ELECTROMECHANICAL MOTION DEVICES

IEEE Press
445 Hoes Lane
Piscataway, NJ 08854

THIRD EDITION

ELECTROMECHANICAL MOTION DEVICES

Rotating Magnetic Field-Based Analysis and Online Animations

PAUL KRAUSE
OLEG WASYNCZUK
STEVEN D. PEKAREK
TIMOTHY O'CONNELL

IEEE PRESS
WILEY

Published by John Wiley & Sons, Inc., Hoboken, New Jersey.

Published simultaneously in Canada.

Library of Congress Cataloging-in-Publication Data is available.

hardback: 9781119489825

Set in 10/12pt Times by SPi Global, Pondicherry, India

Printed in the United States of America.

V10017228_012920

CONTENTS

PREFACE

The list of modern applications requiring high-performance electric machines and drives, already lengthy, is growing rapidly. Originally facilitated by the advent of electronic switching devices in the mid-20th century and accelerated by these devices' ever-increasing performance, this diverse and exciting field has grown into a major industry. Yet, despite the seeming maturity of the field, new applications that require innovative solutions are being developed every day. These include a diverse array of hybrid and electric vehicles – terrestrial- and sea-based crafts; more- and all-electric aircraft, including electrified propulsion systems and unmanned aerial vehicles and drones – large-scale systems for wind energy conversion, and more efficient power generation systems to enable a cleaner grid. Compared to the state-of-the-art, the future machines and drives required to enable these new applications will have higher power, torque, and speed ratings, as well as ever-increasing dynamic performance requirements, all while being subjected to harsher and more demanding environments. The field is no longer limited – nor has it been for decades – to machines and drives that simply sit peacefully on a factory floor driving pumps, fans, or conveyor belts. *Electromechanical Motion Devices Rotating Magnetic Field-Based Analysis and Online Animations Third Edition* has been written specifically to acknowledge this fact and a student who masters the material herein is well on his/her way to contributing to the exciting and dynamic field of high-performance electric machines and drives.

Like its first and second edition predecessors, this introductory-level textbook covers magnetic and magnetically coupled circuits (Chapter 1), electromechanical energy conversion (Chapter 2), and the electric machines in widest use, including dc (Chapter 3), permanent-magnet ac (including the brushless drive) (Chapter 6), synchronous (Chapter 7), induction (Chapter 8), and stepper (Chapter 9), providing sufficient theory, examples, and practice problems to establish a solid base from which to delve deeper into advanced courses and/or texts. In this new edition, Chapters 1, 2, and 9 are unchanged. All of the other chapters have been thoroughly revised and/or rewritten with particular care given to the machine-focused Chapters 6 through 8 to ensure that neither reference frame theory nor control theory is required to understand the material. For each machine, the reference frame derivation which establishes the steady-state phasor equations can be skipped, allowing the instructor to go from the machine variable equations directly to the phasor voltage equations if he or she so chooses. In the case of the drives systems considered, it is assumed that the control is functioning perfectly. Thus, the emphasis is placed on the goal and result of the control, rather than its implementation, and control theory is not needed nor is it recommended as a pre- or co-requisite. Further, the reference frame derivations that establish the phasor voltage equations for the brushless dc and field-orientated drives in Chapters 6 and 8, respectively, can be skipped without loss of cohesiveness. Here, again, the current is assumed to be tightly controlled, whereupon the electric transients are essentially nonexistent and can be neglected. This allows the student to become familiar with the performance of the drives from the steady-state equations without being bogged down with details of the controllers.

Chapter 5 remains devoted to reference frame theory, but it has been completely revised. The subtitle of this third edition, "Rotating Magnetic Field-based Analysis with Online Animations," emphasizes the importance of this revision. It is shown that the basic reference frame transformation is based on and contained in the expression of Tesla's rotating magnetic field and can be obtained using only trigonometry. Applying the transformation to a machine's equations amounts to nothing more than establishing the variables that correctly portray the expression for Tesla's rotating magnetic field as viewed from the given reference frame. This material is not required in order to use the rest of the book; however, it is recommended that at least a part of it be covered since we believe this new way of presenting reference frame theory provides a much clearer and intuitive view of the transformations, allowing the student to gain a deep understanding of the concepts without being bogged down with math and matrix manipulations.

Because this text is intended to serve both the machines and drives areas, there is far more material and detail than can be covered in a three-credit hour course. This is intentional and allows the instructor to pick from a fairly diverse array of material. For example, those interested in the drives area may choose to cover most of Chapter 10 on power electronics while omitting most, if not all, of Chapter 7 on synchronous machines. Alternatively, those interested in the traditional machines area may do the opposite. Moreover, this arrangement of the material allows the text to be used either as a first undergraduate course in machines and drives or as a first semester graduate-level course, where the depth of coverage would be more extensive (or used in both roles at the same school).

Those who are familiar with previous editions will be comfortable using this edition. However, there are important changes, some of which have already been touched upon. The material on reference frame theory has been revised and presented in a new way, and the transformation for rotating circuits has been added. Throughout the text, control theory has been de-emphasized/removed in order to emphasize the goal and result of the controllers, rather than their implementation. The order of Chapters 6 (permanent-magnet ac machine and brushless dc drive) and 8 (induction machines and field orientation) has been switched from the previous editions, and the field-orientation description in Chapter 8 has been revised to follow the new rotating magnetic field-based treatment of reference frame theory in Chapter 5. Chapter 10 on power electronics is new to this edition, and the material on unbalanced operation and single-phase induction machines has been removed.

Finally, as indicated by this edition's new subtitle, detailed online interactive animations that supplement the material in Chapters 3–8 are available. Listed in Appendix B, these animations provide the student and instructor with hands-on visuals and help to portray some of the more difficult concepts. Also, videos and/or PowerPoint presentations of each section of the text are available to the instructor upon request to the publisher.

Paul Krause
Oleg Wasynczuk
Steven D. Pekarek
Timothy O'Connell

CHAPTER 1

MAGNETIC AND MAGNETICALLY COUPLED CIRCUITS

1.1 INTRODUCTION

Before diving into the analysis of electromechanical motion devices, it is helpful to review briefly some of our previous work in physics and in basic electric circuit analysis. In particular, the analysis of magnetic circuits, the basic properties of magnetic materials, and the derivation of equivalent circuits of stationary, magnetically coupled circuits are topics presented in this chapter. Much of this material will be a review for most, since it is covered either in a sophomore physics course for engineers or in introductory electrical engineering courses in circuit theory. Nevertheless, reviewing this material and establishing concepts and terms for later use sets the appropriate stage for our study of electromechanical motion devices.

Perhaps the most important new concept presented in this chapter is the fact that in all electromechanical devices, mechanical motion must occur, either translational or rotational, and this motion is reflected into the electric system either as a change of flux linkages in the case of an electromagnetic system or as a change of charge in the case of an electrostatic system. We will deal primarily with electromagnetic systems. If the magnetic system is linear, then the change in flux linkages results, owing to a change in the inductance. In other words, we will find that the inductances of the electric circuits associated with electromechanical motion devices are functions of the mechanical motion. In this chapter, we shall learn to express the self- and mutual inductances for simple translational and rotational electromechanical devices, and to handle these changing inductances in the voltage equations describing the electric circuits associated with the electromechanical system.

Electromechanical Motion Devices: Rotating Magnetic Field-Based Analysis and Online Animations, Third Edition. Paul Krause, Oleg Wasynczuk, Steven D. Pekarek, and Timothy O'Connell.

Published 2020 by John Wiley & Sons, Inc.

Throughout this text, we will give short problems (SPs) with answers following most sections. If we have done our job, each SP should take less than ten minutes to solve. Also, it may be appropriate to skip or de-emphasize some material in this chapter depending upon the background or interest of the students. At the close of each chapter, we shall take a moment to look back over some of the important aspects of the material that we have just covered and mention what is coming next and how we plan to fit things together as we go along.

1.2 PHASOR ANALYSIS

Phasors are used to analyze steady-state performance of ac circuits and devices. This concept can be readily established by expressing a steady-state sinusoidal variable as

$$F_a = F_p \cos \theta_{ef} \tag{1.2-1}$$

where capital letters are used to denote steady-state quantities and F_p is the peak value of the sinusoidal variation, which is generally voltage or current but could be any electrical or mechanical sinusoidal variable. For steady-state conditions, θ_{ef} may be written as

$$\theta_{ef} = \omega_e t + \theta_{ef}(0) \tag{1.2-2}$$

where ω_e is the electrical angular velocity and $\theta_{ef}(0)$ is the time-zero position of the electrical variable. Substituting (1.2-2) into (1.2-1) yields

$$F_a = F_p \cos\left[\omega_e t + \theta_{ef}(0)\right] \tag{1.2-3}$$

Since

$$e^{j\alpha} = \cos\alpha + j\sin\alpha \tag{1.2-4}$$

equation (1.2-3) may also be written as

$$F_a = Re\left\{F_p e^{j\left[\omega_e t + \theta_{ef}(0)\right]}\right\} \tag{1.2-5}$$

where *Re* is shorthand for the "real part of." Equations (1.2-3) and (1.2-5) are equivalent. Let us rewrite (1.2-5) as

$$F_a = Re\left\{F_p e^{j\theta_{ef}(0)} e^{j\omega_e t}\right\} \tag{1.2-6}$$

Thus, we need to take a moment to define what is referred to as the root mean square (rms) of a sinusoidal variation. In particular, the rms value is defined as

$$F = \left(\frac{1}{T}\int_0^T F_a^2(t)dt\right)^{\frac{1}{2}} \tag{1.2-7}$$

where F is the rms value of $F_a(t)$ and T is the period of the sinusoidal variation. It is left to the reader to show that the rms value of (1.2-3) is $F_p/\sqrt{2}$. Therefore, we can express (1.2-6) as

$$F_a = Re\left[\sqrt{2}Fe^{j\theta_{ef}(0)}e^{j\omega_e t}\right] \tag{1.2-8}$$

By definition, the phasor representing F_a, which is denoted with a raised tilde, is

$$\tilde{F}_a = Fe^{j\theta_{ef}(0)} \tag{1.2-9}$$

which is a complex number. The reason for using the rms value as the magnitude of the phasor will be addressed later in this section. Equation (1.2-6) may now be written as

$$F_a = Re\left[\sqrt{2}\tilde{F}_a e^{j\omega_e t}\right] \tag{1.2-10}$$

A shorthand notation for (1.2-9) is

$$\tilde{F}_a = F\underline{/\theta_{ef}(0)} \tag{1.2-11}$$

Equation (1.2-11) is commonly referred to as the polar form of the phasor. The Cartesian form is

$$\tilde{F}_a = F\cos\theta_{ef}(0) + jF\sin\theta_{ef}(0) \tag{1.2-12}$$

When using phasors to calculate steady-state voltages and currents, we think of the phasors as being stationary at $t = 0$. On the other hand, a phasor is related to the instantaneous value of the sinusoidal quantity it represents. Let us take a moment to consider this aspect of the phasor and, thereby, give some physical meaning to it. From (1.2-4), we realize that $e^{j\omega_e t}$ is a constant-amplitude line of unity length rotating counterclockwise at an angular velocity of ω_e. Therefore,

$$\sqrt{2}\tilde{F}_a e^{j\omega_e t} = \sqrt{2}F\left\{\cos\left[\omega_e t + \theta_{ef}(0)\right] + j\sin\left[\omega_e t + \theta_{ef}(0)\right]\right\} \tag{1.2-13}$$

is a constant-amplitude line $\sqrt{2}F$ in length rotating counterclockwise at an angular velocity of ω_e with a time-zero displacement from the positive real axis of $\theta_{ef}(0)$. Since $\sqrt{2}F$ is the peak value of the sinusoidal variation, the instantaneous value of F_a is the real part of (1.2-13). In other words, the real projection of the phasor $\tilde{F}_a$ is the instantaneous value of $F_a/\sqrt{2}$ at time zero. As time progresses, $\tilde{F}_a e^{j\omega_e t}$ rotates at ω_e in the counterclockwise direction, and its real projection, in accordance with (1.2-10), is the instantaneous value of $F_a/\sqrt{2}$. Thus, for

$$F_a = \sqrt{2}F\cos\omega_e t \tag{1.2-14}$$

the phasor representing F_a is

$$\tilde{F}_a = Fe^{j0} = F\underline{/0^\circ} = F + j0 \tag{1.2-15}$$

For

$$F_a = \sqrt{2}F \sin \omega_e t \tag{1.2-16}$$

the phasor is

$$\widetilde{F}_a = Fe^{-j\pi/2} = F\underline{/-90^\circ} = 0 - jF \tag{1.2-17}$$

Although there are several ways to arrive at (1.2-17) from (1.2-16), it is helpful to ask yourself where must the rotating phasor be positioned at time zero so that, when it rotates counterclockwise at ω_e, its real projection is $(1/\sqrt{2})F_p \sin \omega_e t$? Is it clear that a phasor of amplitude F positioned at $\frac{\pi}{2}$ represents $-\sqrt{2}F \sin \omega_e t$?

In order to show the facility of the phasor in the analysis of steady-state performance of ac circuits and devices, it is useful to consider a series circuit consisting of a resistance, an inductance, and a capacitance. Thus,

$$v_a = Ri_a + L\frac{di_a}{dt} + \frac{1}{C}\int i_a dt \tag{1.2-18}$$

For steady-state operation, let

$$V_a = \sqrt{2}V \cos\left[\omega_e t + \theta_{ev}(0)\right] \tag{1.2-19}$$

$$I_a = \sqrt{2}I \cos\left[\omega_e t + \theta_{ei}(0)\right] \tag{1.2-20}$$

where the subscript a is used to distinguish the instantaneous value from the rms value of the steady-state variable. The steady-state voltage equation may be obtained by substituting (1.2-19) and (1.2-20) into (1.2-18), whereupon we can write

$$\begin{aligned}\sqrt{2}V \cos\left[\omega_e t + \theta_{ev}(0)\right] &= R\sqrt{2}I \cos\left[\omega_e t + \theta_{ei}(0)\right] + \omega_e L\sqrt{2}I \cos\left[\omega_e t + \frac{1}{2}\pi + \theta_{ei}(0)\right] \\ &\quad + \frac{1}{\omega_e C}\sqrt{2}I \cos\left[\omega_e t - \frac{1}{2}\pi + \theta_{ei}(0)\right]\end{aligned} \tag{1.2-21}$$

The second term on the right-hand side of (1.2-21), which is $L\frac{dI_a}{dt}$, can be written as

$$\omega_e L\sqrt{2}I \cos\left[\omega_e t + \frac{1}{2}\pi + \theta_{ei}(0)\right] = \omega_e L Re\left[\sqrt{2}Ie^{j\frac{1}{2}\pi}e^{j\theta_{ei}(0)}e^{j\omega_e t}\right] \tag{1.2-22}$$

Since $\widetilde{I}_a = Ie^{j\theta_{ei}(0)}$, we can write

$$\widetilde{L\frac{dI_a}{dt}} = \omega_e L e^{j\frac{1}{2}\pi}\widetilde{I}_a \tag{1.2-23}$$

Since $e^{j\frac{1}{2}\pi} = j$, (1.2-23) may be written as

$$L\widetilde{\frac{dI_a}{dt}} = j\omega_e L\widetilde{I}_a \tag{1.2-24}$$

If we follow a similar procedure, we can show that

$$\frac{1}{C}\widetilde{\int I_a dt} = -j\frac{1}{\omega_e C}\widetilde{I}_a \tag{1.2-25}$$

It is interesting that differentiation of a steady-state sinusoidal variable rotates the phasor counterclockwise by $\frac{1}{2}\pi$, whereas integration rotates the phasor clockwise by $\frac{1}{2}\pi$.

The steady-state voltage equation given by (1.2-21) can be written in phasor form as

$$\widetilde{V}_a = \left[R + j\left(\omega_e L - \frac{1}{\omega_e C}\right)\right]\widetilde{I}_a \tag{1.2-26}$$

We can express (1.2-26) compactly as

$$\widetilde{V}_a = Z\widetilde{I}_a \tag{1.2-27}$$

where Z, the impedance, is a complex number; it is not a phasor. It is often expressed as

$$Z = R + j(X_L - X_C) \tag{1.2-28}$$

where $X_L = \omega_e L$ is the inductive reactance and $X_C = \frac{1}{\omega_e C}$ is the capacitive reactance.

The instantaneous power is

$$\begin{aligned} P &= V_a I_a \\ &= \sqrt{2}V\cos[\omega_e t + \theta_{ev}(0)]\sqrt{2}I\cos[\omega_e t + \theta_{ei}(0)] \end{aligned} \tag{1.2-29}$$

After some manipulation, we can write (1.2-29) as

$$P = VI\cos[\theta_{ev}(0) - \theta_{ei}(0)] + VI\cos[2\omega_e t + \theta_{ev}(0) + \theta_{ei}(0)] \tag{1.2-30}$$

Therefore, the average power P_{ave} may be written as

$$P_{\text{ave}} = |\widetilde{V}_a||\widetilde{I}_a|\cos[\theta_{ev}(0) - \theta_{ei}(0)] \tag{1.2-31}$$

where $|\widetilde{V}_a|$ and $|\widetilde{I}_a|$ are the magnitude of the phasors (rms value), $\theta_{ev}(0) - \theta_{ei}(0)$ is the power factor angle φ_{pf}, and $\cos[\theta_{ev}(0) - \theta_{ei}(0)]$ is referred to as the power factor. If current is positive in the direction of voltage drop then (1.2-31) is positive if power is consumed and negative if power is generated. It is interesting to point out that in going from (1.2-29) to (1.2-30), the coefficient of the two right-hand terms is $\frac{1}{2}(\sqrt{2}V\sqrt{2}I)$ or one-half the product of the peak values of the sinusoidal variables. Therefore, it was considered more convenient to use the rms values for the phasors, whereupon average power could be calculated by the product of the magnitude of the voltage and current phasors as given by (1.2-31).

We see from (1.2-30) that the instantaneous power of a single-phase ac circuit oscillates at $2\omega_e t$ about an average value. Let us take a moment to calculate the steady-state power of a two-phase ac system. Balanced, steady-state, two-phase variables (a and b phase) may be expressed as

$$V_a = \sqrt{2}V \cos\left[\omega_e t + \theta_{ev}(0)\right] \tag{1.2-32}$$

$$I_a = \sqrt{2}I \cos\left[\omega_e t + \theta_{ei}(0)\right] \tag{1.2-33}$$

$$V_b = \sqrt{2}V \cos\left[\omega_e t - \frac{1}{2}\pi + \theta_{ev}(0)\right] \tag{1.2-34}$$

$$I_b = \sqrt{2}I \cos\left[\omega_e t - \frac{1}{2}\pi + \theta_{ei}(0)\right] \tag{1.2-35}$$

The total instantaneous power is

$$P = V_a I_a + V_b I_b \tag{1.2-36}$$

Substituting (1.2-32) through (1.2-35) into (1.2-36) and after some trigonometric manipulation, the total power for a balanced two-phase system becomes

$$P = 2 \mid \tilde{V}_a \mid\mid \tilde{I}_a \mid \cos\varphi_{pf} \tag{1.2-37}$$

It is important to note that the $2\omega_e t$ oscillation is not present. In other words, the total instantaneous steady-state power is constant. In the case of a three-phase balanced system, the phasors of the three voltages or currents are displaced 120° and the instantaneous steady-state power is also constant and three times the average power of one phase. In other words, the 2 in (1.2-37) becomes 3 when considering a three-phase system.

Example 1A Phasor Diagram

It is often instructive to construct a phasor diagram. For example, let us consider a voltage equation of the form

$$\tilde{V} = Z\tilde{I} + \tilde{E} \tag{1A-1}$$

where Z is given by (1.2-28). Let us assume that $\tilde{V}$ and $\tilde{I}$ are known and that we are to calculate $\tilde{E}$. The phasor diagram may be used as a rough check on these calculations. Let us construct this phasor diagram by assuming that $\mid X_L \mid > \mid X_C \mid$ and $\tilde{V}$ and $\tilde{I}$ are known as shown in Fig. 1A-1. Solving (1A-1) for $\tilde{E}$ yields

$$\tilde{E} = \tilde{V} - [R + j(X_L - X_C)]\tilde{I} \tag{1A-2}$$

To perform this graphically, start at the origin in Fig. 1A-1 and walk to the terminus of $\tilde{V}$. Now, we want to subtract $R\tilde{I}$. To achieve the proper orientation to do this, stand at the terminus of $\tilde{V}$, turn, and look in the $\tilde{I}$ direction which is at the angle

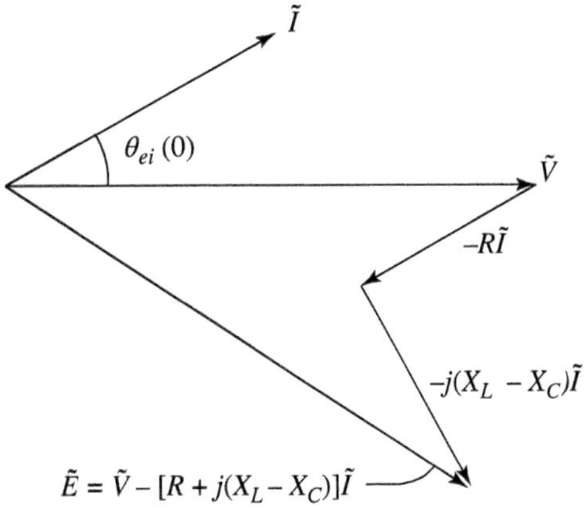

Figure 1A-1 Phasor diagram for (1A-2).

$\theta_{ei}(0)$. But we must subtract $R\tilde{I}$; hence, $-\tilde{I}$ is 180° from $\tilde{I}$, so do an about-face and now we are headed in the $-\tilde{I}$ direction, which is $\theta_{ei}(0) - 180°$. Start walking in the direction of $-\tilde{I}$ for the distance $R\,|\,\tilde{I}\,|$ and then stop. While still facing in the $-\tilde{I}$ direction, let us consider the next term. Now, since we have assumed that $|\,X_L\,| > |\,X_C\,|$, we must subtract $j(X_L - X_C)\tilde{I}$, so let us face in the direction of $-j\tilde{I}$. We are still looking in the $-\tilde{I}$ direction, so we need only to j ourselves. Thus, we must rotate 90° in the counterclockwise direction, whereupon we are standing at the end of $\tilde{V} - R\tilde{I}$ looking in the direction of $\theta_{ei}(0) - 180° + 90°$. Start walking in this direction for the distance of $(X_L - X_C)\,|\,\tilde{I}\,|$, whereupon we are at the terminus of $\tilde{V} - [R + j(X_L - X_C)]\tilde{I}$. According to (1A-2), $\tilde{E}$ is the phasor drawn from the origin of the phasor diagram to where we are.

The average steady-state power for a single-phase circuit may be calculated using (1.2-31). We will mention in passing that the reactive power is defined as

$$Q = |\,\tilde{V}\,||\,\tilde{I}\,|\,\sin\,[\theta_{ev}(0) - \theta_{ei}(0)] \tag{1A-3}$$

The units of Q are var (voltampere reactive). An inductance is said to absorb reactive power and thus, by definition, Q is positive for an inductor and negative for a capacitor. Actually, Q is a measure of the interchange of energy stored in the electric (capacitor) and magnetic (inductance) fields.

SP1.2-1 If $\tilde{V} = 1\underline{/0°}$ and $\tilde{I} = 1\underline{/180°}$ in the direction of the voltage drop, calculate Z and P_{ave}. Is power generated or consumed? [$(-1 + j0)$ ohms, 1 watt, generated]

SP1.2-2 For SP1.2-1, express instantaneous voltage, current, and power if the frequency is 60 Hz. [$V = \sqrt{2}\cos 377t$, $I = \sqrt{2}\cos(377t + \pi)$, $P = -1 + 1\cos(754t + \pi)$]

SP1.2-3 $A = \sqrt{2}\underline{/0°}$, $B = \sqrt{2}\underline{/90°}$. Calculate $A + B$ and $A \times B$. [$1\,\underline{/45°}$, $2\,\underline{/90°}$]

SP1.2-4 In Example 1A, $X_L > X_C$ and yet $\tilde{I}$ was given as leading $\tilde{V}$. How can this be? [$\tilde{E}$]

1.3 MAGNETIC CIRCUITS

An elementary magnetic circuit is shown in Fig. 1.3-1. This system consists of an electric conductor wound N times about the magnetic member, which is generally some type of ferromagnetic material. In this example system, the magnetic member contains an air gap of uniform length between points a and b. We will assume that the magnetic system (circuit) consists only of the magnetic member and the air gap. Recall that Ampere's law states that the line integral of the field intensity **H** about a closed path is equal to the net current enclosed within this closed path of integration. That is,

$$\oint \mathbf{H} \cdot d\mathbf{L} = i_n \tag{1.3-1}$$

where i_n is the net current enclosed. Let us apply Ampere's law to the closed path depicted as a dashed line in Fig. 1.3-1. In particular,

$$\int_a^b H_i dL + \int_b^a H_g dL = Ni \tag{1.3-2}$$

where the path of integration is assumed to be in the clockwise direction. This equation requires some explanation. First, we are assuming that the field intensity exists only in the direction of the given path, hence we have dropped the vector notation. The subscript i denotes the field intensity (H_i) in the ferromagnetic material (iron or steel) and g denotes the field intensity (H_g) in the air gap. The path of integration is taken as the mean length about the magnetic member, for purposes we shall explain later. The right-hand side of (1.3-2) represents the net current enclosed. In particular, we have enclosed the current i, N times. This has the units of amperes but is commonly referred to as ampere-turns (A·t) or magnetomotive force (mmf). We will find that the mmf in magnetic circuits is analogous to the electromotive force (emf) in electric circuits. Note that the current enclosed is positive in (1.3-2) if the current i is positive. The sign of the right-hand side of (1.3-2)

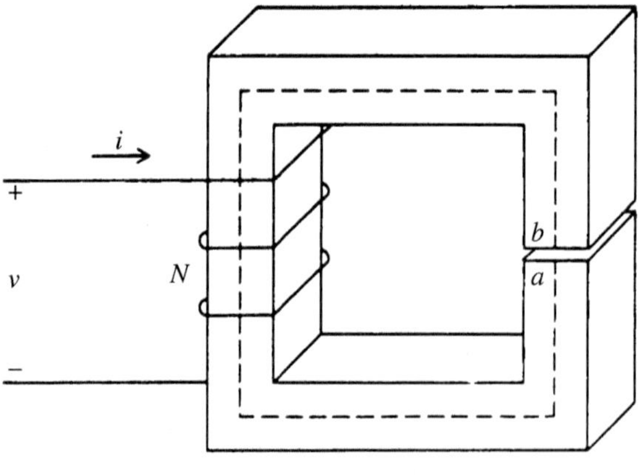

Figure 1.3-1 Elementary magnetic circuit.

may be determined by the so-called "corkscrew" rule. That is, the current enclosed is positive if its assumed positive direction is in the same direction as the advance of a right-hand screw if it were turned in the direction of the path of integration, which in Fig. 1.3-1 is clockwise. Before continuing, it should be mentioned that we refer to **H** as the field intensity; however, some authors prefer to call **H** the field strength.

If we carry out the line integration, (1.3-2) can be written as

$$H_i l_i + H_g l_g = Ni \tag{1.3-3}$$

where l_i is the mean length of the magnetic material and l_g is the length across the air gap. Now, we have some explaining to do. We have assumed that the magnetic circuit consists only of the ferromagnetic material and the air gap, and that the magnetic field intensity is always in the direction of the path of integration or, in other words, perpendicular to a cross section of the magnetic material taken in the same sense as the air gap is cut through the material. The assumed direction of the magnetic field intensity is valid except in the vicinity of the corners. The direction of the field intensity changes gradually rather than abruptly at the corners. Nevertheless, the "mean length approximation" is widely used as an adequate means of analyzing this type of magnetic circuit.

Let us now take a cross section of the magnetic material as shown in Fig. 1.3-2. From our study of physics, we know that for linear, isotropic magnetic materials the flux density **B** is related to the field intensity as

$$\mathbf{B} = \mu \mathbf{H} \tag{1.3-4}$$

where μ is the permeability of the medium. Hence, we can write (1.3-3) in terms of flux density as

$$\frac{B_i}{\mu_i} l_i + \frac{B_g}{\mu_g} l_g = Ni \tag{1.3-5}$$

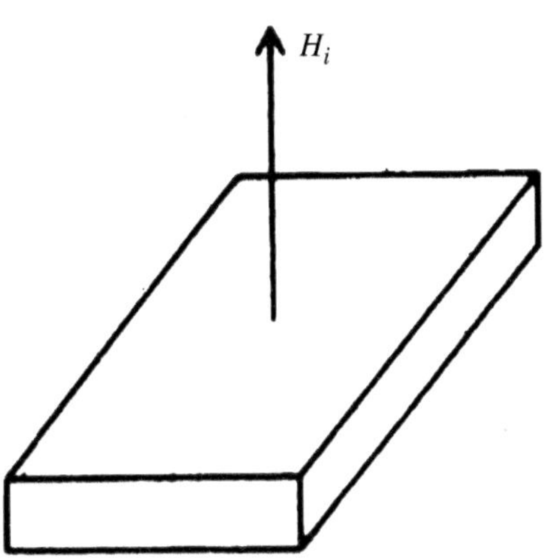

Figure 1.3-2 Cross section of magnetic material.

The surface integral of the flux density is equal to the flux Φ, thus

$$\Phi = \int_A \mathbf{B} \cdot d\mathbf{S} \tag{1.3-6}$$

If we assume that the flux density is uniform over the cross-sectional area, then

$$\Phi_i = B_i A_i \tag{1.3-7}$$

where Φ_i is the total flux in the magnetic material and A_i is the associated cross-sectional area. In the air gap,

$$\Phi_g = B_g A_g \tag{1.3-8}$$

where A_g is the cross-sectional area of the gap. From physics, it is known that the streamlines of flux density **B** are closed; hence, the flux in the air gap is equal to the flux in the core. That is, $\Phi_i = \Phi_g$, and, if the air gap is small, $A_i \cong A_g$, and, therefore, $B_i \cong B_g$. However, the effective area of the air gap is larger than that of the magnetic material, since the flux will tend to balloon or spread out (fringing effect), covering a maximum area midway across the air gap. Generally, this is taken into account by assuming that $A_g = kA_i$, where k, which is greater than unity, is determined primarily by the length of the air gap. Although we shall keep this in mind, it is sufficient for our purposes to assume $A_g = A_i$. If we let $\Phi_i = \Phi_g = \Phi$ and substitute (1.3-7) and (1.3-8) into (1.3-5), we obtain

$$\frac{l_i}{\mu_i A_i}\Phi + \frac{l_g}{\mu_g A_g}\Phi = Ni \tag{1.3-9}$$

The analogy to Ohm's law is at hand. Ni (mmf) is analogous to the voltage (emf), and the flux Φ is analogous to the current. We can complete this analogy if we recall that the resistance of a conductor is proportional to its length and inversely proportional to its conductivity and cross-sectional area. Similarly, $l_i/\mu_i A_i$ and $l_g/\mu_g A_g$ are the reluctances of the magnetic material and air gap, respectively. Generally, the permeability is expressed in terms of relative permeability as

$$\mu_i = \mu_{ri}\mu_0 \tag{1.3-10}$$

$$\mu_g = \mu_{rg}\mu_0 \tag{1.3-11}$$

where μ_0 is the permeability of free space ($4\pi \times 10^{-7}$ Wb/A · m or $4\pi \times 10^{-7}$ H/m, since Wb/A is a henry) and μ_{ri} and μ_{rg} are the relative permeability of the magnetic material and the air gap, respectively. For all practical purposes, $\mu_{rg} = 1$; however, μ_{ri} may be as large as 500–4000 depending upon the type of ferromagnetic material. We will use $\mathcal{R}$ to denote reluctance so as to distinguish reluctance from resistance, which will be denoted by r or R. We can now write (1.3-9) as

$$(\mathcal{R}_i + \mathcal{R}_g)\Phi = Ni \tag{1.3-12}$$

where $\mathcal{R}_i$ and $\mathcal{R}_g$ are the reluctance of the iron and air gap, respectively.

Example 1B Magnetic Circuit – dc Source

A magnetic system is shown in Fig. 1B-1. The total number of turns is 100, the relative permeability of the iron is 1000, and the current is 10 A. Calculate the total flux in the center leg.

Let us draw the electric circuit analog of this magnetic system for which we will need to calculate the reluctance of the various paths:

$$\mathcal{R}_{ab} = \frac{l_{ab}}{\mu_{ri}\mu_0 A_i} = \frac{0.22}{1000(4\pi \times 10^{-7})(0.04)^2} = 109,419\ \mathrm{H}^{-1} \tag{1B-1}$$

Similarly,

$$\mathcal{R}_{bcda} = \frac{0.25 + 0.22 + 0.25}{(1000)(4\pi \times 10^{-7})(0.04)^2} = 358,099\ \mathrm{H}^{-1} \tag{1B-2}$$

Neglecting the air gap length,

$$\mathcal{R}_{bef} = \mathcal{R}_{gha} = \frac{1}{2}\mathcal{R}_{bcda} = 179,049\ \mathrm{H}^{-1} \tag{1B-3}$$

The reluctance of the air gap is

$$\mathcal{R}_{fg} = \frac{0.002}{(4\pi \times 10^{-7})(0.04)^2} = 994,718\ \mathrm{H}^{-1} \tag{1B-4}$$

The electric circuit analog is given in Fig. 1B-2. The polarity of the mmf is determined by the right-hand rule. That is, if we grasp one of the turns of the

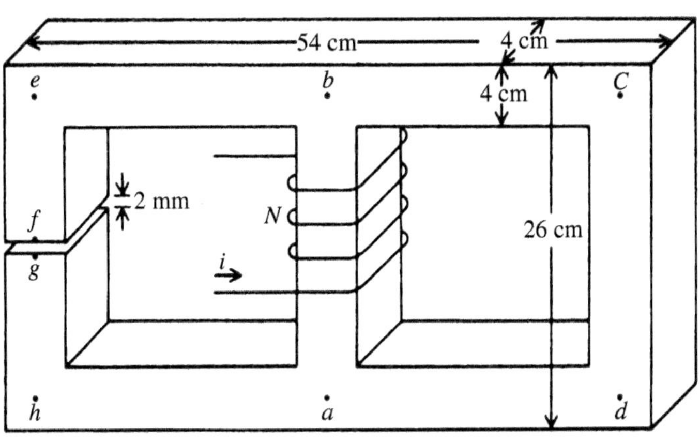

Figure 1B-1 Single-winding magnetic system.

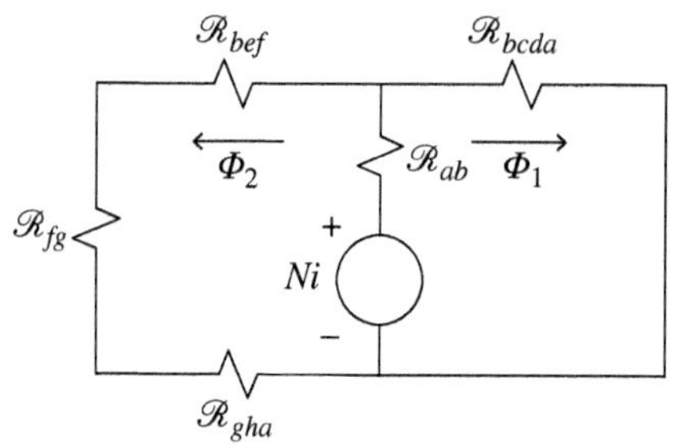

Figure 1B-2 Electric-circuit analog of Fig. 1B-1.

winding with our right hand with the thumb pointed in the direction of positive current, then our fingers will point in the direction of positive flux which flows in the direction of an mmf rise. Or, if we grasp the winding (center leg) with the fingers of our right hand in the direction of positive current, then our thumb will be in the direction of positive flux and in the direction of a rise in mmf.

We can now apply dc circuit theory to solve for the total flux, $\Phi_1 + \Phi_2$, flowing in the center leg. For example, we can use loop equations or, as we will do here, reduce the series–parallel circuit to an equivalent reluctance. The equivalent reluctance of the parallel combination is

$$\begin{aligned} \mathcal{R}_{eq} &= \frac{(\mathcal{R}_{bcda})(\mathcal{R}_{bef} + \mathcal{R}_{fg} + \mathcal{R}_{gha})}{\mathcal{R}_{bcda} + \mathcal{R}_{bef} + \mathcal{R}_{fg} + \mathcal{R}_{gha}} \\ &= \frac{(358,099)(179,049 + 994,718 + 179,049)}{358,099 + 179,049 + 994,718 + 179,049} \\ &= \frac{(358,099)(1,352,816)}{1,710,915} = 283,148\ \mathrm{H}^{-1} \end{aligned} \tag{1B-5}$$

$$\begin{aligned} \Phi_1 + \Phi_2 &= \frac{Ni}{\mathcal{R}_{ab} + \mathcal{R}_{eq}} \\ &= \frac{(100)(10)}{109,419 + 283,148} = 2.547 \times 10^{-3}\ \mathrm{Wb} \end{aligned} \tag{1B-6}$$

Example 1C Magnetic Circuit – ac Source

Consider the magnetic system shown in Fig. 1C-1. The windings are supplied from ac sources and, in the steady state, $I_1 = \sqrt{2}\cos\omega_e t$ and $I_2 = \sqrt{2}\,0.3\cos(\omega_e t + 45°)$, where capital letters are used to denote steady-state conditions. $N_1 = 150$ turns, $N_2 = 90$ turns, and $\mu_r = 3000$. Calculate the flux in the center leg.

The electric circuit analog is given in Fig. 1C-2. The reluctance $\mathcal{R}_x$ is the reluctance of the center leg and $\mathcal{R}_y$ is the reluctance of one of the two parallel paths

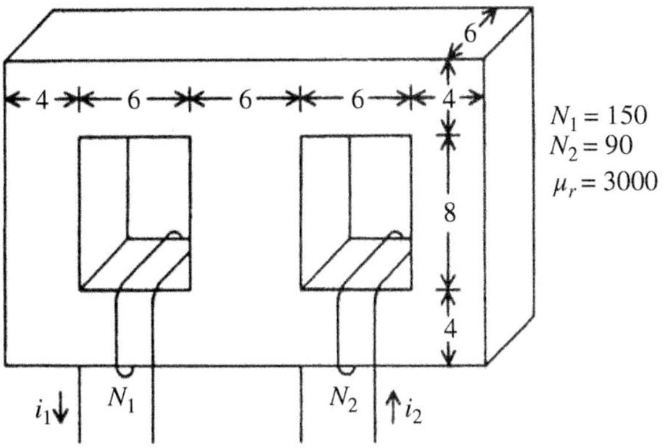

Figure 1C-1 A two-winding magnetic system with dimensions in centimeters.

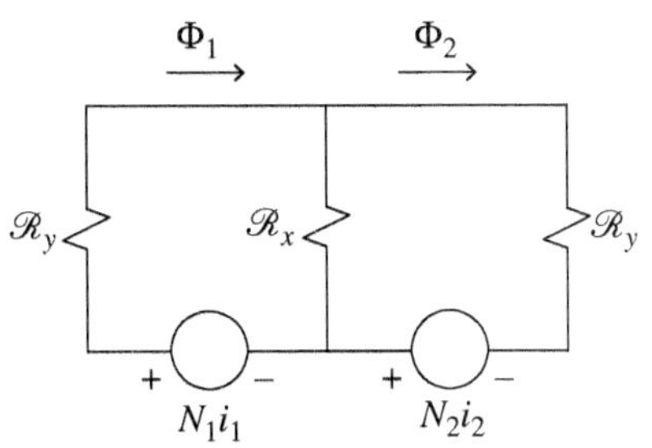

Figure 1C-2 Electric-circuit analog of Fig. 1C-1.

from the top of the center leg through an outside leg to the bottom of the center leg. In particular,

$$\mathcal{R}_y = \frac{2(0.03 + 0.06 + 0.02) + 0.12}{3000(4\pi \times 10^{-7})(0.06)(0.04)} = 37,578\ \text{H}^{-1} \tag{1C-1}$$

$$\mathcal{R}_x = \frac{0.12}{3000(4\pi \times 10^{-7})0.06^2} = 8842\ \text{H}^{-1} \tag{1C-2}$$

Since the currents are sinusoidal, the mmfs will be sinusoidal. Thus, it is convenient to use phasors to solve for Φ_1 and Φ_2. The loop equations are

$$\tilde{\text{mmf}}_1 = \mathcal{R}_y\tilde{\Phi}_1 + \mathcal{R}_x(\tilde{\Phi}_1 - \tilde{\Phi}_2) \tag{1C-3}$$

$$\tilde{\text{mmf}}_2 = \mathcal{R}_x(\tilde{\Phi}_2 - \tilde{\Phi}_1) + \mathcal{R}_y\tilde{\Phi}_2 \tag{1C-4}$$

which may be written in matrix form as

$$\begin{bmatrix} \tilde{\text{mmf}}_1 \\ \tilde{\text{mmf}}_2 \end{bmatrix} = \begin{bmatrix} \mathcal{R}_x + \mathcal{R}_y & -\mathcal{R}_x \\ -\mathcal{R}_x & \mathcal{R}_x + \mathcal{R}_y \end{bmatrix} \begin{bmatrix} \tilde{\Phi}_1 \\ \tilde{\Phi}_2 \end{bmatrix} \tag{1C-5}$$

Now,

$$\tilde{\text{mmf}}_1 = N_1\tilde{I}_1 = (150)(1\underline{/0^\circ}) = 150\underline{/0^\circ}\,\text{At} \tag{1C-6}$$

$$\tilde{\text{mmf}}_2 = N_2\tilde{I}_2 = (90)(0.3\underline{/45^\circ}) = 27\underline{/45^\circ}\,\text{At} \tag{1C-7}$$

Solving (1C-5) yields

$$\tilde{\Phi}_1 = (3.434 + j0.081) \times 10^{-3}\ \text{Wb} \tag{1C-8}$$

$$\tilde{\Phi}_2 = (1.065 + j0.427) \times 10^{-3}\ \text{Wb} \tag{1C-9}$$

The flux flowing down through the center leg is

$$\begin{aligned}\tilde{\Phi}_1 - \tilde{\Phi}_2 &= (2.369 - j0.346) \times 10^{-3} \\ &= 2.39 \times 10^{-3}\underline{/-8.3^\circ}\ \text{Wb}\end{aligned} \tag{1C-10}$$

SP1.3-1 Calculate Φ_1 in Example 1B. $[\Phi_1 = 2.014 \times 10^{-3}\ \text{Wb}]$

SP1.3-2 Calculate $\tilde{\Phi}_1 + \tilde{\Phi}_2$ in Example 1B when $I = \sqrt{2}\,10\cos(\omega_e t - 30^\circ)$. $[\tilde{\Phi}_1 + \tilde{\Phi}_2 = 2.547 \times 10^{-3}\underline{/-30^\circ}\ \text{Wb, rms}]$

SP1.3-3 Remove the center leg of the magnetic system shown in Fig. 1C-1. Calculate the total flux when $I_1 = 9$ A and $I_2 = -15$ A. [Zero]

SP1.3-4 Express the sinusoidal variation represented by $\tilde{\Phi}_2$ given by (1C-9). $[\sqrt{2}\,1.147 \times 10^{-3}\cos(\omega_e t + 21.8^\circ)]$

1.4 PROPERTIES OF MAGNETIC MATERIALS

We may be aware from our study of physics that, when ferromagnetic materials such as iron, nickel, cobalt, or alloys of these elements, such as various types of steels, are placed in a magnetic field, the flux produced is markedly larger (500–4000 times, for example) than that which would be produced when a non-magnetic material is subjected to the same magnetic field. We must take some time to review briefly the basic properties of ferromagnetic materials and to establish terminology for later use.

Let us begin by considering the relationship between B and H shown in Fig. 1.4-1 which is typical of silicon steel used in transformers. We will assume that the ferromagnetic core is initially completely demagnetized (both B and H are zero). As we apply an external H field by increasing the current in a winding wound around the core, the flux density B also increases, but nonlinearly, as shown in Fig. 1.4-1. After H reaches a value of approximately 150 A·t/m, the flux density rises more slowly and the material begins to saturate.

In ferromagnetic materials, the combination of the magnetic moments produced by the electrons orbiting the nucleus of an atom and the electron itself spinning on its axis produce a net magnetic moment of the atom that is not canceled

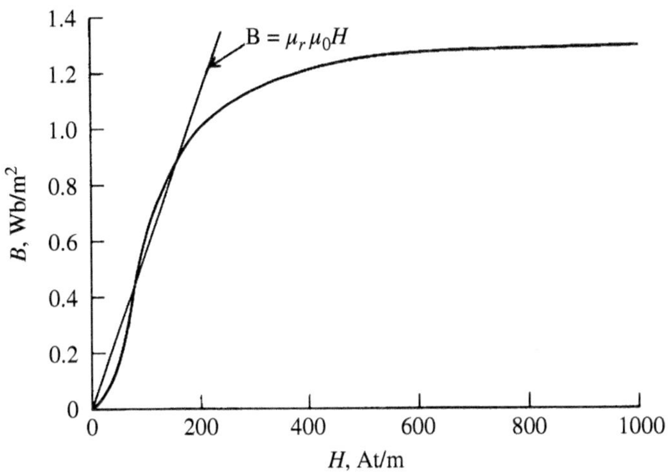

Figure 1.4-1 *B–H* curve for typical silicon steel used in transformers.

by an opposing magnetic moment of a neighboring atom. Ferromagnetic materials have been found to be divided into magnetic domains wherein all magnetic moments (dipoles) are aligned. Although the magnetic moments are all aligned within a magnetic domain, the direction of this alignment will differ from one domain to another.

When a ferromagnetic material is subjected to an external magnetic field, those domains, which originally tend to be aligned with the applied magnetic field, grow at the expense of those domains with magnetic moments that are less aligned. Thereby, the flux is increased from that which would occur with a nonmagnetic material. This is known as *domain-wall motion* [1]. As the strength of the magnetic field increases, the aligned domains continue to grow in nearly a linear fashion. Thus, a nearly linear *B–H* curve results ($B \cong \mu_r \mu_0 H$) until the ability of the aligned domains to take from the unaligned domains starts to slow. This gives rise to the knee of the *B–H* curve and the beginning of saturation. At this point, the displacements of the domain walls are complete. That is, there are no longer unaligned domains from which to take. However, the remaining domains may still not be in perfect alignment with the external *H* field. A further increase in *H* will cause a rotation of the atomic dipole moments within the remaining domains toward a more perfect alignment. However, the marginal increase in *B* due to rotation is less than the original increase in *B* due to domain-wall motion, resulting in a decrease in slope of the *B–H* curve. The magnetic material is said to be completely saturated when the remaining domains are perfectly aligned. In this case, the slope of the *B–H* curve becomes μ_0 [1]. If it is assumed that the magnetic flux is uniform through most of the magnetic material, then *B* is proportional to Φ and *H* is proportional to mmf. Hence, a plot of flux versus current is of the same shape as the *B–H* curve.

A transformer is generally designed so that some saturation occurs during normal operation. Electric machines are also designed similarly in that a machine generally operates slightly in the saturated region during normal, rated operating conditions. Since saturation causes the coefficients of the differential equations describing the behavior of an electromagnetic device to be functions of the winding currents, a transient analysis is difficult without the aid of a computer. However, it is not our purpose to set forth methods of analyzing nonlinear magnetic systems.

In the previous discussion, we have assumed that the ferromagnetic material is initially demagnetized and that the applied field intensity is gradually increased from zero. However, if a ferromagnetic material is subjected to an alternating field intensity, the resulting B–H curve exhibits hysteresis. For example, let us assume that a ferromagnetic material is subjected to an alternating field intensity (alternating current flowing in the winding) and initially the flux density and field intensity are both zero. As H increases from zero, B increases along the initial B–H curve, as shown in Fig. 1.4-2. However, the field intensity varies sinusoidally and, when H decreases from a maximum, B does not follow back down the original B–H curve. After several cycles, the magnetic system will reach a steady-state condition and the plot of B versus H will form a hysteresis loop or a double-valued function, as shown in Fig. 1.4-2. What is happening is very complex. In simple terms, the growth of aligned domains for an incremental change in H in one direction is not equal to the growth of oppositely aligned domains if this change in H were suddenly reversed. We could become quite involved by discussing minor hysteresis loops which would occur if, during the sinusoidal variation of H, it were suddenly stopped at some nonzero value then reversed, stopped, and reversed again [1]. We shall only mention this phenomenon in passing.

A family of hysteresis loops is shown in Fig. 1.4-3. In each case, the applied H is sinusoidal; however, the amplitude of the H field is varied to give the family of

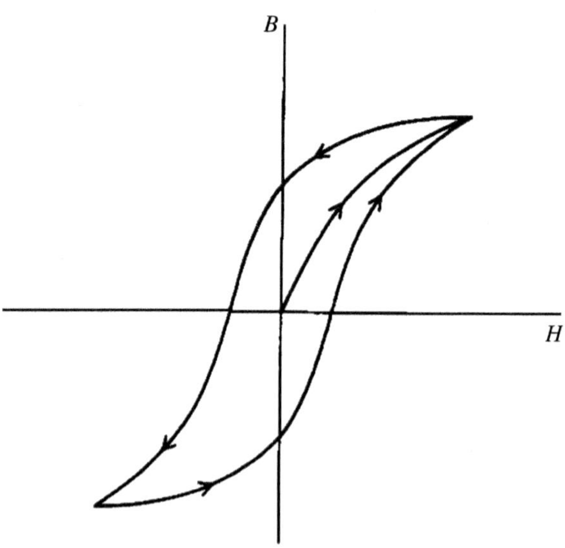

Figure 1.4-2 Hysteresis loop.

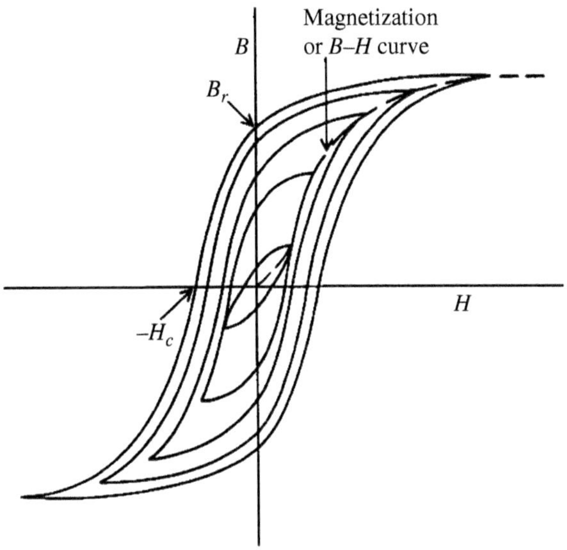

Figure 1.4-3 Family of steady-state hysteresis loops.

loops shown in Fig. 1.4-3. A magnetization or *B–H* curve for a given material is obtained by connecting the tips of the hysteresis loops, as shown by the dashed line in Fig. 1.4-3. The locus of the tips of the hysteresis curves is about the same as the original *B–H* curve in Fig. 1.4-1, which corresponds to a gradual increase of *H* in an initially demagnetized material. If *H* were suddenly stopped at zero, the flux density remaining in the ferromagnetic material is called the residual flux density (B_r). The negative field intensity necessary to bring this residual flux density to zero is called the coercive force (H_c). These two quantities are indicated in Fig. 1.4-3 for the largest hysteresis loop shown.

Energy is required to increase the size of the magnetic domains of the ferromagnetic material. It can be shown that the energy necessary to align alternately the magnetic domains is equal to the area enclosed by the hysteresis loop. This energy causes a rise in the temperature of the magnetic material, and the power associated with this energy loss is called the hysteresis loss.

When a solid block of magnetic material such as that shown in Fig. 1.3-1 is subjected to an alternating field intensity, the resulting alternating flux induces current in the solid magnetic material, which will circulate in a loop perpendicular to the flux density (**B**) inducing it. These so-called eddy currents have two undesirable effects. First, the mmf established by these circulating currents opposes the mmf produced by the winding, and this opposition is greatest at the center of the material because that tends to be also the center of the current loops. Thus, the flux would tend not to flow through the center of the solid magnetic member, thereby not utilizing the full benefits of the ferromagnetic material. Second, there is an i^2r loss associated with these currents, called *eddy current loss*, which is dissipated as heat. These two adverse effects can be minimized in several ways, but the most common is to build the ferromagnetic core of laminations (thin strips) insulated from each other and oriented in the direction of the magnetic field (**B** or **H**). These thin strips offer a much smaller area in which the eddy currents can flow; hence, smaller currents and smaller losses result.

The core losses associated with ferromagnetic materials are the combination of the hysteresis and eddy current losses. Electromagnetic devices are designed to minimize these losses; however, they are always present and are often taken into account in a linear system analysis by assuming that their effects on the electric system can be represented by a resistance.

SP1.4-1 The magnetic circuit of Fig. 1.3-1 is constructed by using silicon sheet steel. Its magnetization curve is given by Fig. 1.4-1. The gap length l_g is 1 mm, the mean core length l_i is 100 cm, $N = 500$, and $A_i = A_g = 25\text{ cm}^2$. Determine the current needed to produce a flux Φ of 2.5×10^{-3} Wb. [*Hint*: First establish H_i, H_g, and use (1.3-3).] [$I = 1.99$ A]

1.5 STATIONARY MAGNETICALLY COUPLED CIRCUITS

Magnetically coupled electric circuits are central to the operation of transformers and electromechanical motion devices. In the case of transformers, stationary circuits are magnetically coupled for the purpose of changing the ac voltage and current levels. In the case of electromechanical devices, circuits in relative motion are magnetically coupled for the purpose of transferring energy between the mechanical and electric systems. Since magnetically coupled circuits play such an important role in energy conversion, it is important to establish the equations that describe their behavior and to express these equations in a form convenient for analysis. Many of these goals may be achieved by considering two stationary electric circuits that are magnetically coupled, as shown in Fig. 1.5-1. The two windings consist of turns N_1 and N_2, and they are wound on a common core, which is a ferromagnetic material with a permeability large relative to that of air. The magnetic core is not illustrated in three dimensions.

Before proceeding, a comment or two is in order. Generally, the concept of an ideal transformer is introduced in a basic circuit course. In the ideal case, v_2 in Fig. 1.5-1 is $(N_2/N_1)v_1$ and i_2 is $-(N_1/N_2)i_1$. Only the turns ratio of the transformer is considered. However, this treatment is often not sufficient for a detailed analysis of transformers, and it is seldom appropriate in the analysis of electromechanical motion devices, since an air gap is necessary for motion to occur; hence, the windings are not as tightly coupled as in the case of transformers and the leakage flux must be taken into account.

In general, the flux produced by each winding can be separated into two components: a leakage component denoted with the subscript l and a magnetizing component denoted by the subscript m. Each of these components is depicted by a single streamline with the positive direction determined by applying the right-hand

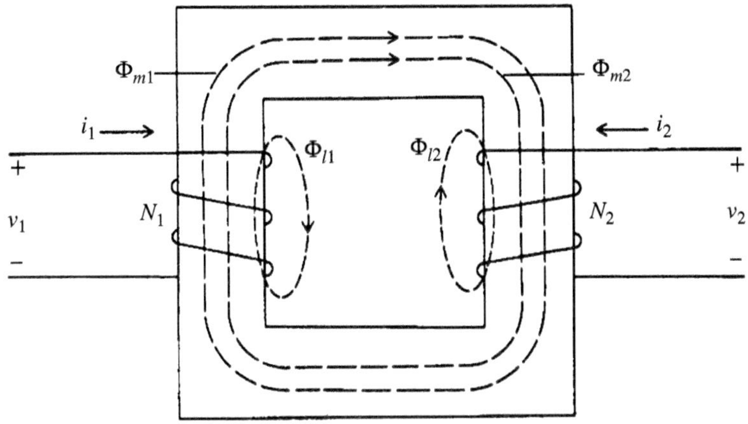

Figure 1.5-1 Magnetically coupled circuits.

rule to the directions of current flow in the winding. (The right-hand rule was reviewed in Example 1B.) The leakage flux associated with a given winding links only that winding, whereas the magnetizing flux, whether it is due to current in winding 1 or winding 2, links both windings. In some cases, i_2 is selected positive out of the top of winding 2 and a dot is placed at that terminal. Although the "dot notation" is convenient for transformers, it is seldom used in the case of electromechanical devices.

The flux linking each winding may be expressed as

$$\Phi_1 = \Phi_{l1} + \Phi_{m1} + \Phi_{m2} \tag{1.5-1}$$

$$\Phi_2 = \Phi_{l2} + \Phi_{m2} + \Phi_{m1} \tag{1.5-2}$$

The leakage flux Φ_{l1} is produced by current flowing in winding 1 and it links only the turns of winding 1. Likewise, the leakage flux Φ_{l2} is produced by current flowing in winding 2 and it links only the turns of winding 2. The flux Φ_{m1} is produced by current flowing in winding 1 and it links all turns of windings 1 and 2. Similarly, the magnetizing flux Φ_{m2} is produced by current flowing in winding 2 and it also links all turns of windings 1 and 2. Both Φ_{m1} and Φ_{m2} are called *magnetizing fluxes*. With the selected positive directions of current flow and the manner in which the windings are wound, magnetizing flux produced by positive current flowing in one winding adds to the magnetizing flux produced by positive current flowing in the other winding. For this case, we will find that the mutual inductance is positive.

It is appropriate to point out that this is an idealization of the actual magnetic system. It seems logical that all of the leakage flux will not link all the turns of the winding producing it; hence, Φ_{l1} and Φ_{l2} are "equivalent" leakage fluxes. Similarly, all of the magnetizing flux of one winding may not link all of the turns of the other winding. To acknowledge this practical aspect of the magnetic system,

N_1 and N_2 are often considered to be the equivalent number of turns rather than the actual number.

The voltage equations may be expressed as

$$v_1 = r_1 i_1 + \frac{d\lambda_1}{dt} \tag{1.5-3}$$

$$v_2 = r_2 i_2 + \frac{d\lambda_2}{dt} \tag{1.5-4}$$

In matrix form,

$$\begin{bmatrix} v_1 \\ v_2 \end{bmatrix} = \begin{bmatrix} r_1 & 0 \\ 0 & r_2 \end{bmatrix} \begin{bmatrix} i_1 \\ i_2 \end{bmatrix} + \frac{d}{dt} \begin{bmatrix} \lambda_1 \\ \lambda_2 \end{bmatrix} \tag{1.5-5}$$

The resistances r_1 and r_2 and the flux linkages λ_1 and λ_2 are related to windings 1 and 2, respectively. Since it is assumed that Φ_1 links the equivalent turns of winding 1 (N_1) and Φ_2 links the equivalent turns of winding 2 (N_2), the flux linkages may be written as

$$\lambda_1 = N_1 \Phi_1 \tag{1.5-6}$$

$$\lambda_2 = N_2 \Phi_2 \tag{1.5-7}$$

where Φ_1 and Φ_2 are given by (1.5-1) and (1.5-2), respectively.

If we assume that the magnetic system is linear, we may apply Ohm's law for magnetic circuits to express the fluxes. Thus, the fluxes may be written as

$$\Phi_{l1} = \frac{N_1 i_1}{\mathcal{R}_{li}} \tag{1.5-8}$$

$$\Phi_{m1} = \frac{N_1 i_1}{\mathcal{R}_m} \tag{1.5-9}$$

$$\Phi_{l2} = \frac{N_2 i_2}{\mathcal{R}_{l2}} \tag{1.5-10}$$

$$\Phi_{m2} = \frac{N_2 i_2}{\mathcal{R}_m} \tag{1.5-11}$$

where $\mathcal{R}_{l1}$ and $\mathcal{R}_{l2}$ are the reluctances of the leakage paths, and $\mathcal{R}_m$ is the reluctance of the path of magnetizing fluxes. Typically, the reluctances associated with leakage paths are much larger than the reluctance of the magnetizing path. The reluctance associated with an individual leakage path is difficult to determine exactly, and it is usually approximated from test data or by using the computer to solve the field equations numerically. On the other hand, the reluctance of the magnetizing path of the core shown in Fig. 1.5-1 may be computed with sufficient accuracy as in Example 1B.

Substituting (1.5-8) through (1.5-11) into (1.5-1) and (1.5-2) yields

$$\Phi_1 = \frac{N_1 i_1}{\mathcal{R}_{l1}} + \frac{N_1 i_1}{\mathcal{R}_m} + \frac{N_2 i_2}{\mathcal{R}_m} \tag{1.5-12}$$

$$\Phi_2 = \frac{N_2 i_2}{\mathcal{R}_{l2}} + \frac{N_2 i_2}{\mathcal{R}_m} + \frac{N_1 i_1}{\mathcal{R}_m} \tag{1.5-13}$$

Substituting (1.5-12) and (1.5-13) into (1.5-6) and (1.5-7) yields

$$\lambda_1 = \frac{N_1^2}{\mathcal{R}_{l1}} i_1 + \frac{N_1^2}{\mathcal{R}_m} i_1 + \frac{N_1 N_2}{\mathcal{R}_m} i_2 \tag{1.5-14}$$

$$\lambda_2 = \frac{N_2^2}{\mathcal{R}_{l2}} i_2 + \frac{N_2^2}{\mathcal{R}_m} i_2 + \frac{N_2 N_1}{\mathcal{R}_m} i_1 \tag{1.5-15}$$

When the magnetic system is linear, the flux linkages are generally expressed in terms of inductances and the currents. We see that the coefficients of the first two terms on the right-hand side of (1.5-14) depend upon N_1 and the reluctance of the magnetic system, independent of the existence of winding 2. An analogous statement may be made regarding (1.5-15) with the roles of winding 1 and winding 2 reversed. Hence, the self-inductances are defined as

$$L_{11} = \frac{N_1^2}{\mathcal{R}_{l1}} + \frac{N_1^2}{\mathcal{R}_m} = L_{l1} + L_{m1} \tag{1.5-16}$$

$$L_{22} = \frac{N_2^2}{\mathcal{R}_{l2}} + \frac{N_2^2}{\mathcal{R}_m} = L_{l2} + L_{m2} \tag{1.5-17}$$

where L_{l1} and L_{l2} are the leakage inductances and L_{m1} and L_{m2} are the magnetizing inductances of windings 1 and 2, respectively. From (1.5-16) and (1.5-17), it follows that the magnetizing inductances may be related as

$$\frac{L_{m2}}{N_2^2} = \frac{L_{m1}}{N_1^2} \tag{1.5-18}$$

The mutual inductances are defined as the coefficient of the third term on the right-hand side of (1.5-14) and (1.5-15). In particular,

$$L_{12} = \frac{N_1 N_2}{\mathcal{R}_m} \tag{1.5-19}$$

$$L_{21} = \frac{N_2 N_1}{\mathcal{R}_m} \tag{1.5-20}$$

We see that $L_{12} = L_{21}$ and, with the assumed positive direction of current flow and the manner in which the windings are wound, the mutual inductances are positive. If, however, the assumed positive directions of current were such that Φ_{m1} opposed Φ_{m2}, then the mutual inductances would be negative.

The mutual inductances may be related to the magnetizing inductances. Comparing (1.5-16) and (1.5-17) with (1.5-19) and (1.5-20), we see that

$$L_{12} = \frac{N_2}{N_1} L_{m1} = \frac{N_1}{N_2} L_{m2} \tag{1.5-21}$$

The flux linkages may now be written as

$$\lambda_1 = L_{11}i_1 + L_{12}i_2 \tag{1.5-22}$$

$$\lambda_2 = L_{21}i_1 + L_{22}i_2 \tag{1.5-23}$$

where L_{11} and L_{22} are defined by (1.5-16) and (1.5-17), respectively, and L_{12} and L_{21}, by (1.5-21). The self-inductances L_{11} and L_{22} are always positive; however, the mutual inductances L_{12} (L_{21}) may be positive or negative, as previously mentioned.

Although the voltage equations given by (1.5-3) and (1.5-4) may be used for purposes of analysis, it is customary to perform a change of variables which yields the well-known equivalent T circuit of two windings coupled by a linear magnetic circuit. To set the stage for this derivation, let us express the flux linkages from (1.5-22) and (1.5-23) as

$$\lambda_1 = L_{l1}i_1 + L_{m1}\left(i_1 + \frac{N_2}{N_1}i_2\right) \tag{1.5-24}$$

$$\lambda_2 = L_{l2}i_2 + L_{m2}\left(\frac{N_1}{N_2}i_1 + i_2\right) \tag{1.5-25}$$

With λ_1 in terms of L_{m1} and λ_2 in terms of L_{m2}, we see two logical candidates for substitute variables, in particular, $(N_2/N_1)i_2$ or $(N_1/N_2)i_1$. If we let

$$i'_2 = \frac{N_2}{N_1}i_2 \tag{1.5-26}$$

then we are using the substitute variable i'_2, which, when flowing through winding 1, produces the same mmf as the actual i_2 flowing through winding 2; $N_1 i'_2 = N_2 i_2$. This is said to be referring the current in winding 2 to winding 1 or to a winding with N_1 turns, whereupon winding 1 becomes the reference winding. On the other hand, if we let

$$i'_1 = \frac{N_1}{N_2}i_1 \tag{1.5-27}$$

then i'_1 is the substitute variable that produces the same mmf when flowing through winding 2 as i_1 does when flowing in winding 1; $N_2 i'_1 = N_1 i_1$. This change of variables is said to refer the current of winding 1 to winding 2 or to a winding with N_2 turns, whereupon winding 2 becomes the reference winding.

We will demonstrate the derivation of the equivalent T circuit by referring the current of winding 2 to a winding with N_1 turns; thus, i'_2 is expressed by (1.5-26). We want the instantaneous power to be unchanged by this substitution of variables. Therefore,

$$v'_2 i'_2 = v_2 i_2 \tag{1.5-28}$$

Hence,

$$v'_2 = \frac{N_1}{N_2}v_2 \tag{1.5-29}$$

Flux linkages, which have the units of V · s, are related to the substitute flux linkages in the same way as voltages. In particular,

$$\lambda_2' = \frac{N_1}{N_2}\lambda_2 \tag{1.5-30}$$

Now, replace $(N_2/N_1)i_2$ with i_2' in the expression for λ_1, given by (1.5-24). Next, solve (1.5-26) for i_2 and substitute it into λ_2 given by (1.5-25). Now, multiply this result by N_1/N_2 to obtain λ_2' and then substitute $(N_2/N_1)^2 L_{m1}$ for L_{m2} in λ_2'. If we do all this, we will obtain

$$\lambda_1 = L_{l1} i_1 + L_{m1}\left(i_1 + i_2'\right) \tag{1.5-31}$$

$$\lambda_2' = L_{l2}' i_2' + L_{m1}\left(i_1 + i_2'\right) \tag{1.5-32}$$

where

$$L_{l2}' = \left(\frac{N_1}{N_2}\right)^2 L_{l2} \tag{1.5-33}$$

The flux linkage equations given by (1.5-31) and (1.5-32) may also be written as

$$\lambda_1 = L_{11} i_1 + L_{m1} i_2' \tag{1.5-34}$$

$$\lambda_2' = L_{m1} i_1 + L_{22}' i_2' \tag{1.5-35}$$

where

$$L_{22}' = \left(\frac{N_1}{N_2}\right)^2 L_{22} = L_{l2}' + L_{m1} \tag{1.5-36}$$

and L_{22} is defined by (1.5-17).

If we multiply (1.5-4) by N_1/N_2 to obtain v_2', the voltage equations become

$$\begin{bmatrix} v_1 \\ v_2' \end{bmatrix} = \begin{bmatrix} r_1 & 0 \\ 0 & r_2' \end{bmatrix} \begin{bmatrix} i_1 \\ i_2' \end{bmatrix} + \frac{d}{dt}\begin{bmatrix} \lambda_1 \\ \lambda_2' \end{bmatrix} \tag{1.5-37}$$

where

$$r_2' = \left(\frac{N_1}{N_2}\right)^2 r_2 \tag{1.5-38}$$

The previous voltage equations, (1.5-37), together with the flux linkage equations, (1.5-34) through (1.5-35), suggest the equivalent T circuit shown in Fig. 1.5-2. This method may be extended to include any number of windings wound on the same core.

Earlier in this section, we mentioned that in the case of an ideal transformer only the turns ratio is considered, that is, $v_2 = (N_2/N_1)v_1$ and $i_2 = -(N_1/N_2)i_1$. We can now more fully appreciate the assumptions that are made in this type of

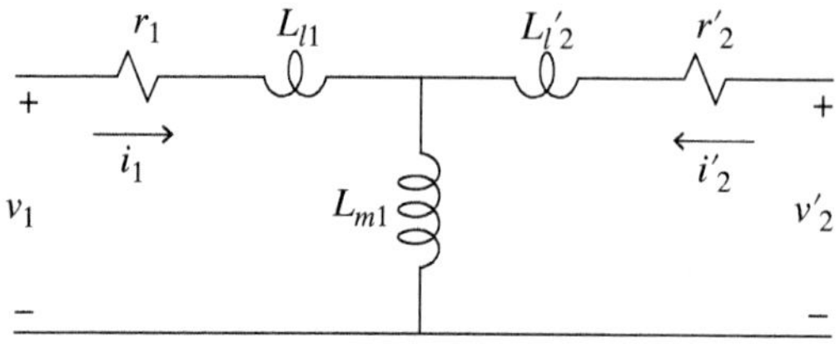

Figure 1.5-2 Equivalent T circuit with winding 1 selected as reference winding.

analysis. In particular, the resistances r_1 and r_2 and the leakage inductances L_{l1} and L_{l2} are neglected, and it is assumed that the magnetizing inductance is large so that the magnetizing current $i_1 + i_2'$ is negligibly small.

The information presented in this section forms the basis of the equivalent circuits for many types of electric machines. Using a turns ratio to refer the voltages and currents of rotor circuits of electric machines to a winding with the same number of turns as the stator windings is common practice.

Example 1D The Equivalent *T* Circuit

It is instructive to illustrate the method of deriving an equivalent T circuit from open- and short-circuit measurements. When winding 2 of the two-winding transformer shown in Fig. 1.5-2 is open circuited and a voltage of 110 V (rms) at 60 Hz is applied to winding 1, the average power supplied to winding 1 is 6.66 W. The measured current in winding 1 is 1.05 A (rms). Next, with winding 2 short-circuited, the current flowing in winding 1 is 2 A (rms) when the applied voltage is 30 V at 60 Hz. The average input power is 44 W. If we assume $L_{l1} = L_{l2}'$, an approximate equivalent T circuit can be determined from these measurements with winding 1 selected as the reference winding.

The average power supplied to winding 1 may be expressed as

$$P_1 = |\tilde{V}_1||\tilde{I}_1| \cos \varphi_{pf} \tag{1D-1}$$

where

$$\varphi_{pf} = \theta_{ev}(0) - \theta_{ei}(0) \tag{1D-2}$$

Here, $\tilde{V}_1$ and $\tilde{I}_1$ are phasors with the positive direction of $\tilde{I}_1$ taken in the direction of the voltage drop, and $\theta_{ev}(0)$ and $\theta_{ei}(0)$ are the phase angles of $\tilde{V}_1$ and $\tilde{I}_1$, respectively. Solving for φ_{pf} during the open-circuit test, we have

$$\varphi_{pf} = \cos^{-1}\frac{P_1}{|\tilde{V}_1||\tilde{I}_1|} = \cos^{-1}\frac{6.66}{(110)(1.05)} = 86.7° \tag{1D-3}$$

Although $\varphi_{pf} = -86.7°$ is also a legitimate solution of (1D-3), the positive solution is taken since $\tilde{V}_1$ leads $\tilde{I}_1$ in an inductive circuit. With winding 2 open-circuited, the input impedance of winding 1 is

$$Z = \frac{\tilde{V}_1}{\tilde{I}_1} = r_1 + j(X_{l1} + X_{m1}) \tag{1D-4}$$

With $\widetilde{V}_1$ as the reference phasor, $\widetilde{V}_1 = 110\angle 0°$, $\widetilde{I}_1 = 1.05\angle -86.7°$. Thus,

$$r_1 + j(X_{l1} + X_{m1}) = \frac{110\angle 0°}{1.05\angle -86.7°} = 6 + j104.6\ \Omega \tag{1D-5}$$

If we neglect core losses, then, from (1D-5), $r_1 = 6\,\Omega$. We also see from (1D-5) that $X_{l1} + X_{m1} = 104.6\,\Omega$.

For the short-circuit test, we will assume that $\widetilde{I}_1 = -I_2'$ since transformers are designed so that at rated frequency $X_{m1} \gg | r_2' + jX_{l2}' |$. Hence, using (1D-1) again,

$$\varphi_{pf} = \cos^{-1}\frac{44}{(30)(2)} = 42.8° \tag{1D-6}$$

In this case, the input impedance is $Z = (r_1 + r_2') + j(X_{l1} + X_{l2}')$. This may be determined as

$$Z = \frac{30\angle 0°}{2\angle -42.8°} = 11 + j10.2\ \Omega \tag{1D-7}$$

Hence, $r_2' = 11 - r_1 = 5\ \Omega$ and, since it is assumed that $X_{l1} = X_{l2}'$, both are $10.2/2 = 5.1\ \Omega$. Therefore, $X_{m1} = 104.6 - 5.1 = 99.5\,\Omega$. In summary, $r_1 = 6\ \Omega$, $L_{l1} = 13.5$ mH, $L_{m1} = 263.9$ mH, $r_2' = 5\,\Omega$, and $L_{l2}' = 13.5$ mH. Make sure we converted from X's to L's correctly.

SP1.5-1 Remove the center leg of the magnetic system shown in Fig. 1C-1. Calculate L_{11}, L_{22}, and L_{12}. Neglect the leakage inductances. [$L_{11} = 299.4$ mH, $L_{22} = 107.8$ mH, $L_{12} = 179.5$ mH]

SP1.5-2 Consider the transformer and parameters calculated in Example 1D. Winding 2 is short-circuited and 12 V (dc) is applied to winding 1. Calculate the steady-state values of i_1 and i_2. Repeat with winding 2 open-circuited. [$I_1 = 2$ A and $I_2 = 0$ in both cases]

SP1.5-3 Calculate the no-load (winding 2 open-circuited) current for the transformer given in Example 1D if $V_1 = \sqrt{2}\ 10\cos 100t$. $[\widetilde{I}_1 = 0.352\ \angle -77.8°\ \text{A}]$

1.6 OPEN- AND SHORT-CIRCUIT CHARACTERISTICS OF STATIONARY MAGNETICALLY COUPLED CIRCUITS

It is instructive to observe the open- and short-circuit characteristics of a transformer with two windings. For this purpose, a transformer with the parameters given in Example 1D was simulated on a computer. The open-circuit characteristics are shown in Figs. 1.6-1 and 1.6-2. The variables plotted are λ, v_1, i_1, v_2', and i_2'. The variable λ is equal to $L_{m1}(i_1 + i_2')$, which is the last term on the right-hand side of (1.5-31) and (1.5-32). This is the flux linkage of winding 1 due to the flux in the transformer iron. It is often referred to as the *magnetizing flux linkage(s)* and denoted λ_m, λ_{mag}, or λ_φ, whereas $i_1 + i_2'$ is called the *magnetizing current.*

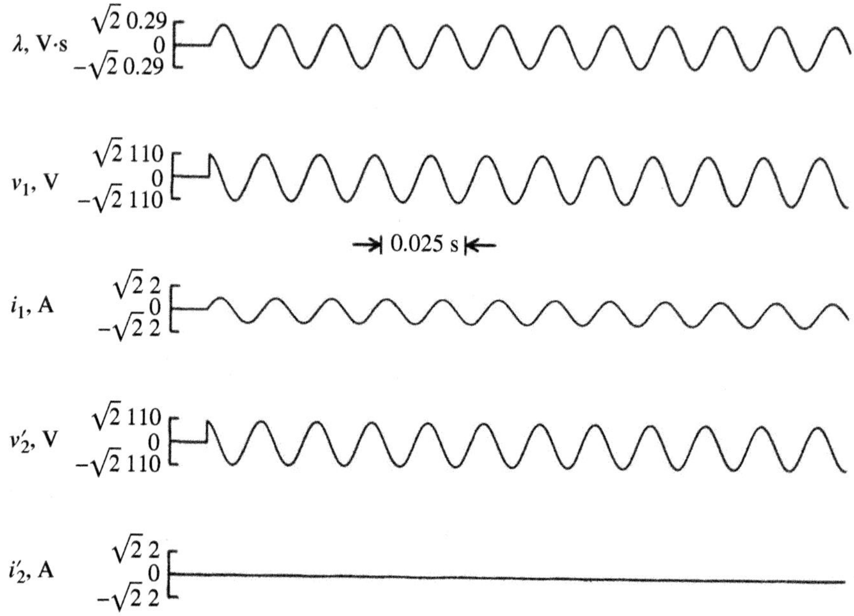

Figure 1.6-1 Open-circuit conditions of a two-winding transformer with $v_1 = \sqrt{2}\ 110 \cos 377t$.

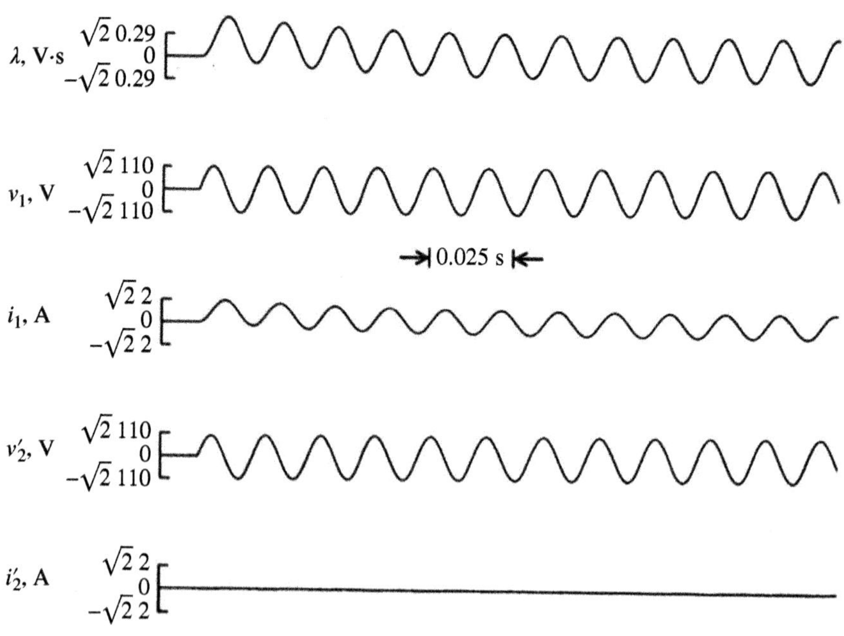

Figure 1.6-2 Open-circuit conditions of a two-winding transformer with $v_1 = \sqrt{2}\ 110 \sin 377t$.

Initially, the windings are unexcited. At time zero ($t = 0$), the voltage applied to winding 1 with winding 2 open-circuited is $v_1 = \sqrt{2}\ 110 \cos 377t$ in Fig. 1.6-1 and $v_1 = \sqrt{2}\ 110 \sin 377t$ in Fig. 1.6-2. The waveforms of the steady-state current i_1 are identical in Figs. 1.6-1 and 1.6-2; however, since the inductive reactance is large, applying a sine wave voltage for v_1 at time zero results in a much larger transient offset in i_1 than when $v_1 = \sqrt{2}\ 110 \cos 377t$. Since $v_1 = \sqrt{2}\ 110 \sin 377t$ causes a larger transient offset, it makes it easier for us to identify the transient period. Therefore, we shall continue with v_1 as a sine wave. Although it is difficult to determine the time constant for the offset of the current i_1 (or λ) to decay to one-third of its original value, it is on the order of 50 ms. The calculated value of the no-load time constant is $\tau_{nl} = (L_{l1} + L_{m1})/r_1 = 46.2$ ms. Before leaving Figs. 1.6-1 and 1.6-2, note that, during steady-state conditions, I_1 lags V_1 by something close to 90° (86.7°, from Example 1D).

Let us now go to the short-circuit characteristics. The transient and steady-state response with $v_1 = \sqrt{2}\ 110 \sin 377t$ and with $v_2' = 0$ are shown in Fig. 1.6-3. There are several things to note. From Fig. 1.6-3, it appears that the time constant associated with the decay of i_1 is small, less than 5 ms. Now let us look at the magnetizing flux linkage λ. We see that it is smaller in amplitude than in the no-load case. We would expect this since during short-circuit conditions $i_1 \cong -i_2'$, whereby the mmfs of the two windings oppose, and the resulting flux

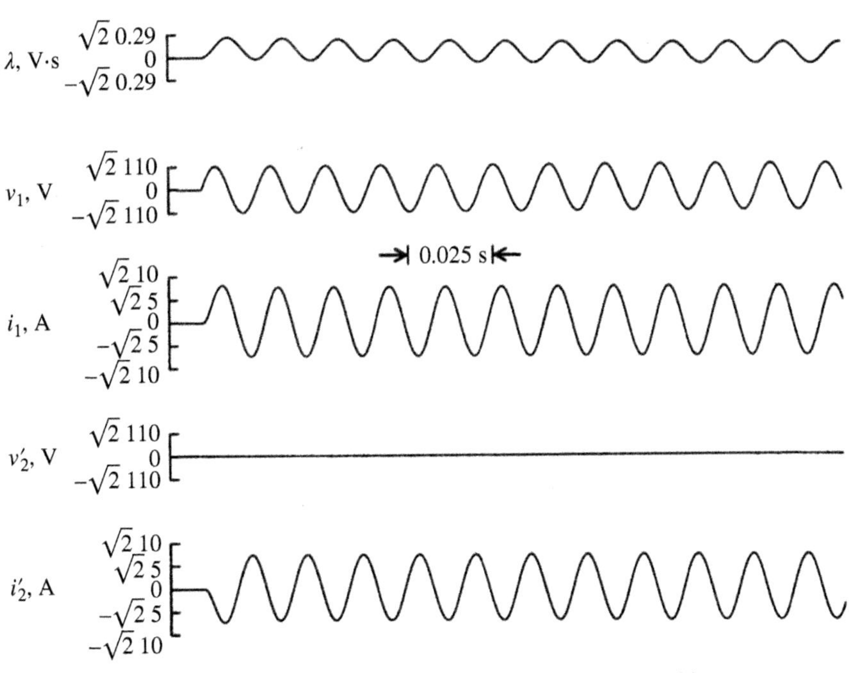

Figure 1.6-3 Short-circuit conditions of a two-winding transformer with $v_1 = \sqrt{2}\ 110 \sin 377t$.

in the transformer iron is less than for the no-load condition where $i_2' = 0$. Looking at this in another way, we realize that i_1 and $-i_2'$ will be essentially equal during short-circuit conditions whenever the impedance of the magnetizing branch $(j\omega_e L_{m1})$ is much larger (say 8–10 times larger) than $r_2' + j\omega_e L_{l2}'$. Here, $\omega_e = 377$ rad/s.

It is interesting to note that the decay of the magnetizing flux linkage λ is much slower than the apparent decay of the currents. As we mentioned, the time constant associated with i_1 is small; however, there is indeed a small difference between i_1 and $-i_2'$, and this small current (magnetizing current), which is actually a small part of i_1, must flow in the large inductance L_{m1}. Hence, the magnetizing current is associated with a longer time constant than the much larger component of the current i_1, which circulates through the series r_2' and L_{l2}'.

SP1.6-1 Use the plot of λ in Fig. 1.6-3 to approximate $|\tilde{I}_1 + \tilde{I}_2'|$. $[|\tilde{I}_1 + \tilde{I}'_2| \cong \frac{1}{2}\text{ A}]$

SP1.6-2 Calculate, using reasonable approximations, the phase angle between the steady-state current $\tilde{I}_1$ and voltage $\tilde{V}_1$ for the conditions of Fig. 1.6-3. Check your answer from the plots. [$\tilde{V}_1$ leads $\tilde{I}_1$ by 42.8°]

SP1.6-3 Consider the transformer given in Example 1D. Assume $V_1 = \sqrt{2}\ 110\cos 1000t$, and a load is connected across winding 2. The impedance of this load referred to winding 1 is $21 + j5\ \Omega$. Calculate $\tilde{I}_2'$. Make valid approximations to reduce your work. $\left[\tilde{I}_2' \cong -2.4\underline{/-45^\circ}\right]$

1.7 MAGNETIC SYSTEMS WITH MECHANICAL MOTION

In Chapter 2, relationships are derived for determining the electromagnetic force or torque established in electromechanical systems. Once this development is completed, three examples of elementary electromechanical systems are considered. It is convenient to introduce these three systems here for the purpose of establishing the voltage equations and expressions for the self- and mutual inductances, thereby setting the stage for the analysis to follow in Chapter 2. The first of these electromechanical systems is an elementary version of an electromagnet. It consists of a magnetic core, part of which is movable. The electric system exerts an electromagnetic force upon this movable member, thereby moving it relative to the stationary member. We shall analyze this device, and in Chapter 2 we shall observe its operating characteristics by computer traces. The second system is a rotational device commonly referred to as a reluctance machine. A large number of stepper motors operate on the reluctance-torque principle. The third device is also a rotational device that has two windings: one on the stationary member and one on the rotational member. This device, although somewhat impracticable, illustrates the concept of windings or magnetic systems in relative motion.

Elementary Electromagnet

An elementary electromagnet that we will consider is shown in Fig. 1.7-1. This system consists of a stationary core with a winding of N turns and a block of magnetic material that is free to slide relative to the stationary member. It is shown in more detail in Chapter 2, wherein a spring, a damper, and an external force are associated with the movable member. We do not need to consider that level of detail here; instead, we will assume that the movable member is at a distance x from the stationary member, which may be a function of time, that is $x = x(t)$.

The voltage equation that describes the electric system is

$$v = ri + \frac{d\lambda}{dt} \tag{1.7-1}$$

where the flux linkages are expressed as

$$\lambda = N\Phi \tag{1.7-2}$$

The flux may be written as

$$\Phi = \Phi_l + \Phi_m \tag{1.7-3}$$

where Φ_l is the leakage flux and Φ_m is the magnetizing flux that is common to both the stationary and movable members. If the magnetic system is considered to be linear (saturation neglected), then, as in the case of the stationary coupled circuits, we can express the fluxes in terms of reluctances. That is,

$$\Phi_l = \frac{Ni}{\mathcal{R}_l} \tag{1.7-4}$$

$$\Phi_m = \frac{Ni}{\mathcal{R}_m} \tag{1.7-5}$$

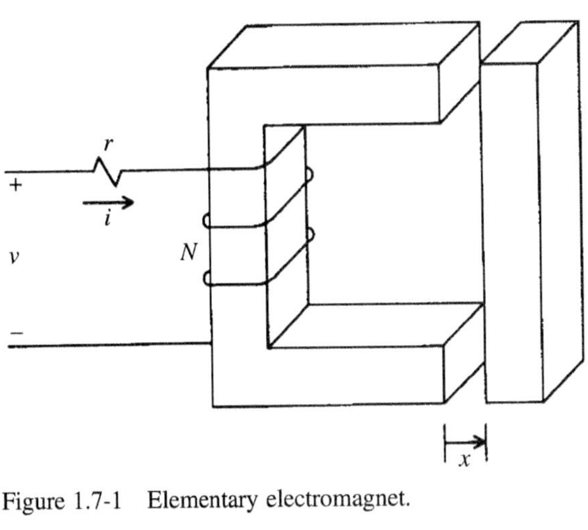

Figure 1.7-1 Elementary electromagnet.

where $\mathcal{R}_l$ and $\mathcal{R}_m$ are the reluctances of the leakage and magnetizing paths, respectively.

The flux linkages may now be written as

$$\lambda = \left(\frac{N^2}{\mathcal{R}_l} + \frac{N^2}{\mathcal{R}_m}\right)i \tag{1.7-6}$$

where the leakage inductance is

$$L_l = \frac{N^2}{\mathcal{R}_l} \tag{1.7-7}$$

and the magnetizing inductance is

$$L_m = \frac{N^2}{\mathcal{R}_m} \tag{1.7-8}$$

The reluctance of the magnetizing path is

$$\mathcal{R}_m = \mathcal{R}_i + 2\mathcal{R}_g \tag{1.7-9}$$

where $\mathcal{R}_i$ is the total reluctance of the magnetic material of the stationary and movable members and $\mathcal{R}_g$ is the reluctance of one of the air gaps. If the cross-sectional area of the stationary and movable members is assumed to be equal and of the same material, the reluctances may be expressed as

$$\mathcal{R}_i = \frac{l_i}{\mu_{ri}\mu_0 A_i} \tag{1.7-10}$$

$$\mathcal{R}_g = \frac{x}{\mu_0 A_g} \tag{1.7-11}$$

We will assume that $A_g = A_i$. Even though, as we have mentioned previously, this may be somewhat of an oversimplification, it is sufficient for our purposes. Hence, $\mathcal{R}_m$ may be written as

$$\mathcal{R}_m = \frac{1}{\mu_0 A_i}\left(\frac{l_i}{\mu_{ri}} + 2x\right) \tag{1.7-12}$$

The magnetizing inductance now becomes

$$L_m = \frac{N^2}{(1/\mu_0 A_i)(l_i/\mu_{ri} + 2x)} \tag{1.7-13}$$

In this analysis, the leakage inductance is assumed to be constant. The magnetizing inductance is clearly a function of displacement. That is, $x = x(t)$ and $L_m = L_m(x)$. Heretofore, when dealing with linear magnetic circuits wherein mechanical motion is not present as in the case of a transformer, the change of flux linkages with respect to time was simply $L(di/dt)$. This is not the case here. When the inductance is a function of $x(t)$,

$$\lambda(i, x) = L(x)i = [L_l + L_m(x)]i \tag{1.7-14}$$

and

$$\frac{d\lambda(i,x)}{dt} = \frac{\partial\lambda}{\partial i}\frac{di}{dt} + \frac{\partial\lambda}{\partial x}\frac{dx}{dt} \tag{1.7-15}$$

With (1.7-15) in mind, we see that the voltage equation (1.7-1) becomes

$$v = ri + [L_l + L_m(x)]\frac{di}{dt} + i\frac{dL_m(x)}{dx}\frac{dx}{dt} \tag{1.7-16}$$

Equation (1.7-16) is a nonlinear differential equation owing to the last two terms on the right-hand side.

In preparation for our work in Chapter 2, let us write (1.7-13) as

$$L_m(x) = \frac{k}{k_0 + x} \tag{1.7-17}$$

where

$$k = \frac{N^2\mu_0 A_i}{2} \tag{1.7-18}$$

$$k_0 = \frac{l_i}{2\mu_{ri}} \tag{1.7-19}$$

When $x = 0$, $L_m(x)$ is determined by the reluctance of the magnetic material. That is, for $x = 0$,

$$L_m(0) = \frac{k}{k_0} = \frac{N^2\mu_0\mu_{ri}A_i}{l_i} \tag{1.7-20}$$

Depending upon the parameters of the magnetic material, $L_m(x)$ may be adequately predicted by

$$L_m(x) = \frac{k}{x} \qquad \text{for } x > 0 \tag{1.7-21}$$

We will use this approximation in Chapter 2.

Elementary Reluctance Machine

An elementary reluctance machine is shown in Fig. 1.7-2. It consists of a stationary core with a winding of N turns and a movable member that rotates at an angular displacement and angular velocity of θ_r and ω_r, respectively. The displacement is defined as

$$\theta_r = \omega_r t + \theta_r(0) \tag{1.7-22}$$

The voltage equation is of the form given by (1.7-1). Similarly, the flux may be divided into a leakage and magnetizing flux, as given by (1.7-3). It is convenient to express the flux linkages as

$$\lambda = (L_l + L_m)i \tag{1.7-23}$$

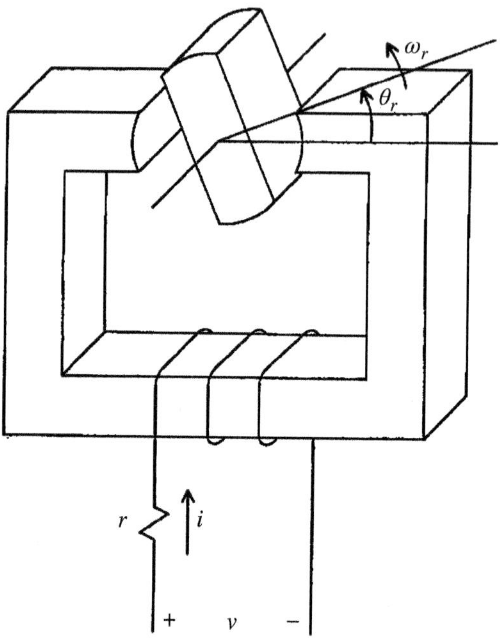

Figure 1.7-2 Elementary reluctance machine.

where L_l is the leakage inductance and L_m is the magnetizing inductance. The leakage inductance is essentially constant, independent of θ_r; however, the magnetizing inductance is a periodic function of θ_r. That is, $L_m = L_m(\theta_r)$. In particular, with $\theta_r = 0$,

$$L_m(0) = \frac{N^2}{\mathcal{R}_m(0)} \tag{1.7-24}$$

Here, the reluctance of the magnetizing path $\mathcal{R}_m$ is maximum due to the large air gap when the rotor is in the vertical (unaligned) position. Hence, L_m is a minimum in this position. Note that this same situation occurs not only at $\theta_r = 0$ but also when $\theta_r = \pi, 2\pi$, and so on.

Now, $\theta_r = \frac{1}{2}\pi$

$$L_m\left(\frac{1}{2}\pi\right) = \frac{N^2}{\mathcal{R}_m\left(\frac{1}{2}\pi\right)} \tag{1.7-25}$$

Here, $\mathcal{R}_m$ is a minimum and, thus, L_m is a maximum. This same situation occurs at $\theta_r = \frac{3}{2}\pi, \frac{5}{2}\pi$, and so on. Hence, the magnetizing inductance varies between maximum and minimum positive values twice per revolution of the rotating member (rotor). Let us make it easy for ourselves and assume that this variation may be adequately approximated by a sinusoidal function. In particular, let $L_m(\theta_r)$ be expressed as

$$L_m(\theta_r) = L_A - L_B \cos 2\theta_r \tag{1.7-26}$$

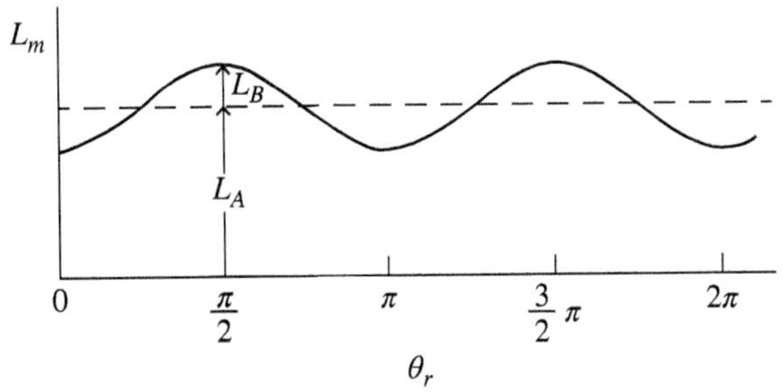

Figure 1.7-3 Approximation of magnetizing inductance of an elementary reluctance machine.

whereupon

$$L_m(0) = L_A - L_B \tag{1.7-27}$$

$$L_m\left(\frac{1}{2}\pi\right) = L_A + L_B \tag{1.7-28}$$

and $L_A > L_B$. The average value is L_A, illustrated in Fig. 1.7-3. The self-inductance may now be expressed as

$$\begin{aligned} L(\theta_r) &= L_l + L_m(\theta_r) \\ &= L_l + L_A - L_B \cos 2\theta_r \end{aligned} \tag{1.7-29}$$

The voltage equation is of the form given by (1.7-16) with x replaced by θ_r.

Windings in Relative Motion

The rotational device shown in Fig. 1.7-4 will be used to illustrate windings in relative motion. This device consists of two windings each containing several turns of a conductor. Winding 1 has N_1 turns and it is on the stationary member (stator); winding 2 has N_2 turns and it is on the rotating member (rotor). The ⊗ indicates that the assumed direction of positive current flow in the conductors is into the paper, whereas ⊙ indicates positive current flow in the conductors is out of the paper. In a practical device, the turns of a winding are distributed over an arc (often 30–60°) of the stator and rotor. However, in this introductory consideration, it is sufficient to assume that the turns are concentrated in one position, as shown in Fig. 1.7-4. Also, the length of the air gap between the stator and rotor is shown exaggerated relative to the inside diameter of the stator.

The voltage equations may be written as

$$v_1 = r_1 i_1 + \frac{d\lambda_1}{dt} \tag{1.7-30}$$

$$v_2 = r_2 i_2 + \frac{d\lambda_2}{dt} \tag{1.7-31}$$

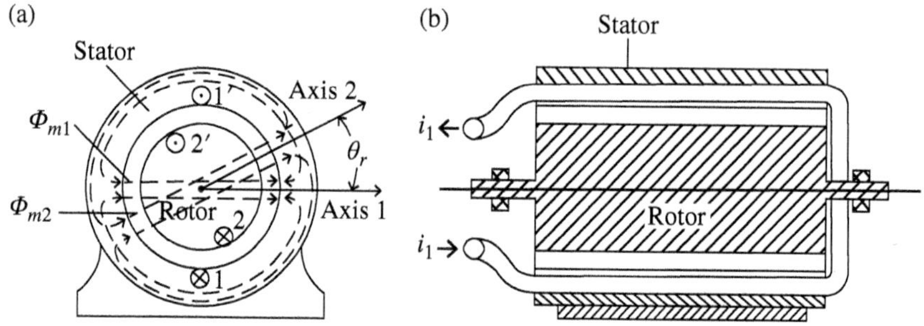

Figure 1.7-4 Elementary rotational electromechanical device. (a) End view; (b) cross-sectional view.

where r_1 and r_2 are the resistances of winding 1 and winding 2, respectively. The magnetic system is assumed linear; therefore, the flux linkages may be expressed as

$$\lambda_1 = L_{11}i_1 + L_{12}i_2 \tag{1.7-32}$$

$$\lambda_2 = L_{21}i_1 + L_{22}i_2 \tag{1.7-33}$$

The self-inductances L_{11} and L_{22} are constants and may be expressed in terms of leakage and magnetizing inductances as

$$\begin{aligned} L_{11} &= L_{l1} + L_{m1} \\ &= \frac{N_1^2}{\mathcal{R}_{l1}} + \frac{N_1^2}{\mathcal{R}_m} \end{aligned} \tag{1.7-34}$$

$$\begin{aligned} L_{22} &= L_{l2} + L_{m2} \\ &= \frac{N_2^2}{\mathcal{R}_{l2}} + \frac{N_2^2}{\mathcal{R}_m} \end{aligned} \tag{1.7-35}$$

where $\mathcal{R}_m$ is the reluctance of the complete magnetic path of Φ_{m1} and Φ_{m2}, which is through the rotor and stator iron and twice across the air gap. Clearly, it is the same for the magnetic system established by either winding 1 or winding 2.

Take a moment to note the designation of axis 1 and axis 2 in Fig. 1.7-4. These axes denote the positive direction of the respective magnetic systems with the assumed positive direction of current flow in the windings (right-hand rule). Now let us consider L_{12}. (Is it clear that $L_{12} = L_{21}$?) When θ_r, which is defined by (1.7-22), is zero, then the coupling between windings 1 and 2 is maximum. In particular, with $\theta_r = 0$ the magnetic system of winding 1 aids that of winding 2 with positive currents assumed. Hence, the mutual inductance is positive and

$$L_{12}(0) = \frac{N_1N_2}{\mathcal{R}_m} \tag{1.7-36}$$

When $\theta_r = \frac{1}{2}\pi$, the windings are orthogonal. The mutual coupling is zero. Hence,

$$L_{12}\left(\frac{1}{2}\pi\right) = 0 \tag{1.7-37}$$

Again let us make it as simple as possible by assuming that the mutual inductance may be adequately predicted by

$$L_{12}(\theta_r) = L_{sr}\cos\theta_r \tag{1.7-38}$$

where L_{sr} is the amplitude of the sinusoidal mutual inductance between the stator and rotor windings as given by (1.7-36).

In writing the voltage equations from (1.7-30) and (1.7-31), the total derivative of the flux linkages is required. This is accomplished by taking the partial derivative of both λ_1 and λ_2 with respect to i_1, i_2, and θ_r.

SP1.7-1 Let $L_m(x) = k/x$, $i = t$, and $x = t$. Express $d[L_m(x)i]/dt$. [Zero]

SP1.7-2 Express $L(\theta_r)$ of the elementary reluctance machine if minimum reluctance occurs at $\theta_r = 0$. $[L(\theta_r) = L_l + L_A + L_b\cos 2\theta_r]$

SP1.7-3 Express L_{11}, L_{22}, and L_{12} if positive i_2 is reversed from that shown in Fig. 1.7-4. [L_{11} and L_{22} are unchanged; $L_{12} = -L_{sr}\cos\theta_r$]

SP1.7-4 Consider Fig. 1.7-4. $I_1 = 1$ A, $L_{sr} = 0.1$ H, $\omega_r = 100$ rad/s, $\theta_r(0) = 0$, and winding 2 is open-circuited. Express V_2. $[V_2 = -10\sin 100t]$

1.8 RECAPPING

We will analyze electromechanical motion devices from the coupled-circuit viewpoint. Although the coupled windings of many electromechanical devices are in relative motion, the equivalent circuit of stationary coupled windings (the transformer) is the beginning of the equivalent circuits that we will develop for these devices in later chapters. We will find the concept of referring variables from one winding to the other very useful as we proceed.

The first step in the analysis of electromechanical motion devices of the electromagnetic type is to express the voltage and flux linkage equations in terms of self- and mutual inductances. We will not consider saturation in our analysis; instead we will restrict our work to linear magnetic systems and leave the analytical treatment of saturation to a more advanced study of these devices. In this chapter, we learned that electromagnetic, electromechanical motion devices are characterized by self- or mutual inductances that vary with displacement of the movable member.

In the next chapter, we will first develop an analytical means of determining the electromagnetic force or torque in electromechanical motion devices. Once we have accomplished this, we will be able to express the electromagnetic force in the elementary electromagnet and the electromagnetic torque in the elementary rotational devices that we have just considered.

1.9 PROBLEMS

1. Consider the magnetic system shown in Fig. 1.3-1. Let $\mu_r = 1500$, $N = 100$ turns, and $I = 2$ A. The cross section of the iron is square, each side 4 cm in length. The air gap is 4 mm in length. The mean length of the iron is 200 times the air gap length. Neglect leakage flux and assume $A_i = A_g$. Calculate the flux.

2. Repeat Example 1B with a second air gap of 2 mm in length cut midway between c and d. Neglect leakage flux and assume $A_i = A_g$.

3. An iron-core transformer that has two windings is shown in Fig. 1.9-1. $N_1 = 50$ turns, $N_2 = 100$ turns, and $\mu_r = 4000$. Calculate L_{12}, L_{m1}, and L_{m2}.

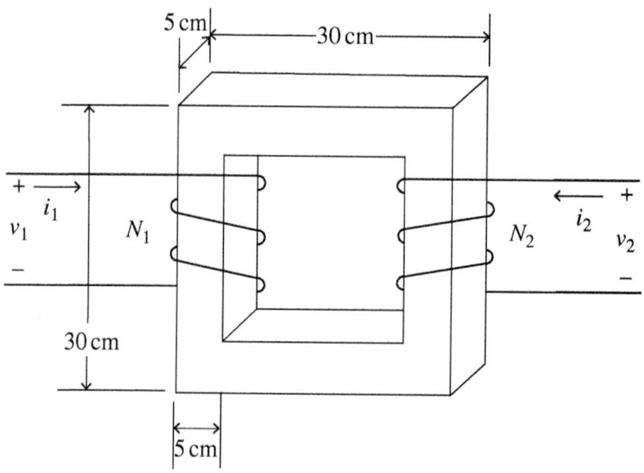

Figure 1.9-1 A two-winding iron-core transformer.

4. An iron "doughnut" (toroid) with two coils is shown in Fig. 1.9-2. $N_1 = 100$ turns and $N_2 = 200$ turns, $\mu_r = 10^4/4\pi$. Calculate L_{12}.

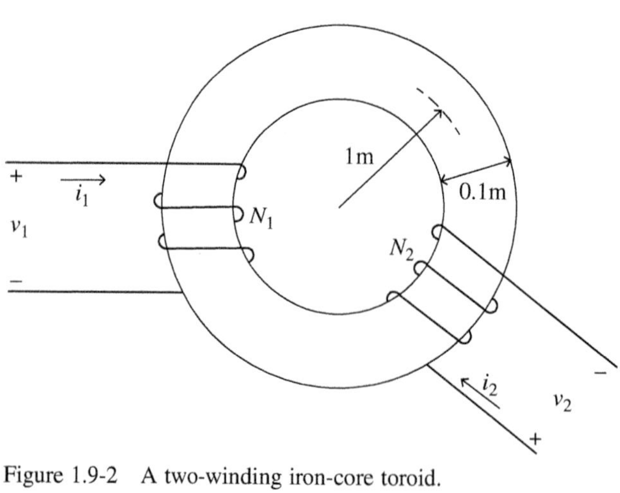

Figure 1.9-2 A two-winding iron-core toroid.

5. An air gap is cut through the left leg of the magnetic system shown in Fig. 1C-1 so that the associated reluctance is $10\mathcal{R}_y$ rather than $\mathcal{R}_y$. Express L_{12} and L_{21} in terms of N_1, N_2, $\mathcal{R}_x$, and $\mathcal{R}_y$.

6. Two coupled coils have the following parameters: $r_1 = 10\,\Omega$, $L_{l1} = 0.1L_{11}$, $r_2 = 2.5\,\Omega$, $L_{l2} = 0.1L_{22}$, $L_{11} = 100$ mH, $N_1 = 100$ turns, $L_{22} = 25$ mH, and $N_2 = 50$ turns. Develop an equivalent T circuit with (a) winding 1 as the reference winding and (b) winding 2 as reference winding.

7. Assume that the direction of positive current is reversed in winding 2 of Fig. 1.5-1. Express (a) L_{12} in terms of N_1, N_2, and $\mathcal{R}_m$; (b) λ_1 and λ_2 in the form of (1.5-22) and (1.5-23); (c) λ_1 and λ_2' in the form of (1.5-31) and (1.5-32); and (d) v_1 and v_2' in the form of (1.5-37).

8. The parameters of a transformer are: $r_1 = r_2' = 10\,\Omega$, $L_{m1} = 300$ mH, and $L_{l1} = L_{l2}' = 30$ mH. A 10-V peak-to-peak 30-Hz sinusoidal voltage is applied to winding 1. Winding 2 is short-circuited. Assume $i_1 = -i_2'$. Calculate the phasor $\tilde{I}_1$ with $\tilde{V}_1$ at zero degrees.

9. A transformer with two windings has the following parameters: $r_1 = r_2 = 1\,\Omega$, $L_{m1} = 1$ H, $L_{l1} = L_{l2} = 0.01$ H, and $N_1 = N_2$. A 2-Ω load resistance R_L is connected across winding 2. $V_1 = 2\cos 400t$. (a) Calculate $\tilde{I}_1$. (b) Express I_1.

10. A transformer with two windings has the following parameters: $r_1 = 1\,\Omega$, $L_{l1} = 0.01$ H, $L_{m1} = 0.2$ H, $N_2 = 2N_1$, $r_2 = 2\,\Omega$, $L_{l2} = 0.04$ H, and $L_{m2} = 0.08$ H. A 4 - Ω resistance R_L is connected across the terminals of winding 2 and a voltage $V_1 = \sqrt{2}\,2\cos 400t$ is applied to winding 1. Calculate and draw the phasor diagram showing $\tilde{V}_1$, $\tilde{I}_1$, $\tilde{V}_2'$, and $\tilde{I}_2'$. Neglect the magnetizing current.

11. For the elementary electromagnet shown in Fig. 1.7-1, assume that the cross-sectional area of the stationary and movable member is the same and $A_i = A_g = 4\text{ cm}^2$. Assume $l_i = 20$ cm, $N = 500$ turns, and $\mu_{ri} = 1000$. Express $L_m(x)$ given by (1.7-17) and the approximation for $x > 0$ given by (1.7-21). Determine the minimum value of x when this approximate expression for $L_m(x)$ is less than 1.1 the value given by (1.7-17).

12. Express the voltage v of the elementary electromagnet given by (1.7-16) for $L_m(x)$ given by (1.7-17), $i = \sqrt{2}I_s \cos\omega_e t$, and $x = t$. Approximate v when t is large.

13. Express the voltage equation for the elementary reluctance machine shown in Fig. 1.7-2. Use (1.7-29) for $L(\theta_r)$.

14. Write the voltage equations for the coils in relative motion shown in Fig. 1.7-4. Use L_{11}, L_{22}, and L_{12} as expressed by (1.7-38). ■

1.10 REFERENCES

1. G. R. Slemon and A. Straughen, *Electric Machines*, Addison-Wesley Publishing Company, Reading, Mass., 1980.
2. P. C. Krause, *Analysis of Electric Machinery*, McGraw-Hill Book Company, New York, 1986.

CHAPTER 2

ELECTROMECHANICAL ENERGY CONVERSION

2.1 INTRODUCTION

The theory of electromechanical energy conversion is the cornerstone for the analysis of electrical motion devices. This theory allows us to express the electromagnetic force or torque in terms of device variables such as the currents and the displacement of the mechanical system. Since numerous types of electromechanical devices are used in motion systems, it is desirable to establish methods of analysis that may be applied to a variety of electromechanical devices rather than just to electric machines. Therefore, the theory of electromechanical energy conversion is set forth in considerable detail for the purpose of providing a background sufficient to analyze electromechanical systems other than just those treated in this text. The first part of this chapter is devoted to establishing analytically the relationships that can be used to express the electromagnetic force or torque. Although one may prefer to perform a separate derivation for each device because it is instructive to do so, a general set of formulas are given in tabular form that are applicable to a variety of electromechanical systems with a single mechanical input.

Once the theory of electromechanical energy conversion is established, a detailed analysis of the elementary electromagnet, which was introduced in Chapter 1, is performed with computer traces included to demonstrate its dynamic performance related to changes in the applied voltage and the external mechanical force. In the final sections, the expressions for the electromagnetic torque are developed for the elementary single-phase reluctance machine and for windings in relative motion. Brief discussions are given of several steady-state modes of operation of these devices, which help to illustrate, in an elementary form, the positioning of stepper motors and the operation of synchronous motors.

Electromechanical Motion Devices: Rotating Magnetic Field-Based Analysis and Online Animations, Third Edition. Paul Krause, Oleg Wasynczuk, Steven D. Pekarek, and Timothy O'Connell.

Published 2020 by John Wiley & Sons, Inc.

The material presented in this chapter is sufficient preparation to study dc machines covered in Chapter 3. However, Chapter 3 is not necessary for the analysis of electromechanical motion devices of the induction and synchronous types, which begins with Chapter 4.

2.2 ENERGY BALANCE RELATIONSHIPS

Electromechanical systems are comprised of an electric system, a mechanical system, and a means whereby the electric and mechanical systems can interact. Interactions can take place through any and all electromagnetic and electrostatic fields that are common to both systems, and energy is transferred from one system to the other as a result of this interaction. Both electrostatic and electromagnetic coupling fields may exist simultaneously and the electromechanical system may have any number of electric and mechanical subsystems. However, before considering an involved system, it is helpful to analyze the electromechanical system in a simplified form. An electromechanical system with one electric subsystem, one mechanical subsystem, and one coupling field is depicted in Fig. 2.2-1. Electromagnetic radiation is neglected, and it is assumed that the electric system operates at a frequency sufficiently low so that the electric system may be considered as a lumped-parameter system.

Heat loss will occur in the mechanical system due to friction, and the electric system will dissipate heat due to the resistance of the current-carrying conductors. Eddy current and hysteresis losses occur in the ferromagnetic materials, whereas dielectric losses occur in all electric fields. If W_E is the total energy supplied by the electric source and W_M the total energy supplied by the mechanical source, then the energy distribution could be expressed as

$$W_E = W_e + W_{eL} + W_{eS} \tag{2.2-1}$$

$$W_M = W_m + W_{mL} + W_{mS} \tag{2.2-2}$$

In (2.2-1), W_{eS} is the energy stored in the electric or magnetic fields, which are not coupled with the mechanical system. The energy W_{eL} is the heat loss associated with the electric system excluding the coupling field losses. This loss occurs due to the resistance of the current-carrying conductors as well as the energy dissipated in the form of heat owing to hysteresis, eddy currents, and dielectric losses external to the coupling field. The energy W_e is the energy transferred to the coupling field by the electric system. The energies common to the mechanical system may be defined in a similar manner. In (2.2-2), W_{mS} is the energy stored in the moving member and compliances of the mechanical system, W_{mL} is the

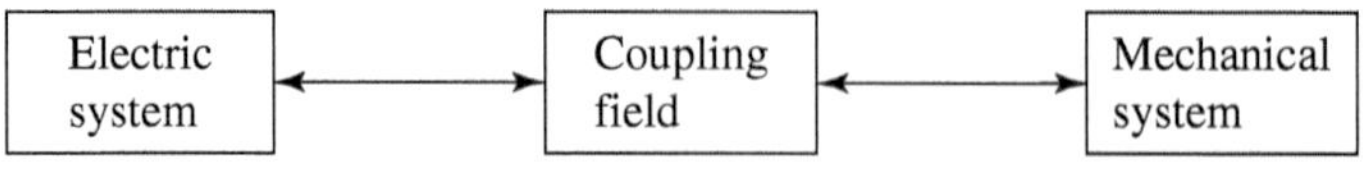

Figure 2.2-1 Block diagram of an elementary electromechanical system.

energy loss in the mechanical system in the form of heat, and W_m is the energy transferred to the coupling field. It is important to note that, with the convention adopted, the energy transferred to the coupling field by either source is considered positive. Also, W_E (W_M) is negative when energy is supplied to the electric source (mechanical source).

If W_F is defined as the total energy transferred to the coupling field, then

$$W_F = W_f + W_{fL} \tag{2.2-3}$$

where W_f is the energy stored in the coupling field and W_{fL} is the energy dissipated in the form of heat due to losses within the coupling field (eddy current, hysteresis, or dielectric losses). In order to comply with convention, we will use W_f to denote the energy stored in the coupling field rather than W_{fS}. The electromechanical system must obey the law of conservation of energy, thus,

$$W_f + W_{fL} = (W_E - W_{eL} - W_{eS}) + (W_M - W_{mL} - W_{mS}) \tag{2.2-4}$$

which may be written as

$$W_f + W_{fL} = W_e + W_m \tag{2.2-5}$$

This energy balance is shown schematically in Fig. 2.2-2.

The actual process of converting electric energy to mechanical energy (or vice versa) is independent of (i) the loss of energy in either the electric or the mechanical systems (W_{eL} and W_{mL}), (ii) the energies stored in the electric or magnetic fields that are not common to both systems (W_{eS}), or (iii) the energies stored in the mechanical system (W_{mS}). If the losses of the coupling field are neglected, then the field is conservative and (2.2-5) becomes

$$W_f = W_e + W_m \tag{2.2-6}$$

Examples of elementary electromechanical systems are shown in Figs. 2.2-3 and 2.2-4. The system shown in Fig. 2.2-3 has a magnetic coupling field, whereas the electromechanical system shown in Fig. 2.2-4 has an electric field as a means of transferring energy between the electric and mechanical systems. In both systems, the space between the movable and stationary members is exaggerated for clarity. In these systems, v is the voltage of the electric source and f is an externally applied

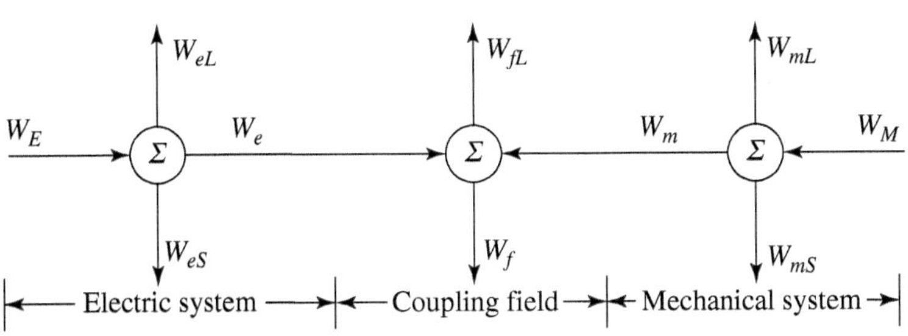

Figure 2.2-2 Energy balance.

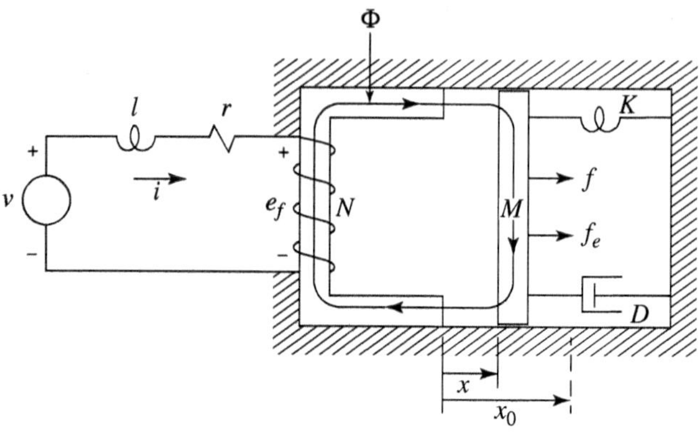

Figure 2.2-3 Electromechanical system with magnetic field.

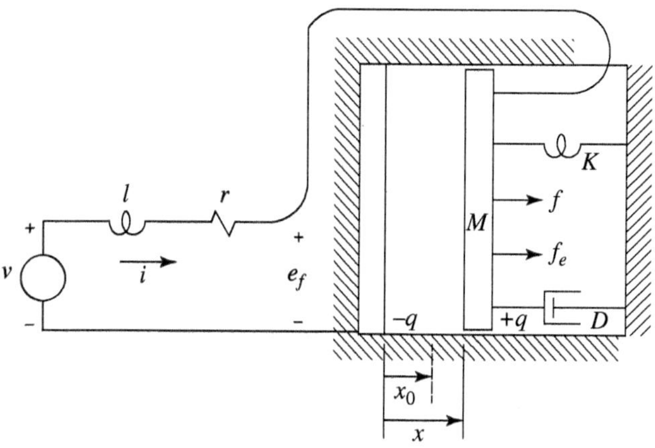

Figure 2.2-4 Electromechanical system with electric field.

mechanical force. The electromagnetic or electrostatic force is denoted f_e. The resistance of the current-carrying conductor is denoted by r, with l denoting the inductance of a linear (conservative) electromagnetic system that does not couple the mechanical system. In the mechanical system, M is the mass of the movable member, and the linear compliance and damper are represented by a spring constant K and a damping coefficient D, respectively. The displacement x_0 is the zero force or equilibrium position of the mechanical system, which is the steady-state position of the mass with f_e and f equal to zero.

The voltage equation that describes the electric systems shown in Figs. 2.2-3 and 2.2-4 may be written as

$$v = ri + l\frac{di}{dt} + e_f \tag{2.2-7}$$

where e_f is the voltage drop due to the coupling field. The dynamic behavior of the translational mechanical systems may be expressed by employing Newton's law of motion. Thus,

$$f = M\frac{d^2x}{dt^2} + D\frac{dx}{dt} + K(x - x_0) - f_e \tag{2.2-8}$$

Since power is the time rate of energy transfer, the total energy supplied by the electric source is

$$W_E = \int vidt \tag{2.2-9}$$

The total energy supplied by the mechanical source is

$$W_M = \int fdx \tag{2.2-10}$$

which may also be expressed as

$$W_M = \int f\frac{dx}{dt}dt \tag{2.2-11}$$

Substituting (2.2-7) into (2.2-9) yields

$$W_E = r\int i^2dt + l\int i\frac{di}{dt}dt + \int e_f idt \tag{2.2-12}$$

The first term on the right-hand side of (2.2-12) represents the energy loss due to the resistance of the conductors (W_{eL}). The second term represents the energy stored in the linear electromagnetic field external to the coupling field (W_{eS}). Therefore, the total energy transferred to the coupling field from the electric system is

$$W_e = \int e_f idt \tag{2.2-13}$$

Similarly, for the mechanical system

$$W_M = M\int\frac{d^2x}{dt^2}dx + D\int\left(\frac{dx}{dt}\right)^2dt + K\int(x - x_0)dx - \int f_e dx \tag{2.2-14}$$

Here, the first and third terms on the right-hand side of (2.2-14) represent the kinetic energy stored in the mass and the potential energy stored in the spring, respectively. The sum of these two stored energies is W_{mS}. You should take a moment to look at the first term on the right-hand side of (2.2-14) and recognize that it can be written as $\frac{1}{2}M(dx/dt)^2$. The second term is the heat loss due to friction (W_{mL}). Thus, the total energy transferred to the coupling field from the mechanical system is

$$W_m = -\int f_e dx \tag{2.2-15}$$

It is important to note, from Figs. 2.2-3 and 2.2-4, that a positive force f_e is assumed to be in the same direction as a positive displacement dx. Substituting (2.2-13) and (2.2-15) into the energy balance relation, (2.2-6), yields

$$W_f = \int e_f i dt - \int f_e dx \tag{2.2-16}$$

The equations set forth may be readily extended to include an electromechanical system with any number of electric and mechanical inputs and any number of coupling fields. Considering the system shown in Fig. 2.2-5, the energy supplied to the coupling field may be expressed as

$$W_f = \sum_{j=1}^{J} W_{ej} + \sum_{k=1}^{K} W_{mk} \tag{2.2-17}$$

wherein J electric and K mechanical inputs exist. The total energy supplied to the coupling field from the electric inputs is

$$\sum_{j=1}^{J} W_{ej} = \int \sum_{j=1}^{J} e_{fj} i_j dt \tag{2.2-18}$$

The total energy supplied to the coupling field from the mechanical inputs is

$$\sum_{k=1}^{K} W_{mk} = -\int \sum_{k=1}^{K} f_{ek} dx_k \tag{2.2-19}$$

In our analysis of electromechanical systems, we will consider devices with only one mechanical input, for example, the shaft of the electric machine or the moving arm of a magnetic relay. On the other hand, since machines may have more than one electric terminal, it is necessary to consider systems with multiple electric inputs. In all cases, however, the multiple electric inputs have a common coupling field. Therefore, we need not become too ambitious in the following derivations. More specifically, hereafter we will restrict our analysis to electromechanical devices with only one mechanical input. Thus, the k subscript will be dropped from

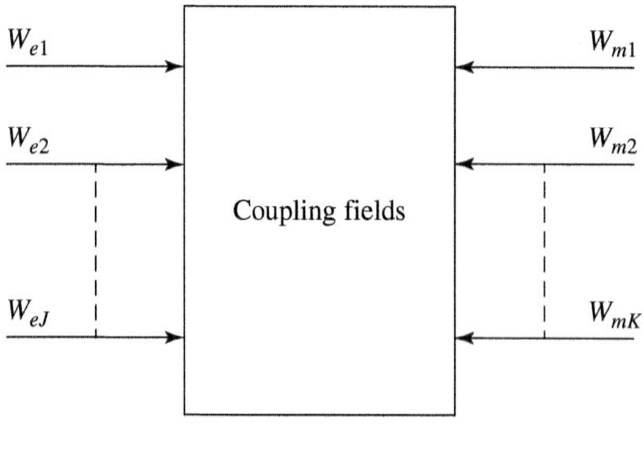

Figure 2.2-5 Multiple electric and mechanical inputs.

f_e, x and W_m. This reduces our work considerably without restricting the practical application of our results. With one mechanical input, the energy balance equation becomes

$$W_f = \int \sum_{j=1}^{J} e_{fj} i_j dt - \int f_e dx \tag{2.2-20}$$

In differential form, which will be the form we will use extensively,

$$dW_f = \sum_{j=1}^{J} e_{fj} i_j dt - f_e dx \tag{2.2-21}$$

SP2.2-1 The current flowing in a 1-H inductor which is external to the coupling field is $i = kt + k_0$. Calculate W_{eS}, the energy stored in the inductor at $t = 1$ s. $\left[W_{eS} = \frac{1}{2}(k + k_0)^2\right]$

SP2.2-2 Express the undamped natural frequency of the mechanical system described by (2.2-8). $\left[\omega_n = (K/M)^{1/2}\right]$

SP2.2-3 Express the instantaneous power delivered to the inductor in SP2.2-1. $\left[P = k^2 t + kk_0\right]$

SP2.2-4 For two electric inputs, $e_{f1} = k_1 t, i_1 = k_0, e_{f2} = k_2 t^2$ and $i_2 = k_3 t$. Express the total energy supplied to the coupling field (W_e) in 2 s. $[W_e = 2k_0 k_1 + 4k_2 k_3]$

2.3 ENERGY IN COUPLING FIELD

Before using (2.2-21) to obtain an expression for the electromagnetic force f_e, it is necessary to derive an expression for the energy stored in the coupling field. Once we have an expression for W_f, we can take the total derivative to obtain dW_f, which can then be substituted into (2.2-21). When expressing the energy in the coupling field, it is convenient to neglect all losses associated with the electric or magnetic coupling field, whereupon the field is assumed to be conservative and the energy stored therein is a function of the state of the electrical and mechanical variables. Although the effects of the core losses of the coupling field may be functionally accounted for by appropriately introducing a resistance in the electric circuit, this refinement is generally not necessary since the ferromagnetic material is selected and arranged in laminations so as to minimize the hysteresis and eddy current losses. Moreover, most of the energy stored in the coupling field is stored in the air gap of the electromechanical device. Since air is a conservative medium, all of the energy stored therein can be returned to the electric or mechanical systems. Therefore, the assumption of a lossless coupling field is not as restrictive as it might first appear.

The energy stored in a conservative field is a function of the state of the system variables and not the manner in which the variables reached that state. It is convenient to take advantage of this feature when developing a mathematical expression for the field energy. In particular, it is convenient to fix mathematically the position of the mechanical system associated with the coupling field and then excite the electric system with the displacement of the mechanical system held fixed. During the excitation of the electric inputs, $dx = 0$, hence, W_m is zero even though electromagnetic or electrostatic forces may occur. Therefore, with the displacement held fixed, the energy stored in the coupling field during the excitation of the electric inputs is equal to the energy supplied to the coupling field by the electric inputs. Thus, with $dx = 0$, the energy supplied from the electric system may be expressed from (2.2-20) as

$$W_f = \int \sum_{j=1}^{J} e_{fj} i_j dt \quad \text{with } dx = 0 \tag{2.3-1}$$

Let us consider a singly excited electromagnetic system similar to that shown in Fig. 2.2-3. In this case, $e_f = d\lambda/dt$, whereupon (2.3-1) becomes

$$W_f = \int i d\lambda \quad \text{with } dx = 0 \tag{2.3-2}$$

Here, $j = 1$; however, the subscript is omitted for the sake of brevity. The area to the left of the λi relationship, shown in Fig. 2.3-1 for a singly excited electromagnetic system, is the area described by (2.3-2). In Fig. 2.3-1, this area represents the energy stored in the field at the instant when $\lambda = \lambda_a$ and $i = i_a$. The λi relationship need not be linear; it need only be single-valued, a property that is characteristic of

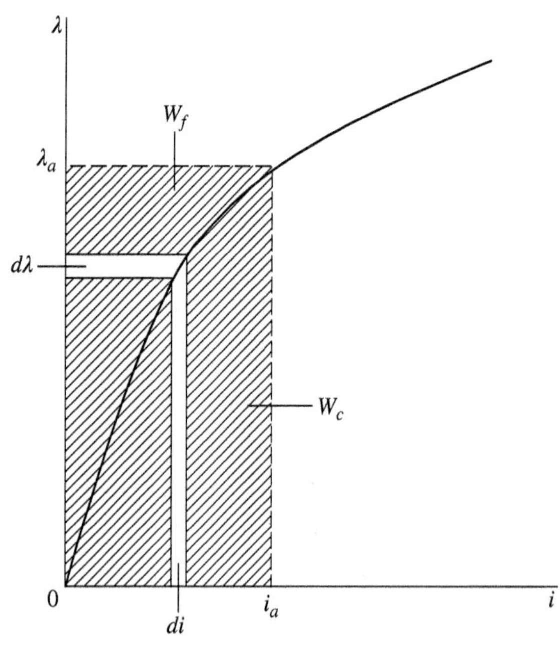

Figure 2.3-1 Stored energy and coenergy in a magnetic field of a singly excited electromagnetic device.

a conservative or lossless field. Moreover, since the coupling field is conservative, the energy stored in the field with $\lambda = \lambda_a$ and $i = i_a$ is independent of the excursion of the electrical and mechanical variables before reaching this state.

The area to the right of the λi curve is called the *coenergy* and can be expressed as

$$W_c = \int \lambda di \qquad \text{with } dx = 0 \tag{2.3-3}$$

Although the coenergy has little or no physical significance, we will find it a convenient quantity for expressing the electromagnetic force. From Fig. 2.3-1, we see that the sum of W_f and W_c is λ times i, that is,

$$\lambda i = W_c + W_f \tag{2.3-4}$$

which is also valid for multiple electric inputs, where λi in (2.3-4) is replaced by $\sum_{j=1}^{J} \lambda_j i_j$. It should be clear that, for a linear magnetic system where the λi plots are straight-line relationships, $W_f = W_c = \frac{1}{2}\lambda i$.

The displacement x defines completely the influence of the mechanical system upon the coupling field; however, since λ and i are related, only one is needed in addition to x in order to describe the state of the electromechanical system. Therefore, we can select either λ and x as independent variables or i and x. If i and x are selected as the independent variables, it is convenient to express the field energy and the flux linkages as

$$W_f = W_f(i, x) \tag{2.3-5}$$

$$\lambda = \lambda(i, x) \tag{2.3-6}$$

With i and x as independent variables, we must express $d\lambda$ in terms of di before substituting into (2.3-2). Thus, from (2.3-6),

$$d\lambda = \frac{\partial \lambda(i,x)}{\partial i} di + \frac{\partial \lambda(i,x)}{\partial x} dx \tag{2.3-7}$$

In the derivation of an expression for the energy stored in the field, dx is set equal to zero. Hence, in the evaluation of field energy where $dx = 0$, $d\lambda$ is equal to the first term on the right-hand side of (2.3-7). Substituting into (2.3-2) yields

$$\begin{aligned} W_f(i,x) &= \int i \frac{\partial \lambda(i,x)}{\partial i} di \\ &= \int_0^i \xi \frac{\partial \lambda(\xi,x)}{\partial \xi} d\xi \qquad \text{with } dx = 0 \end{aligned} \tag{2.3-8}$$

where ξ is a dummy variable of integration. Evaluation of (2.3-8) gives the energy stored in the field of the singly excited system. The coenergy in terms of i and x may be evaluated from (2.3-3) as

$$\begin{aligned} W_c(i,x) &= \int \lambda(i,x)di \\ &= \int_0^i \lambda(\xi,x)d\xi \end{aligned} \tag{2.3-9}$$

With λ and x as independent variables

$$W_f = W_f(\lambda, x) \tag{2.3-10}$$

$$i = i(\lambda, x) \tag{2.3-11}$$

The field energy may be evaluated from (2.3-2) as

$$\begin{aligned} W_f(\lambda,x) &= \int i(\lambda,x)d\lambda \\ &= \int_0^\lambda i(\xi,x)d\xi \end{aligned} \tag{2.3-12}$$

To evaluate the coenergy with λ and x as independent variables, we need to express di in terms of $d\lambda$. Thus, from (2.3-11),

$$di = \frac{\partial i(\lambda,x)}{\partial \lambda}d\lambda + \frac{\partial i(\lambda,x)}{\partial x}dx \tag{2.3-13}$$

Since $dx = 0$ in this evaluation, (2.3-3) becomes

$$\begin{aligned} W_c(\lambda,x) &= \int \lambda \frac{\partial i(\lambda,x)}{d\lambda}d\lambda \\ &= \int_0^\lambda \xi \frac{\partial i(\xi,x)}{\partial \xi}d\xi \end{aligned} \tag{2.3-14}$$

For a linear electromagnetic system, the λi plots are straight-line relationships. Thus, for the singly excited magnetically linear system,

$$\lambda(i,x) = L(x)i \tag{2.3-15}$$

Or

$$i(\lambda,x) = \frac{\lambda}{L(x)} \tag{2.3-16}$$

where $L(x)$ is the inductance. Let us evaluate $W_f(i,x)$. With $dx = 0$ and since $\dfrac{\partial \lambda(i,x)}{\partial i} = L(x)$, (2.3-7) becomes

$$d\lambda = L(x)di \tag{2.3-17}$$

Hence, from (2.3-8),

$$W_f(i,x) = \int_0^i \xi L(x)d\xi = \frac{1}{2}L(x)i^2 \tag{2.3-18}$$

It is left to the reader to show that by a similar procedure $W_f(\lambda, x)$, $W_c(i, x)$, and $W_c(\lambda, x)$ are equivalent to (2.3-18) for this magnetically linear system.

The field energy is a state function and the expression describing the field energy in terms of system variables is valid regardless of the variations in the system variables. For example, (2.3-18) expresses the field energy regardless of the variations in $L(x)$ and i. The fixing of the mechanical system so as to obtain an expression for the field energy is a mathematical convenience and not a restriction upon the result.

In the case of a multiexcited electromagnetic system, an expression for the field energy may be obtained by evaluating the following relation with $dx = 0$:

$$W_f = \int \sum_{j=i}^{J} i_j d\lambda_j \quad \text{with } dx = 0 \tag{2.3-19}$$

Since the coupling field is considered conservative, (2.3-19) may be evaluated independent of the order in which the flux linkages or currents are brought to their final values. To illustrate evaluation of (2.3-19) for a multiexcited system, we will allow the currents to establish their final states one at a time while all other currents are mathematically fixed either in their unexcited or final states. This procedure may be illustrated by considering a doubly excited electric system with one mechanical input. An electromechanical system of this type could be constructed by placing a second winding, supplied from a second electric system, on either the stationary or movable member of the system shown in Fig. 2.2-3. In this evaluation, it is convenient to use currents and displacement as the independent variables. Hence, for a doubly excited electric system,

$$W_f(i_1, i_2, x) = \int [i_1 d\lambda_1(i_1, i_2, x) + i_2 d\lambda_2(i_1, i_2, x)] \quad \text{with } dx = 0 \tag{2.3-20}$$

In this determination of an expression for W_f, the mechanical displacement is held constant ($dx = 0$); thus (2.3-20) becomes

$$\begin{aligned} W_f(i_1, i_2, x) = \int i_1 &\left[\frac{\partial\lambda_1(i_1, i_2, x)}{\partial i_1} di_1 + \frac{\partial\lambda_1(i_1, i_2, x)}{\partial i_2} di_2\right] \\ + i_2 &\left[\frac{\partial\lambda_2(i_1, i_2, x)}{\partial i_1} di_1 + \frac{\partial\lambda_2(i_1, i_2, x)}{\partial i_2} di_2\right] \end{aligned} \tag{2.3-21}$$

We will evaluate the energy stored in the field by employing (2.3-21) twice. First, we will mathematically increase the current i_1 from zero to its desired final value while holding i_2 at zero. Thus, i_1 is the variable of integration and $di_2 = 0$. Energy is supplied to the coupling field from the source connected to winding 1. As the second evaluation of (2.3-21), i_2 is increased from zero to its desired final value while maintaining i_1 at the value attained in the preceding step. Hence, i_2 is the variable of integration and $di_1 = 0$. During this time, energy is supplied from both sources to the coupling field since $i_1 d\lambda_1$ is, in general, nonzero. The total energy stored in the coupling field is the sum of the two evaluations. Following this two-step procedure, the evaluation of (2.3-21) for the total field energy becomes

$$\begin{aligned} W_f(i_1, i_2, x) = &\int i_1 \frac{\partial \lambda_1(i_1, 0, x)}{\partial i_1} di_1 \\ &+ \int \left[i_1 \frac{\partial \lambda_1(i_1, i_2, x)}{\partial i_2} di_2 + i_2 \frac{\partial \lambda_2(i_1, i_2, x)}{\partial i_2} di_2 \right] \end{aligned} \tag{2.3-22}$$

which should be written as

$$\begin{aligned} W_f(i_1, i_2, x) = &\int_0^{i_1} \xi \frac{\partial \lambda_1(\xi, 0, x)}{\partial \xi} d\xi \\ &+ \int_0^{i_2} \left[i_1 \frac{\partial \lambda_1(i_1, \xi, x)}{\partial \xi} + \xi \frac{\partial \lambda_2(i_1, \xi, x)}{\partial \xi} d\xi \right] \end{aligned} \tag{2.3-23}$$

The first integral on the right-hand side of (2.3-22) or (2.3-23) results from the first step of the evaluation with i_1 as the variable of integration and with $i_2 = 0$ and $di_2 = 0$. The second integral comes from the second step of the evaluation with i_1 equal to its final value ($di_1 = 0$) and i_2 as the variable of integration. The order of allowing the currents to reach their final state is irrelevant; that is, as our first step, we could have made i_2 the variable of integration while holding i_1 at zero ($di_1 = 0$) and then let i_1 become the variable of integration while holding i_2 at its final value. The results would be the same. For three electric inputs, the evaluation procedure would require three steps, one for each current to be brought mathematically to its final state.

Let us now evaluate the energy stored in a magnetically linear system with two electric inputs and one mechanical input. For this, let

$$\lambda_1(i_1, i_2, x) = L_{11}(x)i_1 + L_{12}(x)i_2 \tag{2.3-24}$$

$$\lambda_2(i_1, i_2, x) = L_{21}(x)i_1 + L_{22}(x)i_2 \tag{2.3-25}$$

where the self-inductances $L_{11}(x)$ and $L_{22}(x)$ include the leakage inductances. With the mechanical displacement held constant ($dx = 0$),

$$d\lambda_1(i_1, i_2, x) = L_{11}(x)di_1 + L_{12}(x)di_2 \tag{2.3-26}$$

$$d\lambda_2(i_1, i_2, x) = L_{12}(x)di_1 + L_{22}(x)di_2 \tag{2.3-27}$$

The coefficients on the right-hand side of (2.3-26) and (2.3-27) are the partial derivatives. For example, $L_{11}(x)$ is the partial derivative of $\lambda_1(i_1, i_2, x)$ with respect to i_1. Appropriate substitution into (2.3-23) gives

$$W_f(i_1, i_2, x) = \int_0^{i_1} \xi L_{11}(x) d\xi + \int_0^{i_2} [i_1 L_{12}(x) + \xi L_{22}(x)] d\xi \tag{2.3-28}$$

which yields

$$W_f(i_1, i_2, x) = \frac{1}{2} L_{11}(x) i_1^2 + L_{12}(x) i_1 i_2 + \frac{1}{2} L_{22}(x) i_2^2 \tag{2.3-29}$$

It follows that the total field energy of a linear electromagnetic system with J electric inputs may be expressed as

$$W_f(i_1, \ldots, i_J, x) = \frac{1}{2}\sum_{p=1}^{J}\sum_{q=1}^{J} L_{pq} i_p i_q \tag{2.3-30}$$

Example 2A Calculation of W_f and W_c
Consider the magnetic system described by

$$\lambda(i, x) = \left(0.1 + kx^{-1}\right)i \tag{2A-1}$$

Calculate W_f and W_c. Here, it is convenient to work with i and x as independent variables. Thus, from (2.3-8),

$$W_f = \int_0^i \xi\left(0.1 + kx^{-1}\right)d\xi = \frac{1}{2}\left(0.1 + kx^{-1}\right)i^2 \tag{2A-2}$$

From (2.3-9),

$$W_c = \int_0^i \left(0.1 + kx^{-1}\right)\xi d\xi = \frac{1}{2}\left(0.1 + kx^{-1}\right)i^2 \tag{2A-3}$$

We see that $W_f = W_c$. Did you recognize the fact that this is a linear magnetic system with

$$L(x) = 0.1 + kx^{-1} \tag{2A-4}$$

where the first term on the right-hand side is analogous to the leakage inductance and kx^{-1} is the magnetizing inductance for $x > 0$ of the electromagnet treated in Chapter 1.

Let us make it a nonlinear system; in particular, let

$$\lambda(i, x) = \left(0.1 + kx^{-1}\right)i^2 \tag{2A-5}$$

From (2.3-8),

$$W_f = \int_0^i \xi 2\left(0.1 + kx^{-1}\right)\xi d\xi = \frac{2}{3}\left(0.1 + kx^{-1}\right)i^3 \tag{2A-6}$$

From (2.3-9),

$$W_c = \int_0^i \left(0.1 + kx^{-1}\right)\xi^2 d\xi = \frac{1}{3}\left(0.1 + kx^{-1}\right)i^3 \tag{2A-7}$$

We see that W_f and W_c are not equal; however, according to (2.3-4),

$$\lambda i = W_f + W_c \tag{2A-8}$$

Let us show that this is true.

$$\left[\left(0.1 + kx^{-1}\right)i^2\right]i = \frac{2}{3}\left(0.1 + kx^{-1}\right)i^3 + \frac{1}{3}\left(0.1 + kx^{-1}\right)i^3$$

$$\left(0.1 + kx^{-1}\right)i^3 = \left(0.1 + kx^{-1}\right)i^3 \tag{2A-9}$$

SP2.3-1 $\lambda = kx^2 i$. Calculate W_f and W_c when $kx^2 = 1$ V · s /A and $i = 2$ A. $[W_f = W_c = 2\ \text{J}]$

SP2.3-2 $\lambda = kxi^2$. Calculate W_f and W_c when $kx = 1$ V · s /A^2 and $i = 2$ A. $\left[W_f = \frac{16}{3}\ \text{J};\ W_c = \frac{8}{3}\ \text{J}\right]$

SP2.3-3 The current is increased from 2 to 3 A in SP2.3-2. Calculate the change in W_f and W_c. $\left[\Delta W_f = \frac{38}{3}\ \text{J};\ \Delta W_c = \frac{19}{3}\ \text{J}\right]$

SP2.3-4 $i = b(x)\lambda^2$. Express $W_f(\lambda, x)$ and $W_c(\lambda, x)$. $\left[W_f(\lambda, x) = \frac{1}{3}b(x)\lambda^3;\ W_c(\lambda, x) = \frac{2}{3}b(x)\lambda^3\right]$

2.4 GRAPHICAL INTERPRETATION OF ENERGY CONVERSION

Before proceeding to the derivation of expressions for the electromagnetic force, it is instructive to consider briefly a graphical interpretation of the energy conversion process. For this purpose, let us again refer to the elementary system shown in Fig. 2.2-3 and assume that as the movable member moves from $x = x_a$ to $x = x_b$, where $x_b < x_a$, the λi characteristics are given by Fig. 2.4-1. Let us also assume that, as the member moves from x_a to x_b, the λi trajectory moves from point A to point B. The exact trajectory from A to B is determined by the combined dynamics of the electrical and mechanical systems, and any variation in v and f which may occur. Now, the area $0AC0$ represents the original energy stored in the coupling field; area $0BD0$ represents the final energy stored in the field. Therefore, the change in field energy is

$$\Delta W_f = \text{area } 0BD0 - \text{area } 0AC0 \tag{2.4-1}$$

Since $e_f = \dfrac{d\lambda}{dt}$

$$W_e = \int i d\lambda \tag{2.4-2}$$

therefore, the change in W_e, denoted as ΔW_e, is

$$\Delta W_e = \int_{\lambda_A}^{\lambda_B} i d\lambda = \text{area } CABDC \tag{2.4-3}$$

From (2.2-6)

$$\Delta W_m = \Delta W_f - \Delta W_e \tag{2.4-4}$$

Hence, by adding and subtracting the appropriate areas in Fig. 2.4-1, we can obtain ΔW_m. In particular,

$$\begin{aligned} \Delta W_m &= \text{area } 0BD0 - \text{area } 0AC0 - \text{area } CABDC \\ &= -\text{area } 0AB0 \end{aligned} \tag{2.4-5}$$

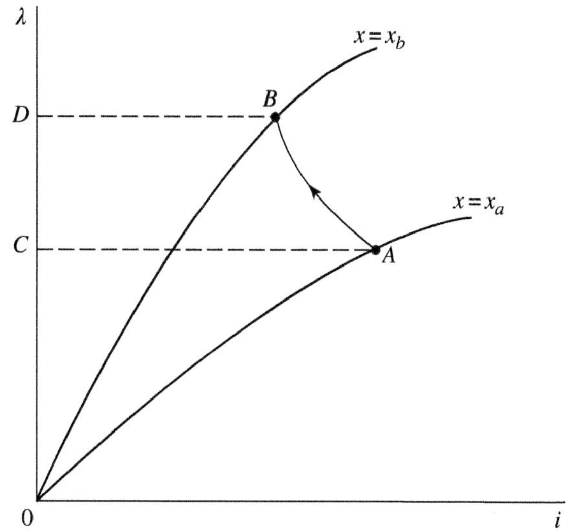

Figure 2.4-1 Graphical representation of electromechanical energy conversion for λi path A to B.

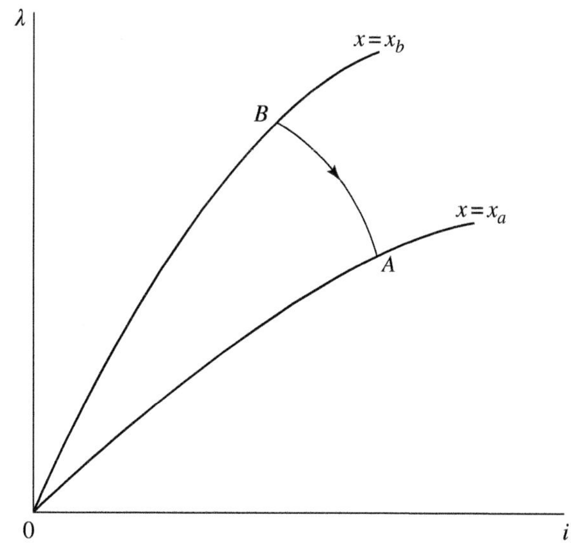

Figure 2.4-2 Graphical representation of electromechanical energy conversion for λi path B to A.

The energy contributed to the coupling field from the mechanical system ΔW_m is negative. Energy has been supplied to the mechanical system from the coupling field, part of which came from the energy stored in the field, and part of which came from the electric system. If the member is now moved back to x_a, the λi trajectory may be as shown in Fig. 2.4-2. Here, ΔW_m is still area $0AB0$, but it is positive, which means that energy was supplied from the mechanical system to the coupling

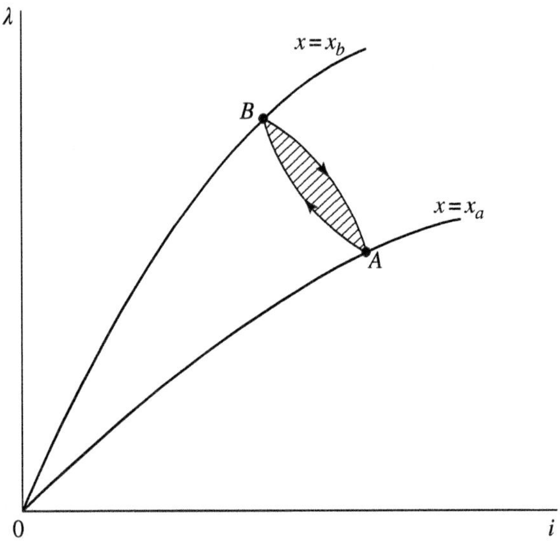

2.4-3 Graphical representation of electromechanical energy conversion for λi path A to B to A.

field, part of which is stored in the field and part of which is transferred to the electric system.

The energy supplied by the mechanical system during the motion from B to A (area $0AB0$ in Fig. 2.4-2) is larger than the energy supplied to the mechanical system during the original motion from A to B (area $0AB0$ in Fig. 2.4-1). Therefore, the net energy supplied by the mechanical system for the complete cycle is positive. The net ΔW_m for the cycle from A to B back to A is the shaded area shown in Fig. 2.4-3. Since the coupling field energy at point A is uniquely determined from the mechanical displacement and current at point A, the net change in field energy is zero as we move from A to B and back to A. Since ΔW_f is zero for this cycle,

$$\Delta W_m = -\Delta W_e \tag{2.4-6}$$

For the cycle shown, the net ΔW_e is negative since the magnitude of the change in W_e is larger when we went from B to A than from A to B, and ΔW_e is negative from B to A. If the trajectory had been in the counterclockwise direction, the net ΔW_e would have been positive and the net ΔW_m negative.

SP2.4-1 $\lambda = a(x)i^2$. The current is held constant at 2 A while $a(x)$ increases from 1 to 2. Calculate ΔW_m. $\left[\Delta W_m = -\frac{8}{3}\text{ J}\right]$

SP2.4-2 Repeat SP2.4-1 if it were possible for $a(x)$ to instantaneously increase from 1 to 2 while λ is held constant at 4 V · s followed by a current increase back to 2 A with $a(x) = 2$ (λ must change accordingly). $\left[\Delta W_m = \frac{-16 + 8\sqrt{2}}{3}\text{ J}\right]$

2.5 ELECTROMAGNETIC AND ELECTROSTATIC FORCES

The stage is set for us to obtain expressions for the electromagnetic and electrostatic forces in electromechanical devices. We will focus on electromagnetic devices. For this purpose, recall that e_{fj} in (2.2-21) may be expressed as

$$e_{fj} = \frac{d\lambda_j}{dt} \tag{2.5-1}$$

If we substitute (2.5-1) into (2.2-21) and if we solve for $f_e dx$, we obtain

$$f_e dx = \sum_{j=1}^{J} i_j d\lambda_j - dW_f \tag{2.5-2}$$

Although we will use (2.5-2), it is helpful to express it in an alternative form. For this purpose, let us first write (2.3-4) for multiple electrical inputs:

$$\sum_{j=1}^{J} \lambda_j i_j = W_c + W_f \tag{2.5-3}$$

If we take the total derivative of (2.5-3), we obtain

$$\sum_{j=1}^{J} \lambda_j di_j + \sum_{j=1}^{J} i_j d\lambda_j = dW_c + dW_f \tag{2.5-4}$$

We realize that when we evaluate the force f_e we must select the independent variables; that is, either the flux linkages and x or the currents and x. The flux linkages and the currents cannot simultaneously be considered independent variables when evaluating the force f_e. Nevertheless, (2.5-4), wherein both $d\lambda_j$ and di_j appear, is valid in general, before a selection of independent variables is made to evaluate f_e. If we solve (2.5-4) for the field energy dW_f and substitute the result into (2.5-2), we obtain

$$f_e dx = -\sum_{j=1}^{J} \lambda_j di_j + dW_c \tag{2.5-5}$$

Either (2.5-2) or (2.5-5) can be used to evaluate the electromagnetic force f_e. If flux linkages and x are selected as independent variables, (2.5-2) is the most direct, whereas (2.5-5) is the most direct if currents and x are selected.

With flux linkages and x as the independent variables, the currents are expressed functionally as

$$i_j(\lambda_1, \ldots, \lambda_j, x) \tag{2.5-6}$$

For the purpose of compactness, we will denote $(\lambda_1, \ldots, \lambda_j, x)$ as $(\boldsymbol{\lambda}, x)$, where $\boldsymbol{\lambda}$ is an abbreviation for the complete set of flux linkages associated with the J windings. Let us write (2.5-2) with flux linkages and x as independent variables:

$$f_e(\boldsymbol{\lambda}, x)dx = \sum_{j=1}^{J} i_j(\boldsymbol{\lambda}, x)d\lambda_j - dW_f(\boldsymbol{\lambda}, x) \tag{2.5-7}$$

If we take the total derivative of the field energy with respect to $\boldsymbol{\lambda}$ and x, and substitute that result into (2.5-7), we obtain

$$f_e(\boldsymbol{\lambda}, x)dx = \sum_{j=1}^{J} i_j(\boldsymbol{\lambda}, x)d\lambda_j - \sum_{j=1}^{J} \frac{\partial W_f(\boldsymbol{\lambda}, x)}{\partial \lambda_j} d\lambda_j - \frac{\partial W_f(\boldsymbol{\lambda}, x)}{\partial x} dx \tag{2.5-8}$$

Equating the coefficients of dx gives

$$f_e(\boldsymbol{\lambda}, x) = -\frac{\partial W_f(\boldsymbol{\lambda}, x)}{\partial x} \tag{2.5-9}$$

A second expression for $f_e(\boldsymbol{\lambda}, x)$ may be obtained by expressing (2.5-3) with flux linkages and x as independent variables, solving for $W_f(\boldsymbol{\lambda}, x)$, and then taking the partial derivative with respect to x. Thus,

$$f_e(\boldsymbol{\lambda}, x) = -\sum_{j=1}^{J} \left[\lambda_j \frac{\partial i_j(\boldsymbol{\lambda}, x)}{\partial x}\right] + \frac{\partial W_c(\boldsymbol{\lambda}, x)}{\partial x} \tag{2.5-10}$$

If we now select $\mathbf{i}$ and x as independent variables, where $(\mathbf{i}, x)$ is the abbreviated notation for $(i_1, \ldots, i_J, x)$, then (2.5-5) can be written as

$$f_e(\mathbf{i}, x)dx = -\sum_{j=1}^{J} \lambda_j(\mathbf{i}, x)di_j + dW_c(\mathbf{i}, x) \tag{2.5-11}$$

If we take the total derivative of $W_c(\mathbf{i}, x)$ with $\mathbf{i}$ and x as independent variables and substitute the result into (2.5-11), we obtain

$$f_e(\mathbf{i}, x)dx = -\sum_{j=1}^{J} \lambda_j(\mathbf{i}, x)di_j + \sum_{j=1}^{J} \frac{\partial W_c(\mathbf{i}, x)}{\partial i_j} di_j + \frac{\partial W_c(\mathbf{i}, x)}{\partial x} dx \tag{2.5-12}$$

Equating coefficients of dx yields

$$f_e(\mathbf{i}, x) = \frac{\partial W_c(\mathbf{i}, x)}{\partial x} \tag{2.5-13}$$

We will make extensive use of this expression. If we now solve (2.5-3) for $W_c(\mathbf{i}, x)$ and then take the partial derivative with respect to x, we can obtain a second expression for $f_e(\mathbf{i}, x)$. That is,

$$f_e(\mathbf{i}, x) = \sum_{j=1}^{J} \left[i_j \frac{\partial \lambda_j(\mathbf{i}, x)}{\partial x}\right] - \frac{\partial W_f(\mathbf{i}, x)}{\partial x} \tag{2.5-14}$$

TABLE 2.5-1 Electromagnetic Forces at Mechanical Input (For Rotational Systems, Replace f_e with T_e and x with θ_r)

$$f_e(\mathbf{i},x) = \sum_{j=1}^{J}\left[i_j\frac{\partial\lambda_j(\mathbf{i},x)}{\partial x}\right] - \frac{\partial W_f(\mathbf{i},x)}{\partial x}$$

$$f_e(\mathbf{i},x) = \frac{\partial W_c(\mathbf{i},x)}{\partial x}$$

$$f_e(\boldsymbol{\lambda},x) = -\frac{\partial W_f(\boldsymbol{\lambda},x)}{\partial x}$$

$$f_e(\boldsymbol{\lambda},x) = -\sum_{j=1}^{J}\left[\lambda_j\frac{\partial i_j(\boldsymbol{\lambda},x)}{\partial x}\right] + \frac{\partial W_c(\boldsymbol{\lambda},x)}{\partial x}$$

We have derived four expressions for the electromagnetic force which are summarized in Table 2.5-1. Since we will generally use currents and x as independent variables, the two expressions for $f_e(\mathbf{i},x)$ are listed first in Table 2.5-1.

Before proceeding to the next section, it is important to take a moment to look back. In order to obtain $f_e(\boldsymbol{\lambda},x)$, we equated the coefficients of dx in (2.5-8). If, however, we equate the coefficients of $d\lambda_j$ in (2.5-8), we obtain

$$\sum_{j=1}^{J}\frac{\partial W_f(\boldsymbol{\lambda},x)}{\partial\lambda_j} = \sum_{j=1}^{J} i_j(\boldsymbol{\lambda},x) \tag{2.5-15}$$

Similarly, if we equate the coefficients of di_j in (2.5-12) we obtain

$$\sum_{j=1}^{J}\frac{\partial W_c(\mathbf{i},x)}{\partial i_j} = \sum_{j=1}^{J}\lambda_j(\mathbf{i},x) \tag{2.5-16}$$

Equations (2.5-15) and (2.5-16) are readily verified by recalling the definitions of W_f and W_c, (2.3-2) and (2.3-3), respectively, which were obtained by holding x fixed $(dx = 0)$.

In Table 2.5-1, the independent variables to be used are designated in each equation by the abbreviated functional notation. Although only translational mechanical systems have been considered, all force relationships developed herein may be modified for the purpose of evaluating the torque in rotational systems. In particular, when considering a rotational system, f_e is replaced with the electromagnetic torque T_e and x with the angular displacement θ_r of the rotating member. These substitutions are justified since the change of mechanical energy in a rotational system is expressed as

$$dW_m = -T_e\ d\theta_r \tag{2.5-17}$$

The force equation for an electromechanical system with electric coupling fields may be derived by following a procedure similar to that used in the case of magnetic coupling fields. These relationships are given in Table 2.5-2 without explanation or proof.

TABLE 2.5-2 Electrostatic Forces at Mechanical Input (For Rotational Systems, Replace f_e with T_e and x with θ_r)

$$f_e(\mathbf{e}_f, x) = \sum_{j=1}^{J}\left[e_{fj}\frac{\partial q_j(\mathbf{e}_f, x)}{\partial x}\right] - \frac{\partial W_f(\mathbf{e}_f, x)}{\partial x}$$

$$f_e(\mathbf{e}_f, x) = \frac{\partial W_c(\mathbf{e}_f, x)}{\partial x}$$

$$f_e(\mathbf{q}, x) = -\frac{\partial W_f(\mathbf{q}, x)}{\partial x}$$

$$f_e(\mathbf{q}, x) = -\sum_{j=1}^{J}\left[q_j\frac{\partial e_{fj}(\mathbf{q}, x)}{\partial x}\right] + \frac{\partial W_c(\mathbf{q}, x)}{\partial x}$$

Example 2B Detailed Calculation of Electromagnetic Force

We have mentioned that one may prefer to determine the electromagnetic force or torque by starting with the relationship $dW_f = dW_e + dW_m$ rather than by selecting a formula from a table. To illustrate this procedure, let

$$\lambda = [1 + a(x)]i^2 \tag{2B-1}$$

First, we must evaluate the field energy. Since losses in the coupling field are neglected, W_f is a function of state. Hence, W_f may be evaluated by fixing the mechanical displacement. This is done by setting $dx = 0$, whereupon

$$dW_f = dW_e = id\lambda \qquad \text{with } dx = 0 \tag{2B-2}$$

where dW_e is obtained from (2.2-13) with $e_f = d\lambda/dt$. From (2B-1) with $dx = 0$,

$$d\lambda = 2[1 + a(x)]idi \tag{2B-3}$$

Substituting (2B-3) into (2B-2) and solving for W_f yields

$$W_f = \int_0^i 2[1 + a(x)]\xi^2 d\xi = \frac{2}{3}[1 + a(x)]i^3 \tag{2B-4}$$

To obtain an expression for f_e, we go back to the basic relationship that $dW_f = dW_e + dW_m$; however, now $dx \neq 0$. Thus, from (2B-4),

$$dW_f = \frac{2}{3}i^3\frac{\partial a(x)}{\partial x}dx + 2[1 + a(x)]i^2 di \tag{2B-5}$$

Now,

$$dW_e = id\lambda = i^3\frac{\partial a(x)}{\partial x}dx + 2[1 + a(x)]idi \tag{2B-6}$$

and from (2.2-15),

$$dW_m = -f_e dx \tag{2B-7}$$

Substituting into $dW_f = dW_e + dW_m$ yields

$$\frac{2}{3}i^3\frac{\partial a(x)}{\partial x}dx + 2[1 + a(x)]i^2di = i^3\frac{\partial a(x)}{\partial x}dx + 2[1 + a(x)]i^2di - f_edx \qquad \text{(2B-8)}$$

The di terms cancel as (2.5-10) tells us they should, and by equating the coefficients of dx,

$$f_e = \frac{1}{3}i^3\frac{\partial a(x)}{\partial x} \qquad \text{(2B-9)}$$

Let us check our work by using information from Table 2.5-1. Since the system is magnetically nonlinear, $W_f \neq W_c$. Thus, knowing W_f, we can use the first entry in Table 2.5-1 or evaluate W_c and use the second entry. We shall do both. From Table 2.5-1 with one electric input ($J = 1$),

$$f_e(i,x) = i\frac{\partial \lambda(i,x)}{\partial x} - \frac{\partial W_f(i,x)}{\partial x} \qquad \text{(2B-10)}$$

Now, $\lambda(i,x)$ is expressed by (2B-1) and $W_f(i,x)$ by (2B-4), thus

$$\begin{aligned} f_e(i,x) &= i\left[i^2\frac{\partial a(x)}{\partial x}\right] - \frac{2}{3}i^3\frac{\partial a(x)}{\partial x} \\ &= \frac{1}{3}i^3\frac{\partial a(x)}{\partial x} \end{aligned} \qquad \text{(2B-11)}$$

which agrees with our previous result. Alternatively, from the second entry of Table 2.5-1,

$$f_e(i,x) = \frac{\partial W_c(i,x)}{\partial x} \qquad \text{(2B-12)}$$

From (2.3-4),

$$W_c(i,x) = \lambda(i,x)i - W_f(i,x) = \frac{1}{3}[1 + a(x)]i^3 \qquad \text{(2B-13)}$$

[Obtain this same expression by using (2.3-3) to evaluate W_c.] Substituting (2B-13) into (2B-12) yields

$$f_e(i,x) = \frac{1}{3}i^3\frac{\partial a(x)}{\partial x} \qquad \text{(2B-14)}$$

SP2.5-1 $\lambda = kx^2i^2$. Express f_e when $i = 2$ A and $x = 1$ m. [$f_e = (16k/3)$ N]

SP2.5-2 $\lambda = ki/x$. Express f_e if $i = \sqrt{2}I_s \cos\omega_e t$. [$f_e = -(ki_s^2/2x^2)(1 + \cos 2\omega_e t)$]

SP2.5-3 $i = a(x)\lambda^3$. Express f_e by using the third entry in Table 2.5-1.
$\left[f_e = -\frac{1}{4}\lambda^4(da(x)/dx)\right]$

SP2.5-4 In a rotational system, $\lambda = ki^2 \sin\theta_r$. Express the torque T_e. [$T_e = \frac{1}{3}ki^3\cos\theta_r$]

2.6 OPERATING CHARACTERISTICS OF AN ELEMENTARY ELECTROMAGNET

From our work in Section 1.7, we established that for $x > 0$ the inductance of the electromagnet shown in Fig. 2.2-3 may be adequately approximated by (1.7-21). That is,

$$L(x) = L_l + L_m(x) = L_l + \frac{k}{x} \tag{2.6-1}$$

Now,

$$\lambda(i,x) = L(x)i \tag{2.6-2}$$

and, since the system is magnetically linear,

$$W_f(i,x) = W_c(i,x) = \frac{1}{2}L(x)i^2 \tag{2.6-3}$$

From the second entry of Table 2.5-1,

$$f_e(i,x) = \frac{\partial W_c(i,x)}{\partial x} = \frac{1}{2}i^2\frac{dL(x)}{dx} \tag{2.6-4}$$

Substituting (2.6-1) into (2.6-4) yields

$$f_e(i,x) = -\frac{ki^2}{2x^2} \tag{2.6-5}$$

The force f_e is always negative in the system shown in Fig. 2.2-3; the electromagnetic force pulls the moving member to the stationary member. In other words, an electromagnetic force is set up so as to minimize the reluctance (maximize the inductance) of the magnetic system.

The differential equations that describe the electromagnet are given by (2.2-7) for the electric system and (2.2-8) for the mechanical system. If the applied voltage v and the applied mechanical force f are both constant, all derivatives with respect to time in (2.2-7) and (2.2-8) are zero during steady-state operation. Thus,

$$v = ri \tag{2.6-6}$$

$$f = K(x - x_0) - f_e \tag{2.6-7}$$

Equation (2.6-7) may be written as

$$-f_e = f - K(x - x_0) \tag{2.6-8}$$

A plot of (2.6-8), with f_e replaced by (2.6-5), is shown in Fig. 2.6-1 for the following system parameters: $r = 10\ \Omega, K = 2667\ \text{N/m}, x_0 = 3$ mm, and $k = 6.283 \times 10^{-5}$ H · m.

In Fig. 2.6-1, the plot of the negative of the electromagnetic force is for an applied voltage of 5 V, whereupon the steady-state current is 0.5 A. The straight lines represent the right-hand side of (2.6-8) with $f = 0$ (lower straight line) and

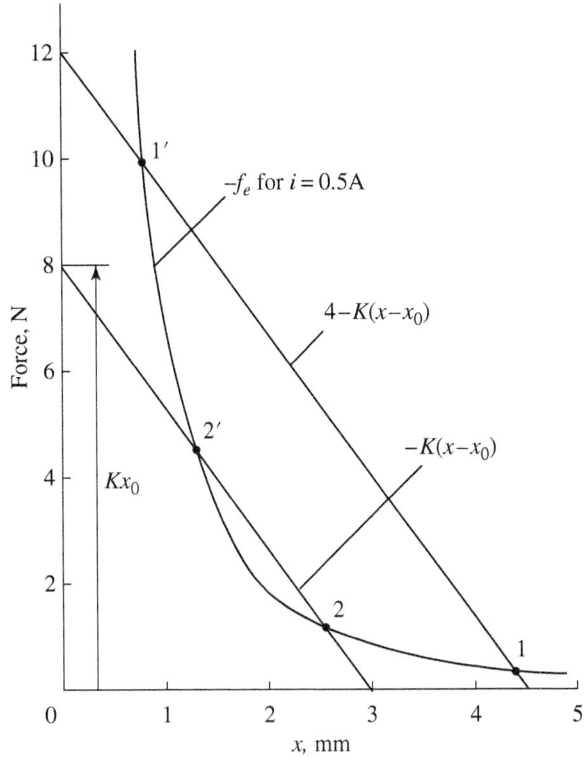

Figure 2.6-1 Steady-state operation of electromechanical system shown in Fig. 2.2-3.

$f = 4$ N (upper straight line). Both lines intersect the $-f_e$ curve at two points. In particular, the upper line intersects the $-f_e$ curve at 1 and 1′; the lower line intersects at 2 and 2′. Stable operation occurs at only points 1 and 2. The system will not operate stably at points 1′ and 2′. This can be explained by assuming the system is operating at one of these points (1′ or 2′) and then show that any system disturbance whatsoever will cause the system to move away from these points. If, for example, x increases slightly from its value corresponding to point 1′, the restraining force, $f - K(x - x_0)$, which is a force to pull the movable member to the right, is larger than the electromagnetic force to pull the movable member to the left. Hence, x will continue to increase until the system reaches operating point 1. If x increases beyond its value corresponding to operating point 1, the restraining force, $f - K(x - x_0)$, is less than the electromagnetic force. Therefore, the system will establish steady-state operation at 1. If, on the other hand, x decreases from point 1′, the electromagnetic force is larger than the restraining force. Therefore, the movable member will move until it comes in contact with the stationary member ($x = 0$). The restraining force that yields a straight line below the $-f_e$ curve will not permit stable operation with $x > 0$. Note also that, at point 2, x is less than x_0 and the spring is extended, exerting a force to the right on the movable member. At point 1, x is greater than x_0 and the spring is in compression, exerting a force to the left on the movable member. In fact, at 1 most of the force $4 - K(x - x_0)$ goes to compress the spring since $x > x_0$. This can be seen graphically by extending the $-K(x - x_0)$ line into the negative-force region and visualizing the intersection of this line and a vertical line through 1.

The dynamic behavior of the system during step changes in the source voltage v is shown in Fig. 2.6-2. Figures 2.6-3 and 2.6-4 are for step changes in the applied force f. The following system parameters were used in addition to those

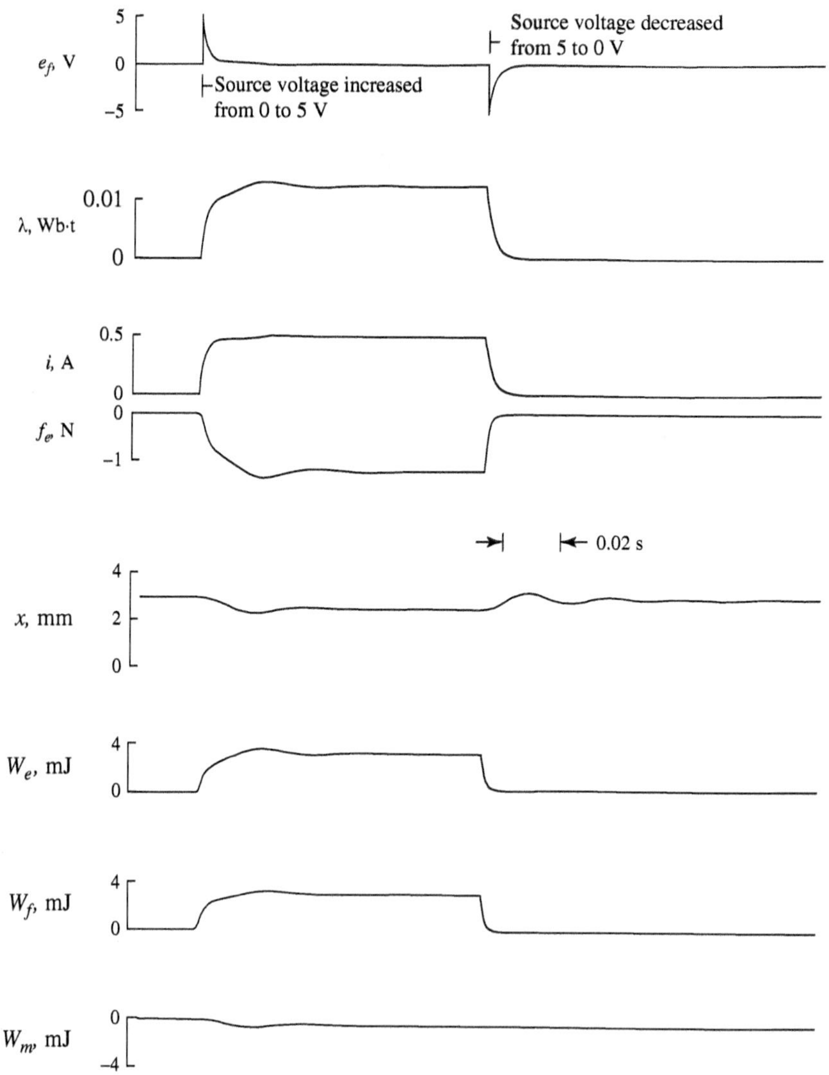

Figure 2.6-2 Dynamic performance of the electromechanical system shown in Fig. 2.2-3 during step changes in the source voltage.

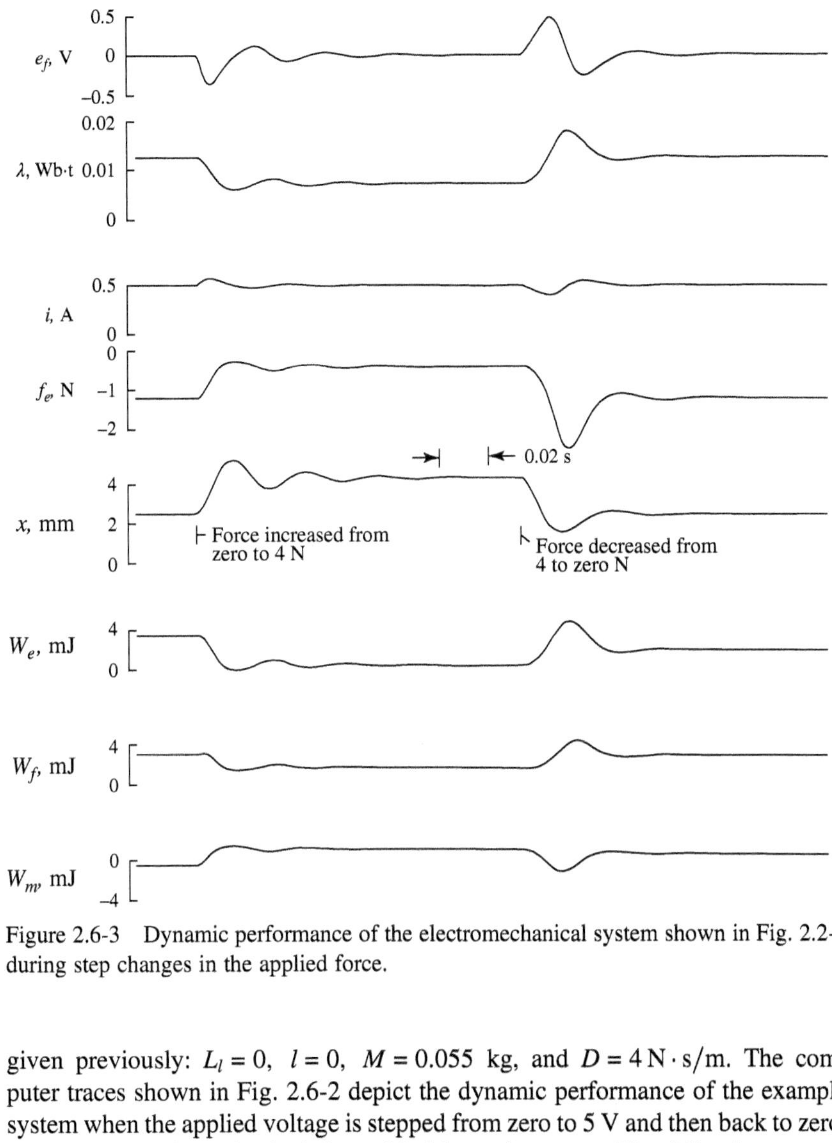

Figure 2.6-3 Dynamic performance of the electromechanical system shown in Fig. 2.2-3 during step changes in the applied force.

given previously: $L_l = 0$, $l = 0$, $M = 0.055$ kg, and $D = 4\,\text{N}\cdot\text{s/m}$. The computer traces shown in Fig. 2.6-2 depict the dynamic performance of the example system when the applied voltage is stepped from zero to 5 V and then back to zero, with the applied mechanical force f held equal to zero. The following system variables are plotted: $e_f, \lambda, i, f_e, x, W_e, W_f$, and W_m. The energies are plotted in millijoules (mJ). Initially, the mechanical system is at rest with $x = x_0$ (3 mm).

When the source voltage is applied, x decreases and, when steady-state operation is established, x is approximately 2.5 mm, which is operating point 2 in Fig. 2.6-1. During the transient period, energy enters the coupling field via W_e. The bulk of this energy is stored in the field (W_f) with a smaller amount transferred to the mechanical system, some of which is dissipated in the damper during the transient period, and the remainder is stored in the spring. When the applied voltage

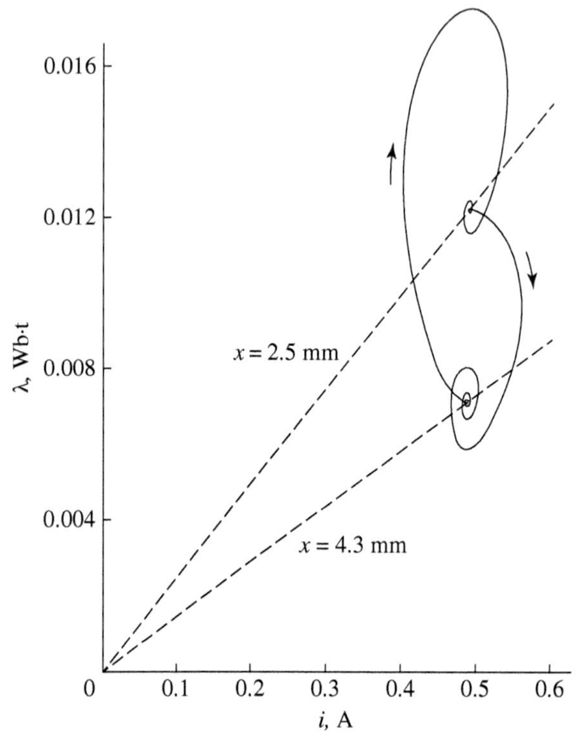

Figure 2.6-4 λ versus i plot of the system response shown in Fig. 2.6-3.

is removed, the electric and mechanical systems return to their original states. The change in W_m is small, increasing only slightly. Hence, during the transient period, there is an interchange of energy between the spring and mass that is finally dissipated in the damper. The net change in W_f during the application and removal of the applied voltage is zero, hence the net change in W_e is positive and equal to the negative of the net change in W_m. The energy transferred to the mechanical system during this cycle is dissipated in the damper since f is fixed at zero and the mechanical system returns to its initial rest position with zero energy stored in the spring.

In Fig. 2.6-3, the initial state is that shown in Fig. 2.6-2 with 5 V applied to the electric system. The mechanical force f is increased from zero to 4 N, whereupon energy enters the coupling field from the mechanical system. Energy is transferred from the coupling field to the electric system and dissipated in the resistor, some coming from the mechanical system and some from the energy originally stored in the magnetic field. We have moved from point 2 to point 1 in Fig. 2.6-1. Next, the force is stepped back to zero from 4 N. The electric and mechanical systems return to their original states. During the cycle, a net energy has been transferred from the mechanical system to the electric system, which is dissipated in the resistance. This cycle is depicted in the λi plot shown in Fig. 2.6-4.

SP2.6-1 In Fig. 2.6-2, e_f jumps to 5 V when the source voltage is stepped from zero to 5 V and jumps from zero to −5 V when the source voltage is stepped from 5 V to zero. Why? [At first $t = 0^+$, $v = L(x)(di/dt)$; at second $t = 0^+$, $L(x)(di/dt) = -ir$]

SP2.6-2 Consider Fig. 2.6-1 with initial operation at point 2. Determine the final operating value of x (a) if f is stepped from zero to −1 N; (b) if f is zero but v is stepped from 5 to 10 V. [(a) and (b) $x = 0$]

SP2.6-3 Assume that the elementary electromagnet shown in Fig. 2.2-3 portrays the λi characteristics shown in Fig. 2.4-1. As the system moves from x_a to x_b, the λi trajectory moves from A to B, as shown in Fig. 2.4-1. Assume steady-state operation exists at A and B. (a) Does the voltage v increase or decrease? (b) Does the applied force f increase or decrease? [(a) and (b) decrease]

SP2.6-4 Why does $i(t)$ not portray an exponential increase when the source voltage is increased from zero to 5 V in Fig. 2.6-2? [$L(x)$]

2.7 SINGLE-PHASE RELUCTANCE MACHINE

An elementary two-pole, single-phase reluctance machine, which was first shown in Fig. 1.7-2, is shown in a slightly different form in Fig. 2.7-1. In particular, the notation has been changed, that is, winding 1 is now winding *as*. We shall see the convenience of this notation as we go along. Next, the stator or stationary member has been changed to depict more accurately the configuration common for this device. The voltage equation may be written as

$$v_{as} = r_s i_{as} + \frac{d\lambda_{as}}{dt} \tag{2.7-1}$$

where r_s is the resistance of the *as* winding and

$$\lambda_{as} = L_{asas} i_{as} \tag{2.7-2}$$

The self-inductance of the *as* winding, which is denoted L_{asas}, can be approximated by an expression of the same form as that for the device shown in Fig. 1.7-2. In particular, from (1.7-29),

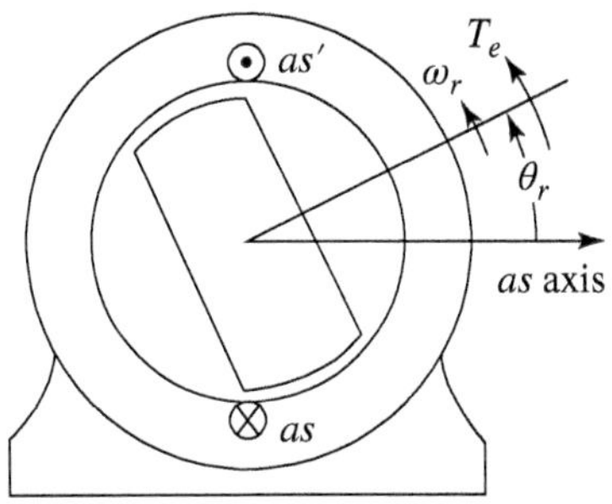

Figure 2.7-1 Elementary two-pole, single-phase reluctance machine.

$$L_{asas} = L_{ls} + L_A - L_B \cos 2\theta_r \tag{2.7-3}$$

where, for constant rotor speed

$$\theta_r = \omega_r t + \theta_r(0) \tag{2.7-4}$$

and L_{ls} is the leakage inductance.

An expression for the electromagnetic torque may be obtained by using the information given in Table 2.5-1. The magnetic system is linear; hence, $W_f = W_c$. We will use the second entry to evaluate T_e; in particular,

$$T_e(i_{as}, \theta_r) = \frac{\partial W_c(i_{as}, \theta_r)}{\partial \theta_r} \tag{2.7-5}$$

where

$$W_c(i_{as}, \theta_r) = \frac{1}{2}(L_{ls} + L_A - L_B \cos 2\theta_r) i_{as}^2 \tag{2.7-6}$$

from which

$$T_e(i_{as}, \theta_r) = L_B i_{as}^2 \sin 2\theta_r \tag{2.7-7}$$

Although the expression for torque given by (2.7-7) is valid for transient and steady-state conditions, we shall consider only elementary modes of steady-state operation in this section.

As a first example, let i_{as} be a constant current. In this case, the torque may be expressed as

$$T_e = K \sin 2\theta_r \tag{2.7-8}$$

where K is a constant equal to $L_B I_{as}^2$. Equation (2.7-8) is plotted in Fig. 2.7-2 with the position of the rotor shown for $\theta_r = 0, \frac{1}{4}\pi, \frac{1}{2}\pi, \ldots, 2\pi$. Let us assume that there

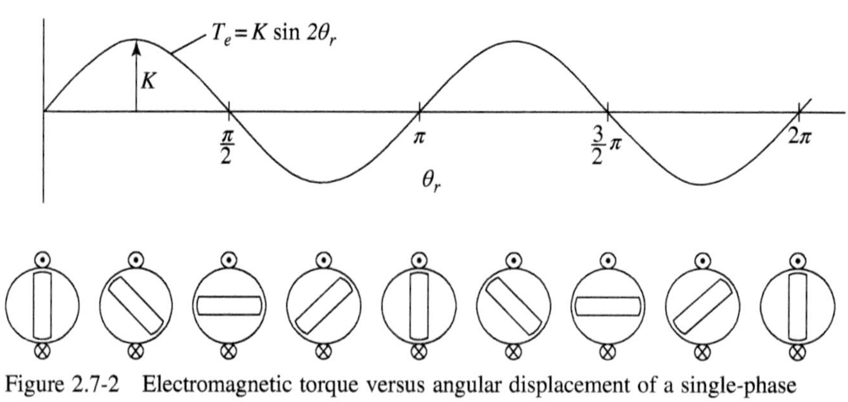

Figure 2.7-2 Electromagnetic torque versus angular displacement of a single-phase reluctance machine with constant stator current.

is no external torque on the shaft; that is, there is no torque to twist the shaft one way or the other. We learned from the study of the electromagnet in the previous section that there is a force created to minimize the reluctance of the magnetic system. This would suggest that the rest position of the rotor with constant I_{as} would be at the minimum reluctance positions; either $\theta_r = \frac{1}{2}\pi$ or $\frac{3}{2}\pi$, depending upon the initial rotor position. For example, if $0 < \theta_r < \frac{1}{2}\pi$ at the time i_{as} was increased from zero to a constant value (transients neglected), the torque T_e would immediately become $K \sin 2\theta_r$, which is a positive value. Recall that the positive assumed direction of the electromagnetic torque is in the positive direction of θ_r (Fig. 2.7-1). Thus, the electromagnetic torque would cause the rotor to rotate to $\theta_r = \frac{1}{2}\pi$. Is this a stable point of operation? Well, at $\theta_r = \frac{1}{2}\pi$, $T_e = 0$. This tells us that electromagnetic torque is not created to move the rotor if θ_r is exactly at $\frac{1}{2}\pi$. However, let us try the stability test that we used on the electromagnet. If we displace the rotor ever so slightly from this operating point, will it return to $\theta_r = \frac{1}{2}\pi$? If so, then $\theta_r = \frac{1}{2}\pi$ is a stable operating point for a constant stator current. Let θ_r decrease very slightly, whereupon T_e becomes positive; hence, an electromagnetic torque is developed to increase θ_r back to $\frac{1}{2}\pi$. Now, let θ_r increase very slightly beyond $\frac{1}{2}\pi$; T_e now becomes negative which forces θ_r back to $\frac{1}{2}\pi$. Certainly, $\theta_r = \frac{1}{2}\pi$ is a stable operating point and we can see that $\theta_r = -\frac{1}{2}\pi$ would also be a stable operating point.

We still have some questions. What would happen if the rotor were initially positioned at $\theta_r = 0$ when we increased i_{as} from zero to a constant value? Well, from our previous discussion we know that the rotor would come to rest at $\theta_r = \frac{1}{2}\pi$ or $-\frac{1}{2}\pi$. However, since there is no external torque on the rotor and since (2.7-8) tells us that $T_e = 0$ at $\theta_r = 0$, why does it not remain at $\theta_r = 0$? The fact that the external torque is satisfied is certainly one condition for stable operation; however, let us try our stability test again. When we increase θ_r slightly from zero, T_e becomes positive, which would cause θ_r to increase even further. The rotor will move away from $\theta_r = 0$ to $\frac{1}{2}\pi$, a stable operating point. If we decrease θ_r slightly from zero, T_e becomes negative, decreasing θ_r even further, and the rotor will come to rest at $-\frac{1}{2}\pi$. Therefore, let us assume that, before increasing the current from zero to a constant value, we took great pains to set θ_r exactly at zero. If then we increased the current from zero to a constant value, the rotor would theoretically remain at $\theta_r = 0$. However, any disturbance, ever so slight, will cause the rotor to rotate away from $\theta_r = 0$ and there is a 50–50 chance as to the direction, counterclockwise or clockwise.

We have established that, with zero external torque on the shaft, $\theta_r = \frac{1}{2}\pi$ and $\frac{3}{2}\pi$ are stable points of operation for a constant stator current, and $\theta_r = 0$ and π are unstable. Let us see what happens when we apply an external torque. For this purpose, let us assume that initially the rotor is positioned at $\theta_r = \frac{1}{2}\pi$ with i_{as} a constant. Assume that we apply the torque to twist the shaft in the counterclockwise direction, increasing θ_r. An electromagnetic torque T_e will be developed to oppose the external torque in an attempt to maintain alignment with the *as* axis. When the

T_e produced, in the attempt to align, is equal and opposite to the external torque, the rotor will stop advancing and come to rest with $\frac{1}{2}\pi < \theta_r < \frac{3}{4}\pi$. Now, the external torque must be less than the peak absolute value of T_e at $\theta_r = \frac{3}{4}\pi$ (K in Fig. 2.7-2). Otherwise, the torque to align cannot satisfy the external torque, whereupon the rotor would continue to advance counterclockwise as long as the external torque were present. If, instead of applying a torque to twist the rotor in the counterclockwise direction, the external torque is applied to twist the rotor in the opposite direction (clockwise), θ_r would decrease. If the electromagnetic torque developed in an attempt to maintain alignment with the *as* axis is large enough to satisfy the external torque, then the rotor will come to rest with $\frac{1}{4}\pi < \theta_r < \frac{1}{2}\pi$. With a little thought, we can convince ourselves that if T_e can satisfy the external torque requirements, then stable operation will occur only with $\frac{1}{4}\pi < \theta_r < \frac{3}{4}\pi$ or $\frac{5}{4}\pi < \theta_r < \frac{7}{4}\pi$, which represents the negative slope portions of the T_e versus θ_r characteristics.

We need to discuss the convention regarding the sign of T and T_e when dealing with rotational devices. In this chapter, it was established in Fig. 2.2-3 that f and f_e were both positive in the direction of positive displacement x. Therefore, for a rotational system, T and T_e would be positive in the direction of positive shaft displacement (counterclockwise θ_r). However, over the years it has become convention when considering motor action to let T_e be positive for counterclockwise shaft displacement but make the load torque, which we will denote T_L rather than T, positive for negative shaft displacement (clockwise θ_r). Starting in Chapter 3, we will follow this convention even when dealing with generator action where the input torque will be a negative T_L. Thus, in the previous paragraph, the torque to twist the shaft in the counterclockwise direction is an input torque or a negative load torque ($-T_L$); the torque to twist the shaft in the clockwise direction is a positive load torque ($+T_L$). Hopefully, this will minimize the confusion that would certainly occur if we were to again change the sense to comply with the convention that is often used when analyzing generator action.

Although operation of a single-phase reluctance machine with a constant stator current is impracticable, it provides a basic understanding of reluctance torque, which is the operating principle of variable-reluctance stepper motors. We will find that, in its simplest form, a variable-reluctance stepper motor consists of three cascaded, single-phase reluctance motors with the rotors on a common shaft and arranged so that their minimum reluctance paths are displaced from each other. We shall talk more about stepper motors in a later chapter. Before leaving this brief encounter with the single-phase reluctance machine, it is instructive to look at an approximation of a practicable mode of operation of this device. To approximate the operation of a single-phase reluctance machine, we will assume that the steady-state stator current is sinusoidal. Thus,

$$I_{as} = \sqrt{2} I_s \cos \theta_{esi} \tag{2.7-9}$$

where, for steady-state operation,

$$\theta_{esi} = \omega_e t + \theta_{esi}(0) \tag{2.7-10}$$

In the above equations, the electrical angular displacement has the subscript *esi*. The *e* denotes association with an electrical variable, *s* with a stator variable, and *i* with current. We know that θ_r, which is given by (2.7-4), may also be written in a form similar to (2.7-10) for steady-state operation. Substituting (2.7-9) into (2.7-7) and after using a few trigonometric identities from Appendix A, we have

$$\begin{aligned} T_e &= L_B I_s^2 \sin 2[\omega_r t + \theta_r(0)] \\ &+ \frac{1}{2} L_B I_s^2 \{ \sin 2[(\omega_r + \omega_e)t + \theta_r(0) + \theta_{esi}(0)] \\ &+ \sin 2[(\omega_r - \omega_e)t + \theta_r(0) - \theta_{esi}(0)]\} \end{aligned} \quad (2.7\text{-}11)$$

We see from the first term that an average steady-state torque is produced when $\omega_r = 0$. More importantly, we see that either the second or the third term produces an average torque whenever $|\omega_r| = \omega_e$, where ω_e is the angular velocity of the electric system (synchronous speed). For this example, let $\omega_r = \omega_e$, then (2.7-11) becomes

$$\begin{aligned} T_e &= L_B I_s^2 \sin 2[\omega_e t + \theta_r(0)] \\ &+ \frac{1}{2} L_B I_s^2 \sin 2[2\omega_e t + \theta_r(0) + \theta_{esi}(0)] \\ &+ \frac{1}{2} L_B I_s^2 \sin 2[\theta_r(0) - \theta_{esi}(0)] \end{aligned} \quad (2.7\text{-}12)$$

Recall that $\theta_r(0)$ and $\theta_{esi}(0)$ are the time-zero values of the displacement of the rotor and the current, respectively.

It appears from (2.7-12) that we should expect the steady-state torque to pulsate at $4\omega_e$ and $2\omega_e$, and the average torque is a double-angle sinusoidal function of $\theta_r(0) - \theta_{esi}(0)$. Although (2.7-12) illustrates that a reluctance machine develops an average torque, it is difficult to explain how this all takes place. Even though you probably have numerous questions regarding the operation of a single-phase reluctance machine, it is perhaps best to delay a more in-depth discussion until we have studied the material on Tesla's rotating magnetic field (rotating air gap mmf) in Chapter 4.

SP2.7-1 Initially, the rotor of a single-phase reluctance motor is positioned with $\theta_r = 0$. Instantaneously, i_{as} is stepped to 1 A and an external torque of 0.01 N · m is applied to rotate the rotor in the clockwise direction. $L_B = 0.02$ H. Neglect transients and determine the final value of θ_r. [$\theta_r = -105°$]

SP2.7-2 The device in SP2.7-1 is operating steadily at $\theta_r = -105°$ when i_{as} is reversed instantaneously ($I_{as} = -1$ A). Neglect the transients and determine the final value of θ_r. [$\theta_r = -105°$]

SP2.7-3 The friction and windage losses of a reluctance motor appear as a retarding torque of 0.01 N · m at $\omega_r = \omega_e$. The steady-state current is $I_{as} = 2\cos\omega_e t$ and $L_B = 0.02$ H. Calculate $\theta_r(0) - \theta_{esi}(0)$ if the motor is rotating steadily at ω_e. [$\theta_r(0) = -105°$]

2.8 WINDINGS IN RELATIVE MOTION

It is instructive to formulate an expression for the electromagnetic torque of the elementary rotational device shown in Fig. 1.7-4. The voltage equations are given by (1.7-30) and (1.7-31), the self-inductances L_{11} and L_{22} are constant as given by (1.7-34) and (1.7-35), and the mutual inductance is given by (1.7-38). Let us rewrite these equations for convenience. The voltage equations may be written in matrix form as

$$\begin{bmatrix} v_1 \\ v_2 \end{bmatrix} = \begin{bmatrix} r_1 & 0 \\ 0 & r_2 \end{bmatrix} \begin{bmatrix} i_1 \\ i_2 \end{bmatrix} + \frac{d}{dt} \begin{bmatrix} \lambda_1 \\ \lambda_2 \end{bmatrix} \tag{2.8-1}$$

which may also be written as

$$\mathbf{v}_{12} = \mathbf{r}\mathbf{i}_{12} + \frac{d}{dt}\boldsymbol{\lambda}_{12} \tag{2.8-2}$$

All terms are defined by comparison with (2.8-1). The flux linkage equations are

$$\begin{bmatrix} \lambda_1 \\ \lambda_2 \end{bmatrix} = \begin{bmatrix} L_{l1} + L_{m1} & L_{sr}\cos\theta_r \\ L_{sr}\cos\theta_r & L_{l2} + L_{m2} \end{bmatrix} \begin{bmatrix} i_1 \\ i_2 \end{bmatrix} \tag{2.8-3}$$

From Table 2.5-1,

$$T_e(i_1, i_2, \theta_r) = \frac{\partial W_c(i_1, i_2, \theta_r)}{\partial \theta_r} \tag{2.8-4}$$

Since the magnetic system is assumed to be linear,

$$W_c(i_1, i_2, \theta_r) = \frac{1}{2}L_{11}i_1^2 + L_{12}i_1i_2 + \frac{1}{2}L_{22}i_2^2 \tag{2.8-5}$$

The self-inductances are constants and the mutual inductance is represented by the off-diagonal terms in (2.8-3). Thus,

$$T_e(i_1, i_2, \theta_r) = -i_1i_2L_{sr}\sin\theta_r \tag{2.8-6}$$

Although (2.8-6) is valid regardless of the form of i_1 and i_2, let us consider for a moment the form of the torque if i_1 and i_2 are both positive and constant. For the positive direction of current shown, the torque would then be of the form

$$T_e = -K\sin\theta_r \tag{2.8-7}$$

where K is a positive constant equal to $i_1i_2L_{sr}$.

A plot of (2.8-7) is shown in Fig. 2.8-1 with the position of the windings illustrated for $\theta_r = 0,\ \frac{1}{2}\pi, \ldots, 2\pi$. Also shown are the positions of the poles of the magnetic systems created by constant positive current flowing in the windings. It may at first appear that the north (N^s) and south (S^s) poles produced by positive current flowing in the stator winding are positioned incorrectly. However, recall that flux issues from a north pole of a magnet into the air. Since the stator and rotor

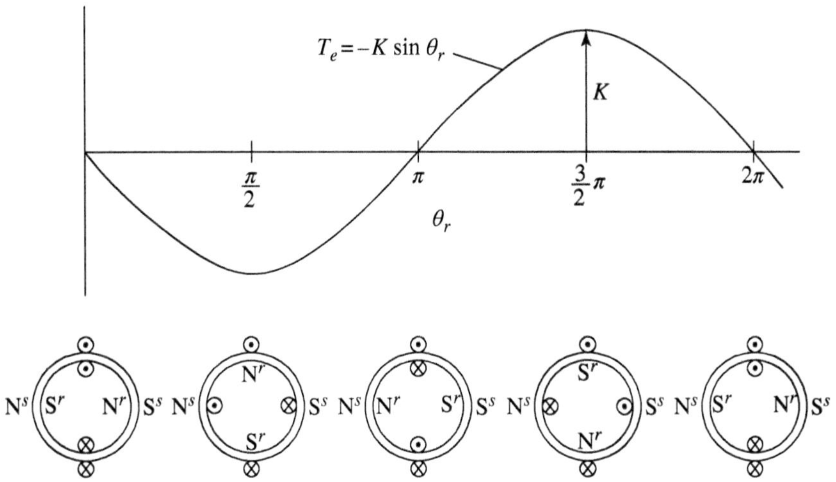

Figure 2.8-1 Electromagnetic torque versus angular displacement with constant winding currents.

windings must each be considered as creating separate magnetic systems, we realize, by the right-hand rule, that flux issues from the north pole of the magnetic system established by the stator winding into the air gap. Similarly, the flux produced by positive current flowing in the rotor winding enters the air gap from the north pole of the magnetic system of the rotor.

Here, we see that an electromagnetic torque is produced in an attempt to align the magnetic systems established by currents flowing in the stator and rotor windings; in other words, to align the 1 and 2 axes. We can use our stability test, as we did in the case of the single-phase reluctance machine, to establish the fact that, with no external torque on the rotor, stable positioning occurs at $\theta_r = 0$, whereas unstable operation occurs at $\theta_r = \pi$. This, of course, assumes that i_1 and i_2 are positive constants. Also, stable operation with an external applied torque occurs over the range of $-\frac{1}{2}\pi < \theta_r < \frac{1}{2}\pi$, the negative slope portion of the T_e versus θ_r characteristic. Although operation with constant winding currents is somewhat impracticable, it does illustrate the principle of positioning of stepper motors.

Before leaving this elementary device for good, let us assume that i_2 is constant ($i_2 = I_2$) and

$$I_1 = \sqrt{2}I \cos \theta_{esi} \tag{2.8-8}$$

where, for steady-state operation

$$\theta_{esi} = \omega_e t + \theta_{esi}(0) \tag{2.8-9}$$

Also,

$$\theta_r = \omega_r t + \theta_r(0) \tag{2.8-10}$$

Substituting into (2.8-6) yields

$$T_e = -\sqrt{2}II_2L_{sr}\cos\left[\omega_e t + \theta_{esi}(0)\right]\sin\left[\omega_r t + \theta_r(0)\right] \tag{2.8-11}$$

which can be expressed as

$$T_e = -\sqrt{2}II_2L_{sr}\left\{\frac{1}{2}\sin\left[(\omega_r + \omega_e)t + \theta_r(0) + \theta_{esi}(0)\right] + \frac{1}{2}\sin\left[(\omega_r - \omega_e)t + \theta_r(0) - \theta_{esi}(0)\right]\right\} \tag{2.8-12}$$

Here, we see that an average torque is produced if $|\,\omega_r\,| = \omega_e$. In particular, if $\omega_r = \omega_e$, (2.8-12) becomes

$$T_e = -\sqrt{2}II_2L_{sr}\left\{\frac{1}{2}\sin\left[2\omega_e t + \theta_r(0) + \theta_{esi}(0)\right] + \frac{1}{2}\sin\left[\theta_r(0) - \theta_{esi}(0)\right]\right\} \tag{2.8-13}$$

From this brief analysis, we can see that an average torque will be produced with $|\omega_r| = \omega_e$. In fact, with i_2 constant and I_1 given by (2.8-8), this device is an elementary single-phase synchronous machine. However, as in the case of the single-phase reluctance machine, we do not really have all the tools at hand to provide an in-depth explanation of its operation. We will be able to do this once we have studied the material in Chapter 4.

SP2.8-1 In the system shown in Fig. 1.7-4, $L_{sr} = 0.1$ H, $i_1 = 2$ A, and $i_2 = 10$ A. (a) A torque of 1 N·m is applied in the clockwise direction. Calculate the steady-state value of θ_r. (b) Repeat with a torque of 2 N·m. [(a) $\theta_r = -30°$; (b) unstable]

SP2.8-2 Reverse the direction of i_1 in Fig. 1.7-4 and determine the range of stable operation for constant, positive currents. $\left[\frac{1}{2}\pi < \theta_r < \frac{3}{2}\pi\right]$

SP2.8-3 In Fig. 1.7-4, winding 1 is moved so that ⊗ is at three o'clock and ⊙ at nine. Express T_e. $[T_e = i_1 i_2 L_{sr}\cos\theta_r]$

SP2.8-4 The currents i_1 and i_2 are positive constants in the device shown in Fig. 1.7-4. (a) An external torque is applied to increase θ_r to 45° and then released. Assume damping exists. What is the final position of the rotor? Repeat for θ_r increased to (b) 90°, (c) 120°, (d) 180°, and (e) 210°. [(a) $\theta_r = 0$; (b) 0; (c) 0; (d) 0 or 2π; (e) 2π]

SP2.8-5 In the system shown in Fig. 1.7-4, $L_{sr} = 0.1$ H, $i_1 = 20\cos\omega_e t$, $i_2 = 2$ A, and $\omega_r = \omega_e$. Calculate $\theta_r(0) - \theta_{esi}(0)$ if an external torque of 1 N·m is applied in the clockwise direction. $[\theta_r(0) = -30°]$

2.9 RECAPPING

The primary purpose of this chapter is to establish a means of expressing the force or torque in electromechanical devices. Therefore, an understanding of all relationships given in Table 2.5-1 is of importance. However, since the magnetic system of the devices that we will consider is assumed to be linear, we will make extensive use of the second entry of Table 2.5-1.

The detailed analysis of the elementary electromagnet and the associated problems at the end of the chapter warrant consideration since this material will not be revisited. You are now prepared to proceed to the study of dc machines (Chapter 3) or jump to Chapter 4 and begin study of the electromechanical motion devices that operate on the principle of Tesla's rotating magnetic field. This includes induction machines, synchronous machines, stepper motors, and the permanent-magnet ac machine.

2.10 PROBLEMS

1. An electromechanical system has two electric inputs. The flux linkages may be expressed as

$$\lambda_1(i_1, i_2, x) = x^2 i_1^2 + x i_2 + i_1$$

$$\lambda_2(i_1, i_2, x) = x^2 i_2^2 + x i_1 + i_2$$

Express $W_f(i_1, i_2, x)$ and $W_c(i_1, i_2, x)$ by first making i_2 the variable of integration with $di_1 = 0$ and $i_1 = 0$. Then, let i_1 be the variable of integration with $di_2 = 0$ and i_2 at its final value.

2. The mechanical system moves from x_1 to x_2 along the λi path 1 to 2 in Fig. 2.10-1. Identify the following by areas: (a) ΔW_f, (b) ΔW_c, (c) ΔW_e, (d) ΔW_m.

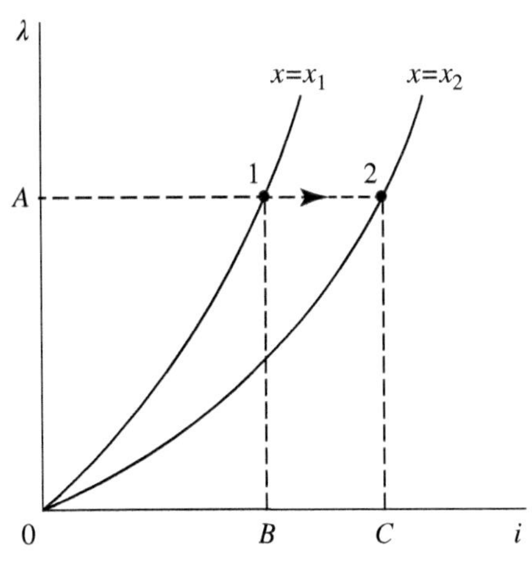

Figure 2.10-1 λi characteristics.

3. Let $i = k\lambda^2 e^{2x}$. Evaluate f_e when $\lambda = 2$ V · s, $x = 1$ m, and $k = 1$ A/(V · s)^2.

4. Express the force of attraction between the iron faces of the air gap in Fig. 1.3-1 in terms of N, the turns of the winding; i, the current flowing in the winding; μ_0; A_g, the

cross-sectional area of the air gap; and g, the air gap length. Neglect the reluctance of the iron. [*Hint*: Express the energy in the air gap in terms of air gap length and allow an infinitesimal change in air gap length dg. This procedure is often referred to as "virtual displacement."]

5. Express $f_e(i_1, i_2, x)$ for λ_1 and λ_2 given in Problem 1.

6. For the electromagnet shown in Fig. 2.2-3, let $L(x) = L_l + L_m(x)$, where L_l is constant and $L_m(x)$ is expressed by (1.7-17). Express f_e for $x = 0$.

7. Following the system transients due to the application of the source voltage in Fig. 2.6-2 ($v = 5$ V), the system assumes steady-state operation. Calculate W_{eS}, W_f, and W_{mS} during steady-state operation. [*Hint*: The external inductance l is set equal to zero in this example and $W_{mS} = K\int_{x_0}^{x}(\xi - x_0)d\xi$.]

8. Consider Fig. 2.6-3 wherein the force f is changed. The source voltage v is constant and the leakage inductance L_l and the external inductance l are both zero. As f is changed, e_f changes during the transient. Show that, for transient and steady-state conditions, the energy dissipated in the resistor may be expressed as

$$W_{eL} = \frac{v^2}{r}\int dt - \frac{2v}{r}\int e_f dt + \frac{1}{r}\int e_f^2 dt$$

   Also express W_E and W_e.

9. The dc source for the electromagnet considered in Section 2.6 is replaced with a 60-Hz source. With $f = 0$ and $i = \sqrt{2}\ 0.5\cos\omega_e t$, it is found that the system operates with x constant at 2.5 mm. (a) Express f_e and justify this observation. (b) Calculate the applied voltage.

10. Assume that the steady-state current supplied to a single-phase reluctance motor is sinusoidal. The rotor speed is $\omega_r = \omega_e$. Determine the harmonics that must be present in the applied voltage V_{as} in order for $I_{as} = \sqrt{2}I_s\cos\omega_e t$.

11. In Fig. 2.10-2, θ_r and ω_r are positive in the clockwise direction. The peak amplitude of the mutual inductance is L_{sr}. Express (a) the mutual inductance L_{ab} and (b) the electromagnetic torque T_e.

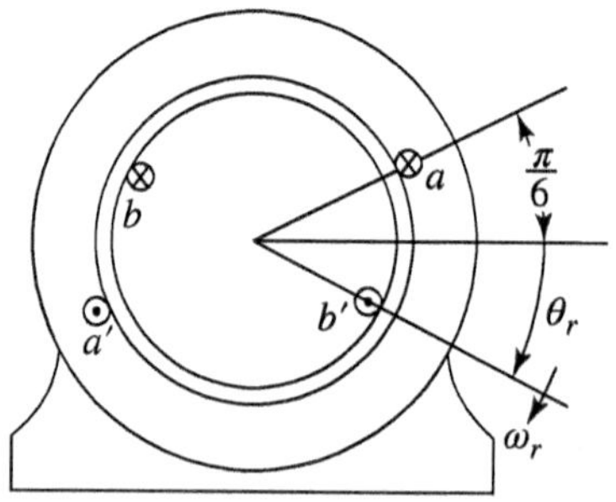

Figure 2.10-2 A two-winding device with clockwise rotation.

12. Consider the electromechanical system shown in Fig. 2.10-3. Assume the peak amplitude of the mutual inductance is L_{sr}. (a) Express the mutual inductance. (b) Show the location of the north and south poles of the stator (N^s and S^s) and of the rotor (N^r and S^r) for positive i_1 and negative i_2. (c) Express the electromagnetic torque T_e.

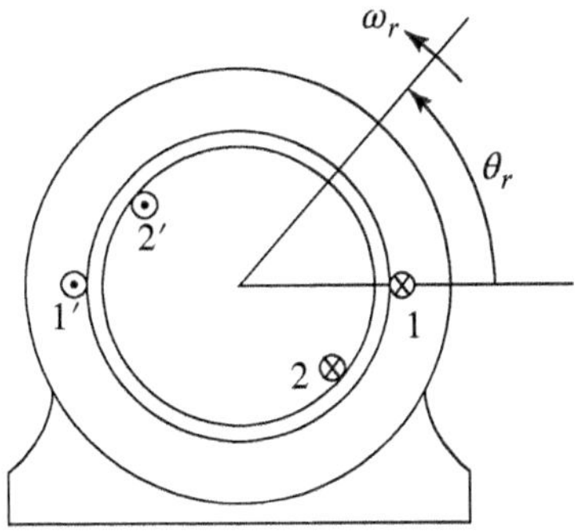

Figure 2.10-3 A two-winding device with counterclockwise rotation.

13. The device shown in Fig. 2.10-4 has two stator windings. Winding 1 is fixed in the position shown. However, we are able to wind winding 2 at any angle α relative to winding 1. (a) Express the mutual inductance between windings 1 and 2 as a function of α. Let the peak amplitude be M. (b) Use the concept of virtual displacement to express the torque between the windings. ■

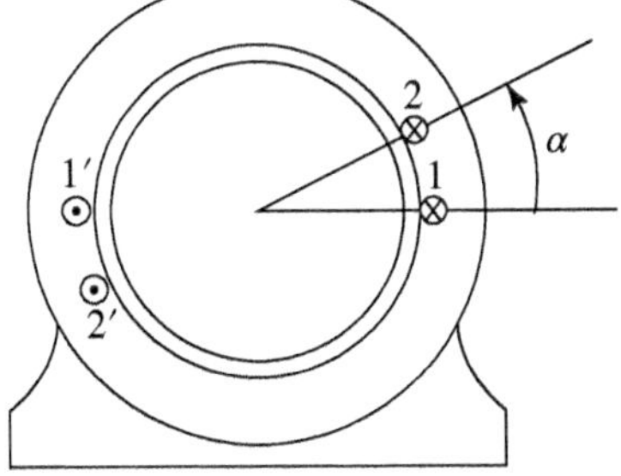

Figure 2.10-4 Two stator windings.

CHAPTER 3

DIRECT-CURRENT MACHINES AND THE DC DRIVE

3.1 INTRODUCTION

The direct-current (dc) machine is not as widely used today as it was in the past. The dc generator has been replaced by solid-state rectifiers which convert alternating current into dc with provisions to control the magnitude of the dc voltage and, in drive applications, the dc motor is being replaced by the voltage-controlled permanent magnet synchronous machine (brushless dc drive) and/or the vector-controlled induction motor. Nevertheless, it is still desirable to devote some time to the dc machine in an introductory course since it is still used as a low-power drive motor, especially in automotive applications. There is another and perhaps more important reason to consider the dc machine. Because of the brushes, the maintenance and environmental issues hamper the use of dc machines; however, this device is the only electric machine that inherently produces maximum torque per ampere. With the advent of electronic switching devices, there has been a large effort to control the permanent-magnet ac synchronous and induction machines so as to emulate the performance characteristics of the dc motor without the maintenance and environmental problems inherent to the dc machines. This chapter is an attempt to treat dc machines sufficient to introduce the reader to the operating principles of dc machines with focus on the shunt-connected and permanent-magnet dc machines, thereby setting the stage for a comparison of the operating characteristics with the voltage-controlled permanent-magnet ac drive and the vector-controlled induction motor.

A simplified method of analysis is presented rather than an analysis wherein commutation is treated in detail. With this type of analytical approach, the dc machine is considered to be the most straightforward to analyze of all electromechanical devices. The dynamic characteristics of the permanent-magnet dc machine are illustrated and the time-domain block diagram and state equations

Electromechanical Motion Devices: Rotating Magnetic Field-Based Analysis and Online Animations, Third Edition. Paul Krause, Oleg Wasynczuk, Steven D. Pekarek, and Timothy O'Connell.

Published 2020 by John Wiley & Sons, Inc.

are developed. In the last section of this chapter, the analysis of dc/dc converters used as a means of voltage control and thus speed control in dc drives systems is presented. This includes converter switching for continuous current operation of two quadrant converters and the derivation of average-value equivalent circuits for two quadrant converters supplying a permanent-magnet dc machine. The material presented in this chapter is not a prerequisite for material in later chapters. Commutation and the ideal dc drive are animated in the online Animation A.

3.2 ELEMENTARY DIRECT-CURRENT MACHINE

It is instructive to discuss the concept of commutation using the elementary machine shown in Fig. 3.2-1 prior to a formal analysis of the performance of a practical dc machine. The two-pole elementary machine is equipped with a field

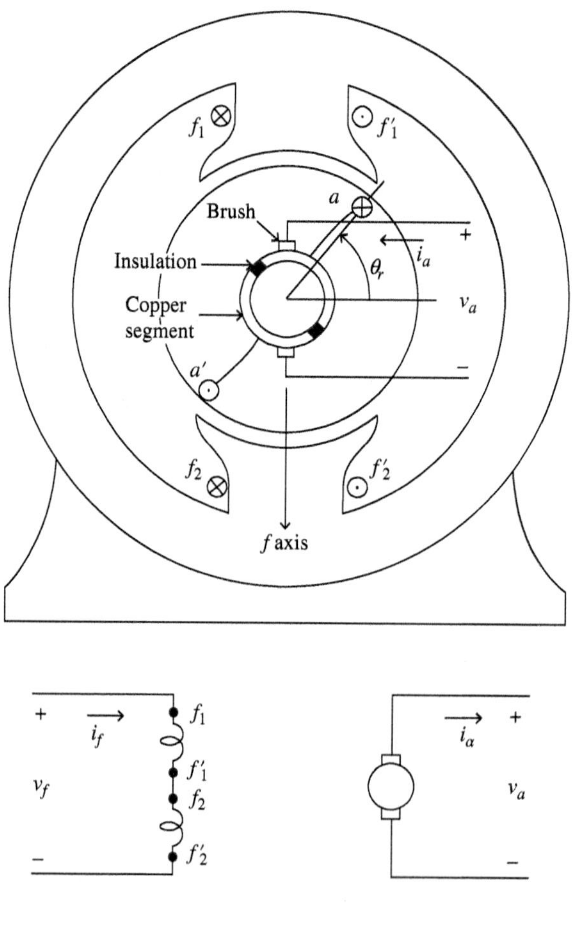

Figure 3.2-1 Elementary two-pole dc machine.

winding wound on the stator poles, a rotor coil $(a-a')$, and a commutator. The commutator is made up of two semicircular copper segments mounted on the shaft at the end of the rotor and insulated from one another as well as from the iron of the rotor. Each terminal of the rotor coil is connected to a copper segment. Stationary carbon brushes ride upon the copper segments whereby the rotor coil is connected to a stationary circuit by a near-frictionless contract.

The voltage equations for the field winding and rotor coil are

$$v_f = r_f i_f + \frac{d\lambda_f}{dt} \tag{3.2-1}$$

$$v_{a-a'} = r_a i_{a-a'} + \frac{d\lambda_{a-a'}}{dt} \tag{3.2-2}$$

where r_f and r_a are the resistance of the field winding and armature coil, respectively. The rotor of a dc machine is commonly referred to as the *armature*; rotor and armature will be used interchangeably. At this point in the analysis it is sufficient to express the flux linkages as

$$\lambda_f = L_{ff} i_f + L_{fa} i_{a-a'} \tag{3.2-3}$$

$$\lambda_{a-a'} = L_{af} i_f + L_{aa} i_{a-a'} \tag{3.2-4}$$

As a first approximation, the mutual inductance between the field winding and an armature coil may be expressed as a sinusoidal function of θ_r as

$$L_{af} = L_{fa} = -L\cos\theta_r \tag{3.2-5}$$

where L is a constant. As the rotor revolves, the action of the commutator is to switch the stationary terminals from one terminal of the rotor coil to the other. For the configuration shown in Fig. 3.2-1, this switching or commutation occurs at $\theta_r = 0, \pi, 2\pi, \ldots$. At the instant of switching, each brush is in contact with both copper segments whereupon the rotor coil is short-circuited. It is desirable to commutate (short-circuit) the rotor coil at the instant the induced voltage is a minimum. The waveform of the voltage induced in the open-circuited armature coil, during constant-speed operation with a constant field winding current, may be determined by setting $i_{a-a'} = 0$ and i_f equal to a constant. Substituting (3.2-4) and (3.2-5) into (3.2-2) yields the following expression for the open-circuit voltage of coil *a–a'* with the field current i_f a constant:

$$v_{a-a'} = \omega_r L I_f \sin\theta_r \tag{3.2-6}$$

where $\omega_r = d\theta_r/dt$ is the rotor speed. The open-circuit coil voltage $v_{a-a'}$ is zero at $\theta_r = 0, \pi, 2\pi, \ldots,$ which is the rotor position during commutation. Commutation is illustrated in Fig. 3.2-2. The open-circuit terminal voltage, v_a, corresponding to the rotor positions denoted as θ_{ra}, $\theta_{rb}(\theta_{rb} = 0)$, and θ_{rc} are indicated. It is important to note that, during one revolution of the rotor, the assumed positive direction of armature current i_a is down coil side a and out of coil side a' for $0 < \theta_r < \pi$. For $\pi < \theta_r < 2\pi$, positive current is down coil side a' and out of coil side a. In

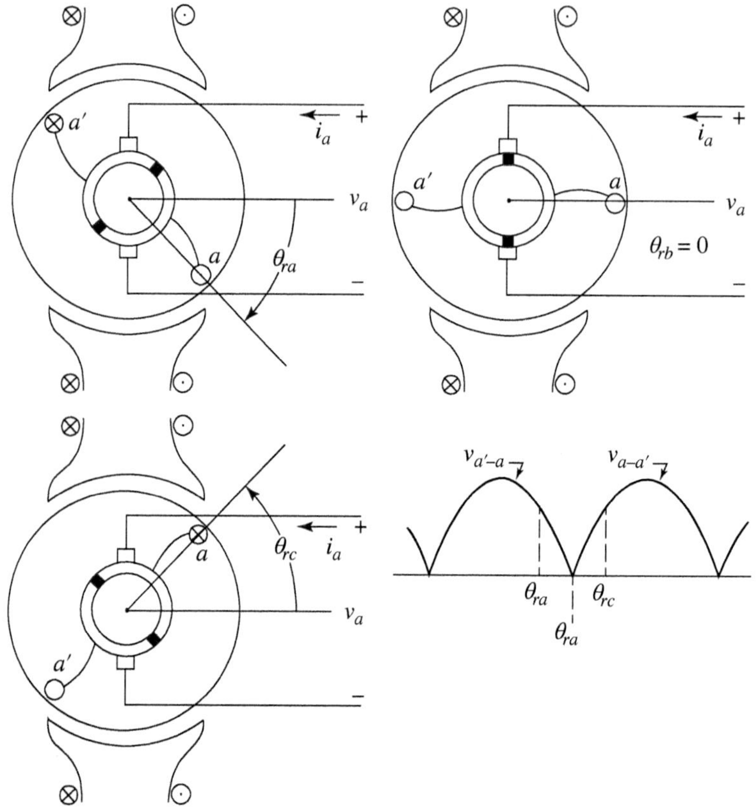

Figure 3.2-2 Commutation of the elementary dc machine.

Chapters 1 and 2 we let positive current flow into the winding denoted without a prime and out the winding denoted with a prime. We will not be able to adhere to this relationship in the case of the armature windings of a dc machine in either the illustrations in this chapter or in the animations since commutation is involved. Although the machine shown in Fig. 3.2-1 could be operated as a generator supplying a resistive load, it could not be operated effectively as a motor supplied from a voltage source owing to the short-circuiting of the armature coil at each commutation. Nevertheless, this impracticable device helps to illustrate two features of commutation; first, commutation takes place when the voltage induced in the rotor winding is ideally zero and secondly, commutation can be thought of as a mechanical means of full-wave rectification. We will see another important feature that commutation provides when we consider the torque produced by the dc machine.

A practicable dc machine, with the rotor equipped with an a winding and an A winding, is shown schematically in Fig. 3.2-3. At the rotor position depicted, coils $a_4 - a_4'$ and $A_4 - A_4'$ are being commutated. The bottom brush short-circuits the $a_4 - a_4'$ coil while the top brush short-circuits the $A_4 - A_4'$ coil. Figure 3.2-3 illustrates the instant when the assumed direction of positive current is into the paper in

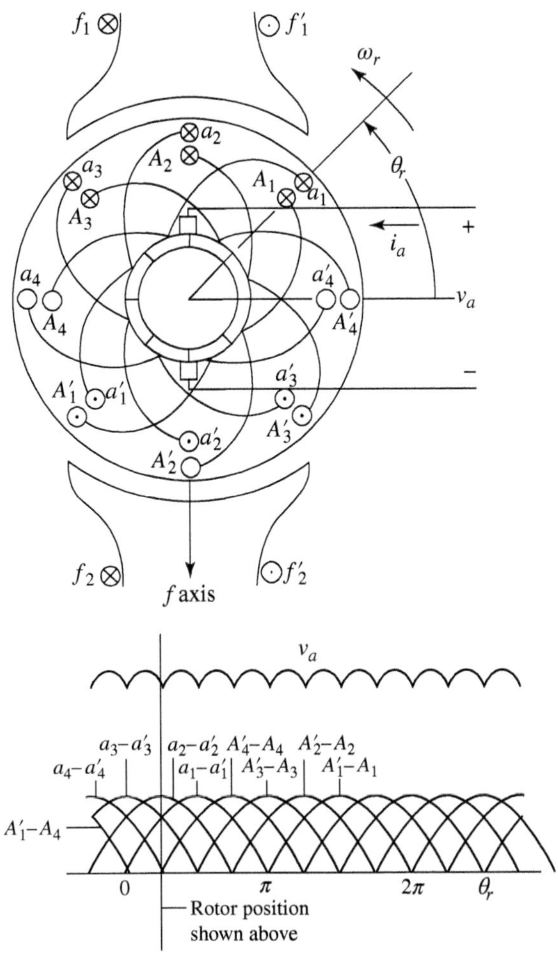

Figure 3.2-3 A dc machine with parallel armature windings.

coil sides $a_1, A_1; a_2, A_2; \ldots$, and out in coil sides $a_1', A_1'; a_2', A_2'; \ldots$. It is instructive to follow the path of current through one of the parallel paths from one brush to the other. For the angular position shown in Fig. 3.2-3, positive current enters the top brush and flows down the rotor via a_1 and back through a_1'; down a_2 and back through a_2'; down a_3 and back through a_3' to the bottom brush. A parallel current path exists through $A_3 - A_3'$, $A_2 - A_2'$, and $A_1 - A_1'$. The open-circuit or induced armature voltage is also shown in Fig. 3.2-3; however, these idealized waveforms require additional explanation. As the rotor advances in the counterclockwise direction, the segment connected to a_1 and A_4 moves from under the top brush, as shown in Fig. 3.2-4. The top brush then rides only on the segment connecting A_3 and A_4'. At the same time the bottom brush is riding on the segment connecting a_4 and a_3'. With the rotor so positioned, current flows

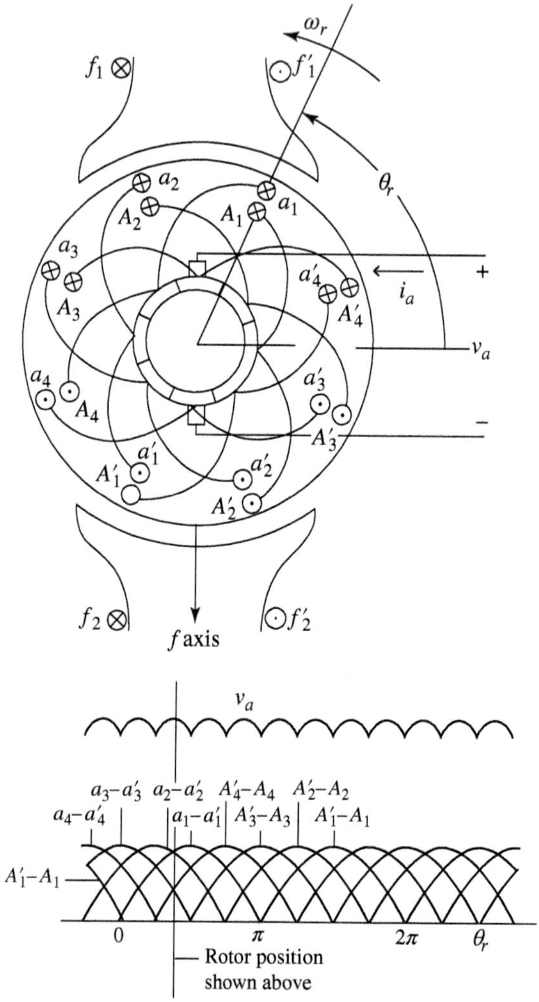

Figure 3.2-4 Same as Fig. 3.2-3 with rotor advanced approximately 22.5° counterclockwise.

in A_3 and A_4' and out a_3' and a_4. In other words, current flows down the coil sides in the upper one half of the rotor and out of the coil sides in the bottom one half. Let us follow the current flow through the parallel paths of the armature windings shown in Fig. 3.2-4. Current now flows through the top brush into A_4' out A_4, into a_1 out a_1', into a_2 out a_2', into a_3 out a_3' to the bottom brush. The parallel path beginning at the top brush is $A_3 - A_3'$, $A_2 - A_2'$, $A_1 - A_1'$, and $a_4' - a_4$ to the bottom brush. The voltage induced in the coils is shown in Figs. 3.2-3 and 3.2-4 for the first parallel path described. It is noted that the induced voltage is plotted only when the coil is in this parallel path.

In Figs. 3.2-3 and 3.2-4, the parallel windings consist of only four coils. Usually the number of rotor coils is substantially more than four, thereby reducing the harmonic content of the open-circuit armature voltage. In this case, the rotor coils may be approximated as a uniformly distributed winding, as illustrated in Fig. 3.2-5. Therein, the rotor winding is considered as current sheets which are fixed in space due to the action of the commutator and which establish a magnetic axis positioned orthogonal to the magnetic axis of the field winding. We will see the importance of this orthogonal positioning of the magnetic axes when we derive the expression of the torque. The brushes are shown positioned on the current sheet for the purpose of depicting commutation. The small angular displacement, denoted by 2γ, designates the region of commutation wherein the coils are short-circuited. However, commutation cannot be visualized from Fig. 3.2-5; one must refer to Figs. 3.2-3 and 3.2-4.

Before proceeding to the development of the equations portraying the operating characteristics of the dc machine, it is instructive to take a brief look at the arrangement of the armature windings and the method of commutation used in many of the low-power permanent-magnet dc motors. Small dc motors used in

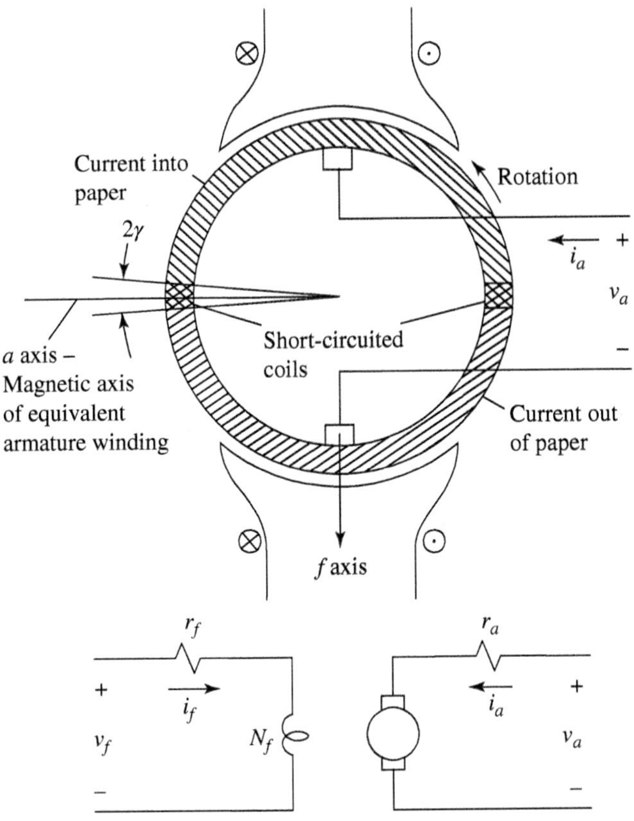

Figure 3.2-5 Idealized dc machine with uniformly distributed rotor winding.

low-power control systems are often the permanent-magnet type, wherein a constant field flux is established by a permanent magnet rather than by a current flowing in a field winding.

Three rotor positions of a typical low-power permanent-magnet dc motor are shown in Fig. 3.2-6. The rotor is turning in the counterclockwise direction and the rotor position advances from (a) to (c). Physically, these devices may be an inch or less in diameter with brushes sometimes as small as a pencil lead. They are mass-produced and relatively inexpensive. Although the brushes actually ride on the outside of the commutator, for convenience they are shown on the inside in Fig. 3.2-6. The armature windings consist of a large number of turns of fine wire; hence, each circle shown in Fig. 3.2-6 represents many conductors. Note that the position of the brushes is shifted approximately 40° relative to a line drawn between the center of the north and south poles. This shift in the brushes was probably determined experimentally by minimizing brush

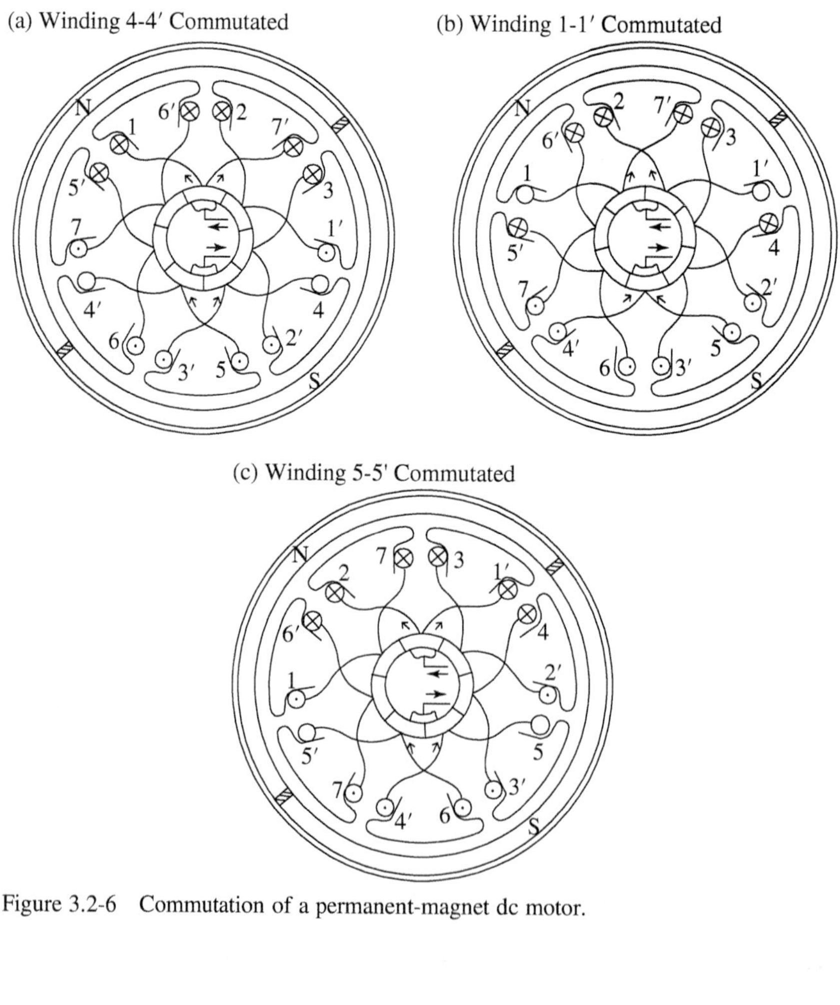

Figure 3.2-6 Commutation of a permanent-magnet dc motor.

Figure 3.2-7 Two-pole 0.1-hp 6-V 12,000-r/min permanent-magnet dc motor. (Courtesy of Vick ElectroMech.)

arcing for normal load conditions. Note also that the windings do not span π radians but more like 140°, and there is an odd number of commutator segments.

In Fig. 3.2-6a, only winding 4 is being commutated. As the rotor advances from the position shown in Fig. 3.2-6a, both windings 4 and 1 are being commutated. In Fig. 3.2-6b, winding 1 is being commutated; then windings 1 and 5, and, finally, in Fig. 3.2-6c, only winding 5 is being commutated. The windings are being commutated when the induced voltage is nonzero and one would expect some arcing to occur at the brushes. We must realize, however, that the manufacture and sale of these devices is very competitive, and one often must compromise when striving to produce an acceptable motor at the least cost possible. Although we realize that in some cases it may be a rather crude approximation, we will consider the permanent-magnet dc motor as having current sheets on the armature with orthogonal armature and field magnetic axes as shown in Fig. 3.2-5. A disassembled two-pole 0.1-hp 6-V 12,000-r/min permanent-magnet dc motor is shown in Fig. 3.2-7. The magnets are samarium cobalt and the device is used to drive hand-held battery-operated surgical instruments. Although some of these terms are new to us, they will be defined as we go along.

SP3.2-1 The peak value of the voltage induced in one coil shown in Fig. 3.2-3 is 1 V. Determine, from Fig. 3.2-3, the maximum and minimum value of v_a. [2.613 V; 2.414 V].

SP3.2-2 Consider Fig. 3.2-3. Indicate the two parallel paths immediately following commutation of $a_3 - a_3'$ and $A_3 - A_3'$. $[A_3' - A_3, A_4' - A_4, a_1 - a_1'$, and $a_2 - a_2'$; $A_2 - A_2'$, $A_1 - A_1'$, $a_4' - a_4$, and $a_3' - a_3]$

3.3 VOLTAGE AND TORQUE EQUATIONS

It is advantageous to first consider the dc machine with a field and armature winding before turning to the permanent-magnet device exclusively. Although rigorous derivation of the voltage and torque equations is possible, it is rather lengthy and little is gained since these relationships may be deduced. The armature coils revolve in a magnetic field established by a current flowing in the field winding. We have established that voltage is induced in these coils by virtue of this rotation. However, the action of the commutator causes the armature coils to appear as a stationary winding with its magnetic axis orthogonal to the magnetic axis of the field winding. In other words, the stator and rotor mmfs are orthogonal. Consequently, voltages are not induced in one winding due to the time rate of change of the current flowing in the other (transformer action). Mindful of these conditions, we can write the field and armature voltage equations in matrix form as

$$\begin{bmatrix} v_f \\ v_a \end{bmatrix} = \begin{bmatrix} r_f + pL_{FF} & 0 \\ \omega_r L_{AF} & r_a + pL_{AA} \end{bmatrix} \begin{bmatrix} i_f \\ i_a \end{bmatrix} \tag{3.3-1}$$

where L_{FF} and L_{AA} are the self-inductances of the field and armature windings, respectively, and p is the short-hand notation for the operator d/dt. The rotor speed is denoted as ω_r, and L_{AF} is the mutual inductance between the field and the rotating armature coils. The above equation suggests the equivalent circuit shown in Fig. 3.3-1. The voltage induced in the armature circuit, $\omega_r L_{AF} i_f$, is commonly referred to as the counter or back emf. It also represents the open-circuit armature voltage.

There are several other forms in which the field and armature voltage equations are often expressed. For example, L_{AF} may also be written as

$$L_{AF} = \frac{N_a N_f}{\mathcal{R}} \tag{3.3-2}$$

where N_a and N_f are the equivalent turns of the armature and field windings, respectively, and $\mathcal{R}$ is the reluctance. Thus,

$$L_{AF} i_f = N_a \frac{N_f i_f}{\mathcal{R}} \tag{3.3-3}$$

If we now replace $N_f i_f / \mathcal{R}$ with Φ_f, the field flux per pole, then $N_a \Phi_f$ may be substituted for $L_{AF} i_f$ in the armature voltage equation.

Another substitute variable often used is

$$k_v = L_{AF} i_f \tag{3.3-4}$$

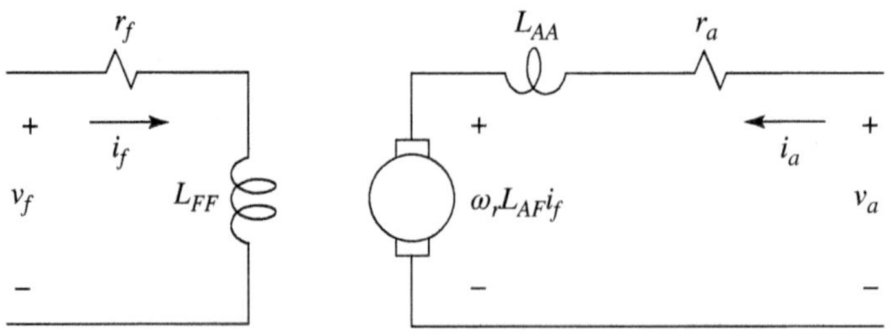

Figure 3.3-1 Equivalent circuit of dc machine.

We will find that this substitute variable is particularly convenient and frequently used. Even though a permanent-magnet dc machine has no field circuit, the constant field flux produced by the permanent magnet is analogous to a dc machine with a constant k_v.

We can take advantage of previous work to obtain an expression for the electromagnetic torque. In particular, the expression for torque given by (2.8-6) may be used directly to express the torque for the dc machine. If we fix θ_r in Fig. 1.7-4 or Fig. 2.8-1 at $-\frac{1}{2}\pi$, the same relationship exists between the magnetic axes of a dc machine (Fig. 3.2-5) and the magnetic axes of the two-coil machine. Hence, (2.8-6) may be written for the dc machine as

$$T_e = L_{AF} i_f i_a \tag{3.3-5}$$

Here again the variable k_v is often substituted for $L_{AF} i_f$. In some instances k_v is multiplied by a factor less than unity when substituted into (3.3-5) so as to approximate the effects of rotational losses. It is interesting that the field winding produces a stationary mmf and, owing to commutation, the armature winding also produces a stationary mmf which is displaced $\frac{1}{2}\pi$ electrical degrees from the mmf produced by the field winding. It follows then that the interaction of these two mmfs produces the electromagnetic torque and, due to the method of commutation, the mmfs are in quadrature thereby producing the maximum torque possible for any field and armature currents. That is, ideally, commutation positions the magnetic fields stationary and orthogonal which yields the maximum possible torque per ampere; a feature that is trying to be emulated by modern-day control of *ac* machines.

The torque and rotor speed are related by

$$T_e = J\frac{d\omega_r}{dt} + B_m \omega_r + T_L \tag{3.3-6}$$

where J is the inertia of the rotor and rigidly connected mechanical load. The units of the inertia are $\text{kg} \cdot \text{m}^2$ or $\text{J} \cdot \text{s}^2$. A positive electromagnetic torque T_e acts to turn the rotor in the direction of increasing θ_r. The load torque T_L is positive for a torque, on the shaft of the rotor, which opposes a positive electromagnetic torque T_e. The constant B_m is a damping coefficient associated with the mechanical rotational system of the machine. It has the units of $\text{N} \cdot \text{m} \cdot \text{s}$ and it is generally small and often neglected.

In the next section we will focus on the permanent-magnet dc motor; however, it is worthwhile to take a moment to mention that we have established the basis for several types of dc machines. In particular, the machine shown in Fig. 3.3-1 is a separately excited dc machine. If we connect the field winding in parallel with the armature winding, it becomes a shunt-connected dc machine. If the field winding is connected in series with the armature winding, it is a series-connected dc machine. If two windings are used, one in parallel with and another in series with the armature, it is referred to as a compound-connected dc machine. Clearly, this is an overly simplistic description and the reader is referred to [1] for a more detailed consideration of these machine types.

Before proceeding, it is appropriate to mention briefly generator action even though the dc generator has been replaced by the ac to dc converter. We see from Fig. 3.3-1 that if $\omega_r L_{AF} i_f$ is greater than v_a, i_a is reversed and we have generator action. In this case T_L is negative since the dc machine is being driven and T_e, (3.3-5), is negative and in the steady state from (3.3-6) $\frac{d\omega_r}{dt}$ is zero and B_m is small.

SP3.3-1 When a 12 V permanent-magnet dc motor is driven at 100 rad/s, the open-circuit voltage is 10 V. Calculate k_v. [$k_v = 0.1\ \text{V}\cdot\text{s/rad}$]

SP3.3-2 The armature applied voltage of a dc motor is 240 V; the rotor speed is constant at 50 rad/s. The steady-state armature current is 15 A, the armature resistance is 1 Ω, and $L_{AF} = 1$ H. Calculate the steady-state field current. [$I_f = 4.5$ A]

SP3.3-3 Calculate the no-load speed ($T_L = 0$) for the permanent-magnet dc motor in SP3.3-1 when rated voltage (12 V) is applied to the armature. [$\omega_r = 120\ \text{rad/s}$]

SP3.3-4 Multiply the expression for v_a given in (3.3-1) by i_a and identify all terms. [$v_a i_a$ – input power to armature, $i_a^2 r_a$ – armature ohmic loss, $L_{AA} i_a p i_a$ – change of energy stored in L_{AA}, $L_{AF} i_f i_a \omega_r = T_e \omega_r$ – output power]

SP3.3-5 Is SP3.3-4 an alternate method of deriving an expression for torque? [Yes; $L_{AF} i_f i_a$ is the coefficient of $p\theta_r$]

3.4 PERMANENT-MAGNET DC MACHINE

In the case of the permanent-magnet dc machine, $L_{AF} I_f$ is replaced with k_v whereupon the steady-state armature voltage equation becomes

$$V_a = r_a I_a + k_v \omega_r \tag{3.4-1}$$

If (3.4-1) is solved for I_a and substituted into (3.3-5) with $L_{AF} I_f$ replaced by k_v, the steady-state torque may be expressed as

$$T_e = \frac{k_v V_a - k_v^2 \omega_r}{r_a} \tag{3.4-2}$$

The steady-state torque-speed characteristic is shown in Fig. 3.4-1.

It is apparent from Fig. 3.4-1, that the stall ($\omega_r = 0$) torque could be made larger for a given armature voltage by reducing r_a. Although the machine may be designed with a smaller armature resistance, there is a problem since, at stall, the steady-state armature current is limited by the armature resistance; hence, for a constant V_a, reducing r_a will result in a larger I_a which can cause damage to the brushes. On the other hand, increasing the starting torque by reducing r_a causes the torque-speed characteristics to have a steeper slope which results in a smaller change in speed for a given change in load torque during normal (near-rated) operation. If, however, the armature voltage is reduced during the starting period to protect the brushes, the desirable characteristic of a small speed change during load torque variations during normal operation could be achieved. In fact, controlled regulation of the armature voltage is generally employed for large

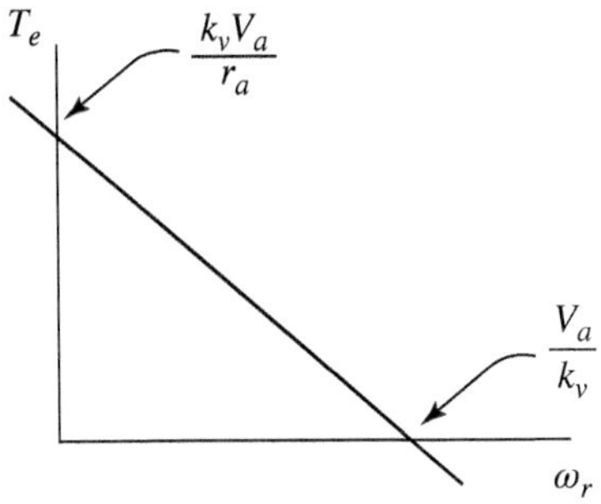

Figure 3.4-1 Steady-state torque-speed characteristic of a permanent-magnet dc machine.

horsepower machines by using a converter; however, low-power permanent-magnet dc machines are generally supplied from a battery as in the case of the automobile and, therefore, a large armature resistance is necessary in order to prevent brush damage during the early part of the starting period. Fortunately, a small speed variation during load torque changes is not required in many applications of the permanent-magnet dc machine; therefore, the steep torque-speed characteristics are not necessary. We will consider the dynamic performance of a permanent-magnet dc motor in the next section. In a later chapter, we will discover that the brushless dc machine can be controlled to have torque-speed characteristics similar to that of the permanent-magnet dc motor.

Example 3A Calculating Machine Parameters

A permanent-magnet dc motor similar to that shown in Fig. 3.2-7 is rated at 6 V with the following parameters: $r_a = 7\,\Omega$, $L_{AA} = 120$ mH, $k_T = 2\,\text{oz}\cdot\text{in}/\text{A}$, and $J = 150\,\mu\,\text{oz}\cdot\text{in}\cdot\text{s}^2$. According to the motor information sheet, the no-load speed is approximately 3350 r/min and the no-load armature current is approximately 0.15 A. Let us attempt to interpret this information.

First, let us convert k_T and J to units which we have been using in this text. In this regard, we will convert the inertia to $\text{kg}\cdot\text{m}^2$ which is the same as $\text{N}\cdot\text{m}\cdot\text{s}^2$. To do this, we must convert ounces to Newtons and inches to meters (Appendix A). Thus,

$$J = \frac{150 \times 10^{-6}}{(3.6)(39.37)} = 1.06 \times 10^{-6}\,\text{kg}\cdot\text{m}^2 \tag{3A-1}$$

We have not seen k_T before. It is the torque constant and, if expressed in the appropriate units, it is numerically equal to k_v. When k_v is used in the expression for $T_e(T_e = k_v i_a)$, it is often referred to as the *torque constant* and denoted k_T. When used in the voltage equation, it is always denoted k_v. Now, we must convert oz·in to N·m, whereupon k_T equals our k_v; hence,

$$k_v = \frac{2}{(3.6)(39.37)} = 1.41 \times 10^{-2}\,\text{N}\cdot\text{m}/\text{A} = 1.41 \times 10^{-2}\,\text{V}\cdot\text{s}/\text{rad} \tag{3A-2}$$

What do we do about the no-load armature current? What does it represent? Well, probably it is a measure of the friction and windage losses. We could neglect it, but we will not. Instead, let us include it as B_m. First, however, we must calculate the no-load speed. We can solve for the no-load rotor speed from the steady-state armature voltage equation (3.4-1)

$$\begin{aligned}\omega_r &= \frac{V_a - r_a I_a}{k_v} = \frac{6-(7)(0.15)}{1.41 \times 10^{-2}} = 351.1 \text{ rad/s} \\ &= \frac{(351.1)(60)}{2\pi} = 3353 \text{ r/min}\end{aligned} \tag{3A-3}$$

Now, at this no-load speed,

$$T_e = k_v i_a = \left(1.41 \times 10^{-2}\right)(0.15) = 2.12 \times 10^{-3}\ \text{N}\cdot\text{m} \tag{3A-4}$$

Since T_L and $J(d\omega_r/dt)$ are zero for this steady-state no-load condition, (3.3-6) tells us that (3A-4) is equal to $B_m\omega_r$; hence,

$$B_m = \frac{2.12 \times 10^{-3}}{\omega_r} = \frac{2.12 \times 10^{-3}}{351.1} = 6.04 \times 10^{-6}\ \text{N}\cdot\text{m}\cdot\text{s} \tag{3A-5}$$

Example 3B Efficiency of Operation

The permanent-magnet dc machine described in Example 3A is operating with rated applied armature voltage and a load torque T_L of 0.5 oz · in. Our task is to determine the efficiency where percent eff = (power output/power input)100.

First, let us convert oz·in to N · m:

$$T_L = \frac{0.5}{(3.6)(39.37)} = 3.53 \times 10^{-3}\ \text{N}\cdot\text{m} \tag{3B-1}$$

In Example 3A we determined k_v to be 1.41×10^{-2} V · s/rad and B_m to be 6.04×10^{-6} N · m · s. During steady-state operation, (3.3-6) becomes

$$T_e = B_m\omega_r + T_L \tag{3B-2}$$

From (3.3-5), with $L_{AF}i_f$ replaced by k_v, the steady-state electromagnetic torque is

$$T_e = k_v I_a \tag{3B-3}$$

Substituting (3B-3) into (3B-2) and solving for ω_r yields

$$\omega_r = \frac{k_v}{B_m} I_a - \frac{1}{B_m} T_L \tag{3B-4}$$

Repeating (3.4-1)

$$V_a = r_a I_a + k_v \omega_r \tag{3B-5}$$

Substituting (3B-4) into (3B-5) and solving for I_a yields

$$\begin{aligned} I_a &= \frac{V_a + (k_v/B_m)T_L}{r_a + \left(k_v^2/B_m\right)} \\ &= \frac{6 + \left[\left(1.41 \times 10^{-2}\right)/\left(6.04 \times 10^{-6}\right)\right]\left(3.53 \times 10^{-3}\right)}{7 + \left(1.41 \times 10^{-2}\right)^2/\left(6.04 \times 10^{-6}\right)} = 0.347 \text{ A} \end{aligned} \tag{3B-6}$$

From (3B-4),

$$\begin{aligned} \omega_r &= \frac{1.41 \times 10^{-2}}{6.04 \times 10^{-6}} 0.357 - \frac{1}{6.04 \times 10^{-6}}\left(3.53 \times 10^{-3}\right) \\ &= 249 \text{ rad/s} \end{aligned} \tag{3B-7}$$

The power input is

$$P_{\text{in}} = V_a I_a = (6)(0.357) = 2.14 \text{ W} \tag{3B-8}$$

The useable power output is

$$P_{\text{out}} = T_L \omega_r = \left(3.53 \times 10^{-3}\right)(249) = 0.88 \text{ W} \tag{3B-9}$$

The efficiency is

$$\begin{aligned} \text{Percent eff} &= \frac{P_{\text{out}}}{P_{\text{in}}} 100 \\ &= \frac{0.88}{2.14} 100 = 41.1 \text{ percent} \end{aligned} \tag{3B-10}$$

This low efficiency is characteristic of low-power dc motors due to the relatively large armature resistance. In this regard, it is interesting to determine the losses due to i^2r, friction, and windage.

$$P_{i^2r} = r_a I_a^2 = (7)(0.357)^2 = 0.89 \text{ W} \tag{3B-11}$$

$$P_{fw} = (B_m \omega_r)\omega_r = \left(6.04 \times 10^{-6}\right)(249)^2 = 0.37 \text{ W} \tag{3B-12}$$

Let us check our work:

$$P_{\text{in}} = P_{i^2r} + P_{fw} + P_{\text{out}} = 0.89 + 0.37 + 0.88 = 2.14 \text{ W} \tag{3B-13}$$

which is equal to (3B-8).

SP3.4-1 A 12-V permanent-magnet dc motor has an armature resistance of 12 Ω and $k_v = 0.01 \text{ V} \cdot \text{s/rad}$. Calculate the steady-state stall torque (T_e with $\omega_r = 0$). $[T_e = 0.01 \text{ N} \cdot \text{m}]$

SP3.4-2 Determine T_e in Example 3B. $[T_e = 0.713 \text{ oz} \cdot \text{in}]$

3.5 TIME-DOMAIN BLOCK DIAGRAM AND STATE EQUATIONS FOR THE PERMANENT-MAGNET DC MACHINE

Although the analysis of control systems is not our intent, it is worthwhile to set the stage for this type of analysis by means of a "first look" at time-domain block diagrams and state equations. Block diagrams, which portray the interconnection of the system equations, are used exclusively in control system analysis and design. We shall work with the time-domain equations, using the p operator to denote differentiation with respect to time and the operator $1/p$ to denote integration. Arranging the equations of a permanent-magnet dc machine into a block diagram representation is straightforward. The armature voltage equation, (3.3-1), and the relationship between torque and rotor speed, (3.3-6), may be written as

$$v_a = r_a(1 + \tau_a p)i_a + k_v\omega_r \tag{3.5-1}$$

$$T_e - T_L = (B_m + Jp)\omega_r \tag{3.5-2}$$

where the armature time constant is $\tau_a = L_{AA}/r_a$. Solving (3.5-1) for i_a and (3.5-2) for ω_r yields

$$i_a = \frac{1/r_a}{\tau_a p + 1}(v_a - k_v\omega_r) \tag{3.5-3}$$

$$\omega_r = \frac{1}{Jp + B_m}(T_e - T_L) \tag{3.5-4}$$

A few comments are in order regarding these expressions. In (3.5-3), we see that $(v_a - k_v\omega_r)$ is multiplied by the operator $(1/r_a)/(\tau_a p + 1)$ to obtain the armature current i_a. The fact that we are multiplying the voltage by an operator to obtain current is in no way indicative of the procedure that we might actually use to calculate the current i_a given the voltage $(v_a - k_v\omega_r)$. We are simply expressing the dynamic relationship between voltage and current in a form convenient for drawing block diagrams. However, to calculate i_a, we may prefer to express (3.5-3) in its equivalent form (3.5-1) and solve the given first-order differential equation using standard techniques. The operator $(1/r_a)/(\tau_a p + 1)$ in (3.5-3) may also be interpreted as a transfer function relating the voltage and current. Those of you who are familiar with Laplace transform methods are likely accustomed to seeing transfer functions expressed in terms of the Laplace operator s instead of the differentiation operator p. In fact, the same transfer functions are obtained by using Laplace transform theory with p replaced by s.

The time-domain block diagram portraying (3.5-3) and (3.5-4) with $T_e = k_v i_a$ is shown in Fig. 3.5-1. The diagram consists of a set of linear blocks, wherein the relationship between the input and corresponding output variable is depicted in transfer function form.

The so-called state equations of a system represent the formulation of the state variables into a matrix form convenient for computer implementation for

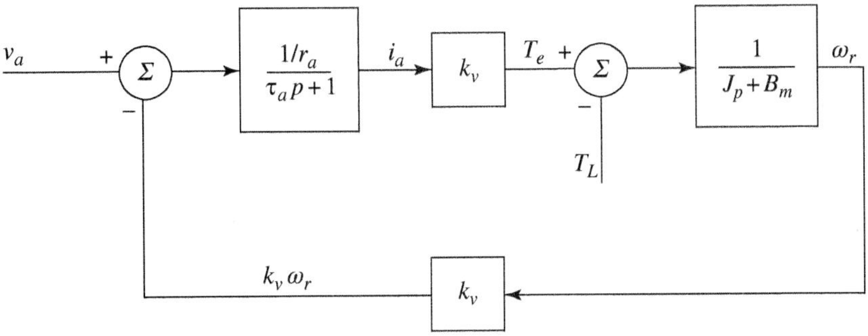

Figure 3.5-1 Time-domain block diagram of a permanent-magnet dc machine.

linear systems. The state variables of a system are defined as a minimal set of variables such that knowledge of these variables at any initial time t_0 plus information on the input excitation subsequently applied is sufficient to determine the state of the system at any time $t > t_0$. In the case of the permanent-magnet dc machines, the armature current i_a, the rotor speed ω_r, and the rotor position θ_r are the state variables. However, since θ_r can be established from ω_r by using

$$\frac{d\theta_r}{dt} = \omega_r \tag{3.5-5}$$

and since θ_r is considered a state variable only when the shaft position is a controlled variable, we will omit θ_r from consideration in this development.

Solving the armature voltage equation, (3.5-1), for di_a/dt yields

$$\frac{di_a}{dt} = \frac{r_a}{L_{AA}} i_a - \frac{k_v}{L_{AA}} \omega_r + \frac{1}{L_{AA}} v_a \tag{3.5-6}$$

From (3.5-2) with $k_v i_a$ substituted for T_e,

$$\frac{d\omega_r}{dt} = -\frac{B_m}{J}\omega_r + \frac{k_v}{J} i_a - \frac{1}{J} T_L \tag{3.5-7}$$

The system is described by a set of linear differential equations. In matrix form, the state equations become

$$p\begin{bmatrix} i_a \\ \omega_r \end{bmatrix} = \begin{bmatrix} -\dfrac{r_a}{L_{AA}} & \dfrac{k_v}{L_{AA}} \\ \dfrac{k_v}{J} & \dfrac{B_m}{J} \end{bmatrix} \begin{bmatrix} i_a \\ \omega_r \end{bmatrix} + \begin{bmatrix} \dfrac{1}{L_{AA}} & 0 \\ 0 & -\dfrac{1}{J} \end{bmatrix} \begin{bmatrix} v_a \\ T_L \end{bmatrix} \tag{3.5-8}$$

The form in which the state equations are expressed in (3.5-8) is called the fundamental form. In particular, the previous matrix equation may be expressed symbolically as

$$p\mathbf{x} = \mathbf{A}\mathbf{x} + \mathbf{B}\mathbf{u} \tag{3.5-9}$$

which is called the fundamental form, where p is the operator d/dt, $\mathbf{x}$ is the state vector (column matrix of state variables), and $\mathbf{u}$ is the input vector (column matrix of inputs to the system). We see that (3.5-8) and (3.5-9) are identical in form. Methods of solving equations of the fundamental form given by (3.5-9) are well known. If we were to consider a dc machine with a field winding as shown in Fig. 3.3-1, the field current would be a state variable. In this case, the state equations become nonlinear due to the product of state variables in $L_{AF}i_f i_a$ and $L_{AF}i_f\omega_r$.

Example 3C Time-Domain Block Diagram of Shunt-Connected DC Machine It is instructive to derive the time-domain block diagram of a shunt-connected dc machine. The equivalent circuit is shown in Fig. 3.3-1. The voltage and torque equations given by (3.3-1), (3.3-5), and (3.3-6) are repeated

$$v_f = r_f i_f + pL_{FF}i_f \tag{3C-1}$$

$$v_a = r_a i_a + pL_{AA}i_a + \omega_r L_{AF}i_f \tag{3C-2}$$

$$L_{AF}i_f i_a = Jp\omega_r + B_m\omega_r + T_L \tag{3C-3}$$

where T_e has been replaced by $L_{AF}i_f i_a$. Solving (3C-1) for i_f, (3C-2) for i_a, and (3C-3) for ω_r yields the time-domain block diagram shown in Fig. 3C-1.

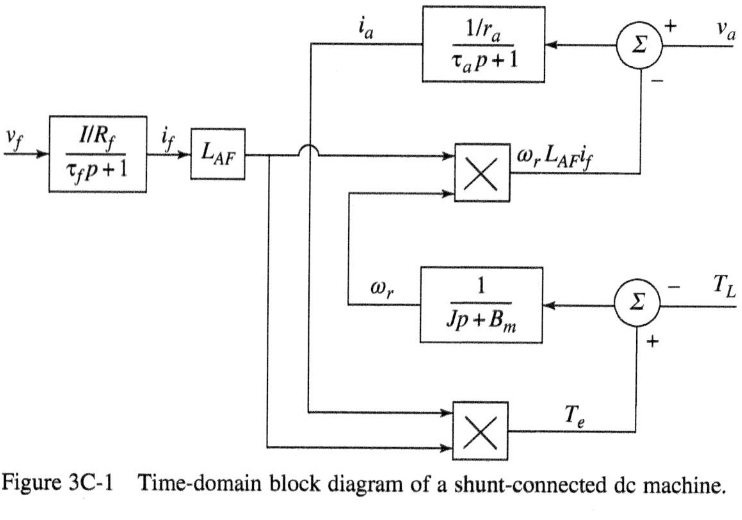

Figure 3C-1 Time-domain block diagram of a shunt-connected dc machine.

SP3.5-1 Can the state equations of the shunt-connected dc machine be written in the form of (3.5-9)? Why? [No, the state equations are nonlinear due to the product of state variables.]

3.6 DYNAMIC CHARACTERISTICS OF PERMANENT-MAGNET DC MOTORS

We can use the material in the previous section to simulate the permanent-magnet dc machine. Two modes of dynamic operation are of interest; starting from stall and changes in load torque with the machine supplied from a constant-voltage source. The permanent-magnet dc machine considered in Examples 3A and 3B is used to demonstrate these modes of operation. It is important that the reader becomes familiar with the material in these examples before proceeding.

Dynamic Performance During Starting

In the previous section, it was pointed out that, if the armature resistance is small, damaging armature current could result if rated voltage is applied to the armature terminals when the machine is stalled ($\omega_r = 0$). With the machine at stall, the counter-emf is zero; therefore, at the beginning of the transient starting period, the armature current is opposed only by the voltage drop across the armature resistance ($r_a i_a$) and the armature inductance ($L_{AA} di_a/dt$). We have mentioned and also noted in Examples 3A and 3B that low-power permanent-magnet dc motors are designed with a large armature resistance making it possible to "directline" start these devices without damaging brushes or the rotor windings. The no-load starting characteristics ($T_L = 0$) of the permanent-magnet dc motor described in Example 3A are shown in Fig. 3.6-1. The armature voltage v_a, the armature current i_a, and the

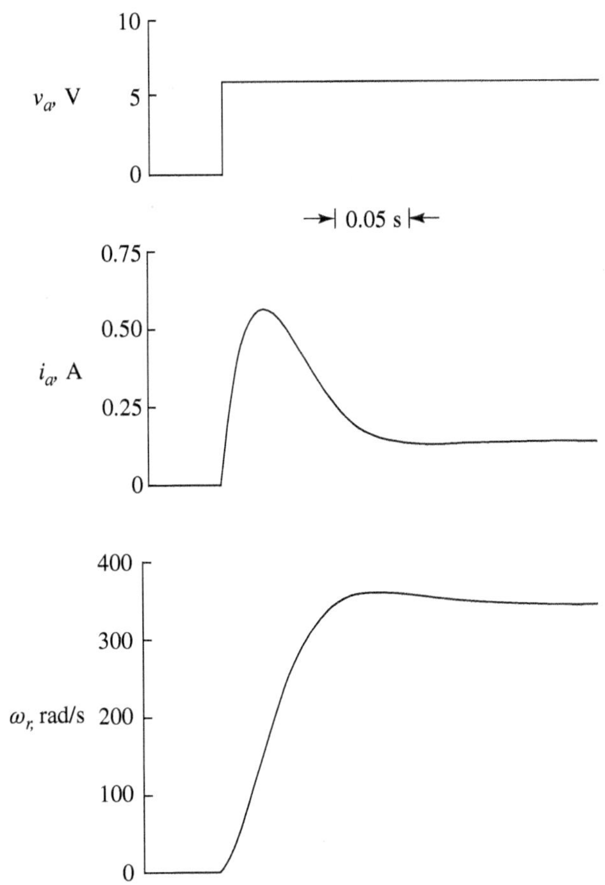

Figure 3.6-1 Starting characteristics of a permanent-magnet dc machine.

rotor speed ω_r are plotted. Initially the motor is at stall and, at time zero, 6 V is applied to the armature terminals. The peak transient armature current is limited to approximately 0.55 A due to the voltage drop across the inductance and resistance of the armature and the fact that the rotor is accelerating from stall, thereby developing the voltage $k_v\omega_r$ which also opposes the applied voltage. After about 0.25 s, steady-state operation is achieved with the no-load armature current of 0.15 A. (From Example 3A, $B_m = 6.04 \times 10^{-6}$ N · m · s.) It is noted that the rotor speed is slightly oscillatory, as illustrated by the small overshoot of the final, steady-state value.

Dynamic Performance During Sudden Changes in Load Torque

In Example 3B, we calculated the efficiency of the permanent-magnet dc motor given in Example 3A with a load torque of 0.5 oz · in (3.53×10^{-3} N·m). Let us assume that this load torque was suddenly applied with the motor initially operating at the no-load condition ($I_a = 0.15$ A). The dynamic characteristics following a step change in load torque T_L from zero to 0.5 oz · in are shown in Fig. 3.6-2. The armature current i_a and the rotor speed ω_r are plotted. Since $T_e = k_v i_a$ and since k_v is constant, T_e differs from i_a by a constant multiplier. It is noted that the system is slightly oscillatory. Also, it is noted that the change in the steady-state rotor speed is quite large. From Example 3A or Fig. 3.6-2, we see that no-load speed is 351.1 rad/s. With $T_L =$ 0.5 oz · in, the rotor speed is 249 rad/s as calculated in Example 3B and noted from Fig. 3.6-2. There has been approximately a 30 percent decrease

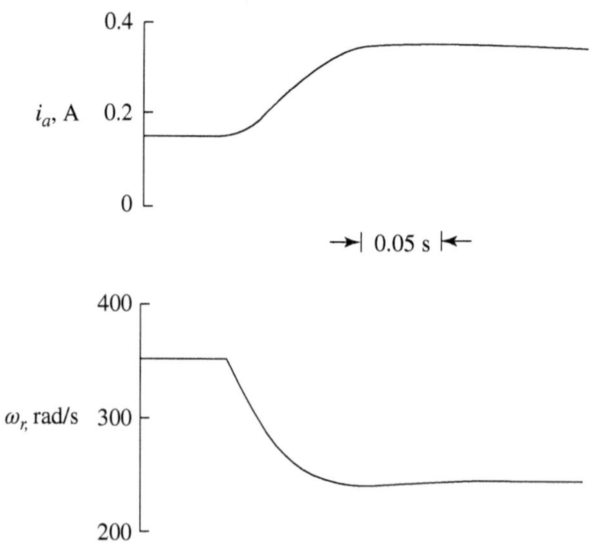

Figure 3.6-2 Dynamic performance of a permanent-magnet dc motor following a sudden increase in load torque from zero to 0.5 oz in.

in speed for this increase in load torque. As we have mentioned, this is characteristic of a low-power permanent-magnet motor owing to the high armature resistance.

SP3.6-1 Plot the steady-state T_e versus ω_r, characteristic for the permanent-magnet dc motor given in Example 3A. [A straight line between $(12.1 \times 10^{-3}, 0)$ and (0, 425.5)]

SP3.6-2 Assume that the peak value of i_a in Fig. 3.6-1 is 0.55 A. Calculate $k_v\omega_r$ at peak i_a. [$k_v\omega_r = 2.15$ V]

3.7 DC DRIVE

Although the dc machine is not used as extensively as in the past, the dc drive still plays a role in some drive applications and a brief look at a method of voltage control is appropriate. Our focus will be on the permanent-magnet dc machine supplied from a two-quadrant dc converter. Steady-state and dynamic performance are illustrated. An average value model is developed and a time-domain block diagram is given. This section is not a prerequisite for material in later chapters.

Since the dc converters used in dc drive systems are often called choppers, we will use dc converter and chopper interchangeably. In this section, we will analyze the operation and establish the average-value model for a two-quadrant chopper drive. A two-quadrant dc converter is depicted in Fig. 3.7-1. Each switch $S1$ or $S2$ is a transistor. It is assumed to be ideal; that is, if $S1$ or $S2$ is closed, current is allowed to flow in the direction of the arrow; current is not permitted to flow opposite to the arrow. If $S1$ or $S2$ is open, current is not allowed to flow in either direction regardless of the voltage across the switch. If $S1$ or $S2$ is closed and the current is positive, the voltage drop across the switch is assumed to be zero. Similarly, each diode $D1$ or $D2$ is ideal. Therefore, if the diode current i_{D1} or i_{D2} is greater than zero, the voltage across the diode is zero. The diode current can never be less than zero.

A voltage control scheme that is often used in dc drives is shown in Fig. 3.7-2. As illustrated, a ramp generator provides a sawtooth waveform of period T that ramps from zero to one. This ramp is compared to k, which is referred to as the duty

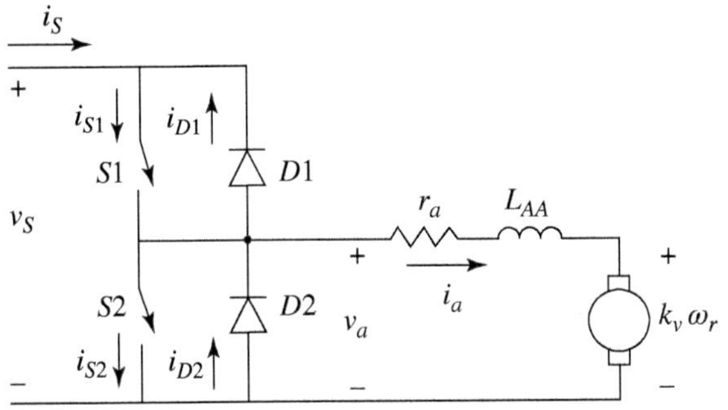

Figure 3.7-1 Two-quadrant chopper drive.

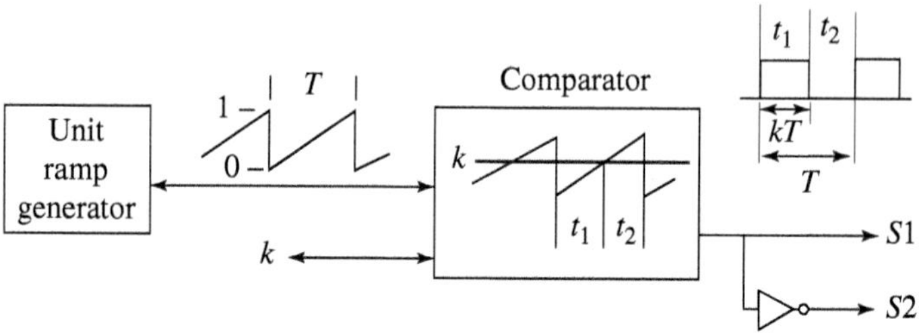

Figure 3.7-2 Pulse-width modulation voltage control.

cycle control signal. As the name implies, k is often the output variable of an open- or closed-loop control. The switches are controlled by the output of the comparator. The duty cycle control signal may vary between zero and one ($0 \leq k \leq 1$). From Fig. 3.7-2, we see that whenever k is greater than the ramp signal, the logic output of the comparator is high. This corresponds to the time interval t_1 in Fig. 3.7-2.

Now, since the ramp signal (sawtooth waveform) varies between zero and one, and since $0 \leq k \leq 1$, we can relate k, T, and t_1 as

$$t_1 = kT \tag{3.7-1}$$

which may be written as

$$t_1 = \frac{k}{f_s} \tag{3.7-2}$$

where f_s is the switching or chopping frequency $\left(f_s = \frac{1}{T}\right)$.

When k is less than the ramp signal, the logic output is low. This corresponds to the time interval t_2. Thus, since

$$t_1 + t_2 = T \tag{3.7-3}$$

we can write

$$t_2 = (1-k)T = (1-k)\frac{1}{f_s} \tag{3.7-4}$$

It follows that if k is fixed at one, $S1$ is always closed ($S2$ is always open) and if k is fixed at zero, $S1$ is always open ($S2$ is always closed).

Waveforms of the converter variables during steady-state operation are shown in Fig. 3.7-3. Therein, the switching period T is large relative to the armature time constant τ_a for the purpose of depicting the transient of the armature current. Normally, the switching period is much smaller than the armature time constant and the switching segments of i_a are nearly sawtooth in shape. This is portrayed later in this section. With a two-quadrant chopper, the armature voltage cannot be negative ($v_a \geq 0$); however, the armature current can be positive or negative. That is, I_1 and I_2 (Fig. 3.7-3) can both be positive, or I_1 can be negative and I_2 positive, or I_1 and I_2 can both be negative. In Fig. 3.7-3, I_1 is negative and I_2 is positive and the average value of i_a is positive. Since the average value of v_a is positive, the mode of operation depicted is motor action if ω_r is positive (ccw).

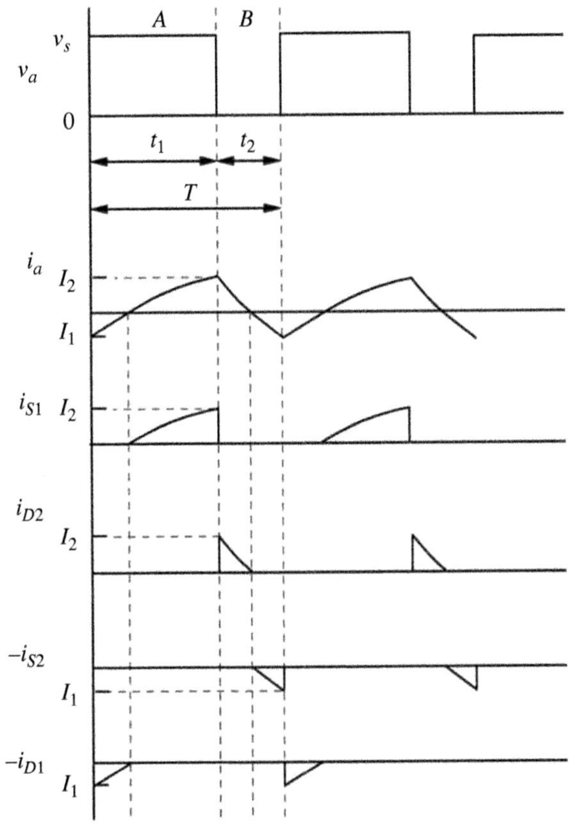

Figure 3.7-3 Steady-state operation of a two-quadrant dc converter drive.

In a two-quadrant chopper, the switching logic is generated from the duty cycle k, as shown in Fig. 3.7-2. When the comparator output signal is high, $S1$ is closed and $S2$ is open (interval A in Fig. 3.7-3); when it is low, $S1$ is open and $S2$ is closed (interval B in Fig. 3.7-3). When $S1$ is closed, current will flow either through $S1$ or $D1$; when $S2$ is closed, current will flow either through $S2$ or $D2$. There is a practical consideration that must be mentioned. Electronic switches have finite turn-off and turn-on times. The turn-off time is generally longer than the turn-on time. Therefore, the switching logic must be arranged so that the turn-on signal is delayed in order to prevent short-circuiting the source, causing "shoot through." Although the delay is very short, it must be considered in the design; however, it does not make our analysis invalid, wherein we will assume instant-on, instant-off operation.

It is important to discuss the mode of operation depicted in Fig. 3.7-3. During interval A, $S1$ is closed and $S2$ is open and, at the start of interval A, $i_a = I_1$, which is negative. Since $S2$ is open, a negative i_a (I_1) can only flow through $D1$. It is important to note that $-i_{D1}$ and $-i_{S2}$ are plotted in Fig. 3.7-3 to allow ready comparison with the waveform of i_a, since they are opposite to positive i_a. Let us go back to the start of interval A. How did i_a become negative? Well, during the interval B in the preceding period, $S2$ was closed with $S1$ open. With $S2$ closed, the armature terminals are short-circuited and the counter-emf has driven i_a negative. Therefore, when $S1$ is closed and $S2$ is opened at the start of interval A, the source voltage has to contend with this negative I_1. We see from Fig. 3.7-3 that the average value of i_a is slightly positive; therefore, v_S is larger than the counter-emf and at the start of interval A when v_S is applied to the machine, the armature current begins to increase toward zero from the negative value of I_1. Once i_a reaches zero, the diode D_1 blocks the current flow. That is, i_{D1} cannot become negative (cannot conduct positive i_a); however, $S1$ has been closed since the start of interval A and since i_{S1} can only be positive, $S1$ is ready to carry the positive i_a. The armature current, which is now i_{S1}, continues to increase until the end of interval A (I_2).

At the beginning of interval B, $S1$ is opened and $S2$ is closed; however, $S2$ cannot conduct a positive armature current. Therefore, the positive current (I_2) is diverted to diode $D2$ which is short-circuiting the armature terminals. Now, the counter-emf has the positive current (I_2) with which to contend. It is clear that if the armature terminals were permanently short-circuited, the counter-emf would drive i_a negative. At the start of interval B, the counter-emf begins to do just that; however, when i_a becomes zero, diode $D2$ blocks i_{D2} and the negative armature current is picked up by $S2$, which has been closed since the beginning of interval B, waiting to be called upon to conduct a negative armature current. This continues until the end of interval B, whereupon we are back to where we started.

Average-Value Time-Domain Block Diagram

The average-value time-domain block diagram for the two-quadrant chopper drive system is shown in Fig. 3.7-4. From Fig. 3.7-3, the average armature voltage may be determined as

$$\overline{v}_a = \frac{1}{T}\left[\int_0^{t_1} v_S\,d\xi + \int_{t_1}^{T} 0\,d\xi\right] \tag{3.7-5}$$

Since $t_1 = kT$, the average armature voltage becomes

$$\overline{v}_a = kv_S \tag{3.7-6}$$

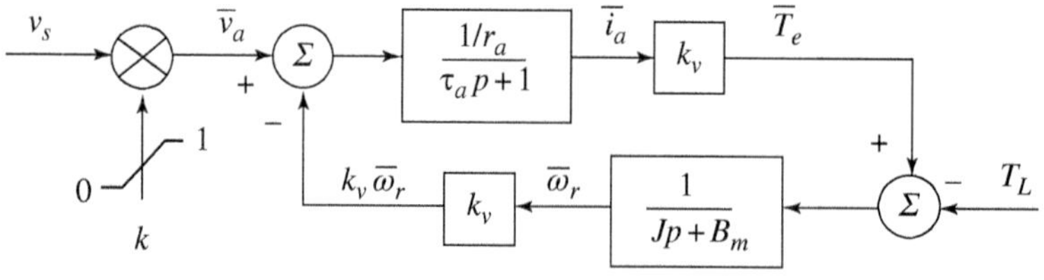

Figure 3.7-4 Average-value model of two-quadrant dc converter drive.

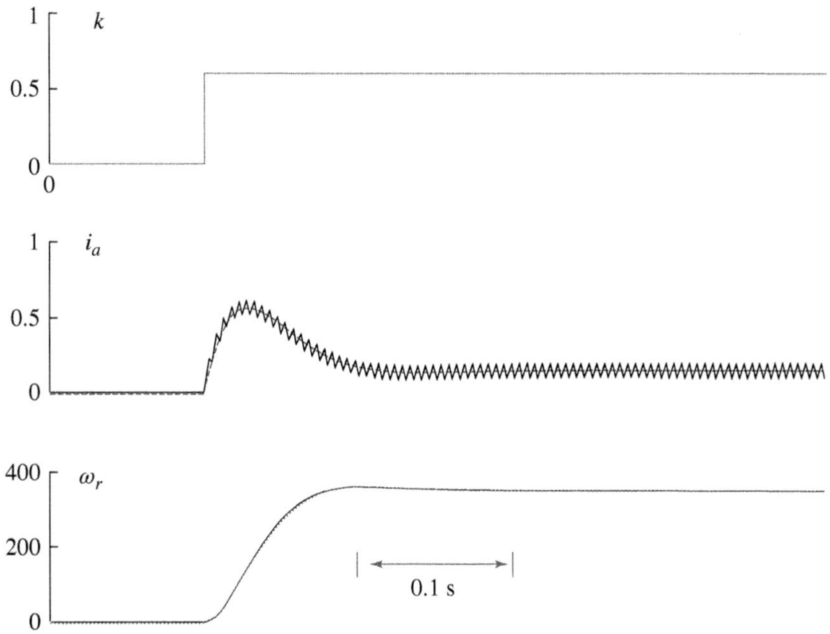

Figure 3.7-5 Starting characteristics of a permanent-magnet dc machine with a two-quadrant dc/dc converter drive.

In Fig. 3.7-4, the bars over the variables denote average values. Other than depiction of (3.7-6) and the bar notation, the block diagram shown in Fig. 3.7-4 is the same as that shown in Fig. 3.5-1. In the steady state

$$\overline{V}_a = kV_s = r_a\overline{I}_a + k_v\omega_r \quad (3.7\text{-}7)$$

The starting characteristics of a permanent-magnet dc machine with a two-quadrant chopper drive are depicted in Fig. 3.7-5. The machine parameters are identical to those in Example 3A. Here, the switching frequency f_s is set to 200 Hz and the source voltage to 10 V. Typically, the switching frequency is much higher, generally greater than 20 kHz. The frequency was selected to illustrate the dynamics introduced by the converter. Even at this low switching frequency, the switching period T is much less than the armature time constant τ_a. Thus, the armature current essentially consists of piecewise linear segments about an average response. In Fig. 3.7-5, the duty cycle is stepped from 0 to 0.6, corresponding to a step increase in average applied voltage from 0 to 6 V. The start-up response established using the average-value model is superimposed for purposes of comparison. As shown, the only salient difference between the two responses is the "sawtooth" behavior of the armature current due to converter switching. The difference in rotor speeds is indistinguishable.

Example 3D Torque Control

The parameters of a permanent-magnet dc machine are $V_a = 6$ V rated, $r_a = 7\ \Omega$, $k_v = 1.41 \times 10^{-2}$ V·s/rad, $L_{AA} = 120$ mH, $J = 1.06$ kg·m^2, and $B_m = 6.04 \times 10^{-6}$ N·m·s. We are to limit the torque T_e^* to 0.423×10^{-2} N·m or $I_a^* = 0.3$ A where the asterisk denotes commanded values. The control is shown symbolically in Fig. 3D-1.

Since the current is controlled, the electric dynamics are neglected and only the mechanical dynamics are considered. The equations involved in Fig. 3D-1 are

$$V_a = r_a I_a^* + \omega_r k_v \tag{3D-1}$$

$$T_e^* = J\frac{d\omega_r}{dt} + B_m\omega_r + T_L \tag{3D-2}$$

where $T_e^* = k_v I_a^* = 0.423 \times 10^{-2}$ N·m. The load line is

$$T_L = K\omega_r^2 \tag{3D-3}$$

where

$$K = 5.529 \times 10^{-8}\ \text{N}\cdot\text{m}\cdot\text{s}^2 \tag{3D-4}$$

This intersects rated V_a torque versus rotor speed plot at Operating Point 1, where $\omega_r = 276.6$ rad/s, as shown in Fig. 3D-2.

The dc machine is operating at Operating Point 1. The commanded torque is suddenly switched $\frac{1}{2}$ the original value which intersects the limiting torque (I_a^*) at Operating Point 2, where $\omega_r = 195.6$ rad/s. The mechanical dynamics are

$$T_e^* = J\frac{d\omega_r}{dt} + B_m\omega_r + T_L \tag{3D-5}$$

Assuming the torque control is functioning perfectly, the rotor slows and steady state is reached at Operating Point 2. The voltage at Operating Point 2 is

$$\begin{aligned} V_a &= r_a I_a^* + k_v\omega_r \\ &= (7)(0.3) + (1.41 \times 10^{-2})(195.6) \\ &= 2.1 + 2.75 = 4.85\ \text{V} \end{aligned} \tag{3D-6}$$

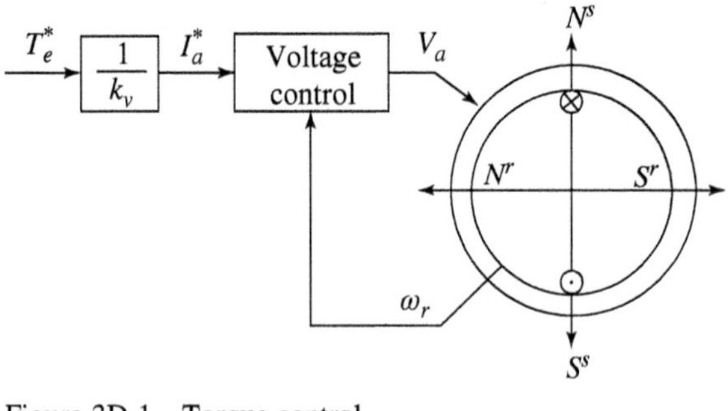

Figure 3D-1 Torque control.

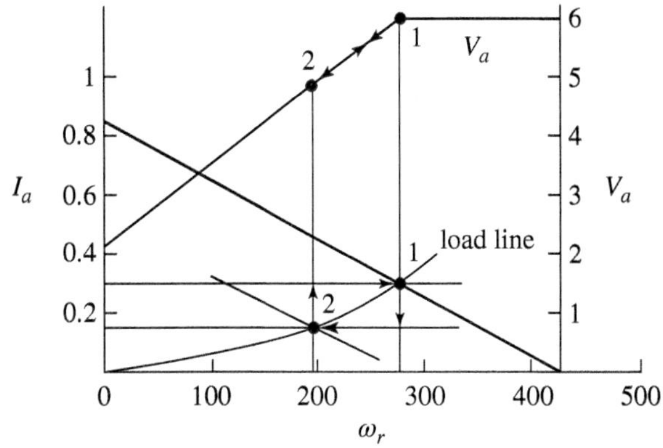

Figure 3D-2 Drive operation during T_e^* switching.

The commanded torque T_e^* is returned to the original value. The rotor speeds up and reaches steady state at Operating Point 1. The trajectory from Operating Point 1 to Operating Point 2 and then back to 1 is shown in Fig. 3D-2.

SP3.7-1 The dc machine given in Example 3A is supplied from a two-quadrant convertor. Determine k to reduce the no-load speed to one-half that when full voltage is applied to the armature. [$k = 0.5$]

SP3.7-2 Sketch Fig. 3.7-3 for only motor action; i.e. $i_{S2}$and i_{D1} equal to zero.

3.8 RECAPPING

The dc machine is unique in that it exerts a torque on the rotating member as a result of the interaction of two stationary, orthogonal magnetic systems. One is produced by current flowing in the windings or a permanent magnet on the stationary member, and the other is caused by the current flowing in the windings of the rotating member. DC machines are being replaced by the converter-controlled permanent-magnet ac machine, which by modern control techniques is made to emulate the maximum torque-per-ampere characteristics of a dc machine.

Since the dynamic characteristics of the permanent-magnet dc machine can be described by linear differential equations, we have also tried to set the stage for those who are interested in control system analysis by formulating the state equations in fundamental form. In the last section of this chapter, a brief analysis of a converter used in dc-drive systems has been presented and an average-value block diagram for this converter supplying a permanent-magnet dc machine was derived.

3.9 PROBLEMS

1. A permanent-magnet dc motor has the following parameters: $r_a = 8\,\Omega$ and $k_v = 0.01\ \text{V}\cdot\text{s/rad}$. The shaft load torque is approximated as $T_L = K\omega_r$, where $K = 5 \times 10^{-6}\ \text{N}\cdot\text{m}\cdot\text{s}$. The applied voltage is 6 V and $B_m = 0$. Calculate the steady-state rotor speed ω_r in rad/s.

2. A permanent-magnet dc motor is driven by a mechanical source at 3820 r/min. The measured open-circuit armature voltage is 7 V. The mechanical source is disconnected, and a 12-V electric source is connected to the armature. With zero-load torque, $I_a = 0.1$ A and $\omega_r = 650$ rad/s. Calculate k_v, B_m, and r_a.

3. Modify the state equations given by (3.5-8) for a permanent-magnet dc machine to include θ_r as a state variable.

4. The parameters of a permanent-magnet dc machine are $r_a = 6\,\Omega$ and $k_v = 2 \times 10^{-2}\,\text{V}\cdot\text{s/rad}$. V_a can be varied from zero to 10 V. The device is to be operated in the constant-torque mode with $T_e = 4 \times 10^{-3}\,\text{N}\cdot\text{m}$. (a) Determine V_a for $\omega_r = 0$. (b) Determine maximum ω_r range of the constant-torque mode of operation, that is, maximum ω_r with $T_e = 4 \times 10^{-3}\ \text{N}\cdot\text{m}$ and $V_a = 10\,\text{V}$.

5. Sketch Fig. 3.7-3 for generator action; i.e. i_{S1} and i_{D2} equal to zero.

6. Obtain (3.3-5) using the mmfs. ■

3.10 REFERENCE

1. P. C. Krause, O. Wasynczuk, and S. D. Sudhoff, *Analysis of Electric Machinery and Drive Systems*, IEEE Press, 2002.

CHAPTER 4

WINDING DISTRIBUTION AND TESLA'S ROTATING MAGNETIC FIELD

4.1 INTRODUCTION

In the previous chapter, we found the dc machine to be a bit involved in that it has windings on the rotating member and in relative motion with the stator magnetic field established by the field winding or a permanent magnet. However, due to the action of the commutator, the resultant mmf produced by currents flowing in the rotating windings is stationary. In other words, the rotor windings appear to be stationary, magnetically. Therefore, with a constant current in the field (stator) winding or a permanent magnet, torque is produced and rotation results owing to the force established between two stationary, orthogonal magnetic fields.

In rotational electromechanical devices other than dc machines and some switched reluctance machines, torque is produced as a result of the interaction of magnetic fields that rotate about the air gap of the device. Induction machines, synchronous machines, and brushless dc motors, which are actually permanent-magnet synchronous machines, all develop torque in this manner. There are features of these devices that are common to all. In particular, the winding arrangement of the stator and the method of producing a rotating magnetic field due to stator currents are the same. Moreover, the rotor of an induction machine produces a rotating magnetic field similar to the stator. It seems logical, therefore, to cover these common features once and for all rather than repeat this material for each electromechanical device covered in subsequent chapters.

The rotating magnetic field is key to the operation of ac machines. In the late 19th century, Tesla showed that a constant amplitude rotating magnetic field, often referred to as a rotating air-gap mmf or rotating magnetic poles, could be

Electromechanical Motion Devices: Rotating Magnetic Field-Based Analysis and Online Animations, Third Edition. Paul Krause, Oleg Wasynczuk, Steven D. Pekarek, and Timothy O'Connell.

Published 2020 by John Wiley & Sons, Inc.

established with balanced multiphase currents flowing in appropriately placed multiphase symmetrical windings. This led to Tesla's invention of the ac machine [1]. Not only is Tesla's rotating magnetic field the backbone of ac-machine operation it is also the key to the analysis of ac machines.

In this chapter, an analytical means of expressing and thus "viewing" Tesla's rotating magnetic field from any frame of reference is established. That is, not only as viewed by a stationary observer but as viewed by an observer rotating at any angular velocity. An animation is available to assist in this visualization.

The importance of grasping the concept of viewing Tesla's rotating magnetic field from any frame of reference cannot be overemphasized. In Chapter 5, we will find that our expression for Tesla's rotating magnetic field leads directly to reference frame theory. This leads to voltage, current, and flux linkage relationships associated with the fictitious circuits that produce the rotating magnetic field as it is viewed from any frame of reference. This establishes an analytical basis for the change of variables used in machine, power systems, and electric drive analyses. In other words, we will find the change of variables used in machine analysis is nothing more than a formulation of the variables that produce Tesla's rotating magnetic field as viewed from any frame of reference. Being aware of this straightforward connection is of utmost importance to the understanding of present-day machine and electric drive analysis.

In this chapter, the focus is primarily the two-pole machine where the stator variables produce a rotating pair of north and south poles. In the final section of this chapter the analysis of multipole machines is considered. It is shown that with a simple substitution, *P*-pole machines (4, 6, 8, …, poles) can be analyzed as two-pole devices. This is a welcomed piece of information.

4.2 WINDING DISTRIBUTION

In Chapters 1 and 2, we considered two windings in relative motion; one on the stator and one on the rotor (Fig. 1.7-4). When introducing this concept in Section 1.7, we assumed the turns of the windings to be concentrated in one position even though, in practice, they would normally be distributed over a 30°–60° arc. Let us redraw Fig. 1.7-4 with the following changes. First, we will disregard the rotor winding and focus our attention on the stator windings, since the stator windings are common to the electromechanical motion devices considered in subsequent chapters. Finally, we will call winding 1 of Fig. 1.7-4 the *as* winding and assume that it is distributed in slots (not shown) over the inner circumference of the stator, as shown in Fig. 4.2-1. Also in Fig. 4.2-1, ϕ_s is a coordinate of positional displacement referenced to the *as* axis; we will refer to ϕ_s as a displacement or a coordinate.

The winding is depicted as a series of individual coils. Each coil is placed in a slot in the stator steel. If we follow the path of positive current i_{as} flowing in the *as* winding, we see that current enters as_1, depicted by ⊗, the tail of an arrow, to indicate that the assumed direction of positive current is down the length of the

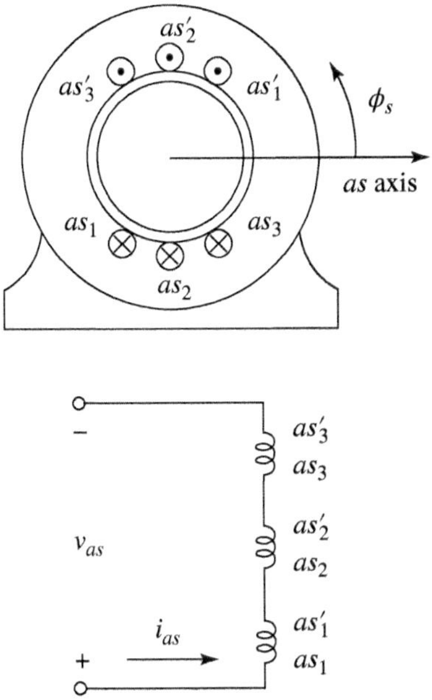

Figure 4.2-1 Elementary two-pole single-phase stator winding.

stator in an axial direction (into the paper). Current flows down the length of the stator, loops at the end (end turn, see Fig. 1.7-4), and flows back down the length of the stator and out at as'_1, depicted by ⊗, the point of an arrow. Note that as_1 and as'_1 are placed in stator slots which span π radians. This is referred to as the *winding* or *pole pitch*. A pole pitch of π radians is characteristic of a two-pole machine; we shall talk more about poles later. Now, as_1 around to as'_1 is referred to as a *coil* and as_1 or as'_1 is a *coil side*. In practice, a coil will contain more than one conductor. That is, current will flow into as_1 in a conductor and out as'_1 via the same conductor. The conductor, which is, of course, insulated, may then be looped back to as_1 and the winding of the conductor around the $as_1 - as'_1$ path repeated, thereby forming a coil with numerous turns. The number of conductors in a stator coil side tells us the number of turns in the coil. This number will be denoted as nc_s. Once we have wound nc_s turns in the $as_1 - as'_1$ coil, we will take the same conductor and repeat this winding process to form the $as_2 - as'_2$ coil. We will assume that the same number of turns (nc_s) make up the $as_2 - as'_2$ coil as the $as_1 - as'_1$ coil and, similarly, for the $as_3 - as'_3$. We could have wound a different number of turns in each coil but we will assume that this was not done. With the same number of turns (nc_s) in each of the coils, the winding is said to be *uniformly distributed* over the stator arc from as_1 to as_3 or 60°. Once the winding is wound, we can use the right-hand rule to give a meaning to the *as* axis shown in Fig. 4.2-1. It is, by definition, the principal direction of the magnetic flux established by positive current flowing in the *as* winding. It is said to denote the positive direction of the magnetic axis of the *as* winding.

In the analysis of electromechanical motion devices, it is generally assumed that the stator windings and, in the case of an induction machine, the rotor windings may be approximated as *sinusoidally distributed* windings. That is, the physical distribution of a phase winding may be approximated as a sinusoidal function of ϕ_s, whereupon, the waveform of the resulting mmf dropped across the air gap (air-gap mmf) of the device may also be approximated as a sinusoidal function of ϕ_s. Actually, most electric machines, particularly large machines, are designed so that the windings, especially the stator windings, are sinusoidally distributed to minimize the voltage and current harmonics. Except in cases where the harmonics due to the winding configuration are of importance, it is typically assumed that all windings may be approximated as sinusoidally distributed windings. We will make this assumption in our analysis of the electromechanical motion devices considered in subsequent chapters.

For the purpose of considering a sinusoidally distributed winding, let us add a few coils to the *as* winding shown in Fig. 4.2-1, whereupon we have Fig. 4.2-2a.

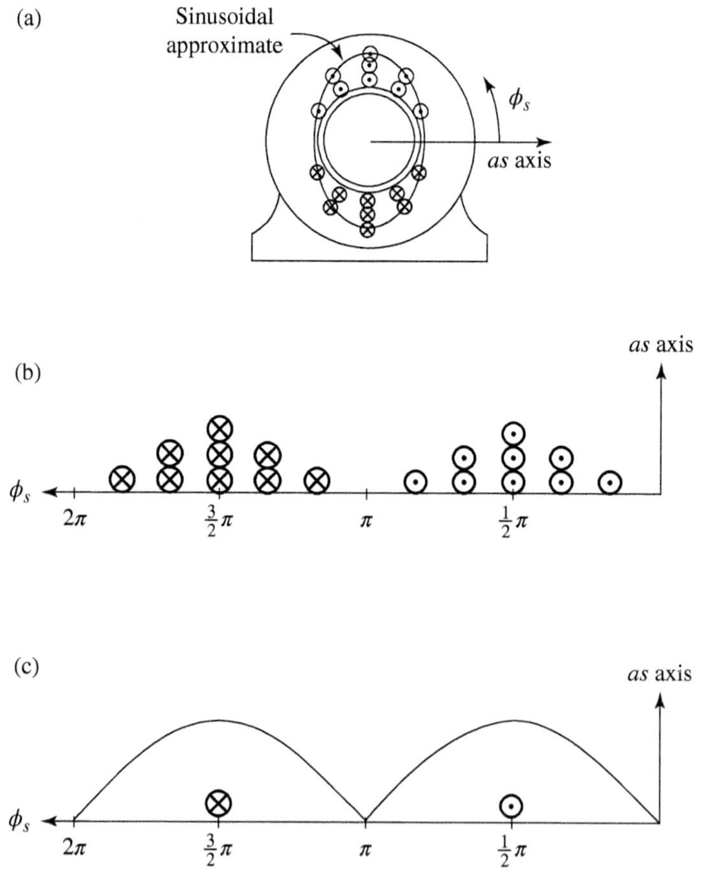

Figure 4.2-2 Sinusoidally distributed stator winding. (a) Approximate sinusoidal distribution. (b) Developed diagram of (a). (c) Sinusoidal approximation of (b).

Although we may be going a bit far by adding coils that now span a 120° arc, it helps to illustrate our point. It is convenient to employ the so-called *developed diagram* of the cross-sectional view of the device shown in Fig. 4.2-2a. This developed diagram is obtained by "flattening out" the rotor and stator as depicted in Fig. 4.2-2b. This linear approximation of the circular surfaces is possible since the air-gap length is very small relative to the radius of the rotor. To relate the developed diagram to the cross-sectional view of the device, it is helpful to define the displacement ϕ_s to the left of the *as* axis when plotting the mmf. This allows us to position the stator above the rotor and to preserve the rules governing the mmfs.

From Fig. 4.2-2a or the developed diagram in Fig. 4.2-2b and from Fig. 4.2-2c, we see that the *winding distribution*, i.e. turn density function, from $0 < \phi_s < \pi$ may be approximated by the expression

$$N_{as} = N_p \sin \phi_s \qquad \text{for } 0 < \phi_s < \pi \tag{4.2-1}$$

and from $\pi < \phi_s < 2\pi$ as

$$N_{as} = -N_p \sin \phi_s \qquad \text{for } \pi < \phi_s < 2\pi \tag{4.2-2}$$

where N_p is the peak turns density in turns/radian which could be determined from a Fourier analysis of the actual winding distribution. If N_s represents the number of turns of the equivalent sinusoidally distributed winding, then, using ξ as a dummy variable of integration,

$$N_s = \int_0^{\pi} N_p \sin \xi d\xi = 2N_p \tag{4.2-3}$$

It is important to note that the number of turns is obtained by integrating the turns density N_{as} from zero to π rather than from zero to 2π. Also, N_s represents not the total turns of the winding, but those of the equivalent sinusoidally distributed winding that corresponds to the fundamental component of the actual winding distribution.

SP4.2-1 The winding shown in Fig. 4.2-2 is repositioned so that the maximum turn density ⊙ is now at $\phi_s = \pi$ and ⊗ is at $\phi_s = 0$. Express the sinusoidal approximation of this winding similar to that given by (4.2-1) and (4.2-2). Determine the position of the positive magnetic axis. $\left[N_{as} = N_p \cos \phi_s \text{ for } -\frac{1}{2}\pi < \phi_s < \frac{1}{2}\pi; N_{as} = -N_p \cos \phi_s \text{ for } \frac{1}{2}\pi < \phi_s < \frac{3}{2}\pi;\right.$ positive magnetic axis at $\left.\phi_s = \frac{1}{2}\pi\right]$.

SP4.2-2 The winding shown in Fig. 4.2-2 is repositioned so that the positive magnetic axis is now at $\phi_s = -\frac{1}{2}\pi$. Express the sinusoidal approximation of this winding. [N_{as} same as SP4.2-1; however, ⊙ maximum at $\phi_s = 0$; ⊗ maximum at $\phi_s = \pi$]

4.3 AIR-GAP MMF

The air-gap magnetomotive force (magnetic field) is the mmf drop in ampere-turns across the air gap of the electromechanical device. Magnetic fields are established by currents flowing in the phase windings of the machine. The interaction of these magnetic fields is the basis of Tesla's rotating magnetic field and, thus, energy conversion. In order to analyze the electromechanical device, it is necessary to express the air-gap mmf. We will begin by considering the *as*-phase winding depicted in Fig. 4.2-2 and establishing an expression for the *as*-phase air-gap mmf (mmf_{as}) using Ampere's law. Ampere's law allows us to establish the mmf that causes the magnetic flux to flow across the air gap. In particular, it is the current enclosed within a closed-line integral of the field intensity $\left(\oint \mathbf{H} \cdot d\mathbf{L}\right)$ around the flux path.

In the case of an electric machine, it is convenient to perform this line integral about a path that spans π radians for a two-pole device or a winding pitch as shown in Fig. 4.3-1a–f where we have not shown the outer machine housing for

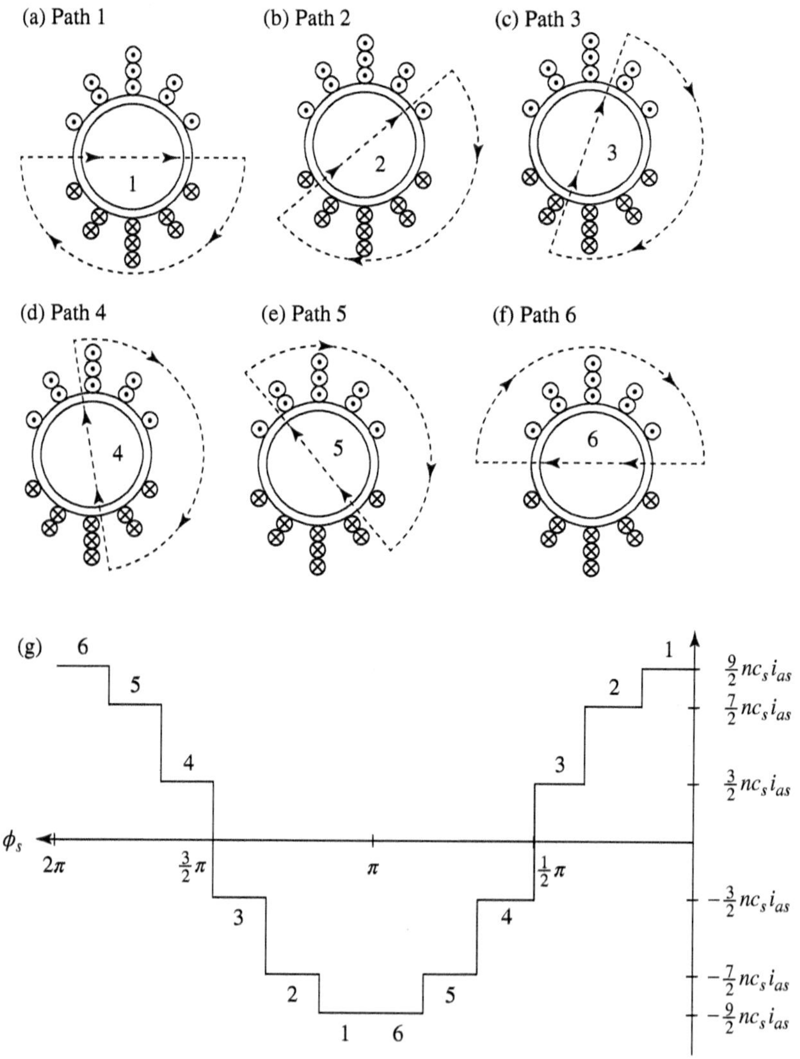

Figure 4.3-1 Closed paths of integration and the plot of mmf_{as}.

convenience. By the "corkscrew rule" the mmf is positive for positive current into the paper. Thus, for path 1, in Fig. 4.3-1a, the mmf is $9nc_s i_{as}$. We will assume that a positive air-gap mmf is a drop from rotor to stator. Also, if we neglect the reluctance of the iron, all of the drop in mmf would be across the two air gaps in the integration path. Therefore, for path 1 shown in Fig. 4.3-1a, $\frac{9}{2}nc_s i_{as}$ is the mmf drop at $\phi_s = 0$ and $-\frac{9}{2}nc_s i_{as}$ at $\phi_s = \pi$. Now, this is true if the reluctance at $\phi_s = 0$ and $\phi_s = \pi$ are the same. This is the case for a two-pole device and we will find that the symmetry repeats every $\left(\frac{P}{2}\right)2\pi$ radians of the displacement about the rotor where P is the total number of north and south poles.

How do we obtain the plot of mmf_{as} for a given current i_{as} shown in Fig. 4.3-1g? Well, we have explained the $\frac{9}{2}nc_s i_{as}$ at $\phi_s = 0$ and $-\frac{9}{2}nc_s i_{as}$ at $\phi_s = \pi$ shown in Fig. 4.3-1a. Now, as we rotate the path of integration counterclockwise we enclose the same mmf until we cross $\phi_s = \frac{\pi}{6}$ where we now enclose 1 ⊙ and drop 1 ⊗. The current enclosed that spans π radians becomes 8⊗ and 1⊙ or a net of 7⊗ as shown in Fig. 4.3-1b. Thus, the mmf is $7nc_s i_{as}$ and mmf_{as} becomes $\frac{7}{2}nc_s i_{as}$ from $\frac{\pi}{6} < \phi_s < \frac{\pi}{3}$ and $-\frac{7}{2}nc_s i_{as}$ from $\frac{7\pi}{6} < \phi_s < \frac{4\pi}{3}$ as shown in Fig. 4.3-1g. It is left to the reader to justify the remainder of the plot shown in Fig. 4.3-1g.

The plot of mmf_{as} shown in Fig. 4.3-1g strongly suggests that the expression for the air-gap mmf or air-gap magnetic field may be approximated as

$$\text{mmf}_{as} = \frac{N_s}{2} i_{as} \cos \phi_s \tag{4.3-1}$$

where N_s is defined following (4.2-3). Equation (4.3-1) is plotted in Fig. 4.3-2. The ⊙ and ⊗ indicate the maximum turn density of the assumed sinusoidally distributed phase winding. Hereafter, we will use the pair of single circles to depict sinusoidal windings. It should not be confused with the same notation used for a concentrated winding in earlier chapters.

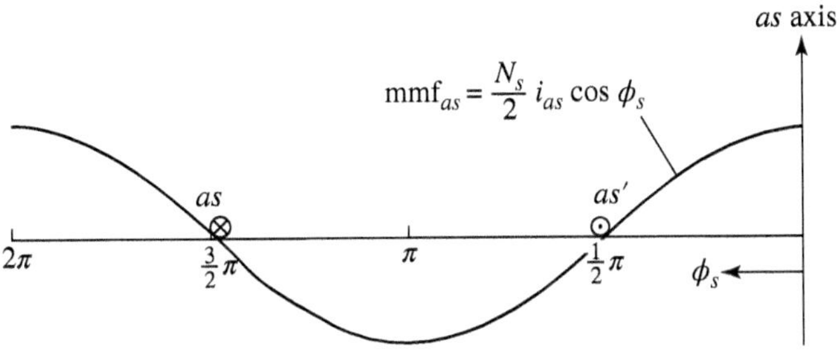

Figure 4.3-2 Plot of sinusoidal approximation of mmf_{as} shown in Fig. 4.3-1g.

Let us take a minute to consider what we have in Fig. 4.3-2. The plot is for a given phase current. If i_{as} is a constant positive value, mmf_{as} would appear as shown in Fig. 4.3-2. If i_{as} is a sinusoidal function of time, mmf_{as} would be pulsating at the frequency of the current. In particular, if we observe this mmf while standing at $\phi_s = 0$, we would see the air-gap mmf pulsating between $\frac{9}{2}nc_sI_p$ when I_{as} is positive maximum and $-\frac{9}{2}nc_sI_p$ when I_{as} is negative maximum where I_p is the peak magnitude of the sinusoidal current. What would we see if we are standing at $\phi_s = \frac{1}{2}\pi$? [Zero; always]. The air-gap mmf_{as} would be viewed as a function of time and displacement ϕ_s.

Example 4A Distributed Stator Winding

Consider the uniformly distributed stator winding shown in Fig. 4.2-1 and shown again here as Fig. 4A-1. Each of the coils is assumed to have nc_s turns. Determine and plot the air-gap mmf versus ϕ_s.

The path of integration shown in Fig. 4A-1 spans π radians and encloses $3nc_si_{as}$ ampere-turns. One half of this mmf is dropped across the air gap at $\phi_s = 0$ (positive) and one half at $\phi_s = \pi$ (negative). The enclosed ampere-turns remain constant as the path of integration is rotated counterclockwise until it encloses as_1' and fails to enclose as_1 whereupon the total ampere-turns enclosed becomes nc_si_{as}. The total enclosed ampere-turns remains at nc_si_{as} until the path of integration rotates to $\frac{1}{2}\pi$, whereupon it becomes zero. When the path of integration rotates beyond $\frac{1}{2}\pi$, one ⊙ is enclosed until as_1' through as_3' are all enclosed. The plot of the air-gap mmf is shown in Fig. 4A-2b.

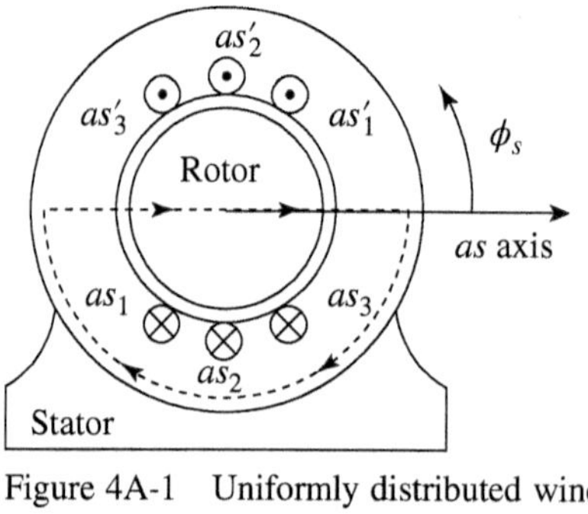

Figure 4A-1 Uniformly distributed winding.

SP4.3-1 Express the sinusoidal approximation of mmf_{as} if the assumed positive direction of i_{as} is reversed in Fig. 4.3-1. [$\text{mmf}_{as} = -(4.3\text{-}1)$]

SP4.3-2 Assume that winding 1 in Fig. 1.7-4 has one turn. Sketch the air-gap mmf due to current i_1 flowing in this winding. Let ϕ_s be the displacement from axis 1. [$\frac{1}{2}i_1$ for $-\frac{1}{2}\pi < \phi_s < \frac{1}{2}\pi$ and $-\frac{1}{2}i_1$ for $\frac{1}{2}\pi < \phi_s < \frac{3}{2}\pi$]

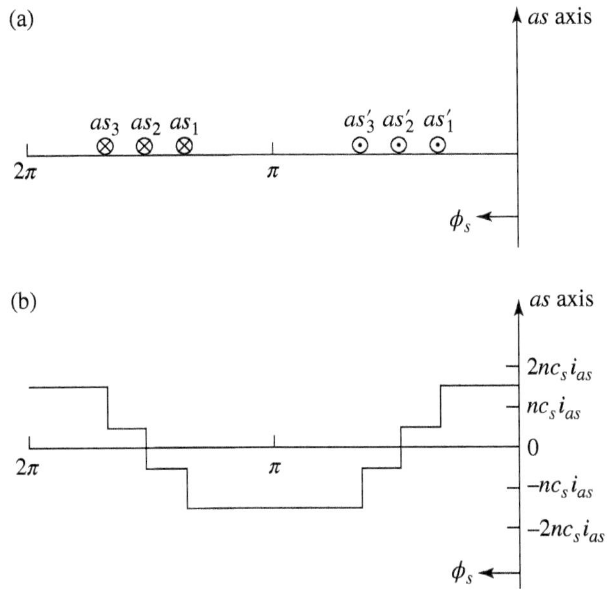

Figure 4A-2 The *as* winding with three coils. (a) Developed diagram; (b) air-gap mmf.

SP4.3-3 Repeat SP4.3-2 for winding 2. Let ϕ_r be the displacement from axis 2. [Same as SP4.3-2 with ϕ_s replaced by ϕ_r]

SP4.3-4 Express the sinusoidal approximation of the mmf$_s$ of the windings described in SP4.2-1 and SP4.2-2. [SP4.2-1, mmf$_{as}$ = $\frac{N_s}{2} i_{as} \sin \phi_s$; SP4.2-2 mmf$_{as}$ = $-$mmf$_{as}$ for SP4.2-1]

4.4 TESLA'S ROTATING MAGNETIC FIELD – SYMMETRICAL STATOR CIRCUITS

In the case of a single-phase stator winding, the sinusoidal approximation of the air-gap mmf is given by (4.3-1). Let us think about this for a minute; (4.3-1) is a product of i_{as} and $\cos \phi_s$; therefore, the air-gap mmf will be nonexistent at all positions around the air gap whenever $i_{as} = 0$. This is not too desirable since we will find that in order to produce a constant steady-state torque we need a constant amplitude rotating magnetic field (rotating north and south poles) produced by the stator currents. Is it possible to place another winding on the stator and arrange its position and the time sequence of the current flowing in it so that its mmf would complement the mmf due the *as* winding resulting in a constant rotating magnetic field? Yes, fortunately, and Tesla showed us how. We are going to talk about Tesla's rotating magnetic field in this section.

Two-Phase Stator

In the case of a symmetrical two-pole two-phase stator, the *bs* winding is located $\frac{1}{2}\pi$ radians from the *as* winding and it is identical in distribution and parameters to the *as* winding. Let us assume it is positioned as illustrated in Fig. 4.4-1. Therein, a circle placed at the position of maximum turn density represents a sinusoidally distributed winding. The air-gap mmf created by i_{bs} can be approximated as

$$\text{mmf}_{bs} = \frac{N_s}{2} i_{bs} \sin \phi_s \tag{4.4-1}$$

The total air-gap mmf due to the two-pole stator windings would be

$$\begin{aligned} \text{mmf}_s &= \text{mmf}_{as} + \text{mmf}_{bs} \\ &= \frac{N_s}{2}(i_{as} \cos \phi_s + i_{bs} \sin \phi_s) \end{aligned} \tag{4.4-2}$$

This relationship is perhaps the most important equation we will run into in this text since it is key to the operation and analysis of ac machines. The currents i_{as} and i_{bs} are sinusoidal and when animating the rotating magnetic field it is necessary to show positive and negative currents. Therefore, when the current is negative, positive current flows into the primed part of the coil shown in Fig. 4.4-1.

A two-phase steady-state balanced set of stator currents (*ab* sequence) may be expressed as

$$I_{as} = \sqrt{2} I_s \cos [\omega_e t + \theta_{esi}(0)] \tag{4.4-3}$$

$$I_{bs} = \sqrt{2} I_s \sin [\omega_e t + \theta_{esi}(0)] \tag{4.4-4}$$

where I_s is the rms value of the phase current, ω_e is the electrical angular velocity in rad/s, and $\theta_{esi}(0)$ is the phase angle of the currents. For this balanced set, $\tilde{I}_{bs} = -j\tilde{I}_{as}$.

The concept of Tesla's rotating field can be explained by assuming the steady-state phase currents are (4.4-3) and (4.4-4) with $\theta_{esi}(0) = 0$. At time zero,

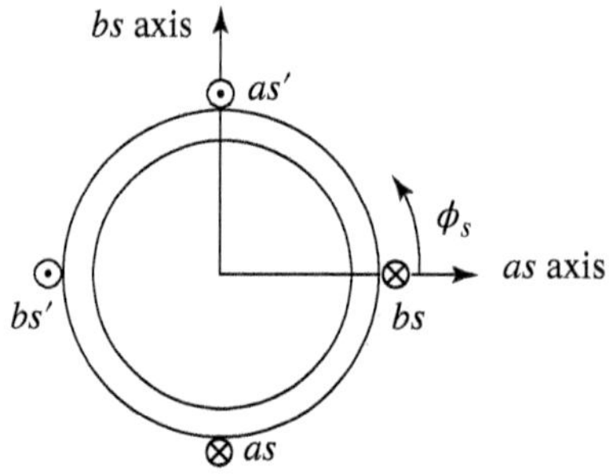

Figure 4.4-1 Elementary symmetrical, two-pole two-phase sinusoidally distributed stator windings. Outer housing not shown.

I_{as} is maximum and I_{bs} is zero. The air-gap mmf is oriented along the *as* axis. When the time has advanced to where $\omega_e t = \frac{1}{2}\pi$, $I_{as} = 0$ and I_{bs} is maximum. The air-gap mmf has rotated from the *as* axis to the *bs* axis with an angular velocity of ω_e. That is, the mmf has rotated counterclockwise relative to an observer on the stator at ω_e.

Now, since Tesla's magnetic field rotates relative to the stator, perhaps, looking at it from a rotating axis will help us visualize what is going on. In Fig. 4.4-2, we have added a third axis, the q axis, to Fig. 4.4-1. As shown in Fig. 4.4-2, the concept of a q axis is no different from the *as* or *bs* axis. In fact, in Chapter 5 we will associate a q winding with the q axis. The q axis can rotate at any angular velocity ω, or it can be stationary ($\omega = 0$). We are free to specify it. The angle from the *as* axis to the q axis is θ and for a constant ω,

$$\theta = \omega t + \theta(0) \tag{4.4-5}$$

The displacement or coordinate ϕ shown in Fig. 4.4-2 is from the q axis just as ϕ_s is the displacement or coordinate from the *as* axis. Let us relate this displacement to an adjacent displacement on the stator. From Fig. 4.4-2,

$$\phi_s = \theta + \phi \tag{4.4-6}$$

If we substitute (4.4-6) into (4.4-2), we obtain

$$\text{mmf}_s = \frac{N_s}{2}[i_{as}\cos(\theta + \phi) + i_{bs}\sin(\theta + \phi)] \tag{4.4-7}$$

Equation (4.4-7) is cornerstone to reference frame theory which we will consider in Chapter 5. Substituting the steady-state currents given by (4.4-3) and (4.4-4), for i_{as} and i_{bs}, in (4.4-7) and (4.4-5) for θ and after some work

$$\text{mmf}_s = \frac{N_s}{2}\sqrt{2}I_s\cos[(\omega_e - \omega)t + \theta_{esi}(0) - \phi] \tag{4.4-8}$$

Now, ω_e is the angular frequency of the electrical system and ω is the angular velocity of the q axis, where ϕ is the displacement from the q axis, and $\theta(0)$ is the

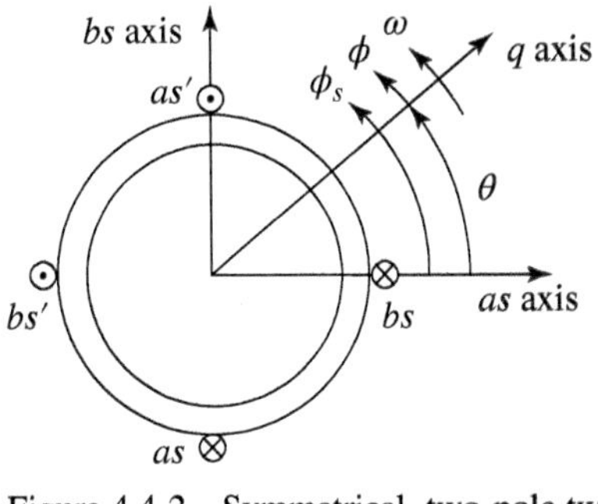

Figure 4.4-2 Symmetrical, two-pole two-phase sinusoidally distributed stator windings with a third magnetic axis (q axis).

time-zero position of the q axis (4.4-5) which we have let be zero in (4.4-8). We will continue to set $\theta(0)$ equal to zero unless otherwise specified.

If now we let the angular velocity of the q axis be zero ($\omega = 0$) and $\theta(0) = 0$, then $\phi = \phi_s$ and we would be viewing Tesla's rotating magnetic field as a stationary observer. In this case, (4.4-8) becomes

$$\text{mmf}_s^s = \frac{N_s}{2}\sqrt{2}I_s \cos\left[\omega_e t + \theta_{esi}(0) - \phi_s\right] \tag{4.4-9}$$

where we have added a superscript s to emphasize that we are observing Tesla's rotating magnetic field from a stationary frame of reference. If $\theta_{esi}(0)$ and ϕ_s are constants, mmf_s^s is viewed as a sinusoidal variation of frequency ω_e. If, for example, ϕ_s is zero, we would be positioned (fixed) at the *as* axis ($\phi_s = 0$ in Fig. 4.4-2) and the mmf_s^s would be pulsating at ω_e and rotating counterclockwise.

Let us think about this for a minute. We are observing Tesla's rotating magnetic field as we stand on the stator. We see the rotating magnetic field pulsating at ω_e. It seems logical that the variables associated with this rotating magnetic field in this frame of reference would vary at ω_e. This is something we already knew. That is, the balanced set of currents that produces this view of the rotating magnetic field is the balanced steady-state set given by (4.4-4) and (4.4-5).

If we now let $\omega = \omega_e$, (4.4-8) becomes

$$\text{mmf}_s^e = \frac{N_s}{2}\sqrt{2}I_s \cos\left[\theta_{esi}(0) - \phi_e\right] \tag{4.4-10}$$

where ϕ_e is the displacement from the q axis which is rotating at ω_e. Since $\theta_{esi}(0)$ is constant for balanced steady-state conditions, (4.4-10) is a sinusoidal function of the spatial variable ϕ_e (the displacement from the q axis). Its maximum value occurs at $\phi_e = \theta_{esi}(0)$ which is the phase angle of the balanced steady-state stator currents. In other words, we are observing Tesla's rotating magnetic field as we run at ω_e in the counterclockwise direction around the air gap and this field appears constant to us for balanced steady-state conditions. Now, guess the form of the variables in this reference frame. We are guessing dc, and we hope you are too. Since ω_e is the angular frequency of I_{as} and I_{bs}, and the q axis is also rotating at ω_e or in *synchronism* with the electrical system, this is referred to as the *synchronously rotating frame of reference*. Therein, since we have 360° vision, Tesla's rotating magnetic field appears to us as a sinusoidal function of ϕ_e (a space sinusoid) with the maximum value at $\phi_e = \theta_{esi}(0)$ from the q axis and stationary relative to us since we are rotating at ω_e counterclockwise. This is depicted in Fig. 4.4-3 where $\theta_{esi}(0)$ is selected negative to correspond to inductive circuits. A positive mmf_s^e is a south pole due to the stator currents (S^s) with the maximum intensity at $\phi_e = \theta_{esi}(0)$. A negative mmf_s^e represents a stator north pole (N^s).

We have just opened the door to a concept of viewing Tesla's rotating magnetic field from frames of reference other than that of a stationary observer. In Fig. 4.4-3, the reference frame is rotating at ω_e in the counterclockwise direction. The importance of Fig. 4.4-3 cannot be overemphasized. It is the connection

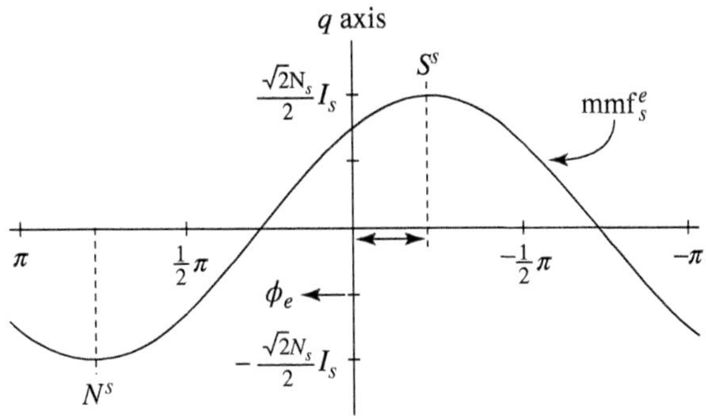

Figure 4.4-3 Tesla's balanced steady-state rotating magnetic field (mmf_s^e) viewed from $-\pi$ to π by an observer rotating counterclockwise about the air gap at ω_e with $\theta(0) = 0$ and $\theta_{esi}(0)$ is negative. The synchronously rotating reference frame.

between displacement and time. We will find that this concept enables a clear visualization of the operation of electric machines and we will use the synchronously rotating reference frame extensively in our analysis of machines and drives. An online animation of Tesla's rotating magnetic field is available.

This is not the first time we have viewed a sinusoidal variation from a reference frame rotating at ω_e. In Section 1.2, it was pointed out that when a phasor is rotated at ω_e in the counterclockwise direction, its real projection is the instantaneous value of the sinusoidal variation. When we consider the phasor as a constant, we are running counterclockwise with it at ω_e or we have stopped rotation of the phasor. Later, we will relate variables in the synchronously rotating reference frame to the instantaneous and steady-state phasors.

Before proceeding, there are a couple things that should be mentioned. It is important not to confuse the magnetic axes of the stator windings (*as* and *bs* axes), which are stationary, with the phasors representing the stator currents ($\tilde{I}_{as}$ and $\tilde{I}_{bs}$). However, it is interesting to note the relative position of the *as* axis and *bs* axis versus the relative position of $\tilde{I}_{as}$ and $\tilde{I}_{bs}$ in a phasor diagram for a constant, counterclockwise mmf$_s$ to occur. The *bs* axis is displaced $\frac{1}{2}\pi$ ahead of the *as* axis, as illustrated in Fig. 4.4-1; however, from (4.4-3) and (4.4-4) we see that $\tilde{I}_{bs}$ lags $\tilde{I}_{as}$ by $\frac{1}{2}\pi$. As we pointed out earlier, the direction of rotation of the air-gap mmf (mmf$_s$) as viewed by a stationary observer may readily be determined by forgetting about $\theta_{esi}(0)$ and locating the position of $\frac{N_s}{2}\sqrt{2}I_s$ when time has progressed from maximum I_{as} to where I_{bs} is maximum and I_{as} is zero $\left(\omega_e t = \frac{1}{2}\pi\right)$. That is, the maximum mmf$_s$ has rotated counterclockwise from the *as* axis to the *bs* axis. In other words, maximum I_{as} followed $\frac{1}{2}\pi$ electrical degrees by maximum I_{bs} produces counterclockwise rotation of mmf$_s$ in the device shown in Fig. 4.4-1.

Three-Phase Stator

The arrangement of the stator windings of a two-pole three-phase device is shown in Fig. 4.4-4. The windings are connected in a wye configuration (Y-connected) and they are identical, sinusoidally distributed with N_s equivalent turns and with their magnetic axes displaced by $\frac{2}{3}\pi$. The positive direction of the magnetic axes is selected so as to achieve counterclockwise rotation of the rotating air-gap mmf with balanced stator currents of the *abc* sequence. We shall see this in just a moment. The expressions for the air-gap mmfs established by the stator phases may be written by inspection of Fig. 4.4-4. In particular,

$$\text{mmf}_{as} = \frac{N_s}{2} i_{as} \cos \phi_s \tag{4.4-11}$$

$$\text{mmf}_{bs} = \frac{N_s}{2} i_{bs} \cos \left(\phi_s - \frac{2}{3}\pi \right) \tag{4.4-12}$$

$$\text{mmf}_{cs} = \frac{N_s}{2} i_{cs} \cos \left(\phi_s + \frac{2}{3}\pi \right) \tag{4.4-13}$$

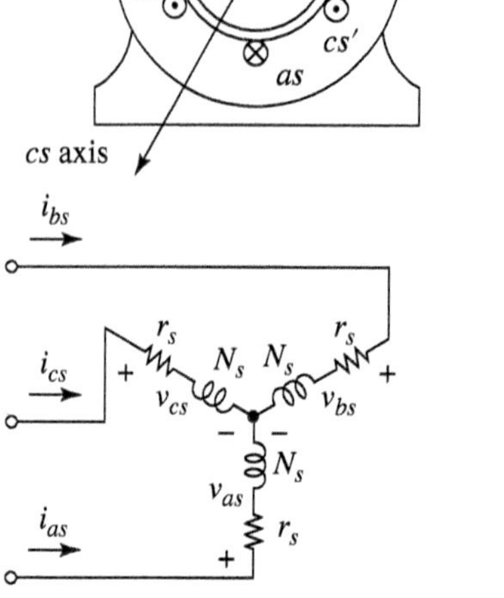

Figure 4.4-4 Elementary two-pole three-phase sinusoidally distributed stator windings.

As before, ϕ_s is the angular displacement about the stator measured from the *as* axis. The total air-gap mmf_s is

$$\text{mmf}_s = \frac{N_s}{2}\left[i_{as}\cos\phi_s + i_{bs}\cos\left(\phi_s - \frac{2}{3}\pi\right) + i_{cs}\cos\left(\phi_s + \frac{2}{3}\pi\right)\right] \tag{4.4-14}$$

For balanced steady-state conditions, the stator currents for an *abc* sequence may be expressed as

$$I_{as} = \sqrt{2}I_s\cos\left[\omega_e t + \theta_{esi}(0)\right] \tag{4.4-15}$$

$$I_{bs} = \sqrt{2}I_s\cos\left[\omega_e t - \frac{2}{3}\pi + \theta_{esi}(0)\right] \tag{4.4-16}$$

$$I_{cs} = \sqrt{2}I_s\cos\left[\omega_e t + \frac{2}{3}\pi + \theta_{esi}(0)\right] \tag{4.4-17}$$

Substituting (4.4-15) through (4.4-17) into (4.4-14) and after some trigonometric manipulations we can obtain an expression for the rotating air-gap mmf (Tesla's rotating field) established by balanced steady-state stator currents as viewed by a stationary observer

$$\text{mmf}_s^s = \frac{N_s}{2}\sqrt{2}I_s\frac{3}{2}\cos\left[\omega_e t + \theta_{esi}(0) - \phi_s\right] \tag{4.4-18}$$

If mmf_s^s for the three-phase device given by (4.4-18) is compared with mmf_s^s for a two-phase device given by (4.4-9), we see that they are identical except that the amplitude for the three-phase device is $\frac{3}{2}$ times that of a two-phase device. Actually, it can be shown that the amplitude for multiphase devices changes from that of a two-phase device in proportion to the number of phases divided by two. It has been pointed out that we will first analyze the two-phase machine since the trigonometry is less involved and once we have considered its "little sister," the extension to the three-phase machine is straightforward. It follows that we need not repeat the work we did regarding introducing the *q* axis and the derivation leading up to (4.4-8). The only difference would be that (4.4-8) would be multiplied by $\frac{3}{2}$. Similarly, Fig. 4.4-3 would apply to a three-phase stator if the amplitude of the magnetic field is multiplied by $\frac{3}{2}$.

It is important to note that, with the assumed positive directions of the magnetic axes, a counterclockwise rotating air-gap mmf is obtained with a three-phase set of balanced stator currents of the *abc* sequence. As in the two-phase case, it is also important to note the relative positions of the magnetic axes versus the relative positions of the phasors representing the currents in order to establish a constant amplitude counterclockwise rotating air-gap mmf. From (4.4-11) through (4.4-13) or Fig. 4.4-4 we see that the *bs* axis is stationary at $\phi_s = 120°$, whereas the *cs* axis is stationary at $\phi_s = -120°$. From (4.4-15) through (4.4-17), $\tilde{I}_{as}, \tilde{I}_{bs}$, and $\widetilde{I}_{cs}$ are 120° out of phase; however, in order for the constant-amplitude air-gap mmf to rotate in the counterclockwise direction, $\widetilde{I}_{bs}$ lags $\widetilde{I}_{as}$ by 120° and $\widetilde{I}_{cs}$ lags $\widetilde{I}_{as}$ by 240°. In other words, maximum positive I_{as} is followed by maximum I_{bs} and then maximum I_{cs} for an *abc* sequence. With the magnetic axis positioned as shown in Fig. 4.4-4, this sequence of maximum currents produces a counterclockwise rotation of the stator air-gap mmf.

Example 4B Components of mmf_{as} and mmf_{bs}

Express mmf_{as}, mmf_{bs}, and mmf_s for balanced steady-state conditions for a two-phase system.

Let us select

$$I_{as} = \sqrt{2}I_s \cos\left[\omega_e t + \theta_{esi}(0)\right] \tag{4B-1}$$

$$I_{bs} = \sqrt{2}I_s \sin\left[\omega_e t + \theta_{esi}(0)\right] \tag{4B-2}$$

Now, from (4.3-1) and (4.4-1)

$$\text{mmf}_{as} = \frac{N_s}{2} i_{as} \cos\phi_s \tag{4B-3}$$

$$\text{mmf}_{bs} = \frac{N_s}{2} i_{bs} \sin\phi_s \tag{4B-4}$$

Substituting (4B-1) and (4B-2) into (4B-3) and (4B-4) yields

$$\text{mmf}_{as} = \frac{N_s}{2}\sqrt{2}I_s \cos\left[\omega_e t + \theta_{esi}(0)\right]\cos\phi_s \tag{4B-5}$$

$$\text{mmf}_{bs} = \frac{N_s}{2}\sqrt{2}I_s \sin\left[\omega_e t + \theta_{esi}(0)\right]\sin\phi_s \tag{4B-6}$$

As we have mentioned, when $\phi_s = 0$ and $\theta_{esi}(0) = 0$, mmf_{as} is pulsating in time between + and – maximums at ω_e; when $\phi_s = \frac{1}{2}\pi$, mmf_{as} is always zero. In the case of mmf_{bs}, it is zero when $\phi_s = 0$ and pulsating when $\phi_s = \frac{1}{2}\pi$.

Since (4B-5) is a product of cosines and (4B-6) is a product of sines, the following trigonometric identities may be used:

$$\cos x \cos y = \frac{1}{2}\cos(x+y) + \frac{1}{2}\cos(x-y) \tag{4B-7}$$

$$\sin x \sin y = \frac{1}{2}\cos(x-y) - \frac{1}{2}\cos(x+y) \tag{4B-8}$$

Now,

$$\text{mmf}_s = \text{mmf}_{as} + \text{mmf}_{bs} \tag{4B-9}$$

Adding (4B-7) and (4B-8), the $\cos(x+y)$ terms cancel and we have

$$\cos x \cos y + \sin x \sin y = \cos(x-y) \tag{4B-10}$$

Therefore, since $x = \omega_e t + \theta_{esi}(0)$ and $y = \phi_s$, (4B-9) becomes

$$\text{mmf}_s = \frac{N_s}{2}\sqrt{2}I_s \cos[\omega_e t + \theta_{esi}(0) - \phi_s] \tag{4B-11}$$

which is mmf_s^s given by (4.4-9).

SP4.4-1 Determine the direction of rotation of mmf_s as viewed by a stationary observer for the winding arrangement shown in Fig. 4.4-1 for (a) $\tilde{I}_{as} = j\tilde{I}_{bs}$, (b) $\tilde{I}_{as} = -j\tilde{I}_{bs}$, (c) $\tilde{I}_{bs} = j\tilde{I}_{as}$, and (d) $\tilde{I}_{bs} = -j\tilde{I}_{as}$. Equations are not needed. [(a) ccw, (b) cw, (c) cw, (d) ccw]

SP4.4-2 In Fig. 4.4-1 assume we are running clockwise at ω_e. Describe our view of the *as* and *bs* axes and the *as* and *bs* windings. [They are all rotating (moving) ccw at ω_e relative to us.]

SP4.4-3 Determine the direction of rotation of mmf_s as viewed by a stationary observer for the winding arrangement shown in Fig. 4.4-4 for (a) an *abc* sequence, (b) an *acb* sequence. Equations are not needed. [(a) ccw, (b) cw]

SP4.4-4 Refer to Fig. 4.4-4. Determine the frequency of the pulsating air-gap mmf if it is viewed from (a) $\omega = 2\omega_e$, (b) $\omega = -\omega_e$. [(a) ω_e, (b) $2\omega_e$]

SP4.4-5 Why is (4B-11) mmf_s^s? [(4B-11) = (4.4-9)].

4.5 TESLA'S ROTATING FIELDS AND TORQUE WITH UNSYMMETRICAL AND SYMMETRICAL ROTOR CIRCUITS

When considering the dc machine, in Chapter 3, we found that a constant torque was produced due to the interaction of the mmf created by the stationary field and the field (mmf) created due to the commutation of the rotor (armature) induced currents. As a result of the commutation, the rotor mmf is also stationary; hence, we have the interaction of two orthogonally positioned mmfs.

In this chapter, we have introduced Tesla's rotating magnetic field. In order to produce a constant torque, the rotor mmf, mmf_r, must rotate in unison with the stator mmf, mmf_s. Although we will deal with torque in detail in later chapters, it is appropriate to take a first look at the means of producing torque in synchronous and induction machines from the standpoint of stator and rotor mmfs. Although the reader has the tools to do it now, he/she may wish to delay study of this section until starting the chapters on synchronous and induction machines.

Unsymmetrical System

An elementary two-phase round-rotor synchronous machine is shown in Fig. 4.5-1. The rotor is equipped with a field winding, labeled *f*–*f*′. The mmf_r^r may now be expressed as

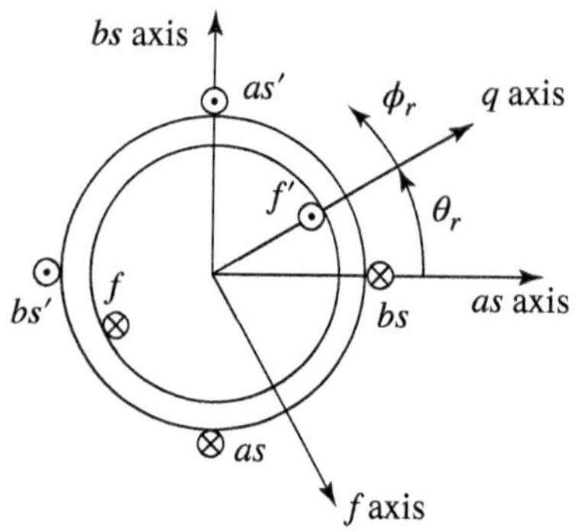

Figure 4.5-1 Elementary two-phase round-rotor synchronous machine.

$$\text{mmf}_r^r = -\frac{N_r}{2} I_{dc} \sin \phi_r \tag{4.5-1}$$

where I_{dc} is the constant current flowing in this winding. Since there is only one winding on the rotating member, the only way to obtain a constant amplitude rotating magnetic field is to hold I_{dc} constant and rotate the rotor. Now, recall that mmf_s^e for balanced steady-state operation is (4.4-14) which is repeated here

$$\text{mmf}_s^e = \frac{N_s}{2} \sqrt{2} I_s \cos [\theta_{esi}(0) - \phi_e] \tag{4.5-2}$$

When are mmf_r^r and mmf_s^e traveling at the same angular velocity? Well it occurs if ω_r is ω_e. That is, if the stator of the machine is supplied from a constant frequency source, say, ω_e and the rotor is rotating in synchronism with it, $\omega_r = \omega_e$ and $\phi_r = \phi_e$. We will find that this is the mode of operation of a synchronous generator connected to a power system (Chapter 7).

There is, however, another mode of operation that will cause mmf_s and mmf_r to rotate in unison. If the source frequency is always equal to the angular velocity of the rotor, ω_r, then ω_e is ω_r. This is the mode of operation of the brushless dc drive, where a power electronic circuit is controlled to always supply the applied stator voltages with the frequency corresponding to electrical angular velocity of the rotor of the permanent magnet ac synchronous machine (Chapter 6).

Symmetrical System

The derivation of the air-gap mmf of symmetrical rotating circuits parallels that of the stationary circuits. As shown in Fig. 4.5-2, the *ar* and *br* windings are orthogonal and if they are identical and sinusoidally distributed, Tesla's rotating magnetic field of these symmetrical windings may be expressed as

$$\text{mmf}_r = \frac{N_r}{2} (i_{ar} \cos \phi_r + i_{br} \sin \phi_r) \tag{4.5-3}$$

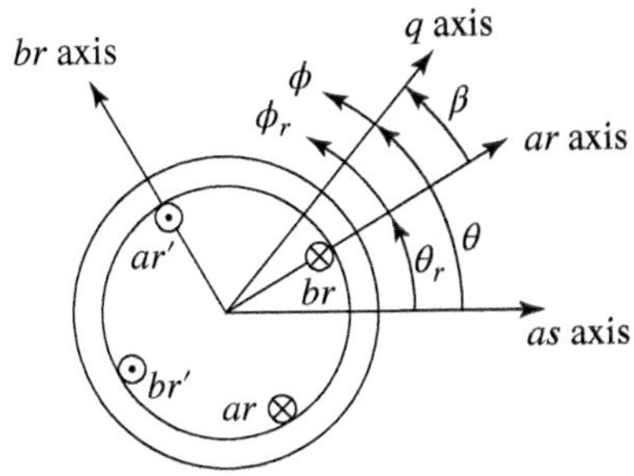

Figure 4.5-2 Two-phase rotating, identical, sinusoidally distributed symmetrical windings.

where N_r is the equivalent number of turns and ϕ_r is the angular displacement from the *ar* axis which is positioned by θ_r from the *as* axis. In Fig. 4.5-2, the positions θ_r and θ are referenced to the *as* axis. Although we will find that rotor windings are generally not sinusoidally distributed, the net rotor mmf is commonly approximated as a sinusoidal function of space and time as is the stator mmf.

Now, the positions are related as

$$\phi_r = \beta + \phi \tag{4.5-4}$$

Substituting (4.5-4) into mmf_r which is similar in form to (4.4-7), that is

$$\begin{aligned}\text{mmf}_r &= \frac{N_r}{2}[i_{ar}\cos\phi_r + i_{br}\sin\phi_r] \\ &= \frac{N_r}{2}[i_{ar}\cos(\beta+\phi) + i_{br}\sin(\beta+\phi)]\end{aligned} \tag{4.5-5}$$

We will return to (4.5-5) in Chapter 5; however, let us continue our focus on the rotating air-gap mmf during balanced steady-state operation.

Now, from Fig. 4.5-2,

$$\beta = \theta - \theta_r \tag{4.5-6}$$

and during steady-state operation

$$\beta = (\omega - \omega_r)t + \theta(0) - \theta_r(0) \tag{4.5-7}$$

We will find that during balanced steady-state operation of an induction motor, the rotor windings are short-circuited and the rotor currents, which are induced by mmf_s, are balanced with the angular frequency of $(\omega_e - \omega_r)$. Thus, for an induction machine,

$$I_{ar} = \sqrt{2}I_r\cos[(\omega_e - \omega_r)t + \theta_{eri}(0)] \tag{4.5-8}$$

$$I_{br} = \sqrt{2}I_r\sin[(\omega_e - \omega_r)t + \theta_{eri}(0)] \tag{4.5-9}$$

If we now substitute (4.5-8), (4.5-9), and (4.5-6) into (4.5-5), we will obtain an expression for mmf_r:

$$\text{mmf}_r = \frac{N_r}{2}\sqrt{2}I_r \cos\left[(\omega_e - \omega_r - \omega + \omega_r)t + \theta_{eri}(0) - \theta(0) + \theta_r(0) - \phi\right] \quad (4.5\text{-}10)$$

Now, we will select $\theta(0)$ and $\theta_r(0)$ to be zero unless otherwise specified; whereupon, (4.5-10) may be written

$$\text{mmf}_r = \frac{N_r}{2}\sqrt{2}I_r \cos\left[(\omega_e - \omega)t + \theta_{eri}(0) - \phi\right] \quad (4.5\text{-}11)$$

Equation (4.5-11) is similar in form to mmf_s given by (4.4-8). This is a very important observation. If the rotor currents are balanced and if the angular frequency of these currents is $(\omega_e - \omega_r)$, which we assumed to derive (4.5-11), then the stator mmf_s and the rotor mmf_r will travel around the air gap at the same angular velocity. If, for example, we observe mmf_s and mmf_r as a stationary observer $(\omega = 0)$, mmf_s^s and mmf_r^s would both be traveling counterclockwise at ω_e relative to us. If we are riding on the rotor $(\omega = \omega_r)$, mmf_s^r and mmf_r^r would be traveling at $(\omega_e - \omega_r)$ counterclockwise relative to us. If we are running at ω_e, mmf_s^e and mmf_r^e would appear to us as constant space sinusoids of ϕ_e. The value of each will depend on ϕ_e. Since mmf_s and mmf_r are traveling in unison, a constant torque (power) is produced. We are starting to uncover the basis of Tesla's contribution to electric power generation and utilization.

Let us summarize by expressing Tesla's rotating magnetic fields for an elementary two-phase induction machine in the stationary, rotor, and synchronously rotating frames of reference by substituting into (4.4-8) for mmf_s and (4.5-11) for mmf_r with $\theta(0) = 0$ and $\theta_r(0) = 0$, and the angular frequency of the balanced stator currents is ω_e and $\omega_e - \omega_r$ is the angular frequency of the rotor currents.

Stationary reference frame; $\omega = 0$ and $\phi = \phi_s$

$$\text{mmf}_s^s = \frac{N_s}{2}\sqrt{2}I_s \cos\left[\omega_e t + \theta_{esi}(0) - \phi_s\right] \quad (4.5\text{-}12)$$

$$\text{mmf}_r^s = \frac{N_r}{2}\sqrt{2}I_r \cos\left[\omega_e t + \theta_{eri}(0) - \phi_s\right] \quad (4.5\text{-}13)$$

Rotor reference frame; $\omega = \omega_r$ and $\phi = \phi_r$

$$\text{mmf}_s^r = \frac{N_s}{2}\sqrt{2}I_s \cos\left[(\omega_e - \omega_r)t + \theta_{esi}(0) - \phi_r\right] \quad (4.5\text{-}14)$$

$$\text{mmf}_r^r = \frac{N_r}{2}\sqrt{2}I_r \cos\left[(\omega_e - \omega_r)t + \theta_{eri}(0) - \phi_r\right] \quad (4.5\text{-}15)$$

Synchronously rotating reference frame; $\omega = \omega_e$ and $\phi = \phi_e$

$$\text{mmf}_s^e = \frac{N_s}{2}\sqrt{2}I_s \cos\left[\theta_{esi}(0) - \phi_e\right] \quad (4.5\text{-}16)$$

$$\text{mmf}_r^e = \frac{N_r}{2}\sqrt{2}I_r\cos\left[\theta_{eri}(0) - \phi_e\right] \tag{4.5-17}$$

We will find that (4.5-16) and (4.5-17) are key to the animation and visualization of induction machine operation. Equations (4.5-12) through (4.5-17) are for two-phase machines. For three-phase machines, multiply each by $\frac{3}{2}$.

A constant torque can be produced only if mmf_s and mmf_r are rotating at the same angular velocity. This is the case for balanced steady-state operation of an induction machine (Chapter 8). The magnitude of the constant torque is determined by the number of turns of stator or rotor windings, the magnitude of the stator and rotor currents, and the relative position of mmf_s and mmf_r. The relative position of the phase between the mmfs, $\theta_{esi}(0)$ and $\theta_{eri}(0)$, is easily discerned in the synchronous reference frame, i.e. mmf_s^e and mmf_r^e given by (4.5-16) and (4.5-17), respectively.

Although we may be getting ahead of ourselves, let us talk a little bit about how torque is produced. In the dc machine torque is produced by the interaction of two stationary, orthogonal mmfs. A synchronous machine develops a constant torque only at synchronous speed, $\omega_r = \omega_e$. This occurs due to the fact that the rotor field which is stationary with respect to the rotor is rotating in unison with Tesla's rotating magnetic field. This of course is necessary to produce an average torque.

Now, the induction machine produces torque at any rotor speed except when $\omega_r = \omega_e$. It is at this rotor speed that current is not induced in the short-circuited rotor circuits, and mmf_r disappears and mmf_s has no other mmf with which to interact.

Please do not misinterpret (4.5-12) through (4.5-17); the only restriction on these equations is that $\omega_r \neq \omega_e$. The induction machine is operating at some rotor speed other than ω_e and we are observing mmf_s and mmf_r from various reference frames. Regardless of from where we observe these mmfs, steady-state mmf_s and mmf_r are stationary relative to each other and torque produced is a function of angle between the mmfs, $\theta_{esi}(0)$ and $\theta_{eri}(0)$. We will talk more of this as we go along; however, our focus will be on the synchronously rotating reference frame.

Example 4C Windings of Elementary Induction Machines

If we place the rotor of Fig. 4.5-2 inside of the stator of Fig. 4.4-2, we would have Fig. 4C-1. If we now short-circuit the rotor windings, we would have an elementary two-pole two-phase induction motor. Equations (4.5-12) through (4.5-17) would then apply to this device. The frequency of the stator currents is ω_e. (a) Justify the angular velocity of (4.5-12) through (4.5-17) by observation. (b) Determine the angular frequency of the voltages and currents associated with the mmfs given by (4.5-12) through (4.5-17).

In the case of part (a), we know that (4.5-12) through (4.5-17) are for balanced steady-state conditions. This will give rise to mmf_s as viewed from the stator to rotate at ω_e counterclockwise which will induce balanced two-phase currents of $(\omega_e - \omega_r)$ angular frequency in the short-circuited rotor windings. These currents

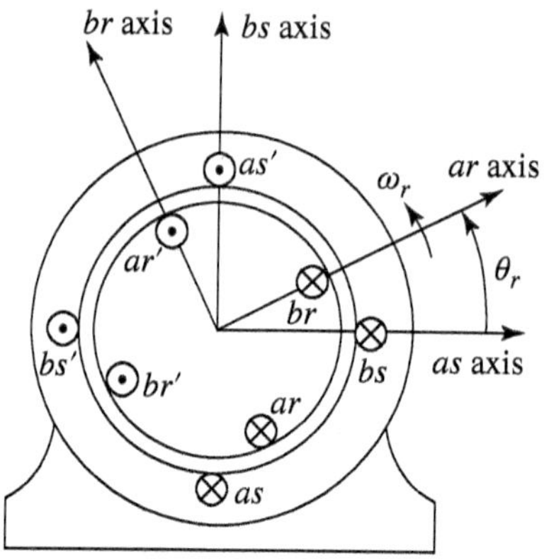

Figure 4C-1 Elementary two-pole two-phase induction machine.

will cause a mmf_r that rotates at $(\omega_e - \omega_r)$ relative to the rotor. If we are on the stator, we would see mmf_s^s varying at ω_e and mmf_r^s would appear to be varying at $(\omega_e - \omega_r + \omega_r)$ or ω_e also. If we are on the rotor, mmf_r^r and mmf_s^r would appear to vary at $(\omega_e - \omega_r)$. If we are running at ω_e, mmf_s^e and mmf_r^e would be stationary relative to us. The reader should verify these angular velocities by appropriate substituting into (4.5-12) through (4.5-17).

In the case of part (b), the mmfs given by (4.5-12) through (4.5-17) have the units of ampere-turns. Therefore, the angular frequency of the currents as viewed from the various frames of reference corresponds to that of the mmfs. It follows that the voltages associated with these balanced currents and symmetrical circuits will also be balanced and of the same angular frequency as the currents.

SP4.5-1 Change Fig. 4.4-3 to portray (4.5-17). [Change all "s" subscripts to "r" and reverse the sense of the poles]

SP4.5-2 Justify $\phi = \phi_s$ in (4.5-12) and (4.5-13), $\phi = \phi_r$ in (4.5-14) and (4.5-15), and $\phi = \phi_e$ in (4.5-16) and (4.5-17). [Use (4.4-8) and (4.5-11) and substitute $\omega = 0$; $\omega = \omega_r$; and $\omega = \omega_e$]

SP4.5-3 Assume balanced steady-state conditions and consider (4.5-12) through (4.5-17). Determine the frequency of the steady-state variables in the stationary, rotor, and synchronously rotating reference frames. $[\omega_e, \omega_e - \omega_r, \text{dc}]$

4.6 *P*-POLE MACHINES

Thus far, we have considered only two-pole electromechanical motion devices; however, electromechanical devices may have any even number of poles (2, 4, 6, 8, ...) up to more than 50 in the case of large hydroturbine synchronous generators. We may at first be a little hesitant to tackle the analysis of machines with more than two poles, since it appears that this may complicate matters before we even get started. Fortunately, this is not the case. We will find that with a simple change of variables we can analyze all machines as if they were two-pole devices. We need only to modify the expression for evaluating torque and show that, physically, the actual rotor speed of a machine with more than two poles will be a multiple less than that which we determine from our two-pole analysis.

The rotating air-gap mmf of a machine with more than two poles can be determined by considering the four-pole device shown in Fig. 4.6-1. In this figure, each phase winding consists of two series-connected windings, each of which is assumed to be sinusoidally distributed. For example, $as1'$ represents a group of conductors sinusoidally distributed over $0 < \phi_s < \frac{1}{2}\pi$, and so on. The phase windings consist of N_s turns, with $N_s/\frac{P}{2}$ turns in each of the two series-connected windings. There may be some confusion in regard to notation. When considering the coils of a winding, notation $as_1, bs_1, \ldots$ was used. In Fig. 4.6-1, the notation $as1$, $bs1$, ... is used to denote sinusoidally distributed windings. Thus, a subscripted number denotes a coil, whereas a number that is not a subscript denotes a sinusoidally distributed winding.

Note that each of the two series windings per phase spans $\frac{1}{2}\pi$ radians and each phase winding establishes two magnetic systems. As shown in Fig. 4.6-1, at the instant when $I_{bs} = 0$ and I_{as} is a positive maximum, the $as1 - as1'$ part of the as winding produces positive flux in the $as1$ axis direction whereas the

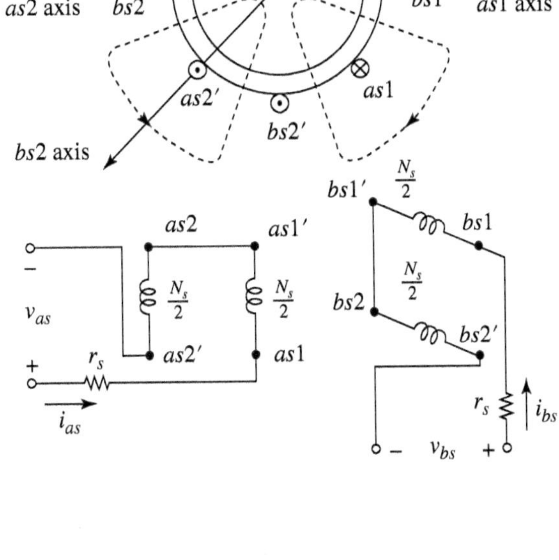

Figure 4.6-1 Stator winding arrangement of a four-pole two-phase symmetrical electromechanical device. Flux streamlines shown for I_{as} maximum positive and I_{bs} zero.

$as2 - as2'$ part of the as winding produces positive flux in the $as2$ axis direction. South poles occur from $-\frac{1}{4}\pi < \phi_s < \frac{1}{4}\pi$ and $\frac{3}{4}\pi < \phi_s < \frac{5}{4}\pi$. Now, as shown in Fig. 4.6-1 by the streamlines, half of the flux that enters the stator steel from $-\frac{1}{4}\pi < \phi_s < \frac{1}{4}\pi$ reenters the air gap between $\frac{5}{4}\pi < \phi_s < \frac{7}{4}\pi$ and the other half between $\frac{1}{4}\pi < \phi_s < \frac{3}{4}\pi$. The flux that enters the stator from $\frac{3}{4}\pi < \phi_s < \frac{5}{4}\pi$ divides similarly. Hence, for $I_{as} = \sqrt{2}I_s$ and $I_{bs} = 0$, two north poles occur; one from $\frac{1}{4}\pi < \phi_s < \frac{3}{4}\pi$ and another from $\frac{5}{4}\pi < \phi_s < \frac{7}{4}\pi$. It should be pointed out that the streamlines should enter and exit the air gap at right angles.

The air-gap mmf established by each phase is a sinusoidal function of $2\phi_s$ for a four-pole machine, or, in general, $(P/2)\phi_s$, for a P-pole, machine. In particular,

$$\text{mmf}_{as} = \frac{2}{P}\frac{N_s}{2} i_{as} \cos \frac{P}{2}\phi_s \tag{4.6-1}$$

$$\text{mmf}_{bs} = \frac{2}{P}\frac{N_s}{2} i_{bs} \sin \frac{P}{2}\phi_s \tag{4.6-2}$$

where N_s is the total equivalent turns per stator phase. For balanced steady-state operation, the stator currents may be expressed as

$$I_{as} = \sqrt{2}I_s \cos\left[\omega_e t + \theta_{esi}(0)\right] \tag{4.6-3}$$

$$I_{bs} = \sqrt{2}I_s \sin\left[\omega_e t + \theta_{esi}(0)\right] \tag{4.6-4}$$

Equations (4.6-3) and (4.6-4) are the same expressions for the stator currents as given for the two-pole case. Also, (4.4-9) would become

$$\text{mmf}_s^s = \left(\frac{2}{P}\right)\frac{N_s}{2}\sqrt{2}I_s \cos\left[\omega_e t + \theta_{esi}(0) - \frac{P}{2}\phi_s\right] \tag{4.6-5}$$

With the stator arranged as in Fig. 4.6-1, balanced steady-state stator currents of frequency ω_e produce a four-pole (P-pole) magnetic system that rotates about the air gap at $(2/4)\omega_e$ or $(2/P)\omega_e$ relative to a stationary observer. How do we know that? Well, in order for (4.6-5) to be a constant, the argument of the cosine must be a constant and this occurs when its time derivative is zero. Thus, $\omega_e - \frac{P}{2}\frac{d\phi_s}{dt} = 0$ which is true if $\frac{d\phi_s}{dt} = \frac{2}{P}\omega_e$. Therefore, the speed of the rotating mmf (rotating poles) is now $(2/P)\omega_e$ which is the synchronous speed of the machine; however, the stator variables are unaware of this. To the electric system, ω_e is synchronous speed; however, the magnetic system of the rotor must "lock in" with the magnetic system established by the stator.

It now seems logical that we could treat any machine regardless of its number of pole pairs, as a cascading of two-pole machines. Thus, we could calculate torque for a device with more than one pole pair by multiplying the expressions given in

Table 2.5-1 by $\frac{P}{2}$, i.e. the number of cascaded two-pole machines and dividing the rotor speed calculated for a two-pole device by $\frac{P}{2}$ to obtain the actual rotor speed. Also, since the electrical system "thinks" the rotor displacement and rotor speed are θ_r and ω_r, respectively, those quantities are referred to as electrical position and electrical angular velocity of the rotor, respectively. If this all seems reasonable, the remainder of this section can be omitted; however, the following development is included in case questions arise.

The four-pole single-phase reluctance machine shown in Fig. 4.6-2 can be used to establish the change of variables necessary to allow us to consider salient-rotor *P*-pole machines as two-pole devices. In Fig. 4.6-2, θ_{rm} and ω_{rm} are the actual angular position and angular velocity, respectively, of the rotor. Recall we used θ_r and ω_r for the two-pole device in our previous work. Now, let us proceed to write the self-inductance L_{asas}. If we follow the same procedure as we did to obtain (1.7-29), we will find that

$$L_{asas} = L_l + L_A - L_B \cos 2\left(\frac{4}{2}\theta_{rm}\right) \tag{4.6-6}$$

Generalizing for a *P*-pole machine,

$$L_{asas} = L_l + L_A - L_B \cos 2\left(\frac{P}{2}\theta_{rm}\right) \tag{4.6-7}$$

It is apparent that if we make the substitution of

$$\theta_r = \frac{P}{2}\theta_{rm} \tag{4.6-8}$$

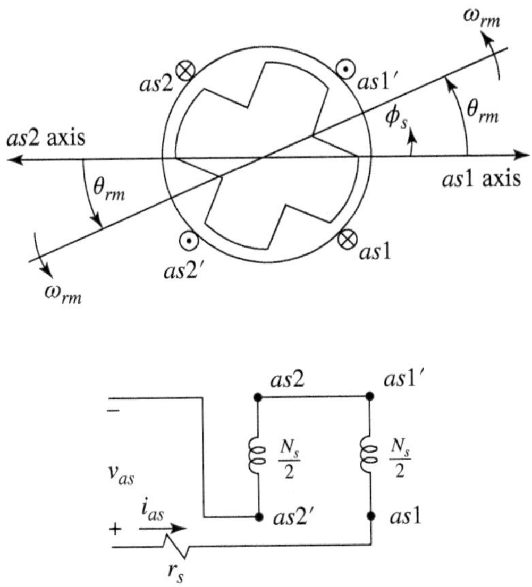

Figure 4.6-2 Elementary four-pole single-phase reluctance machine.

then L_{asas} for the P-pole machine, (4.6-7), is identical in form to L_{asas} for the two-pole machine, (1.7-29). Also, (4.6-8) leads to

$$\omega_r = \frac{P}{2}\omega_{rm} \tag{4.6-9}$$

With the substitution of θ_r and ω_r for $(P/2)\theta_{rm}$ and $(P/2)\omega_{rm}$, respectively, the voltage equations for a P-pole machine are identical to those for a two-pole machine. Actually, the electric system can tell no difference. To the electric system, the angular position of the rotor is θ_r and its velocity is ω_r. As we have mentioned, it is for this reason that θ_r and ω_r are often called the electrical angular position of the rotor and the electrical angular velocity of the rotor, respectively.

Let us now consider a four-pole stator winding and a four-pole rotor winding shown in Fig. 4.6-3. The mutual inductance between the *as* and *ar* windings may be expressed

$$L_{asar} = L_{sr}\cos 2\theta_{rm} \tag{4.6-10}$$

where θ_{rm} is the actual rotor position. In the two-pole case we used θ_r. The argument of the cosine in (4.6-10) may be rewritten in terms of P, so that

$$L_{asar} = L_{sr}\cos\frac{P}{2}\theta_{rm} \tag{4.6-11}$$

If we substitute (4.6-8) into (4.6-11), it becomes what we had for the two-pole case. Now, as in the case of the salient-pole device, θ_{rm} is the actual rotor displacement and θ_r is what is referred to as the electrical angular position of the rotor.

Although we see that if we substitute (4.6-8) into (4.6-11), the result is the mutual inductance for a two-pole device; however, what about the mutual inductance between stator windings as shown in Fig. 4.6-1? In the case of the two-pole machine, $L_{asbs} = 0$, since the magnetic axes are orthogonal. Note, however, that in the four-pole machine in Fig. 4.6-1 the *bs* winding is displaced 45° counterclockwise from the *as* winding. If we assume, for a minute, that in Fig. 4.6-1, we rotated the *bs* winding clockwise 45°, it would be atop the *as* winding and L_{asbs} would be similar to (4.6-11), except L_{asar} would be replaced by L_{asbs}, L_{sr} would be replaced by L_{ms}, and θ_{rm} by the angle between *as* and *bs* windings. If now we rotate the *bs* winding back counterclockwise by an angle of 45°, then L_{asbs} would be zero from our modified (4.6-11). Thus, we conclude that we can safely neglect mutual

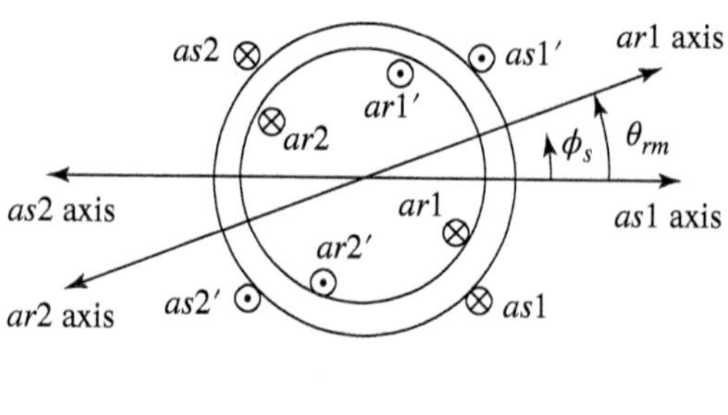

Figure 4.6-3 Mutual coupling between four-pole stator and rotor windings.

coupling between stator windings of a P-pole machine. From here on out, we will treat all electric machines as two-pole devices, knowing full well that (4.6-8) and (4.6-9) are the relationship between θ_r and θ_{rm} and ω_r and ω_{rm}.

Now, since we are going to treat all machines as two-pole devices and use θ_r in all the voltage equations, we have to do one last thing in order to make things work out. Back in Section 2.5 we expressed the change in the energy entering the coupling field from the mechanical system as

$$dW_m = -T_e d\theta_r \tag{4.6-12}$$

and the torque T_e as

$$T_e(\mathbf{i}, \theta_r) = \frac{\partial W_c(\mathbf{i}, \theta_r)}{\partial \theta_r} \tag{4.6-13}$$

Equation (4.6-12) must now be written in terms of θ_{rm}:

$$dW_m = -T_e d\theta_{rm} \tag{4.6-14}$$

and since $\theta_{rm} = \frac{2}{P}\theta_r$, (4.6-13) becomes

$$T_e(\mathbf{i}, \theta_r) = \frac{P}{2}\frac{\partial W_c(\mathbf{i}, \theta_r)}{\partial \theta_r} \tag{4.6-15}$$

Analyzing a P-pole machine has turned out to be not as troublesome as we might have first thought. We are fortunate.

SP4.6-1 If, during steady-state operation, $\omega_r = \omega_e$, calculate the actual rotor speed in revolutions per minute of a 60-pole 60-Hz hydroturbine synchronous generator. [$\omega_{rm} = 120\,\text{r/min}$]

SP4.6-2 Suppose $\theta_{esi}(0) = 45°$ in (4.6-3) and (4.6-4). For a six-pole two-phase stator, determine the location of the positive and negative maximum values of mmf$_s$ at time zero. [S^s at $\phi_s = 15°, 135°, 255°$; N^s at $\phi_s = 75°, 195°, 315°$]

4.7 RECAPPING

The primary focus of this chapter has been the introduction of Tesla's rotating magnetic field which makes ac machines possible. Moreover, in Chapter 5 we will see that Tesla's rotating magnetic field also provides the analytical basis for reference frame theory which is the foundation of machine analysis and simulation. Being able to view Tesla's rotating magnetic field from various frames of reference helps to understand the concept of reference frame analysis wherein substitute variables replace the actual machine variables. The visualization of Tesla's rotating magnetic field shown in the online animation is the first step towards setting the stage of Tesla's based approach to machine analysis.

4.8 PROBLEMS

1. A winding with five coils, each with nc_s conductors, is distributed over $\frac{1}{2}\pi$ of the stator. Sketch the configuration and indicate the direction of the positive magnetic axis of this single-phase stator winding.

2. The four coils, each with nc_s conductors, of the windings of a two-pole two-phase low-power device are concentrated so that four coil sides are placed in one slot per phase. Sketch the winding arrangement and show the *as* and *bs* axes.

3. Consider the two-phase device shown in Fig. 4.8-1. The windings are sinusoidally distributed, each with N_s equivalent turns. Express (a) N_{as} and N_{bs} and (b) mmf_{as} and mmf_{bs}.

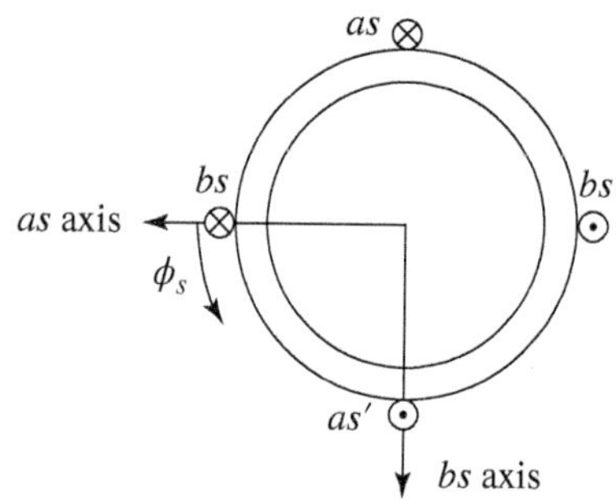

Figure 4.8-1 Elementary two-pole two-phase electromechanical device.

4. Each stator coil in Fig. 4.8-2 contains nc_s turns. Using the information given in Example 4A as a guide, sketch the developed diagram for the *as* and *bs* windings and mmf_{as} and mmf_{bs}.

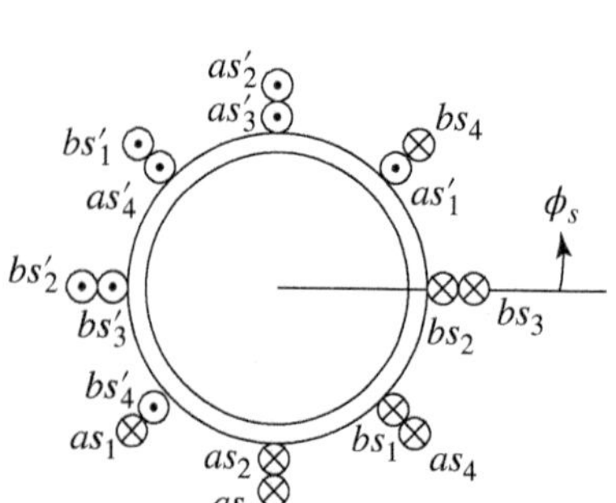

Figure 4.8-2 A two-pole, two-phase electromechanical device.

5. Consider Fig. 4.8-3, where a winding is formed by *N* conductors uniformly distributed over the regions shown. Each conductor carries the current *i*. Sketch the air-gap mmf.

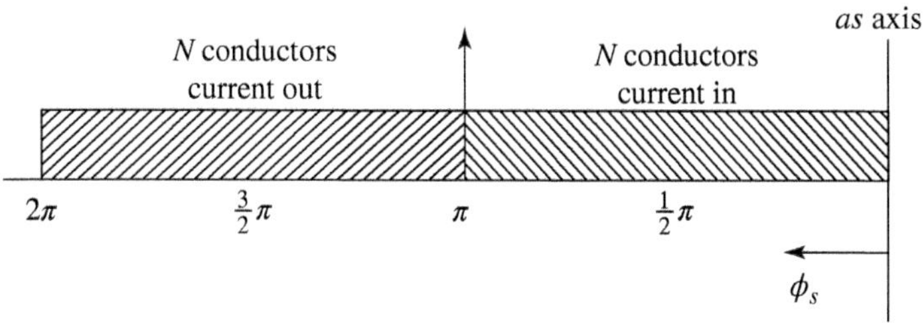

Figure 4.8-3 Uniformly distributed windings.

6. Consider Fig. 4.8-1. (a) Express mmf_s. (b) Determine the phase relationship between $\tilde{I}_{as}$ and $\tilde{I}_{bs}$ that will produce counterclockwise rotation of mmf_s. That is, does $\tilde{I}_{as}$ lead or lag $\tilde{I}_{bs}$ by 90°? (c) Locate the positions of N^s and S^s when $\omega_e t = 30°$ with $I_{as} = \sqrt{2}I_s \cos \omega_e t$ and $I_{bs} = \sqrt{2}I_s \sin \omega_e t$.

7. Consider the device shown in Fig. 4.4-1. Express mmf_s for (a) $I_{as} = \sqrt{2}I_s \cos \omega_e t$, $I_{bs} = \sqrt{2}I_s \cos \omega_e t$; (b) $I_{as} = I_a \cos \omega_e t$, $I_{bs} = I_b \sin \omega_e t$, where $I_a \neq I_b$. Both (a) and (b) are unbalanced sets. Set (a) is unbalanced since I_{as} and I_{bs} are not orthogonal; set (b) is unbalanced due to unequal amplitudes.

8. The windings of the device shown in Fig. 4.8-4 are sinusoidally distributed, each with N_s equivalent turns. The *bs* winding may be arranged so that the *bs* axis is at an arbitrary angle α with the *as* axis. Express i_{as} and i_{bs} so that mmf_s has a constant amplitude and rotates in the counterclockwise direction regardless of the value of α; in particular, $\text{mmf}_s = (N_s/2)\sqrt{2}I_s \cos(\omega_e t - \phi_s)$.

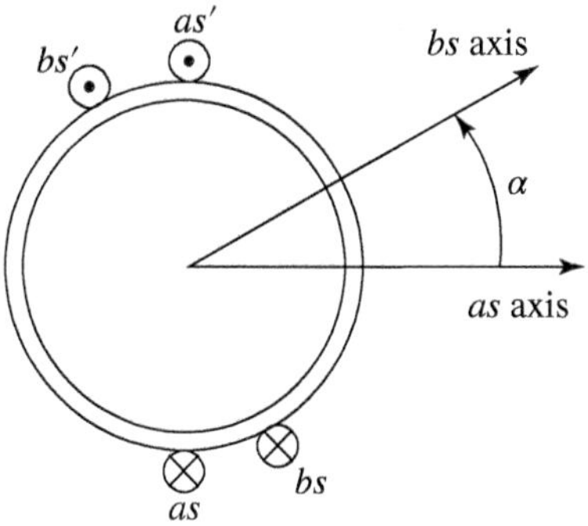

Figure 4.8-4 Magnetic axes displaced by a mechanical angle α.

9. Consider the stator configurations shown in Fig. 4.8-5. For each of the configurations A, B, and C, express I_{as} and I_{bs} for ccw rotation and then express mmf_s^e.

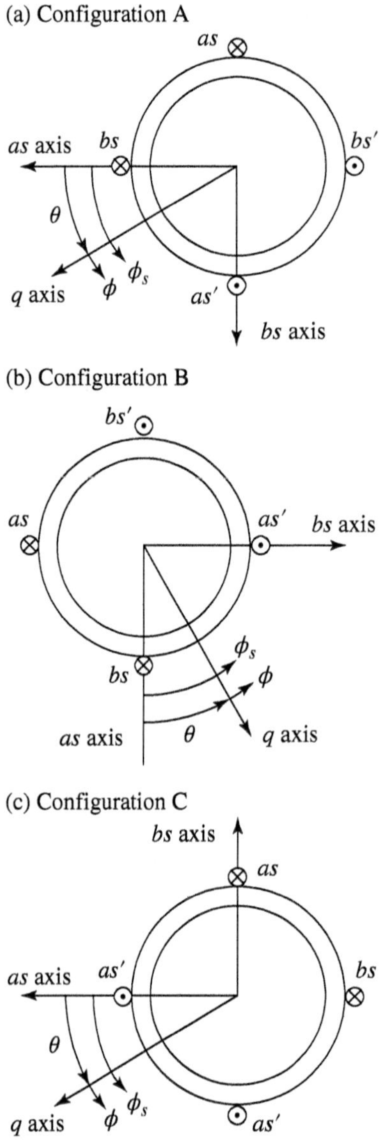

Figure 4.8-5 Stator configurations.

10. Sketch the *as*-phase stator winding for a six-pole stator.

11. A two-phase induction motor is operating at rated-load torque with balanced stator currents. The steady-state rotor speed is $\omega_r = 0.9\omega_e$. It is an eight-pole, 50-Hz device.

Determine (a) the actual rotor speed, (b) the angular velocity of the rotor mmf relative to the rotor, and (c) the frequency of the rotor currents.

12. Determine the steady-state rotor speeds (ω_{rm}) in revolutions per minute of the following devices: (a) a four-pole three-phase 100-hp 60-Hz salient-pole synchronous machine, (b) an eight-pole two-phase 1-hp 60-Hz reluctance machine supplied from a 5-Hz voltage source, and (c) a six-pole three-phase $\frac{1}{2}$-hp permanent-magnet motor supplied from a 15-Hz source.

4.9 REFERENCE

1. Nikola Tesla, *My Inventions and Other Writings*, Penguin Books, 2011.

CHAPTER 5

INTRODUCTION TO REFERENCE FRAME THEORY

5.1 INTRODUCTION

In recent years, there has been a steady evolution of the electric drive industry which began with the advent of the controlled rectifier. Controlled electric drives are being used in a wide range of applications such as brushless dc motors used in appliances, doubly fed induction machines used in wind turbines, drive motors in electric and hybrid vehicles, and now electric drives are replacing hydraulic drives in aircraft. As the applications for converter-controlled electric drives began to emerge, it became evident that the traditional steady-state approach to electric machine analysis and the machine characteristics as portrayed by this type of analysis were inadequate to analyze and understand most modern electric drive systems. In particular, rapid electronic switching enabled control techniques which altered the performance of the electric machine beyond what one could envision from the traditional steady-state analysis. It became obvious that the transient characteristics of the electric machine had to be considered and that computer simulations of the dynamic performance of converter-controlled electric drives were useful, if not necessary, in the design of these drives. Concurrently, it was shown that these needs could only be addressed through the appropriate application of reference frame theory. Unfortunately, reference frame theory has been considered more or less an abstract approach to electric machine analysis. This has made it difficult to convey the concept of reference frame theory to the undergraduate student. Fortunately, the connection between Tesla's rotating magnetic field and reference frame theory provides a direct and straightforward analytical basis for viewing the physical variables of symmetrical windings from any frame of reference [1, 2]. We have already set the stage for this derivation. In particular, reference frame theory is nothing more than establishing the change of variables between the

Electromechanical Motion Devices: Rotating Magnetic Field-Based Analysis and Online Animations, Third Edition. Paul Krause, Oleg Wasynczuk, Steven D. Pekarek, and Timothy O'Connell.

Published 2020 by John Wiley & Sons, Inc.

physical and substitute variables associated with the fictitious circuits that produce the rotating magnetic field as viewed from a given reference frame. Fortunately, the transformation that establishes the substitute variables (voltage equations and flux linkage equations) associated with the fictitious windings is determined directly from the expression of Tesla's rotating magnetic field. Please refer to Animation C of Appendix B.

This chapter can be skipped without hindering the reader's ability to follow the analysis of steady-state machine operation. This, however, is not recommended since reference frame theory is a straightforward extension of Tesla's rotating magnetic field and essential to the derivation of the machine equations and the fundamentals of machine control in modern drive systems. Therefore, it is strongly recommended that at least the basic sections presented in this chapter be covered in order to gain an awareness of reference frame theory even if the derivations are de-emphasized in the following chapters and only steady-state performance considered.

5.2 BACKGROUND

In the late 1920s, R. H. Park [3] introduced a new approach to electric machine analysis. He formulated a change of variables which, in effect, replaced the physical variables (voltages, currents, and flux linkages) associated with the stator windings of a synchronous machine with substitute variables rotating at the electrical angular velocity of the rotor. This change of variables is often described as transforming or referring the stator variables to a frame of reference fixed in the rotor. Park's transformation revolutionized electric machine analysis.

Following Park's work, several changes of variables convenient for induction machine analysis have been developed over the years, each of which appeared to be unique. Consequently, each transformation was derived and treated separately in literature until it was reported in 1965 [4] that all known real transformations used in induction machine analysis are contained in one general transformation that eliminates all position-varying inductances by referring the stator and the rotor variables to a common frame of reference which may rotate at any angular velocity or remain stationary. Our purpose is to derive this transformation for stationary and rotating circuits from Tesla's rotating magnetic field, thereby establishing the concept of viewing the physical variables associated with the actual windings from any frame of reference and thus establishing the basis of reference frame theory.

5.3 CHANGE OF VARIABLES FOR SYMMETRICAL STATOR CIRCUITS

For the purpose of convenience, we will repeat some equations derived in Chapter 4. Tesla's rotating magnetic field for the two-pole two-phase symmetrical, stator windings shown in Fig. 5.3-1 is expressed by (4.4-2) as

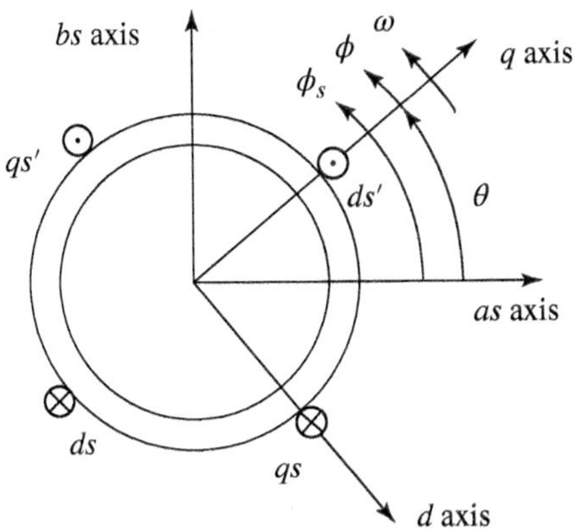

Figure 5.3-1 Two-pole two-phase sinusoidally distributed fictitious windings.

$$\text{mmf}_s = \frac{N_s}{2}(i_{as}\cos\phi_s + i_{bs}\sin\phi_s) \tag{5.3-1}$$

Please note that in Fig. 5.3-1 we have not shown the ⊗ and ⊙ notation of the *as* and *bs* windings for purposes of clarity. From (4.4-6), the displacement coordinates ϕ and ϕ_s are related as

$$\phi_s = \theta + \phi \tag{5.3-2}$$

where θ is the position of the q axis, ϕ is the displacement from the q axis, and ϕ_s is the displacement about the stator periphery measured from the *as* axis. We have also added a *ds* winding and axis. We will talk about this in more detail shortly.

Substituting (5.3-2) into (5.3-1) yields (4.4-7), that is

$$\text{mmf}_s = \frac{N_s}{2}[i_{as}\cos(\theta + \phi) + i_{bs}\sin(\theta + \phi)] \tag{5.3-3}$$

In Chapter 4 we substituted steady-state currents, I_{as} and I_{bs}, into (4.4-7), which is (5.3-3), and viewed the mmf_s from the q axis which could rotate at any angular velocity or remain stationary. The purpose now is to determine the change of variables that will portray Tesla's rotating magnetic field from any frame of reference, the *arbitrary reference frame*. To do this, we will first express (5.3-3) as

$$\text{mmf}_s = \frac{N_s}{2}\cos\phi(i_{as}\cos\theta + i_{bs}\sin\theta) + \frac{N_s}{2}\sin\phi(-i_{as}\sin\theta + i_{bs}\cos\theta) \tag{5.3-4}$$

Since mmf is in ampere-turns, the terms in the parentheses are currents, then $\frac{N_s}{2}\cos\phi$ and $\frac{N_s}{2}\sin\phi$ can be interpreted as the result of fictitious orthogonal sinusoidally distributed windings. From Chapter 4 we realize that the winding distribution is orthogonal to the mmf distribution. When ϕ is zero, $\frac{N_s}{2}\cos\phi$ is symmetrical about the q axis and the current i_{qs} flowing in this q-axis fictitious winding labeled *qs* in Fig. 5.3-1. The current i_{qs} is

$$i_{qs} = i_{as} \cos\theta + i_{bs} \sin\theta \tag{5.3-5}$$

When $\phi = \pm \frac{\pi}{2}$, the second fictitious winding, labeled ds in Fig 5.3-1, is symmetrical about an axis orthogonal to the q axis. We will call this axis the d axis, as indicated in Fig. 5.3-1, and we will select a positive d axis at $\phi = -\frac{\pi}{2}$. From (5.3-4), we see that the current flowing in this winding is

$$i_{ds} = i_{as} \sin\theta - i_{bs} \cos\theta \tag{5.3-6}$$

We can now write (5.3-4) as

$$\text{mmf}_s = \frac{N_s}{2}\left(i_{qs} \cos\phi - i_{ds} \sin\phi\right) \tag{5.3-7}$$

Let us consider (5.3-1) and (5.3-7) for a minute. Equation (5.3-1) is an expression for Tesla's rotating magnetic field in terms of physical windings and variables. The actual currents i_{as} and i_{bs} are flowing in sinusoidally distributed real-life stationary windings. On the other hand, the currents i_{qs} and i_{ds} in (5.3-7) are arbitrary reference frame substitute currents flowing in fictitious windings. These substitute variables are related to the physical variables, i.e. the substitute currents are the currents as viewed from the arbitrary reference frame defined by q and d axes. The fictitious windings, which are rotating at an arbitrary angular velocity of ω, are orthogonal and sinusoidally distributed with an identical number of turns and resistance as the actual windings. The substitute currents, i_{qs} and i_{ds}, that are flowing in these fictitious windings are related to i_{as} and i_{bs} by (5.3-5) and (5.3-6), respectively. We can think of the qs and ds windings as fictitious windings moving "through" the stator iron with i_{qs} and i_{ds} flowing in them, respectively.

The as and bs currents are transformed by the change of variables given by (5.3-5) and (5.3-6) to i_{qs} and i_{ds}, respectively. Since the voltages, currents, and flux linkages are functionally related, the same change of variables applies to all. Thus, the transformation to the arbitrary reference frame for the voltages, currents, and flux linkages may be written in matrix form as

$$\begin{bmatrix} f_{qs} \\ f_{ds} \end{bmatrix} = \begin{bmatrix} \cos\theta & \sin\theta \\ \sin\theta & -\cos\theta \end{bmatrix} \begin{bmatrix} f_{as} \\ f_{bs} \end{bmatrix} \tag{5.3-8}$$

Symbolically

$$\mathbf{f}_{qds} = \mathbf{K}_s \mathbf{f}_{abs} \tag{5.3-9}$$

where

$$(\mathbf{f}_{qds})^T = [f_{qs} \;\; f_{ds}] \tag{5.3-10}$$

$$(\mathbf{f}_{abs})^T = [f_{as} \;\; f_{bs}] \tag{5.3-11}$$

$$\mathbf{K}_s = \begin{bmatrix} \cos\theta & \sin\theta \\ \sin\theta & -\cos\theta \end{bmatrix} \tag{5.3-12}$$

and T denotes the transpose of a matrix. The angular position of the arbitrary reference frame is

$$\frac{d\theta}{dt} = \omega \tag{5.3-13}$$

Let us take a minute to pull things together. First, f can denote either voltage, current, or flux linkage. That is, the same transformation is applied to all. The subscripts, as and bs, respectively, denote the variables associated with the as and bs phases of the stationary (stator) circuits; thus, qs and ds are the substitute variables that are the as and bs variables as viewed from an arbitrary frame of reference. Or, in other words, the substitute variables are the variables associated with the fictitious circuits that produce Tesla's rotating mmf as viewed from a given frame of reference. This interpretation is key to the understanding of reference frame theory. The substitute variables are related to the physical as and bs variables by the transformation matrix $\mathbf{K}_s$. The subscript s is used in qs, ds, and $\mathbf{K}_s$ to distinguish stator physical variables from the variables associated with the rotor circuits. Later, we will designate rotor variables with an r in the subscript. Finally, the angular velocity and angular position are related by (5.3-13). It is interesting that for this choice of the q- and d-axis locations, $\mathbf{K}_s$ is equal to $(\mathbf{K}_s)^{-1}$.

Note that we have not assigned a value to ω or θ. The angular position θ must be continuous; however, the angular velocity associated with the change of variables is unspecified. That is, the frame of reference from which we wish to view Tesla's rotating magnetic field may rotate at any constant or varying angular velocity or it may remain stationary. In other words, the angular velocity of the transformation is unspecified (arbitrary) and can be selected to specify the frame of reference that will expedite the solution of the system equations or to satisfy system constraints. Although we have used Tesla's rotating magnetic field to derive the change of variables, the transformation may be applied to variables of any waveform and time sequence and may have no relation to an electric machine. In an ab sequence, the a-phase variables lead the b-phase variables by $\frac{\pi}{2}$ for balanced steady-state conditions; however, as we just mentioned, this does not imply that the variables must be a balanced steady-state set nor must the set be associated with a machine.

We have set forth the transformation for variables associated with the stator circuits. It is important to understand that the change of variables is nothing more than a transformation of the stator variables to a selected frame of reference so that the substitute variables flowing in the fictitious reference frame windings provide the correct view of Tesla's rotating magnetic field from that frame of reference. In a later section, we will consider rotor circuits and we will find that the transformation for symmetrical rotating circuits is very similar. In fact, we could have set forth just one transformation to a reference frame rotating at an arbitrary angular velocity which could be used for either stationary or rotating symmetrical circuits. That is, the transformation for stationary circuits can then be obtained by fixing the angular velocity of rotating circuits at zero. Although this approach is valid, it tends to be confusing. Therefore, the transformation of stationary and rotor circuits are treated separately.

It is interesting that when we substituted $\theta + \phi$ for ϕ_s into (5.3-1), we leave the real world and entered the world of Tesla's rotating magnetic field where the variables are no longer *as* and *bs* but *qs* and *ds* for compactness. The analysis becomes based on the rotating magnetic field and we must think in terms of rotating with it and yet at any time, we can refer our analyses back to the real world. Once one becomes familiar with this concept, reference frame theory becomes straightforward.

The instantaneous power of a two-phase system is

$$P = v_{as}i_{as} + v_{bs}i_{bs} \tag{5.3-14}$$

In substitute variables, (5.3-14) becomes

$$\begin{aligned} P = &\left(v_{qs}\cos\theta + v_{ds}\sin\theta\right)\left(i_{qs}\cos\theta + i_{ds}\sin\theta\right) \\ &+ \left(v_{qs}\sin\theta - v_{ds}\cos\theta\right)\left(i_{qs}\sin\theta - i_{ds}\cos\theta\right) \end{aligned} \tag{5.3-15}$$

Equation (5.3-15) reduces to

$$P = v_{qs}i_{qs} + v_{ds}i_{ds} \tag{5.3-16}$$

Now, (5.3-16) is in arbitrary-reference-frame variables and it is equal to (5.3-14). Therefore, we have just shown that power is the same in all reference frames. Although the frequency of the voltage and current is reference frame dependent, since these variables must portray the view of Tesla's rotating magnetic field from a specific frame of reference, the average power is not a function of frequency and therefore not altered by the frame of reference from which we view Tesla's rotating magnetic field.

Example 5A Balanced Set Transformed to the Arbitrary Reference Frame Let us assume a balanced set with $f_{as} = \cos\theta_{esf}$ and $f_{bs} = \sin\theta_{esf}$. From (5.3-9),

$$\begin{bmatrix} f_{qs} \\ f_{ds} \end{bmatrix} = \begin{bmatrix} \cos\theta & \sin\theta \\ \sin\theta & -\cos\theta \end{bmatrix} \begin{bmatrix} \cos\theta_{esf} \\ \sin\theta_{esf} \end{bmatrix} \tag{5A-1}$$

This balanced set is viewed from the arbitrary reference frame as

$$f_{qs} = \cos\left(\theta_{esf} - \theta\right) \tag{5A-2}$$

$$f_{ds} = -\sin\left(\theta_{esf} - \theta\right) \tag{5A-3}$$

Note that the amplitude of the balanced set is not altered by the change of variables. If, for example, $\theta = \omega t + \theta(0)$ and $\theta_{esf} = \omega_e t + \theta_{esf}(0)$ where f can be voltage, current, or flux linkage, then f_{qs} and f_{ds} also form a balanced two-phase sinusoidal set if $\omega \neq \omega_e$. If $\omega < \omega_e$, then f_{qs} lags f_{ds} by $\frac{\pi}{2}$; if, however, $\omega > \omega_e$, then f_{qs} leads f_{ds} by $\frac{\pi}{2}$. We will talk more of this later.

If $\omega = \omega_e$, then f_{qs} and f_{ds} are constants; in particular,

$$f_{qs}^e = \cos\left[\theta_{esf}(0) - \theta(0)\right] \tag{5A-4}$$

$$f_{ds}^e = -\sin\left[\theta_{esf}(0) - \theta(0)\right] \tag{5A-5}$$

Be careful here, when f_{qs} and f_{ds} are without superscript these are arbitrary-reference-frame variables. A superscript denotes a specific frame; here, the e denotes viewing f_{as} and f_{bs} from the synchronously rotating reference frame. If we think about this for just a minute, we can see that if in one reference frame a balanced set appears in the steady state, then there is another reference frame where this balanced set is viewed as constants and vice versa. Actually, we are not surprised by all these since we could have predicted it from (4.5-12) through (4.5-17). In particular, since Tesla's rotating magnetic field appears as a constant in the ω_e reference frame during balanced steady-state operation, we would expect that the steady-state substitute variables in this reference frame would also be constants (dc).

Before leaving this example, let us assume that $\theta = 0$, then f_{qs} and f_{ds} in the stationary reference frame, (5A-2) and (5A-3), become

$$f_{qs}^s = \cos\theta_{esf} \tag{5A-6}$$

$$f_{ds}^s = -\sin\theta_{esf} \tag{5A-7}$$

where the s superscript denotes the stationary reference frame variables. Recall that at the beginning of this example, we set $f_{as} = \cos\theta_{esf}$ and $f_{bs} = \sin\theta_{esf}$. It is important to note that f_{ds}^s in the stationary reference frame ($\omega = 0$) is the negative of f_{bs} if $\theta(0)$ is selected to be zero which is generally the case. Moreover, in the stationary reference frame ($\omega = 0$) the fictitious windings for the qs^s and ds^s variables become the physical windings associated with the as and bs variables; however, the positive direction of i_{ds}^s (5.3-6) has been selected opposite to that of i_{bs}.

SP5.3-1 Show that if we placed the positive d axis at $\frac{\pi}{2}$ counterclockwise from the q axis rather than $-\frac{\pi}{2}$, $(\mathbf{K}_s)^{-1}$ would not be equal to $\mathbf{K}_s$.

SP5.3-2 Express i_{ds} for SP5.3-1. [From (5.3-4), $i_{ds} = -i_{as}\sin\theta + i_{bs}\cos\theta$]

SP5.3-3 If $f_{as} = \cos\omega_e t$ and $f_{bs} = \sin\omega_e t$ and $f_{qs} = 1$ and $f_{ds} = 0$. Use (5.3-9) or Example 5A to determine $\theta_{esf}(0)$, ω, and $\theta(0)$. $[0, \omega_e, 0]$.

SP5.3-4 Let $V_{as} = \cos\omega_e t$, $I_{as} = \cos\omega_e t$, $V_{bs} = \sin\omega_e t$, and $I_{bs} = \sin\omega_e t$ and $\theta(0) = 0$. Use (5.3-9) or Example 5A to determine V_{qs}^e, V_{ds}^e, I_{qs}^e, and I_{ds}^e. [(1, 0, 1, 0)].

5.4 TRANSFORMATION OF TWO-PHASE STATOR VARIABLES TO THE ARBITRARY REFERENCE FRAME

In the case of symmetrical two-phase resistive circuits, the voltage equations may be written as

$$\mathbf{v}_{abs} = \mathbf{r}_s \mathbf{i}_{abs} \tag{5.4-1}$$

where $\mathbf{r}_s$ is an equal element diagonal matrix and the *as* and *bs* circuits have the same resistance. We are assuming that the inductance of the windings is negligible which is not possible but it is helpful academically to consider r_s and L_{ss} separately. From (5.3-9)

$$\mathbf{v}_{qds} = \mathbf{K}_s \mathbf{v}_{abs} \tag{5.4-2}$$

Substituting (5.4-1) into (5.4-2) yields

$$\mathbf{v}_{qds} = \mathbf{K}_s \mathbf{r}_s \mathbf{i}_{abs} \tag{5.4-3}$$

Now, since $\mathbf{i}_{abs} = (\mathbf{K}_s)^{-1}\mathbf{i}_{qds}$, we can write (5.4-3) as

$$\mathbf{v}_{qds} = \mathbf{K}_s \mathbf{r}_s (\mathbf{K}_s)^{-1} \mathbf{i}_{qds} \tag{5.4-4}$$

Since $\mathbf{r}_s = r_s \mathbf{I}$, where $\mathbf{I}$ is the 2×2 identity matrix,

$$\begin{aligned} \mathbf{K}_s \mathbf{r}_s (\mathbf{K}_s)^{-1} &= r_s \mathbf{K}_s \mathbf{I} (\mathbf{K}_s)^{-1} \\ &= \mathbf{r}_s \end{aligned} \tag{5.4-5}$$

Thus, as viewed from the arbitrary reference frame

$$\mathbf{v}_{qds} = \mathbf{r}_s \mathbf{i}_{qds} \tag{5.4-6}$$

Therefore, if the phase resistances are equal, Ohm's law is viewed the same regardless of the frame of reference. This, however, is not the case if the phase resistances are unequal, for example, let

$$\mathbf{r}_s = \begin{bmatrix} r_a & 0 \\ 0 & r_b \end{bmatrix} \tag{5.4-7}$$

where $r_a \neq r_b$. Evaluating the right-hand side of (5.4-4)

$$\mathbf{K}_s \mathbf{r}_s (\mathbf{K}_s)^{-1} \mathbf{i}_{qds} = \begin{bmatrix} \cos\theta & \sin\theta \\ \sin\theta & -\cos\theta \end{bmatrix} \begin{bmatrix} r_a & 0 \\ 0 & r_b \end{bmatrix} \begin{bmatrix} \cos\theta & \sin\theta \\ \sin\theta & -\cos\theta \end{bmatrix} \begin{bmatrix} i_{qs} \\ i_{ds} \end{bmatrix} \tag{5.4-8}$$

After some work, we can express $\mathbf{v}_{qds}$ as

$$\begin{bmatrix} v_{qs} \\ v_{ds} \end{bmatrix} = \begin{bmatrix} \dfrac{r_a + r_b}{2} + \dfrac{r_a - r_b}{2}\cos 2\theta & \dfrac{r_a - r_b}{2}\sin 2\theta \\ \dfrac{r_a - r_b}{2}\sin 2\theta & \dfrac{r_a + r_b}{2} - \dfrac{r_a - r_b}{2}\cos 2\theta \end{bmatrix} \begin{bmatrix} i_{qs} \\ i_{ds} \end{bmatrix} \tag{5.4-9}$$

Unfortunately, the change of variables has complicated the situation beyond repair. If, however, $\omega = 0$ for the stationary reference frame, then the resistance matrix in (5.4-9) is independent of time and if $\theta(0)$, which is the time-zero position of the reference frame, is selected to be zero, which is usually the case, then the resistance matrix becomes (5.4-7). This says that if the circuits are unsymmetrical, the simplest formulation of the circuit equations occurs when the reference frame is selected where the asymmetry exists.

Let us now consider two-phase inductive circuits where r_s is neglected. In this case, the voltage equations become

$$\mathbf{v}_{abs} = p\boldsymbol{\lambda}_{abs} \tag{5.4-10}$$

For a magnetically linear system, one is tempted to express $\boldsymbol{\lambda}_{abs}$ in terms of inductances and currents; however, it is more convenient at this point in the derivation to transform flux linkages. Thus,

$$\mathbf{v}_{qds} = \mathbf{K}_s p\boldsymbol{\lambda}_{abs} \tag{5.4-11}$$

In terms of qs and ds variables, (5.4-11) becomes

$$\mathbf{v}_{qds} = \mathbf{K}_s p\left[(\mathbf{K}_s)^{-1}\boldsymbol{\lambda}_{qds}\right] \tag{5.4-12}$$

using the chain rule

$$\mathbf{v}_{qds} = \mathbf{K}_s p\left[(\mathbf{K}_s)^{-1}\right]\boldsymbol{\lambda}_{qds} + \mathbf{K}_s(\mathbf{K}_s)^{-1}p\left[\boldsymbol{\lambda}_{qds}\right] \tag{5.4-13}$$

The last term on the right-hand side of (5.4-13) is $p\boldsymbol{\lambda}_{qds}$. The first term may be written as

$$\begin{aligned}\mathbf{K}_s p\left[(\mathbf{K}_s)^{-1}\right]\boldsymbol{\lambda}_{qds} &= \begin{bmatrix} \cos\theta & \sin\theta \\ \sin\theta & -\cos\theta \end{bmatrix}\omega\begin{bmatrix} -\sin\theta & \cos\theta \\ \cos\theta & \sin\theta \end{bmatrix}\begin{bmatrix} \lambda_{qs} \\ \lambda_{ds} \end{bmatrix} \\ &= \omega\begin{bmatrix} 0 & 1 \\ -1 & 0 \end{bmatrix}\begin{bmatrix} \lambda_{qs} \\ \lambda_{ds} \end{bmatrix}\end{aligned} \tag{5.4-14}$$

Therefore, the matrix voltage equations of the two-phase inductive circuit in the arbitrary reference frame become

$$\mathbf{v}_{qds} = \omega\begin{bmatrix} \lambda_{ds} \\ -\lambda_{qs} \end{bmatrix} + p\boldsymbol{\lambda}_{qds} \tag{5.4-15}$$

The first term on the right-hand side of (5.4-15) is often written as $\omega\boldsymbol{\lambda}_{dqs}$. In expanded form, (5.4-15) becomes

$$v_{qs} = \omega\lambda_{ds} + p\lambda_{qs} \tag{5.4-16}$$

$$v_{ds} = -\omega\lambda_{qs} + p\lambda_{ds} \tag{5.4-17}$$

We have yet to illustrate that the transformation eliminates position-varying $\mathbf{L}_s$. Therefore, we have not illustrated the reason for Park's transformation; however, there is a simple example that will help demonstrate this feature of the transformation. For this purpose let us consider the two-pole two-phase reluctance machine shown in Fig. 5.4-1.

The self-inductance matrix of the stator windings of the two-phase reluctance machine shown in Fig. 5.4-1 is

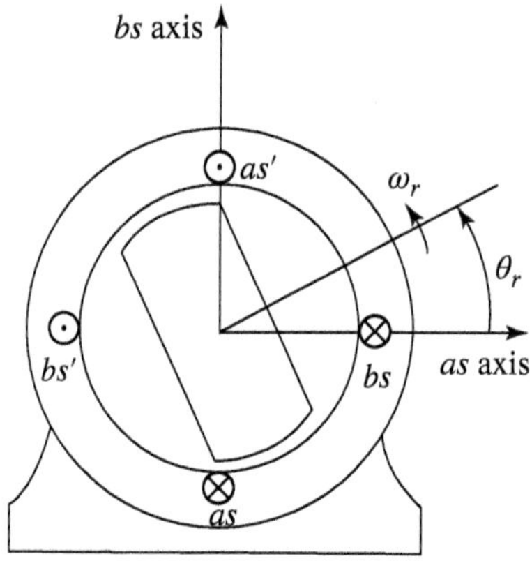

Figure 5.4-1 Elementary two-pole two-phase reluctance machine.

$$\mathbf{L}_s = \begin{bmatrix} L_l + L_A - L_B \cos 2\theta_r & -L_B \sin 2\theta_r \\ -L_B \sin 2\theta_r & L_l + L_A + L_B \cos 2\theta_r \end{bmatrix} \tag{5.4-18}$$

where θ_r is the rotor angular position. The self-inductances follow that given by (2.7-3); however, due to the salient-pole rotor, there is a mutual coupling between the *as* and *bs* windings which is given by the off-diagonal elements in (5.4-18). This occurs due to the low reluctance path of the rotor channeling flux caused by one winding through the turns of the other as it rotates. It is not necessary to follow this explanation of (5.4-18); we will cover it in more detail later. The purpose here is to illustrate the facility of the change of variables. If (5.4-18) is substituted into (5.4-19) and if we set θ in $\mathbf{K_s}$ equal to θ_r and $\omega = \omega_r$, we are using the reference frame rotating at the angular velocity of the rotor. It can be shown that

$$\begin{aligned} \boldsymbol{\lambda}_{qds}^r &= \mathbf{K}_s^r \mathbf{L}_s (\mathbf{K}_s^r)^{-1} \mathbf{i}_{qds}^r \\ &= \begin{bmatrix} L_l + L_A - L_B & 0 \\ 0 & L_l + L_A + L_B \end{bmatrix} \begin{bmatrix} i_{qs}^r \\ i_{ds}^r \end{bmatrix} \end{aligned} \tag{5.4-19}$$

where $\mathbf{K}_s^r$ denotes the transformation to the rotor reference frame. Equation (5.4-19) becomes

$$\lambda_{qs}^r = (L_{ls} + L_A - L_B) i_{qs}^r \tag{5.4-20}$$

$$\lambda_{ds}^r = (L_{ls} + L_A + L_B) i_{ds}^r \tag{5.4-21}$$

where the superscript r denotes variables in the rotor reference frame and $(L_{ls} + L_A - L_B)$ and $(L_{ls} + L_A + L_B)$ are generally referred to as the q-axis inductance and the d-axis inductance, respectively. We feel much better, because the change of variables to the rotor reference frame has eliminated the dependency of the inductances upon rotor position and $p\lambda_{qs}^r$ and $p\lambda_{ds}^r$ become zero for balanced steady-state conditions where i_{qs}^r and i_{ds}^r are constant. We now have hope for the transformation. Moreover, the *qs* and *ds* fictitious circuits are not coupled magnetically. We will talk more about all this in Chapter 7 when we analyze the synchronous machine. Be careful, this elimination of the position-varying inductances occurs only when $\theta = \theta_r$.

It is interesting to mention in passing that Park used ω_r in the speed voltage terms in (5.4-16) and (5.4-17) since this was the only reference frame wherein the position-varying inductances were eliminated. Although he was correct in doing so, reference frame theory had not been set forth and the speed voltages became misinterpreted as voltages induced due to the rotor speed of the machine instead of the reference frame speed. This, of course, is understandable since the speeds are the same if, but only if, $\omega = \omega_r$.

Recall from (5.4-9) that when the resistances of the circuits are unequal, the transformation does not provide an advantage and it is best to perform the analysis in the frame of reference where the parameter asymmetry exists. Now, we see that when a magnetic asymmetry exists, the frame of reference to use for purposes of machine analysis is the reference frame where the magnetic asymmetry exists. These are important findings.

Example 5B Two-Phase Stator Circuits in the Arbitrary Reference Frame Consider the device given in Fig. 5.3-1. (a) Write the voltage equations in *as* and *bs* variables. (b) Write the voltage equations in arbitrary reference frame variables.

For part (a):

$$v_{as} = r_s i_{as} + p\lambda_{as} \tag{5B-1}$$

$$v_{bs} = r_s i_{bs} + p\lambda_{bs} \tag{5B-2}$$

where r_s is the resistance of each winding and

$$\lambda_{as} = L_{ss} i_{as} \tag{5B-3}$$

$$\lambda_{bs} = L_{ss} i_{bs} \tag{5B-4}$$

where L_{ss} is the sum of the leakage inductance L_{ls} and magnetizing inductance L_{ms}.

For part (b):

From (5.4-6) and (5.4-15) the voltage equations for the fictitious circuits in arbitrary reference frame variables may be written as

$$v_{qs} = r_s i_{qs} + \omega\lambda_{ds} + p\lambda_{qs} \tag{5B-5}$$

$$v_{ds} = r_s i_{ds} - \omega\lambda_{qs} + p\lambda_{ds} \tag{5B-6}$$

From (5B-3) and (5B-4) and since

$$\boldsymbol{\lambda}_{qds} = \mathbf{K}_s \boldsymbol{\lambda}_{abs} \tag{5B-7}$$

$$\lambda_{qs} = L_{ss} i_{qs} \tag{5B-8}$$

$$\lambda_{ds} = L_{ss} i_{ds} \tag{5B-9}$$

Substituting (5B-8) and (5B-9) into (5B-5) and (5B-6) yields the arbitrary reference frame voltage equations for the device shown in Fig. 5.3-1. In particular,

$$v_{qs} = r_s i_{qs} + \omega L_{ss} i_{ds} + L_{ss} p i_{qs} \tag{5B-10}$$

$$v_{ds} = r_s i_{ds} - \omega L_{ss} i_{qs} + L_{ss} p i_{ds} \tag{5B-11}$$

SP5.4-1 Consider Fig. 5.3-1. Write the voltage equations in the stationary reference frame. $\left[v_{qs}^s = r_s i_{qs}^s + L_{ss} p i_{qs}^s;\ v_{ds}^s = r_s i_{ds}^s + L_{ss} p i_{ds}^s\right]$

SP5.4-2 Draw the equivalent qs and ds circuits in the arbitrary reference frame. [(5B-10) and (5B-11)]

5.5 BALANCED STEADY-STATE STATOR VARIABLES VIEWED FROM ANY REFERENCE FRAME

A balanced two-phase set may be expressed as

$$f_{as} = \sqrt{2} f_s \cos \theta_{esf} \tag{5.5-1}$$

$$f_{bs} = \sqrt{2} f_s \sin \theta_{esf} \tag{5.5-2}$$

The above equations form an ab-sequence, balanced, two-phase set where f_{as} and f_{bs} may be voltages, currents, or flux linkages and f_s may be a function of time. For steady-state conditions

$$F_{as} = \sqrt{2} F_s \cos \left[\omega_e t + \theta_{esf}(0)\right] \tag{5.5-3}$$

$$F_{bs} = \sqrt{2} F_s \sin \left[\omega_e t + \theta_{esf}(0)\right] \tag{5.5-4}$$

where F_s, ω_e, and $\theta_{esf}(0)$ are all constants. The subscript f is used to denote the zero displacement of the voltage $\theta_{esv}(0)$, current $\theta_{esi}(0)$, and flux linkage $\theta_{es\lambda}(0)$. In phasor form

$$\tilde{F}_{as} = F_s \underline{/\theta_{esf}(0)} \tag{5.5-5}$$

$$\begin{aligned} \tilde{F}_{bs} &= F_s \underline{/\theta_{esf}(0) - \frac{\pi}{2}} \\ &= -j\tilde{F}_{as} \end{aligned} \tag{5.5-6}$$

For balanced steady-state conditions it is not necessary to consider both the as and bs phasors since the phasors of the two-phase set are orthogonal.

Transforming the balanced set given by (5.5-1) and (5.5-2) to the arbitrary reference frame using (5.3-9) yields

$$f_{qs} = \sqrt{2} f_s \cos\left(\theta_{esf} - \theta\right) \tag{5.5-7}$$

$$f_{ds} = -\sqrt{2}f_s \sin\left(\theta_{esf} - \theta\right) \quad (5.5\text{-}8)$$

Equations (5.5-7) and (5.5-8) are how (5.5-1) and (5.5-2), the sinusoidal physical (real-life) variables, are viewed from the arbitrary reference frame.

Asynchronously Rotating Reference Frames

For balanced steady-state conditions $\theta_{esf} = \omega_e t + \theta_{esf}(0)$ and since $\theta = \omega t + \theta(0)$, in the steady state, (5.5-7) and (5.5-8) become

$$F_{qs} = \sqrt{2}\,F_s \cos\left[(\omega_e - \omega)t + \theta_{esf}(0) - \theta(0)\right] \quad (5.5\text{-}9)$$

$$F_{ds} = -\sqrt{2}\,F_s \sin\left[(\omega_e - \omega)t + \theta_{esf}(0) - \theta(0)\right] \quad (5.5\text{-}10)$$

We now understand that the change of variables only changes the frequency of the balanced set. It does not change the amplitude of the balanced set; it only changes the speed at which the magnetic field is rotating relative to us. This makes sense, since the transformation is nothing more than changing the speed, we are rotating when we view Tesla's rotating magnetic field.

An asynchronously rotating reference frame occurs when $\omega \neq \omega_e$. When $\omega < \omega_e$, $(\omega_e - \omega)$ is positive and the mmf$_s$ is moving counterclockwise for $\widetilde{I}_{as} = j\widetilde{I}_{bs}$. Now, when $\omega > \omega_e$, $(\omega_e - \omega)$ becomes negative and the mmf$_s$ is moving clockwise for $\widetilde{I}_{as} = j\widetilde{I}_{bs}$; however, with $(\omega_e - \omega)$ negative, $e^{j(\omega_e - \omega)}$ is rotating clockwise and phasors are being rotated clockwise rather than counterclockwise which occurs when $(\omega_e - \omega)$ is positive, i.e., when $\omega < \omega_e$. In this case, because the sign of $\theta_{esi}(0)$ does not change, current that appears lagging when viewed from a reference frame where $\omega < \omega_e$ becomes leading when viewed from a reference frame where $\omega > \omega_e$.

We will deal exclusively with $\omega < \omega_e$ other than to show that, in a developed diagram, the mmf$_s$ moves from left to right for $\omega > \omega_e$. So, for $\omega < \omega_e$ and $\theta(0) = 0$, which is generally the case, we can write,

$$\widetilde{F}_{as} = \widetilde{F}_{qs} \quad (5.5\text{-}11)$$

$$\widetilde{F}_{bs} = -\widetilde{F}_{ds} \quad (5.5\text{-}12)$$

Therefore, the impedance is the same for all reference frames with $\omega < \omega_e$. This does not seem correct. We see from (5.5-9) and (5.5-10) that the frequency of the substitute variables changes as we move from one reference frame to another. Why, then, does not the reactive component of the impedance change for $\omega < \omega_e$? Well, we are viewing the physical or actual stator variables from different reference frames wherein we have formulated fictitious circuits and substitute variables. The phase relation between the steady-state phasors is not going to change depending on how fast we are running as we view these variables. In other words, let us consider the steady-state variables given by (5.5-9) and (5.5-10) for $\omega < \omega_e$. The quantities given by (5.5-11) and (5.5-12) are actually rotating at $(\omega_e - \omega)$

and when we run with them, we view the quantities as being stationary; however, regardless of how fast we run, our view of the phase relation between phasors ($\tilde{V}$ and $\tilde{I}$, for example) will not change for $\omega < \omega_e$. This phase angle is determined entirely by the physical variables. Clearly, fictitious windings and substitute variables are not independent circuits and variables; rather, they are linked to their corresponding physical counter parts.

The direction of rotation of mmf, as viewed from a reference frame rotating at an angular velocity counterclockwise greater than ω_e, appears to be rotating clockwise at $|\omega_e - \omega|$ relative to us. For $\omega < \omega_e$, mmf$_s$ appears to be rotating counterclockwise at $|\omega_e - \omega|$. We wonder what causes this to happen. Well, as we have said, the reference frame variables (substitute variables) associated with the fictitious windings will provide us the correct view of mmfs that we would see from a given frame of reference. Let us see if this is the case. When $\omega < \omega_e$, F_{qs} (5.5-9) is a positive cosine and F_{ds} (5.5-10) is a negative sine wave. In this case, mmf$_s$ appears to be rotating counterclockwise. Check this by assuming I_{qs} is a positive cosine and I_{ds} is a negative sine in Fig. 5.3-1. We see that the mmf$_s$ is rotating counterclockwise. Now, when $\omega > \omega_e$, I_{qs} is still a cosine but now I_{ds} is a positive sine wave. Using Fig. 5.3-1 again, we see that mmf$_s$ is moving clockwise relative to us when we are moving faster counterclockwise than ω_e. Equation (4.4-8) tells us that the mmf$_s$ is rotating clockwise; however, now we see that the qs and ds variables provide the appropriate view of Tesla's rotating magnetic field for $\omega < \omega_e$ and $\omega > \omega_e$.

Synchronously Rotating Reference Frame

In the synchronous reference frame $\omega = \omega_e$ and with $\theta(0) = 0$, $\theta = \omega_e t$. For steady-state operation, (5.5-9) and (5.5-10) become

$$F_{qs}^e = \sqrt{2}\, F_s \cos \theta_{esf}(0) \tag{5.5-13}$$

$$F_{ds}^e = -\sqrt{2}\, F_s \sin \theta_{esf}(0) \tag{5.5-14}$$

The raised index e signifies variables in the synchronously reference frame.

The as-phasor given by (5.5-5) can be written as

$$\tilde{F}_{as} = F_s \cos \theta_{esf}(0) + jF_s \sin \theta_{esf}(0) \tag{5.5-15}$$

which can be expressed in terms of F_{qs}^e and F_{ds}^e as

$$\sqrt{2}\,\tilde{F}_{as} = F_{qs}^e - j\,F_{ds}^e \tag{5.5-16}$$

We must be careful here; F_{qs}^e and F_{ds}^e are constants in the steady state. Clearly, $\tilde{F}_{as}$ is a complex number and if $\omega = \omega_e$ and $\theta(0) = 0$, then $\tilde{F}_{as}$ can be expressed by (5.5-16). This is summarized in Fig. 5.5-1 where we have shown the phasors of F_{qs} and F_{ds} and their relationship for $\omega < \omega_e$ and $\omega > \omega_e$. The rotating magnetic field viewed from any reference frame appears to be rotating towards synchronous speed.

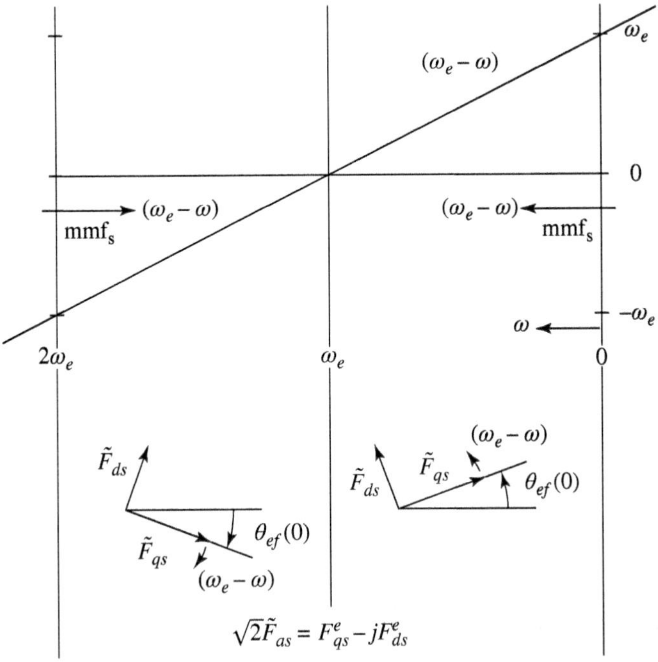

Figure 5.5-1 Direction of rotation of mmf_s and the phase relation between $\widetilde{F}_{qs}$ and $\widetilde{F}_{ds}$ as viewed from the arbitrary reference frame when $\omega < \omega_e$ and $\omega > \omega_e$ for balanced steady-state conditions. For winding axes shown in Fig. 5.3-1 and $\widetilde{I}_{as} = j\widetilde{I}_{bs}$.

For steady-state balanced conditions the qs and ds variables in the synchronously rotating reference frame are constants; therefore, since pi_{qs}^e and pi_{ds}^e are zero and with $\omega = \omega_e$, (5B-10) and (5B-11) become

$$V_{qs}^e = r_s I_{qs}^e + \omega_e L_{ss} I_{ds}^e \tag{5.5-17}$$

$$V_{ds}^e = r_s I_{ds}^e - \omega_e L_{ss} I_{qs}^e \tag{5.5-18}$$

From (5.5-17)

$$\begin{aligned} \widetilde{V}_{as} &= \frac{1}{\sqrt{2}}\left(V_{qs}^e - jV_{ds}^e\right) \\ &= (r_s + j\omega_e L_{ss})\widetilde{I}_{as} \end{aligned} \tag{5.5-19}$$

We will talk more about phasors when we introduce the instantaneous phasor in Section 5.7.

SP5.5-1 If $\widetilde{V}_{as} = r_s\widetilde{I}_{as}$ and $\widetilde{F}_{as} = j\widetilde{F}_{bs}$, express (a) $\widetilde{V}_{qs}$, (b) $\widetilde{V}_{ds}$, (c) V_{qs}^e, and (d) V_{ds}^e. Are there any restrictions on your answer for (a) and (b)? [(a) $\widetilde{V}_{qs} = r_s\widetilde{I}_{qs}$, (b) $\widetilde{V}_{ds} = r_s\widetilde{I}_{ds}$, (c) $V_{qs}^e = r_s I_{qs}^e$, (d) $V_{ds}^e = r_s I_{ds}^e$; $\theta(0) = 0$ and $\omega \neq \omega_e$]

SP5.5-2 Express F_{qs}^e and F_{ds}^e for balanced steady-state conditions if $\theta(0) = \frac{\pi}{2}$. $\left[F_{qs}^e = \sqrt{2}F_s \cos\left(\theta_{esf}(0) - \frac{\pi}{2}\right), F_{ds}^e = -\sqrt{2}F_s \sin\left(\theta_{esf}(0) - \frac{\pi}{2}\right)\right]$

SP5.5-3 Express $\widetilde{F}_{as}$ in terms of F_{qs}^e and F_{ds}^e in SP5.5-2 if $F_{as} = \sqrt{2}F_s \cos\left[\omega_e t + \theta_{esf}(0)\right]$. $[\sqrt{2}\widetilde{F}_{as} = F_{ds}^e + jF_{qs}^e]$.

5.6 STATOR VARIABLES OBSERVED FROM DIFFERENT REFERENCE FRAMES

It is instructive to observe the waveform of the variables of a stationary, two-phase, series rL circuit from commonly used reference frames and the arbitrary reference frame. For this purpose we will assume that both $\mathbf{r}_s$ and $\mathbf{L}_s$ are diagonal matrices each with equal nonzero elements and the voltages applied at time zero are

$$V_{as} = \sqrt{2}V_s \cos \omega_e t \tag{5.6-1}$$

$$V_{bs} = \sqrt{2}V_s \sin \omega_e t \tag{5.6-2}$$

where ω_e is the angular velocity and $\theta_{esv}(0) = 0$. Assuming that the initial currents are zero at time zero, the instantaneous currents may be expressed as

$$i_{as} = \frac{\sqrt{2}V_s}{|Z_s|}\left\{-e^{-\frac{t}{\tau}}\cos\theta_{esi}(0) + \cos\left[\omega_e t - \theta_{esi}(0)\right]\right\} \tag{5.6-3}$$

$$i_{bs} = \frac{\sqrt{2}V_s}{|Z_s|}\left\{-e^{-\frac{t}{\tau}}\sin\theta_{esi}(0) + \sin\left[\omega_e t - \theta_{esi}(0)\right]\right\} \tag{5.6-4}$$

where

$$Z_s = r_s + j\omega_e L_{ss} \tag{5.6-5}$$

$$\tau = \frac{L_{ss}}{r_s} \tag{5.6-6}$$

$$\theta_{esi}(0) = \tan^{-1}\frac{\omega_e L_{ss}}{r_s} \tag{5.6-7}$$

Once a solution is known in one reference frame, it is known in all reference frames. It is only necessary to transform the currents to the arbitrary reference frame. In this case

$$\begin{bmatrix} i_{qs} \\ i_{ds} \end{bmatrix} = \begin{bmatrix} \cos\theta & \sin\theta \\ \sin\theta & -\cos\theta \end{bmatrix}\begin{bmatrix} i_{as} \\ i_{bs} \end{bmatrix} \tag{5.6-8}$$

$$\theta = \omega t + \theta(0) \tag{5.6-9}$$

Substituting (5.6-3) and (5.6-4) into (5.6-8) yields

$$i_{qs} = \frac{\sqrt{2}V_s}{|Z_s|}\left\{-e^{-\frac{t}{\tau}}\cos\left[\omega t - \theta_{esi}(0)\right] + \cos\left[(\omega_e - \omega)t - \theta_{esi}(0)\right]\right\} \tag{5.6-10}$$

$$i_{ds} = \frac{\sqrt{2}V_s}{|Z_s|}\left\{-e^{-\frac{t}{\tau}}\sin\left[\omega t - \theta_{esi}(0)\right] - \sin\left[(\omega_e - \omega)t - \theta_{esi}(0)\right]\right\} \qquad (5.6\text{-}11)$$

Clearly, the state of the electric system is independent of the frame of reference from which it is observed. Although the variables will appear differently in each reference frame, they will exhibit the same mode of operation (transient or steady state) regardless of the reference frame. Except for $\omega = 0$ and $\omega = \omega_e$, (5.6-10) and (5.6-11) contain two balanced sets during balanced operation. When $\omega = 0$, the transient mmf$_s^s$ is exponential decay; when $\omega = \omega_e$, the transient mmf$_s^e$ is a sinusoidal decay. We are already familiar with the steady-state response that is represented by the last term in (5.6-10) and (5.6-11); however, we have not considered the transient response before.

We have talked about the fact that the phase between the voltage and current is not altered by the reference frame from which the steady-state substitute variables are being viewed. This is apparent from the last term of (5.6-10) and (5.6-11) that the phase angle of the current, $\theta_{esi}(0)$, is the same in all asynchronous reference frames. In the case for the transient response as illustrated by the first term of (5.6-10) and (5.6-11), the frequency of the decaying sinusoidal variations corresponds to the electrical angular velocity of the reference frame.

The waveforms of the system variables observed from various reference frames are shown in Figs. 5.6-1 through 5.6-4. The voltages of the form given by (5.6-1) and (5.6-2) are applied to the two-phase series rL circuit with $V_s = 10/\sqrt{2}\,V$, $r_s = 0.216\,\Omega$, and $\omega_e L_{ss} = 1.09\,\Omega$ with $\omega_e = 377$ rad/s. The response, for $t > 0$, of the electric variables in the stationary reference frame is shown in Fig. 5.6-1. Since we have selected $\theta(0) = 0$, $f_{as} = f_{qs}^s$ and the plots of v_{qs}^s and i_{qs}^s are v_{as} and i_{as}, respectively. The raised s is used to denote qs and ds variables in the stationary reference frame. Note the offset present in $i_{ds}^s(-i_{bs})$ and to a lesser extent in $i_{qs}^s(i_{as})$.

The variables for the same mode of operation are shown in the synchronously rotating reference frame in Fig. 5.6-2. We have selected the time-zero position of the two-phase voltages to be zero; thus, from (5.5-7) and (5.5-8) with $\theta(0) = 0$, $v_{qs}^e = 10\text{V}$, and $v_{ds}^e = 0$. The raised e is used to denote variables in the synchronously rotating reference frame.

There are several features worthy of note. The waveform of the instantaneous electric power is the same in all reference frames; power is invariant. This was shown by (5.3-14) through (5.3-16). We are not changing the amplitude of the variables by the change of variables. The electric transient is very evident in the waveforms of the instantaneous electric power and the currents in the synchronously rotating reference frame in Fig. 5.6-2.

In Fig. 5.6-3, the voltages were applied and the solution of the differential equations observed in the reference frame rotating counterclockwise at $2\omega_e$. Therein, the mmf$_s$ is moving from left to right and the current leads the voltage. In Fig. 5.6-4, we first view the steady-state variables from the stationary reference frame ($\omega = 0$). The reference frame is stepped from $\omega = 0$ to $\omega = 2\omega_e$. When switching from one reference frame to another, the variables must be continuous.

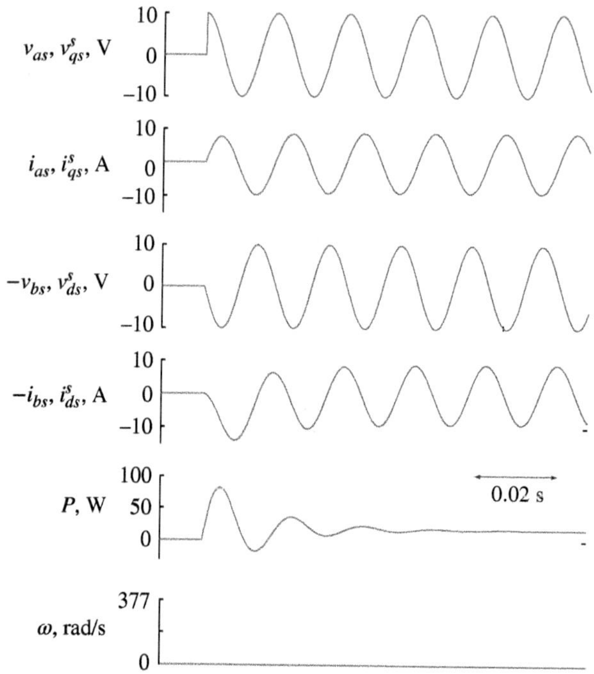

Figure 5.6-1 Variables of a stationary two-phase system viewed from the stationary reference frame.

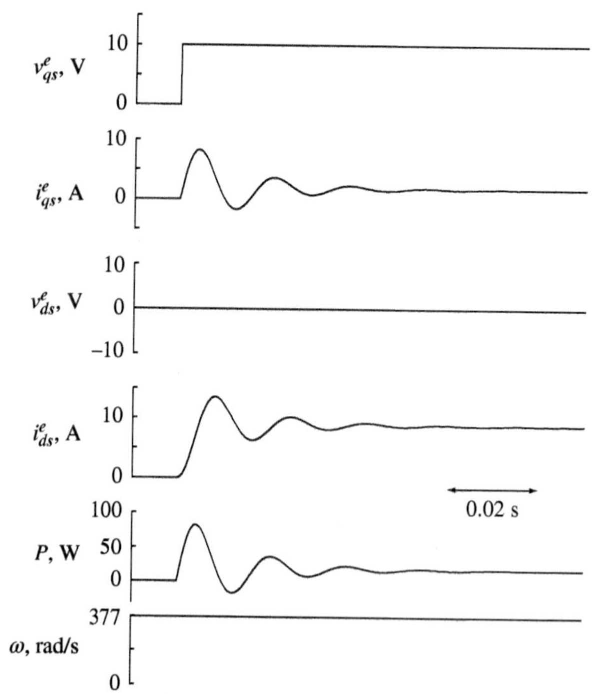

Figure 5.6-2 Variables of a stationary two-phase system viewed from the synchronous reference frame.

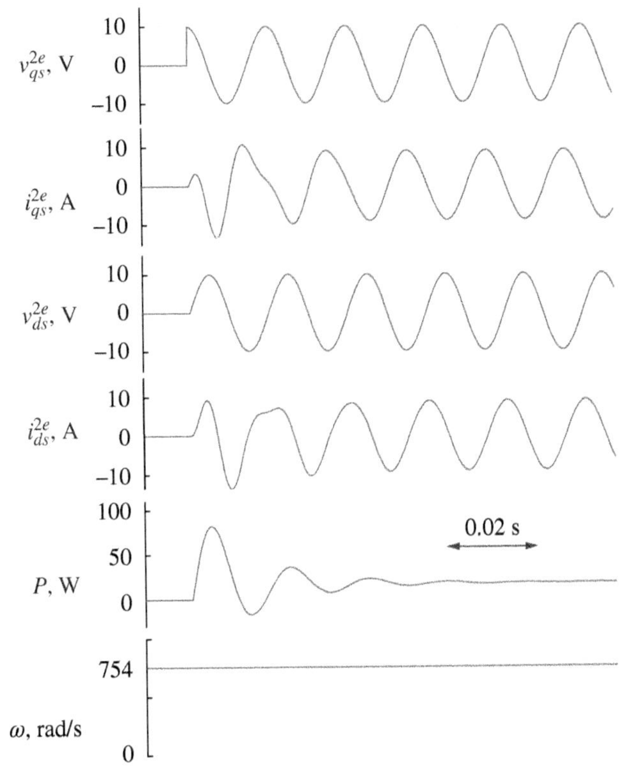

Figure 5.6-3 Variables of a stationary two-phase system viewed from a reference frame rotating at $2\omega_e$ counterclockwise.

A superscript is not attached to the variables since two reference frames are involved. One may question the correctness of Fig. 5.6-4 since $\cos \omega_e t = \cos(-\omega_e t)$; however, we are plotting versus $+\omega_e t$ and not $-\omega_e t$.

SP5.6-1 In Fig. 5.6-1, *sqs* and *sds* variables are plotted and also labeled *as* and −*bs* variables. What restrictions are necessary for this correspondence? [$\theta(0) = 0$, $\omega = 0$]

SP5.6-2 Why is $v_{ds}^e = 0$ in Fig. 5.6-2? [$\theta_{esv}(0) = 0$, $\theta(0) = 0$]

SP5.6-3 The power pulsates negatively during the transient period. Why? How is this portrayed in Fig. 5.6-2? [Energy returns to source from inductance; i_{qs}^e pulsates negatively]

SP5.6-4 In Fig. 5.6-4 there is a smooth transition in the variables between the $\omega = 0$ and $\omega = 2\omega_e$ variables. Why? [The *q* or *d* variables must be continuous regardless of the time of switching from one reference frame to another]

SP5.6-5 Determine the frequency and direction of rotation of the steady-state mmf$_s$ viewed from the reference frames in Fig. 5.6-4. [ω_e ccw, ω_e cw]

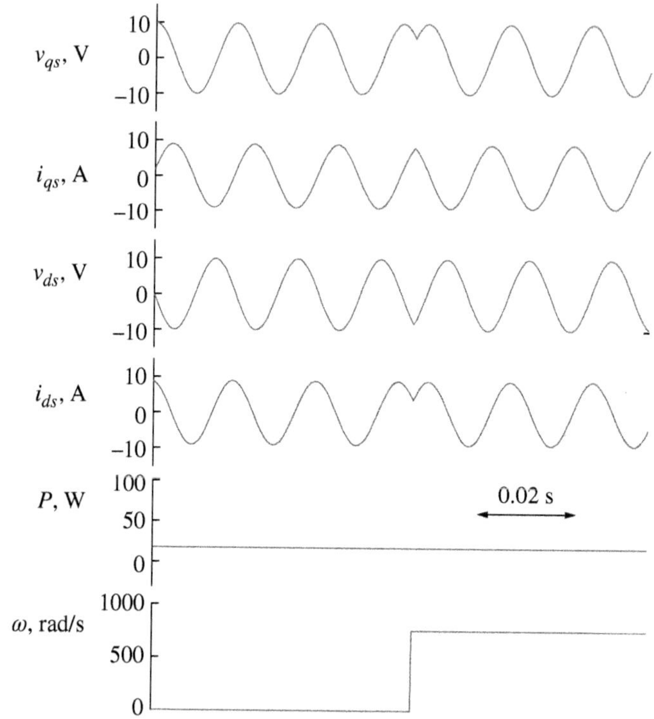

Figure 5.6-4 Variables of a stationary two-phase system viewed first from the stationary reference frame ($\omega = 0$), then switched to $\omega = 2\omega_e$.

5.7 INSTANTANEOUS PHASOR

The phasor was introduced in Section 1.2 for the purpose of analyzing the steady-state performance of an *RLC* circuit. Now, we are focusing on two- and three-phase symmetrical stator circuits and in Section 5.5 we viewed a balanced set of two-phase steady-state variables in all reference frames. This revealed that for balanced steady-state conditions, the synchronously rotating reference frame offered a means of relating q and d quantities to the phasor of the actual phase variables. In particular, from (5.5-16) with $\theta(0) = 0$,

$$\sqrt{2}\tilde{F}_{as} = F^e_{qs} - jF^e_{ds} \tag{5.7-1}$$

In Park's original paper [3], he likened the orthogonal q and d axes to a complex plane and expressed the instantaneous q and d variables in complex form similar to (5.7-1). In the synchronous reference frame, this expression would be

$$\tilde{f}_{as}(t) = f^e_{qs}(t) - jf^e_{ds}(t) \tag{5.7-2}$$

where $\widetilde{f}_{as}(t)$ is the *instantaneous phasor* of the *as*-phase voltage, current, or flux linkage and since it is in the synchronous reference frame, it is rotating at ω_e counterclockwise as a phasor.

From Section 1.2, a steady-state sinusoidal variable was expressed as

$$F_a = F_p \cos \theta_{esf} \tag{5.7-3}$$

which can be written as

$$F_a = \text{Re}\left[\sqrt{2}\widetilde{F}_a e^{j\omega_e t}\right] \tag{5.7-4}$$

where $\widetilde{F}_a$ is a constant complex expression that is related to steady-state q_s^e and d_s^e quantities, F_{qs}^e and F_{ds}^e, by (5.7-1).

Since $f_{qs}^e(t)$ and $f_{ds}^e(t)$ are rotating at ω_e, we can write $f_{as}(t)$ in a form similar to (5.7-4) as

$$\begin{aligned} f_{as}(t) &= \text{Re}\left[\widetilde{f}_{as}(t)e^{j\omega_e t}\right] \\ &= \text{Re}\left\{\left[f_{qs}^e(t) - jf_{ds}^e(t)\right]e^{j\omega_e t}\right\} \end{aligned} \tag{5.7-5}$$

We have referred to $\widetilde{f}_{as}(t)$ as an instantaneous phasor and since we are dealing with instantaneous quantities, the $\sqrt{2}$ is not included as it was in (5.7-4) when using the rms values of steady-state sinusoidal variables.

The q- and d-axis voltage equations for a series rL circuit are given in the arbitrary reference frame by (5B-10) and (5B-11). In the synchronous reference frame, these equations become

$$v_{qs}^e = r_s i_{qs}^e + \omega_e L_{ss} i_{ds}^e + L_{ss} p i_{qs}^e \tag{5.7-6}$$

$$v_{ds}^e = r_s i_{ds}^e - \omega_e L_{ss} i_{qs}^e + L_{ss} p i_{ds}^e \tag{5.7-7}$$

Substituting (5.7-6) and (5.7-7) into (5.7-2), we can write

$$v_{qs}^e - jv_{ds}^e = r_s\left(i_{qs}^e - ji_{ds}^e\right) + \omega_e L_{ss}\left(i_{ds}^e + ji_{qs}^e\right) + L_{ss}\left(pi_{qs}^e - jpi_{ds}^e\right) \tag{5.7-8}$$

We have dropped the functional notation, i.e., the dependence on t, for compactness. Equation (5.7-8) may be written in instantaneous phasor form as

$$\widetilde{v}_{as} = r_s \widetilde{i}_{as} + j\omega_e L_{ss} \widetilde{i}_{as} + L_{ss} p \widetilde{i}_{as} \tag{5.7-9}$$

Equations (5.7-8) and (5.7-9) are complex visualization of v_{as} in the synchronously rotating reference frame. The q_s^e and d_s^e variables shown in Fig. 5.7-1 are portrayed on the complex plane and in the steady state the last term on the right-hand side of (5.7-8) or (5.7-9) becomes zero and $v_{qs}^e - jv_{ds}^e$ becomes $\sqrt{2}\widetilde{V}_{as}$. The mmf$_s^e$ for Fig. 5.7-1 are shown in Fig. 5.7-2.

In Figs. 5.7-1 and 5.7-2, we have identified three different methods of portraying the performance of electromechanical devices: the plots of the qs and ds

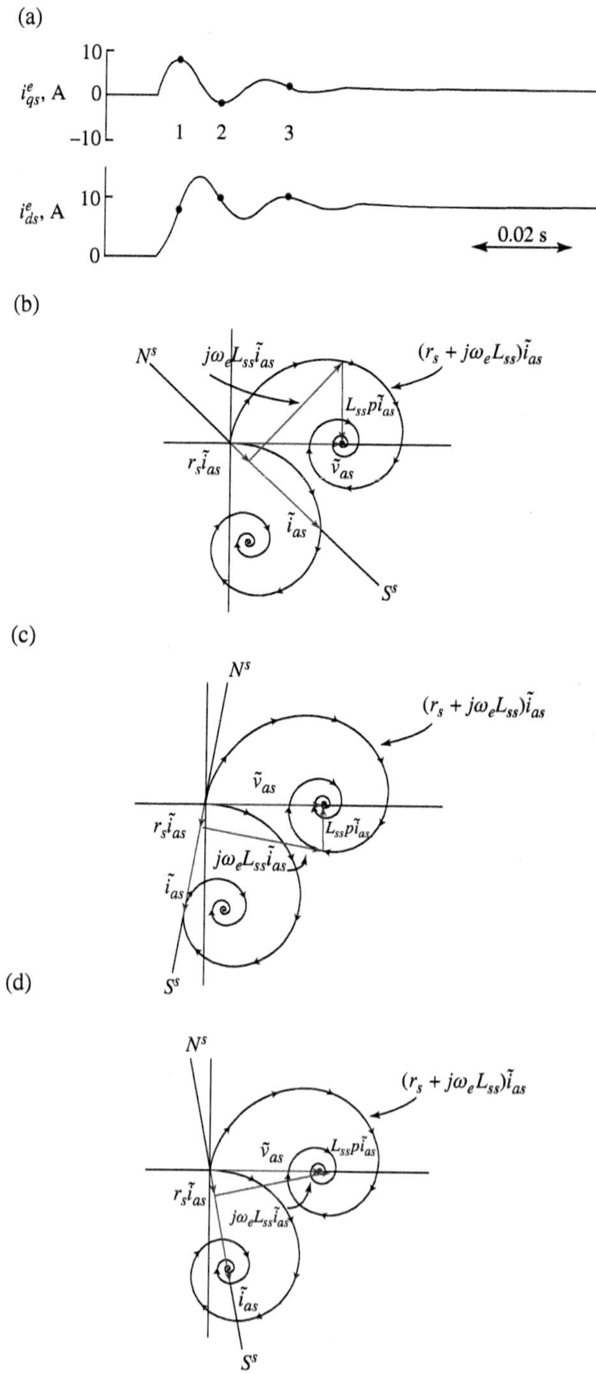

Figure 5.7-1 Plots of $(r_s + j\omega_e L_{ss})\tilde{i}_{as}$ and $\tilde{i}_{as}$ with instantaneous phasors at Points 1, 2, and 3 superimposed. (a) Plots from Fig. 5.6-2 showing Points 1, 2, and 3. (b) Plot of $(r_s + j\omega_e L_{ss})\tilde{i}_{as}$ and $\tilde{i}_{as}$ for Point 1. (c) For Point 2. (d) For Point 3.

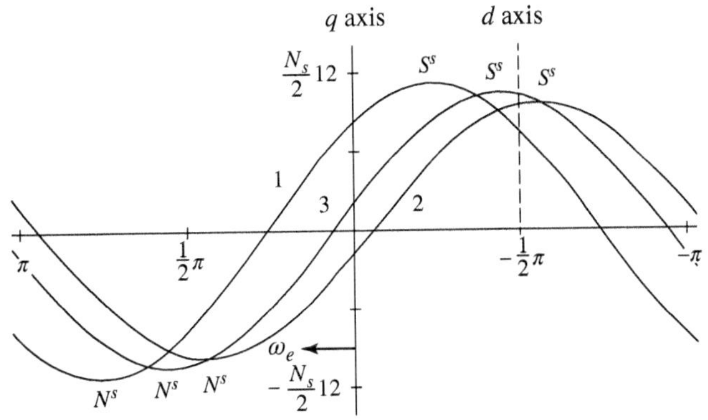

Figure 5.7-2 mmf_s^e for Fig. 5.7-1.

variables in any reference frame, the instantaneous phasor, and the plots of mmf_s^e. The instantaneous phasor is particularly instructive since it allows us to locate the dynamic and steady-state poles.

SP5.7-1 Write (5.7-5) for F_{as}. $\left\{F_{as} = \text{Re}\left[\sqrt{2}\,\tilde{F}_{as}e^{j\omega_e t}\right];\ = \text{Re}\left[\left(F_{qs}^e - jF_{ds}^e\right)e^{j\omega_e t}\right]\right\}$

SP5.7-2 Why does $p\tilde{i}_{as}$ become zero in the steady state? $\left[i_{qs}^e \text{ and } i_{ds}^e \text{ become constant}\right]$

SP5.7-3 In Fig. 5.7-1, we see that when $\tilde{I}_{as}$ is in the third quadrant, i_{qs}^e is negative and power is returned to the source. What is happening? [Some of the energy stored in the field is being returned to the source]

5.8 TRANSFORMATION OF THREE-PHASE STATOR VARIABLES TO THE ARBITRARY REFERENCE FRAME

The mmf_s for a three-phase stator may be expressed by substituting (5.3-2) for ϕ_s into (4.4-14) and using trigonometric identities:

$$\begin{aligned}\text{mmf}_s = {} & \frac{N_s}{2}\cos\phi\left[i_{as}\cos\theta + i_{bs}\cos\left(\theta - \frac{2}{3}\pi\right) + i_{cs}\cos\left(\theta + \frac{2}{3}\pi\right)\right] \\ & - \frac{N_s}{2}\sin\phi\left[i_{as}\sin\theta + i_{bs}\sin\left(\theta - \frac{2}{3}\pi\right) + i_{cs}\sin\left(\theta + \frac{2}{3}\pi\right)\right]\end{aligned} \tag{5.8-1}$$

If we let $\phi = 0$, we obtain an expression for mmf_s along the q axis and if $\phi = -\pi/2$, we obtain mmf_s along the d axis. Thus, the terms inside "[ ]" in the first line of (5.8-1) are i_{qs} and the terms inside "[ ]" on the second line are i_{ds}. This forms the basis for the transformation of stator variables in three-phase machines. Because we are transforming three variables, we need a third variable in addition to the q and d variables. It is referred to as the *zero variable*, which is independent of θ and is denoted as f_{0s}, where $f_{0s} = \frac{1}{3}(f_{as} + f_{bs} + f_{cs})$.

The commonly used three-phase transformation for the stator variables is given by Park [3] as

$$\mathbf{f}_{qd0s} = \mathbf{K}_s \mathbf{f}_{abcs} \tag{5.8-2}$$

where, for an *abc* sequence

$$(\mathbf{f}_{qd0s})^T = [f_{qs} \;\; f_{ds} \;\; f_{0s}] \tag{5.8-3}$$

$$(\mathbf{f}_{abcs})^T = [f_{as} \;\; f_{bs} \;\; f_{cs}] \tag{5.8-4}$$

$$\mathbf{K}_s = \frac{2}{3}\begin{bmatrix} \cos\theta & \cos\left(\theta - \frac{2}{3}\pi\right) & \cos\left(\theta + \frac{2}{3}\pi\right) \\ \sin\theta & \sin\left(\theta - \frac{2}{3}\pi\right) & \sin\left(\theta + \frac{2}{3}\pi\right) \\ \frac{1}{2} & \frac{1}{2} & \frac{1}{2} \end{bmatrix} \tag{5.8-5}$$

$$\frac{d\theta}{dt} = \omega \tag{5.8-6}$$

It can be shown that for the inverse transformation we have

$$(\mathbf{K}_s)^{-1} = \begin{bmatrix} \cos\theta & \sin\theta & 1 \\ \cos\left(\theta - \frac{2}{3}\pi\right) & \sin\left(\theta - \frac{2}{3}\pi\right) & 1 \\ \cos\left(\theta + \frac{2}{3}\pi\right) & \sin\left(\theta + \frac{2}{3}\pi\right) & 1 \end{bmatrix} \tag{5.8-7}$$

Park chose to use a $\frac{2}{3}$ factor in $\mathbf{K}_s$. Tesla's rotating magnetic field (5.8-1) does not suggest the $\frac{2}{3}$ factor. Although it is unclear why Park used the $\frac{2}{3}$ factor, it is more or less irrelevant. We only need to remember that when evaluating power and torque using Park's variables, we must multiply by $\frac{3}{2}$ to compensate for the $\frac{2}{3}$ factor in the three-phase transformation.

Perhaps the best way to justify this is if we calculate power for a single-phase system, it is $v_{as}i_{as}$ which becomes $V_sI_s\cos\phi_{pf}$, for a balanced two-phase system it is $v_{as}i_{as} + v_{bs}i_{cs}$ or $v_{qs}i_{qs} + v_{ds}i_{ds}$ which becomes $2V_sI_s\cos\phi_{pf}$. Now since Park's $\frac{2}{3}$ factor makes $v_{qs}i_{qs} + v_{ds}i_{ds}$ the three-phase power $2V_sI_s\cos\phi_{pf}$ when it should be $3V_sI_s\cos\phi_{pf}$, we must multiply by $\frac{3}{2}$.

Let us assume that

$$v_{as} = r_s i_{as} + p\lambda_{as} \tag{5.8-8}$$

$$v_{bs} = r_s i_{bs} + p\lambda_{bs} \tag{5.8-9}$$

$$v_{cs} = r_s i_{cs} + p\lambda_{cs} \tag{5.8-10}$$

If (5.8-8) through (5.8-10) are transformed to the arbitrary reference frame [5], [6], we obtain

$$v_{qs} = r_s i_{qs} + \omega\lambda_{ds} + p\lambda_{qs} \tag{5.8-11}$$

$$v_{ds} = r_s i_{ds} - \omega\lambda_{qs} + p\lambda_{ds} \tag{5.8-12}$$

$$v_{0s} = r_s i_{0s} + p\lambda_{0s} \tag{5.8-13}$$

Although the $0s$ voltage equation has been added, the v_{qs} and v_{ds} equations are identical in form to those for two-phase systems. This is a result of the $\frac{2}{3}$ factor.

The $0s$ variables are zero for a balanced three-phase system. Also, three-phase electric machines are often connected in wye (Y) without a neutral conductor. In this case, the phase currents sum to zero. Since i_{0s} is zero and if the system is symmetrical, v_{0s} becomes zero.

Let us consider a balanced, *abc*-sequence three-phase set of variables:

$$f_{as} = \sqrt{2}f_s \cos\theta_{esf} \tag{5.8-14}$$

$$f_{bs} = \sqrt{2}f_s \cos\left(\theta_{esf} - \frac{2}{3}\pi\right) \tag{5.8-15}$$

$$f_{cs} = \sqrt{2}f_s \cos\left(\theta_{esf} + \frac{2}{3}\pi\right) \tag{5.8-16}$$

Using (5.8-2) and trigonometric identities to transform to the arbitrary reference frame yields

$$f_{qs} = \sqrt{2}f_s \cos\left(\theta_{esf} - \theta\right) \tag{5.8-17}$$

$$f_{ds} = -\sqrt{2}f_s \sin\left(\theta_{esf} - \theta\right) \tag{5.8-18}$$

$$f_{0s} = 0 \tag{5.8-19}$$

Please note that if f_{0s} is zero, the three-phase system becomes a two-phase system and (5.8-17) and (5.8-18) are (5.5-7) and (5.5-8), respectively. Therefore, the material in Sections 5.5 and 5.7 also applies to a balanced three-phase system, except when calculating torque and power in terms of q and d variable, we must multiply by $\frac{3}{2}$ to make up for the $\frac{2}{3}$ factor in (5.8-5).

SP5.8-1 If a three-phase system had the same elements as the two-phase system in Section 5.6 and if balanced three-phase voltages were applied, the voltages and currents of all figures would be the same for the three-phase system. Why? [(5.5-7) = (5.8-17) and (5.5-8) = (5.8-18)]

SP5.8-2 A three-wire three-phase system is symmetrical and magnetically linear. Do (5.8-17) through (5.8-19) apply if the applied voltages are unbalanced? [Only (5.8-19)]

SP5.8-3 In SP5.8-1 it was shown that the voltages and currents in the figures in Section 5.6 can apply to a balanced three-phase system; however, the trace of power in the figures in Section 5.6 must be multiplied by $\frac{3}{2}$. Why? [The $\frac{2}{3}$ factor in (5.8-5)]

SP5.8-4 Is Park's transformation power invariant? Is Tesla's transformation power invariant? [No, No]

5.9 SUBSTITUTE VARIABLES FOR SYMMETRICAL ROTATING CIRCUITS

Tesla's rotating magnetic field for rotating symmetrical sinusoidally distributed windings was set forth in Section 4.5. We will repeat some of the figures and equations for convenience. Figure 5.9-1 is a repeat of Fig. 4.5-2. Also (4.5-3), which is an expression of Tesla's rotating magnetic field due to the rotating circuits is

$$\text{mmf}_r = \frac{N_r}{2}(i_{ar}\cos\phi_r + i_{br}\sin\phi_r) \tag{5.9-1}$$

In (5.9-1), N_r is the equivalent number of turns and ϕ_r is the angular displacement from the *ar* axis which is displaced by θ_r from the *as* axis. In Fig. 5.9-1, θ_r and θ are referenced to the *as* axis. Note that the q axis in Fig. 5.9-1 is the same q axis as in Fig. 5.3-1. The displacements are related as

$$\phi_r = \beta + \phi \tag{5.9-2}$$

where

$$\beta = \theta - \theta_r \tag{5.9-3}$$

and θ and ϕ are the same as in Fig. 5.3-1.

Substituting (5.9-2) into (5.9-1) yields

$$\text{mmf}_r = \frac{N_r}{2}[i_{ar}\cos(\beta+\phi) + i_{br}\sin(\beta+\phi)] \tag{5.9-4}$$

which is (4.5-5). If $\theta_r(0)$ and $\theta(0)$ are selected to be zero and if the rotor windings are short-circuited whereupon the frequency of the rotor currents is $\omega_e - \omega_r$, then the steady-state mmf_r as viewed from the stator ($\theta = 0$), rotor ($\theta = \omega_r t$), and synchronously rotating ($\theta = \omega_e t$) reference frames are given by (4.5-13), (4.5-15), and (4.5-17), respectively. These expressions are repeated below for convenience:

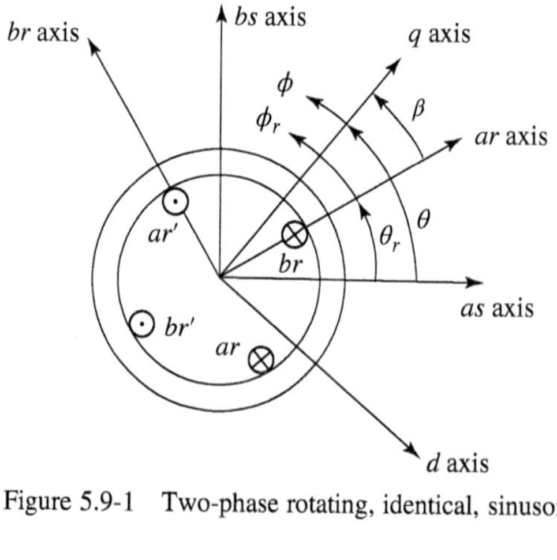

Figure 5.9-1 Two-phase rotating, identical, sinusoidally distributed windings.

$$\text{mmf}_r^s = \frac{N_r}{2}\sqrt{2}I_r \cos\left[\omega_e t + \theta_{eri}(0) - \phi_s\right] \tag{5.9-5}$$

$$\text{mmf}_r^r = \frac{N_r}{2}\sqrt{2}I_r \cos\left[(\omega_e - \omega_r)t + \theta_{eri}(0) - \phi_r\right] \tag{5.9-6}$$

$$\text{mmf}_r^e = \frac{N_r}{2}\sqrt{2}I_r \cos\left[\theta_{eri}(0) - \phi_e\right] \tag{5.9-7}$$

An observer running counterclockwise at ω_e with 360° vision would see a stationary mmf_r^e which would be similar to mmf_s^e shown in Fig. 4.4-3; differing only by the current phase angle.

The transformation for the rotor variables to the arbitrary reference frame follows that set forth in Section 5.3 for the stationary circuits. In particular, (5.9-4) may be written as

$$\text{mmf}_r = \frac{N_r}{2}\cos\phi(i_{ar}\cos\beta + i_{br}\sin\beta) + \frac{N_r}{2}\sin\phi(-i_{ar}\sin\beta + i_{br}\cos\beta) \tag{5.9-8}$$

If we let $\phi = 0$ we obtain i_{qr}, and $\phi = -\frac{\pi}{2}$ we obtain i_{dr}. Equation (5.9-8) may now be written as

$$\text{mmf}_r = \frac{N_r}{2}\left(i_{qr}\cos\phi - i_{dr}\sin\phi\right) \tag{5.9-9}$$

Figure 5.9-2 depicts the fictitious qr and dr windings for the rotor. It is important to mention that if a device has symmetrical stator windings and symmetrical rotor windings, as is the case of an induction machine, the stator and rotor windings have symmetrical fictitious windings in the arbitrary reference frame. Moreover, in the arbitrary reference frame the fictitious windings are all stationary, relative to each other. This is significant since we will find that stator and rotor position-dependent mutual inductances are eliminated in all reference frames in

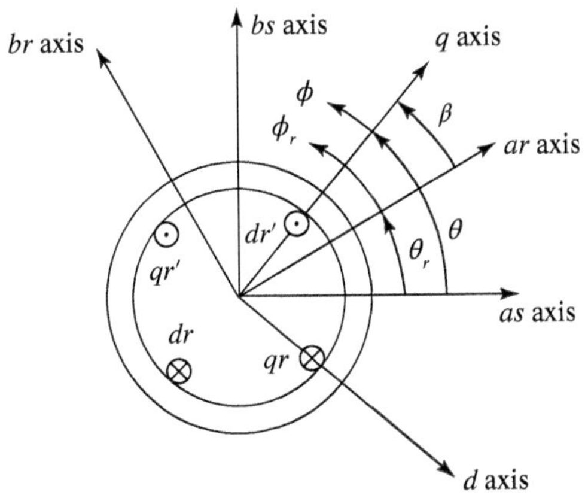

Figure 5.9-2 The fictitious qr and dr windings.

the case of the symmetrical induction machine. Conversely, we will find that in the case of the synchronous machine, the θ_r dependency of the mutual inductances is eliminated only in the rotor reference frame due to the magnetic asymmetry of a salient-pole machine as well as the fact that the rotor is not equipped with symmetrical windings.

The transformation for the rotating symmetrical circuits becomes

$$\begin{bmatrix} f_{qr} \\ f_{dr} \end{bmatrix} = \begin{bmatrix} \cos\beta & \sin\beta \\ \sin\beta & -\cos\beta \end{bmatrix} \begin{bmatrix} f_{ar} \\ f_{br} \end{bmatrix} \tag{5.9-10}$$

or

$$\mathbf{f}_{qdr} = \mathbf{K}_r \mathbf{f}_{abr} \tag{5.9-11}$$

and

$$(\mathbf{K}_r)^{-1} = \mathbf{K}_r \tag{5.9-12}$$

The substitute voltage equations for a resistive-inductive circuit become

$$v_{qr} = r_r i_{qr} + (\omega - \omega_r)\lambda_{dr} + p\lambda_{qr} \tag{5.9-13}$$

$$v_{dr} = r_r i_{dr} - (\omega - \omega_r)\lambda_{qr} + p\lambda_{dr} \tag{5.9-14}$$

These substitute voltage equations are very similar in form to those for stationary circuits given by (5B-5) and (5B-6). We will revisit this material when analyzing the induction machine in Chapter 8.

The extension to three-phase rotating circuits is straightforward and will not be given since we need only to replace θ in (5.8-5) and (5.8-7) with β; however, before proceeding, it is interesting to note that if $\theta_r = 0$ in (5.9-3), it becomes the transformation to the arbitrary reference frame for the stationary circuits. Therefore, (5.9-10) is actually the only transformation that is needed for stationary or rotating symmetrical circuits.

SP5.9-1 Modify Fig. 5.5-1 to apply to mmf_r and F_{qr} and F_{dr}. [Change s subscripts to r]

5.10 RECAPPING

The connection between Tesla's rotating magnetic field and reference frame theory establishes the concept of the change of variables for symmetrical windings to any frame of reference. That is, once the expression for the rotating magnetic field as viewed from any frame of reference is known, the transformation of the actual physical variables to the substitute variables in the arbitrary frame of reference can be readily established from this expression. Tesla's work is the key; moreover, once we have the expression in the arbitrary reference frame, we have it in all reference frames; we need only to change the angular velocity of the reference frame from which the rotating magnet field is to be viewed. Since the substitute variables cause this view of the rotating magnetic field, it follows that if we can visualize the rotating magnetic field, we can visualize the form of the associated substitute variables. For example, if for balanced conditions and symmetrical windings, the rotating magnetic field appears to be varying in time at a frequency, then the expression of the rotating magnetic field as viewed from this reference frame provides the change of variables necessary to transform the physical variables to substitute variables of the same frequency. If, however, in a reference frame, the rotating magnetic field appears constant in time in the steady state, the substitute variables associated with this constant magnetic field will also be constants in the steady state and the expression for the rotating magnetic field viewed from this reference frame provides the necessary change of variables for the physical variables. Although reference frame theory has much broader applications than can be covered in an introductory text, our focus has been and will continue to be on the change of variables for symmetrical, two- and three-phase stationary and rotating circuits and for balanced operating conditions with emphasis on the stationary, rotor, and synchronously rotating reference frames for the induction machine and the rotor reference frame for the synchronous machine.

5.11 PROBLEMS

1. If $f_{as} = \cos \omega_e t$ and $f_{bs} = -\sin \omega_e t$ in Fig. 5.3-1, (a) express the relationship between $\tilde{F}_{as}$ and $\tilde{F}_{bs}$. (b) Obtain f_{qs}^e and f_{ds}^e using (5.3-8) with $\theta(0) = 0$. For f_{as} and f_{bs} given the rotating magnetic field is rotating cw. (c) What do we need to do to make $f_{qs} = 1$ and $f_{ds} = 0$?.

2. The inductance $L_{ss} = L_{ls} + L_{ms}$. Draw the equivalent circuit for inductive stator circuits in the arbitrary reference frame.

3. Consider Fig. 5.3-1, $V_{as} = \sqrt{2}V_s \cos \omega_e t$ and $V_{bs} = \sqrt{2}V_s \sin \omega_e t$. (a) Express the voltage equations in the synchronous reference frame. (b) Express the steady-state currents I_{qs}^e and I_{ds}^e.

4. Show that (5.5-19) results from (5.5-16) through (5.5-18).

5. Show that I_{qs}^e and I_{ds}^e obtained in Problem 3 is $\sqrt{2}\tilde{I}_{as}$ obtained using (5.5-19).

6. If $\mathbf{f}_{qds}^a$ are the *as*- and *bs*-variables expressed in the *a*-reference frame, and $\mathbf{f}_{qds}^b$ are the *as*- and *bs*-variables expressed in the *b*-reference frame, determine ${}^a\mathbf{K}^b$ in terms of $\mathbf{K}_s^a$ and $\mathbf{K}_s^b$, where
$$\mathbf{f}_{qds}^b = {}^a\mathbf{K}^b \mathbf{f}_{qds}^a$$

7. Consider Fig. 5.7-1. Add point 4, a steady-state mode of operation and show plot of $\sqrt{2}\tilde{I}_{as}$ and $j\omega_e L_{ss}\sqrt{2}\tilde{I}_{as}$ for this operating point.

8. Assume a two-phase system with $\mathbf{r}_s$ and $\mathbf{L}_s$ equal-element diagonal matrices. $V_{as} = \sqrt{2}V_s \cos \theta_{esv}$ and $V_{bs} = \sqrt{2}V_s \sin \theta_{esv}$, where $\theta_{esv} = \omega_e t + \theta_{esv}(0)$. Construct the $q\overset{e}{s}$ and $d\overset{e}{s}$ equivalent circuits where $\theta = \omega_e t + \theta(0)$ in $\mathbf{K}_s^e$. It is desirable to investigate variable-frequency operation. Thus, ω_e is a function of time. (a) What changes are

necessary in the $q\overset{e}{s}$ and $d\overset{e}{s}$ equivalent circuits? (b) Are the equivalent circuits valid for $\omega_e = 0$?

9. The inductance $L_{rr} = L_{lr} + L_{mr}$. Draw the equivalent circuit for inductive rotor circuits in the arbitrary reference frame.

10. Derive the qs and ds voltage equations if we selected the d axis at $\phi = \frac{\pi}{2}$ rather than $\phi = -\frac{\pi}{2}$.

11. Assume $V_{as} = \sqrt{2}V_s \cos\theta_e$ and $I_{as} = \sqrt{2}I_s \cos\theta_e$. Show that we must multiply Park's three-phase calculation of power by $\frac{3}{2}$. ■

5.12 REFERENCES

1. P. C. Krause, O. Wasynczuk, T. C. O'Connell, and M. Hasan, "Tesla's Contribution to Electric Machine Analysis," IEEE Trans. on Energy Conversion, Vol. 32, no. 2, 2016, pp. 591–598.
2. P. C. Krause, O. Wasynczuk, T. C. O'Connell, and M. Hasan, *Introduction to Power and Drive Systems*, John Wiley & Sons, IEEE Press, New York, 2017.
3. R. H. Park, "Two-Reaction Theory of Synchronous Machines – Generalized Method of Analysis – Part I," AIEE Trans., Vol. 48, July 1929, pp. 716–727.
4. P. C. Krause and C. H. Thomas, "Simulation of Symmetrical Induction Machinery," IEEE Trans. Power Apparatus and Systems, Vol. 84, November 1965, pp. 1038–1053.
5. P. C. Krause, O. Wasynczuk, and S. D. Pekarek, *Electromechanical Motion Devices*, 2nd Edition, John Wiley & Sons, IEEE Press, New York, 2012.
6. P. C. Krause, O. Wasynczuk, S. D. Sudhoff, and S. D. Pekarek, *Analysis of Electric Machinery and Drive Systems*, 3rd Edition, John Wiley & Sons, IEEE Press, New York, 2013.

CHAPTER 6

PERMANENT-MAGNET AC MACHINE AND FIELD ORIENTATION OF A BRUSHLESS DC DRIVE

6.1 INTRODUCTION

The permanent-magnet ac machine supplied from a controlled inverter is widely used as a low-power control motor and in hybrid and electric vehicles. Although it may seem as a contradiction of terms, this inverter–machine combination is often referred to as a brushless dc drive. Depending upon the control strategy and inverter used, the performance of this brushless dc drive can be made to (a) emulate the performance of a permanent-magnet dc motor, (b) operate in a maximum torque per volt mode, (c) or in a maximum torque per ampere mode. Fortunately, we are able to become quite familiar with the operating features of these various modes without becoming overly involved with the actual inverter control. In particular, if we assume that the stator variables (voltages and currents) are sinusoidal and balanced with the same angular velocity as the rotor speed, we are able to predict the dominant operating features of all common modes of operation without becoming involved with the actual switching or control of the inverter. Therefore, focus on the performance of the brushless dc drive assuming that the inverter is designed and controlled appropriately and not the details of how this is done. We will use the rotor reference frame to derive the equations used in the analysis and for the control of the machine; however, the derivations may be de-emphasized or bypassed if the focus is steady-state performance.

Electromechanical Motion Devices: Rotating Magnetic Field-Based Analysis and Online Animations, Third Edition. Paul Krause, Oleg Wasynczuk, Steven D. Pekarek, and Timothy O'Connell.

Published 2020 by John Wiley & Sons, Inc.

6.2 TWO-PHASE PERMANENT-MAGNET AC MACHINE

A two-pole two-phase permanent-magnet ac machine is shown in Fig. 6.2-1. The stator windings are identical, sinusoidally distributed windings each with N_s equivalent turns and resistance r_s as described in Chapter 4. The angular displacement ϕ_s about the stator is referenced to the *as* axis. The angular displacement about the rotor ϕ_r is referenced to the q axis. The angular velocity of the rotor is ω_r and θ_r is the angular position of the rotor measured from the *as* axis to the q axis. Thus, a given point on the rotor surface at the angular displacement ϕ_r may be related to an adjacent point on the inside stator surface with angular displacement ϕ_s as

$$\phi_s = \theta_r + \phi_r \tag{6.2-1}$$

The d axis (direct axis) is fixed at the center of the north pole of the permanent-magnet rotor and the q axis (quadrature axis) is displaced 90° counterclockwise from the d axis. In a synchronous-type machine, the rotor is unsymmetrical electrically (a field winding or permanent magnet only in the d axis), and the rotor may also be unsymmetrical magnetically (salient pole). Thus, transformation to a reference frame other than the rotor reference frame offers no advantage; and we will make the rotor angular position, θ_r, and the angular position of the q axis of the reference frame, θ, be one in the same. The electromagnetic torque T_e and the load torque T_L are also indicated in Fig. 6.2-1. As defined in Chapter 2, T_e is assumed

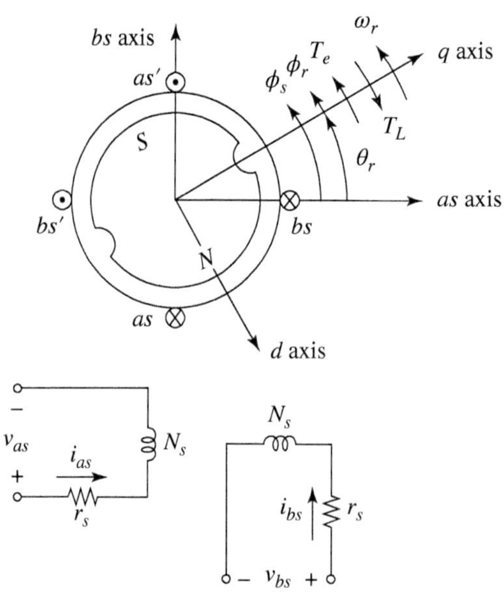

Figure 6.2-1 Two-pole two-phase permanent-magnet ac machine.

positive in the direction of increasing θ_r; T_L is positive in the opposite direction. It is assumed that:

1. The magnetic system is linear.
2. The q- and d-axis reluctances are equal. The slight rotor indentation in the q axis is shown in Fig. 6.2-1 to indicate separation of the N and S poles on the rotor.
3. The open-circuit stator voltages, induced by rotating the permanent-magnet rotor, are sinusoidal.
4. Demagnetization of the permanent magnet is neglected.

Although these assumptions may be somewhat oversimplifying, the analysis is made less involved while still portraying the main operating features of the device.

Even though most brushless dc drives are inverter-controlled three-phase permanent-magnet ac machines, from an analytical standpoint, it is convenient to first consider the two-phase device. In Fig. 6.2-1, the magnetic axes of the stator windings are denoted as the *as* and *bs* axes. Electromagnetic torque is produced by the interaction of the poles of the permanent-magnet rotor (N^r and S^r) and the poles resulting from the rotating air-gap mmf (Tesla's rotating magnetic field) established by currents flowing in the stator windings. The rotating mmf as viewed by a stationary observer (mmf_s^s) established by symmetrical two-phase stator windings carrying balanced two-phase currents is given by (4.4-9).

A four-pole three-phase 28-V $\frac{1}{3}$ – hp permanent-magnet ac machine is shown in Fig. 6.2-2. The disassembled motor is shown in Fig. 6.2-2a. The stator housing is shown in Fig. 6.2-2b wherein the stator windings are visible. Housed therein are the Hall-effect sensors, which are used to determine rotor position, the drive inverter, the filter capacitor, and the logic circuitry.

The stator and rotor of a ten-pole three-phase 28-V 0.63-hp 4500-r/min permanent-magnet ac machine is shown in Fig. 6.2-3. The magnets are samarium cobalt and the drive inverter is supplied from a 28-V dc source. The magnetic end

(a) (b)

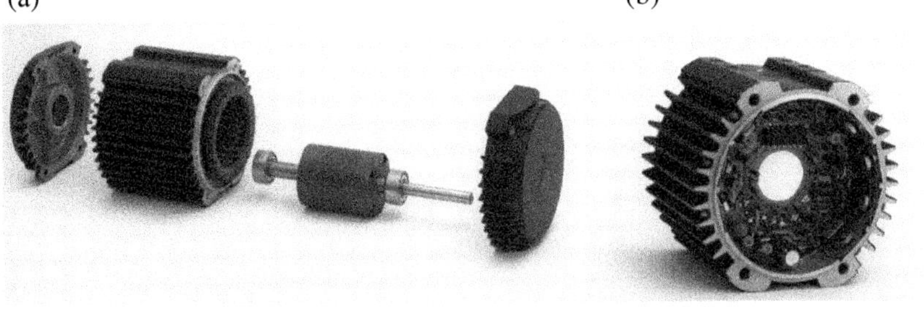

Figure 6.2-2 (a) Four-pole three-phase 28-V $\frac{1}{3}$ – hp permanent-magnet ac machine. (Courtesy of EG and G Rotron.) (b) Stator housing including electronics.

Figure 6.2-3 Two-pole three-phase 28-V 0.63-hp 4500-r/min permanent-magnet ac machine. (Courtesy of Vickers Electromech.)

cap is used in conjunction with Hall-effect sensors mounted in the stator housing (not shown) to determine the rotor position.

SP6.2-1 Express the mmf created by the permanent-magnet rotor $\left(\mathrm{mmf}_r^r\right)$ as viewed from the rotor for the two-pole two-phase permanent-magnet ac machine shown in Fig. 6.2-1. Let F_p denote the peak value. [$\mathrm{mmf}_r^r = -F_p \sin \phi_r$]

6.3 VOLTAGE EQUATIONS AND WINDING INDUCTANCES

The instantaneous voltage equations for the two-pole two-phase permanent-magnet ac machine shown in Fig. 6.2-1 may be expressed as

$$v_{as} = r_s i_{as} + \frac{d\lambda_{as}}{dt} \tag{6.3-1}$$

$$v_{bs} = r_s i_{bs} + \frac{d\lambda_{bs}}{dt} \tag{6.3-2}$$

In matrix form,

$$\mathbf{v}_{abs} = \mathbf{r}_s \mathbf{i}_{abs} + p\boldsymbol{\lambda}_{abs} \tag{6.3-3}$$

where p is the operator d/dt, and for voltages, currents, and flux linkages,

$$(\mathbf{f}_{abs})^T = [f_{as} \;\; f_{bs}] \tag{6.3-4}$$

with

$$\mathbf{r}_s = \begin{bmatrix} r_s & 0 \\ 0 & r_s \end{bmatrix} \tag{6.3-5}$$

The flux linkage equations may be expressed as

$$\lambda_{as} = L_{asas}i_{as} + L_{asbs}i_{bs} + \lambda_{asm} \tag{6.3-6}$$

$$\lambda_{bs} = L_{bsas}i_{as} + L_{bsbs}i_{bs} + \lambda_{bsm} \tag{6.3-7}$$

In matrix form,

$$\boldsymbol{\lambda}_{abs} = \mathbf{L}_s\mathbf{i}_{abs} + \boldsymbol{\lambda}'_m \tag{6.3-8}$$

where $\boldsymbol{\lambda}'_m$ is the column vector

$$\boldsymbol{\lambda}'_m = \begin{bmatrix} \lambda_{asm} \\ \lambda_{bsm} \end{bmatrix} = \lambda'_m \begin{bmatrix} \sin\theta_r \\ -\cos\theta_r \end{bmatrix} \tag{6.3-9}$$

In (6.3-9), λ'_m is the amplitude of the flux linkages established by the permanent magnet as viewed from the stator phase windings. In other words, the magnitude of λ'_m is proportional to the amplitude of the open-circuit sinusoidal voltage induced in each stator phase winding. The rotor position θ_r is expressed as

$$\frac{d\theta_r}{dt} = \omega_r \tag{6.3-10}$$

As mentioned, assume that the reluctance of the rotor of the permanent-magnet ac machine is the same in the q and d axes. Actually, this assumption may be an oversimplification in some cases; however, it markedly reduces our work and allows us to establish directly the basic principles of the controlled permanent-magnet ac machine. Unequal q- and d-axis reluctances are considered briefly in a later section and in detail in [1]. With the assumption of a uniform air gap, the mutual inductance between the as and bs windings is zero. Since the windings are identical, the self-inductances L_{asas} and L_{bsbs} are equal and denoted as L_{ss}. As in the case of a transformer, the self-inductance is made up of a leakage L_{ls} and a magnetizing inductance L_{ms}. Thus,

$$L_{ss} = L_{ls} + L_{ms} \tag{6.3-11}$$

The machine is designed to minimize the leakage inductance; it generally makes up approximately 10 percent of L_{ss}. The magnetizing inductances may be expressed in terms of turns and reluctance. In particular,

$$L_{ms} = \frac{N_s^2}{\mathcal{R}_m} \tag{6.3-12}$$

The magnetizing reluctance $\mathcal{R}_m$ is an equivalent reluctance due to the stator steel, the permanent magnet, and the air gap. Assume that $\mathcal{R}_m$ is independent of rotor position θ_r, whereupon, the self-inductance matrix $\mathbf{L}_s$ may be expressed as the constant matrix:

$$\mathbf{L}_s = \begin{bmatrix} L_{ss} & 0 \\ 0 & L_{ss} \end{bmatrix} \tag{6.3-13}$$

Example 6A Rotor Flux Linkage Referred to the Stator
The parameters of a four-pole two-phase permanent-magnet ac machine are $r_s = 3.4\,\Omega$, $L_{ls} = 1.1$ mH, and $L_{ms} = 11$ mH. When the device is driven at 1000 r/ min, the open-circuit phase voltage is sinusoidal with a peak-to-peak value of 34.6 V. Determine λ_m'. The actual rotor speed at which the measurement was taken is

$$\begin{aligned}\omega_{rm} &= \frac{(\text{r/ min})(\text{rad/r})}{\text{s/ min}}\\ &= \frac{(1000)(2\pi)}{60} = \frac{100}{3}\pi\ \text{rad/s}\end{aligned} \tag{6A-1}$$

From (4.6-9), the electrical angular velocity is

$$\begin{aligned}\omega_r &= \frac{P}{2}\omega_{rm}\\ &= \frac{4}{2}\frac{100\pi}{3} = \frac{200}{3}\pi\ \text{rad/s}\end{aligned} \tag{6A-2}$$

With the phases open-circuited, $i_{as} = i_{bs} = 0$. Thus, from (6.3-1) and (6.3-9),

$$v_{as} = \frac{d(\lambda_m' \sin\theta_r)}{dt} = \lambda_m'\omega_r\cos\theta_r \tag{6A-3}$$

Now, the peak-to-peak voltage is 34.6 V; hence, from (6A-3), with the peak-to-peak voltage divided by 2, we have

$$\frac{34.6}{2} = \lambda_m'\left(\frac{200}{3}\pi\right) \tag{6A-4}$$

Solving for λ_m' yields

$$\lambda_m' = \frac{(34.6)(3)}{(2)(200\pi)} = 0.0826\ \text{V}\cdot\text{s/rad} \tag{6A-5}$$

SP6.3-1 The stator windings of the permanent-magnet ac machine shown in Fig. 6.2-1 are open-circuited. The rotor is driven clockwise and $V_{as} = -10\sin 100t$. Determine V_{bs}. $[V_{bs} = -10\cos 100t]$

SP6.3-2 Determine λ_m' for SP6.3-1. $[\lambda_m' = 0.1\ \text{V}\cdot\text{s/rad}]$

SP6.3-3 During steady-state operation, one can use (6.3-10) to show $\theta_r = \omega_r t + \theta_r(0)$. What is $\theta_r(0)$ in SP6.3-1? $\left[\theta_r(0) = -\frac{1}{2}\pi\right]$

6.4 TORQUE

An expression for the electromagnetic torque may be obtained by using the second entry in Table 2.5-1. Since we are assuming a linear magnetic system, the coenergy may be expressed as

$$W_c = \frac{1}{2}L_{ss}\left(i_{as}^2 + i_{bs}^2\right) + \lambda_m' i_{as} \sin\theta_r - \lambda_m' i_{bs} \cos\theta_r + W_{pm} \tag{6.4-1}$$

where W_{pm} is the energy associated with the permanent magnet, which is assumed constant for the device shown in Fig. 6.2-1. Taking the partial derivative with respect to θ_r and accounting for the number of poles by multiplying by $P/2$ yields

$$T_e = \frac{P}{2}\lambda_m'(i_{as}\cos\,\theta_r + i_{bs}\sin\,\theta_r) \tag{6.4-2}$$

The above expression is positive for motor action. The torque and speed may be related as

$$T_e = J\left(\frac{2}{P}\right)\frac{d\omega_r}{dt} + B_m\left(\frac{2}{P}\right)\omega_r + T_L \tag{6.4-3}$$

where J is in $\text{kg}\cdot\text{m}^2$; it is the inertia of the rotor and any rigidly connected mechanical load. We will be concerned primarily with motor action; therefore, the load torque T_L is assumed positive in the direction indicated in Fig. 6.2-1. The constant B_m is a damping coefficient associated with the rotational system of the machine and mechanical load. It has the units $\text{N}\cdot\text{m}\cdot\text{s}/\text{rad}$ and it is generally small and often neglected in the case of the machine but may be considerable for the mechanical load.

SP6.4-1 Calculate T_e for a two-pole permanent-magnet ac motor if $I_{as} = \cos\,\theta_r$, and $I_{bs} = \sin\,\theta_r$. $\lambda_m' = 0.1\ \text{V}\cdot\text{s/rad}$. $[T_e = 0.1\ \text{N}\cdot\text{m}]$

SP6.4-2 Repeat SP6.4-1 with $\theta_r = \omega_r t$ and $I_{as} = \cos\left(\omega_r t + \frac{1}{2}\pi\right)$ and $I_{bs} = \sin\left(\omega_r t + \frac{1}{2}\pi\right)$. $[T_e = 0]$

6.5 MACHINE EQUATIONS IN THE ROTOR REFERENCE FRAME

The reference frame derivation of the voltage and torque equations is straightforward; however, if the focus is on steady-state performance, this material may be bypassed. Study may then be resumed with (6.6-5), (6.6-6), and (6.6-9) which are the expressions for steady-state $\widetilde{V}_{as}$, $\widetilde{E}_a$, and T_e, respectively.

Since the rotor of the permanent-magnet ac machine is magnetically unsymmetrical with the magnetic field fixed in the d axis, the reference frame is fixed in the rotor; however, the voltage equations may be written directly from the voltage equations in the arbitrary reference frame. In particular, if we select $\theta = \theta_r$ and $\omega = \omega_r$ and combine (5.4-6), (5.4-16), and (5.4-17), the stator voltage equations transformed to the rotor reference frame, expressed in expanded form, become

$$v_{qs}^r = r_s i_{qs}^r + \omega_r \lambda_{ds}^r + p\lambda_{qs}^r \tag{6.5-1}$$

$$v_{ds}^r = r_s i_{ds}^r - \omega_r \lambda_{qs}^r + p\lambda_{ds}^r \tag{6.5-2}$$

where the raised r is used to denote the variables in the rotor reference frame.

For a magnetically linear system, the stator flux linkages are expressed by (6.3-8). Substituting the change of variables into (6.3-8) yields

$$(\mathbf{K}_s^r)^{-1}\boldsymbol{\lambda}_{qds}^r = \mathbf{L}_s(\mathbf{K}_s^r)^{-1}\mathbf{i}_{qds}^r + \boldsymbol{\lambda}_m' \tag{6.5-3}$$

where, from (5.3-12) with $\theta = \theta_r$:

$$\mathbf{K}_s^r = \begin{bmatrix} \cos\theta_r & \sin\theta_r \\ \sin\theta_r & -\cos\theta_r \end{bmatrix} \tag{6.5-4}$$

Substituting (6.3-9) for $\boldsymbol{\lambda}_m'$ and (6.3-13) for $\mathbf{L}_s$ and then premultiplying by $\mathbf{K}_s^r$ we have

$$\begin{bmatrix} \lambda_{qs}^r \\ \lambda_{ds}^r \end{bmatrix} = \begin{bmatrix} L_{ss} & 0 \\ 0 & L_{ss} \end{bmatrix} \begin{bmatrix} i_{qs}^r \\ i_{ds}^r \end{bmatrix} + \lambda_m'^r \begin{bmatrix} 0 \\ 1 \end{bmatrix} \tag{6.5-5}$$

To be consistent, we have added the superscript r to λ_m'. We see from (6.5-5) that, in our new system of variables, the flux linkage created by the permanent magnet is constant. Hence, the fictitious circuits are fixed relative to the permanent magnet and, therefore, fixed in the rotor reference frame. We have accomplished the goal of eliminating flux linkages that vary with θ_r. From (6.5-5)

$$\lambda_{qs}^r = L_{ss} i_{qs}^r \tag{6.5-6}$$

$$\lambda_{ds}^r = L_{ss} i_{ds}^r + \lambda_m'^r \tag{6.5-7}$$

and, as before,

$$L_{ss} = L_{ls} + L_{ms} \tag{6.5-8}$$

Substituting (6.5-6) and (6.5-7) into (6.5-1) and (6.5-2) and, since $\lambda_m'^r$ is constant, $p\lambda_m'^r = 0$ yields

$$v_{qs}^r = (r_s + pL_{ss})i_{qs}^r + \omega_r L_{ss} i_{ds}^r + \omega_r \lambda_m'^r \tag{6.5-9}$$

$$v_{ds}^r = (r_s + pL_{ss})i_{ds}^r - \omega_r L_{ss} i_{qs}^r \tag{6.5-10}$$

Equations (6.5-9) and (6.5-10) suggest the equivalent circuits given in Fig. 6.5-1.

For balanced conditions, the stator variables may be expressed as

$$f_{as} = \sqrt{2} f_s \cos\theta_{esf} \tag{6.5-11}$$

$$f_{bs} = \sqrt{2} f_s \sin\theta_{esf} \tag{6.5-12}$$

where

$$\frac{d\theta_{esf}}{dt} = \omega_e \tag{6.5-13}$$

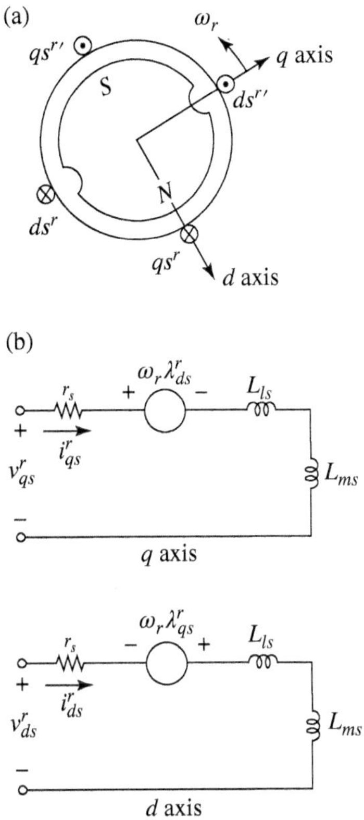

Figure 6.5-1 (a) The q_s^r- and d_s^r-axis equivalent diagram of a permanent-magnet ac machine, (b) Quadrature- and direct-axis equivalent circuits.

Transforming the stator variables to the rotor reference frame by (6.5-4) yields

$$f_{qs}^r = \sqrt{2} f_s \cos\left(\theta_{esf} - \theta_r\right) \tag{6.5-14}$$

$$f_{ds}^r = -\sqrt{2} f_s \sin\left(\theta_{esf} - \theta_r\right) \tag{6.5-15}$$

where

$$\frac{d\theta_r}{dt} = \omega_r \tag{6.5-16}$$

Later, we will find it convenient to relate the angular positions of the applied voltage of the *as* phase and the rotor as

$$\phi_v = \theta_{esv} - \theta_r \tag{6.5-17}$$

where θ_{esv} is the position of the positive peak value of v_{as} and θ_r the displacement of the q axis both referenced to the *as* axis. We will talk more of this as we go along; however, it is mentioned here to prevent confusion between (6.5-17) and (6.2-1).

Equation (6.2-1) relates a displacement on the rotor (ϕ_r) to a displacement on the stator (ϕ_s). Equation (6.5-17) relates the angular position of the q axis (θ_r) to the angular position of the maximum positive value of v_{as}.

The expression for electromagnetic torque is obtained by expressing i_{as} and i_{bs} in (6.4-2) in terms of i^r_{qs} and i^r_{ds}. In particular, the terms within the parentheses of (6.4-2) is i^r_{qs}, therefore

$$T_e = \frac{P}{2}\lambda'^r_m i^r_{qs} \tag{6.5-18}$$

which is positive for motor action.

For balanced, steady-state conditions $\omega_r = \omega_e$, the q^r_s and d^r_s are constants and the voltage equations given by (6.5-9) and (6.5-10) become

$$V^r_{qs} = r_s I^r_{qs} + \omega_r L_{ss} I^r_{ds} + \omega_r \lambda'^r_m \tag{6.5-19}$$

$$V^r_{ds} = r_s I^r_{ds} - \omega_r L_{ss} I^r_{qs} \tag{6.5-20}$$

where pI^r_{qs} and pI^r_{ds} are zero and capital letters denote steady-state (in this case, constant) quantities, except for λ'^r_m which is assumed to always be constant. Since we are assuming $\omega_r = \omega_e$, we could use either e or r as the superscript; we will use r. Equations (6.5-17) through (6.5-20) are given again in Table 6.6-1.

If (6.5-19) and (6.5-20) are solved for I^r_{qs} and the result substituted into the expression for electromagnetic torque given by (6.5-18), we obtain

$$T_e = \frac{P}{2}\frac{\lambda'^r_m r_s}{r_s^2 + \omega_r^2 L_{ss}^2}\left(V^r_{qs} - \frac{\omega_r L_{ss} V^r_{ds}}{r_s} - \omega_r \lambda'^r_m\right) \tag{6.5-21}$$

TABLE 6.6-1 Voltage and Torque Equations for Steady-State Operation of a Two-Phase Permanent-Magnet AC Machine

q^r_s and d^r_s Variables	Phasor Variables
$V^r_{qs} = r_s I^r_{qs} + \omega_r L_{ss} I^r_{ds} + \omega_r \lambda'^r_m$	$\tilde{V}_{as} = (r_s + j\omega_r L_{ss})\tilde{I}_{as} + \tilde{E}_a$
$V^r_{ds} = r_s I^r_{ds} - \omega_r L_{ss} I^r_{qs}$	$\tilde{E}_a = \frac{1}{\sqrt{2}}\omega_r \lambda'^r_m e^{j0°}$
$V^r_{qs} = \sqrt{2}\, V_s \cos\phi_v$	$\sqrt{2}\tilde{F}_{as} = F^r_{qs} - jF^r_{ds}$
$V^r_{ds} = -\sqrt{2}\, V_s \sin\phi_v$	$T_e = \frac{P}{2}\lambda'^r_m \sqrt{2}\, I_s \cos[\theta_{esi}(0) - \theta_r(0)]$
$T_e = \frac{P}{2}\lambda'^r_m I^r_{qs}$	

$$\omega_r = \omega_e$$
$$\phi_v = \theta_{esv} - \theta_r = \theta_{esv}(0)$$
$$L_{ss} = L_{ls} + L_{ms}$$

SP6.5-1 Let $f_{as} = -\cos\theta_r$ and $f_{bs} = -\sin\theta_r$. Determine f_{qs}^r and f_{ds}^r. [$f_{qs}^r = -1, f_{ds}^r = 0$]

SP6.5-2 Let $f_{as} = \cos\theta_{esf}$, $f_{bs} = \sin\theta_{esf}$. If $\omega_e = \omega_r$ and $\theta_r(0) = 0$, determine θ_{esf} for $f_{qs}^r = 0$ and $f_{ds}^r = 1$. $\left[\theta_{esf} = \omega_e t - \frac{1}{2}\pi\right]$

SP6.5-3 If the speed ω_r is constant and if $V_{as} = \cos\theta_e$ and $V_{bs} = \sin\theta_e$ and $\theta_e = \theta_r$, what will be the type of waveform of I_{qs}^r and I_{ds}^r in the steady state? [dc]

SP6.5-4 Show that (6.5-18) is equal to $T_e = \frac{P}{2}\left(\lambda_{ds}^r i_{qs}^r - \lambda_{qs}^r i_{ds}^r\right)$. Would this be the case if the reluctances in the q and d axes are different? [No]

SP6.5-5 Use (6.5-18) to justify the answer to SP6.4-1. [Express i_{qs}^r in terms of i_{as} and i_{bs} given in SP6.4-1]

6.6 INSTANTANEOUS AND STEADY-STATE PHASORS

We will derive the instantaneous phasor voltage equation by following the derivation used in Section 5.7. From (5.7-2),

$$\tilde{f}_{as}(t) = f_{qs}^r(t) - jf_{ds}^r(t) \tag{6.6-1}$$

In the case of the permanent-magnet ac machine used as the motor in a brushless dc drive, the electrical angular velocity (frequency) of the applied stator voltages is controlled by the inverter to be the same as the electrical angular velocity of the rotor; in other words, ω_e is controlled to always be ω_r. Therefore, (6.6-1) will apply as long as $\omega_e = \omega_r$. We will find that this is an important observation since the instantaneous and steady-state phasors may be used to portray transient and steady-state machine and drive operation on a phasor diagram if the mode of operation can be expressed in the synchronous reference frame and if steady-state operation yields constant synchronous reference frame variables. If the rotor and stator are symmetrical as in the case of an induction machine (Chapter 8), all balanced modes of operation regardless of rotor speed yield constant steady-state synchronous reference frame variables. In the case of the permanent-magnet ac machine, which is an unsymmetrical machine due to the rotor, constant synchronous reference frame variables in the steady state exist only if the rotor speed is fixed at synchronous speed. This is not the case in general; however, we will find that when the permanent-magnet ac machine is controlled as a brushless dc drive, the fundamental frequency of the applied stator voltages is controlled nearly instantaneously to be equal to the rotor speed. Therefore, ideally all modes of operation are essentially fixed in the synchronous reference frame and the phasor representation is valid if we neglect the harmonics introduced in the stator voltages due to the switching of the drive inverter. We are getting a little ahead of ourselves; but, we will talk more of all this later.

Substituting (6.5-9) and (6.5-10) into (6.6-1) yields

$$\left(v_{qs}^r - jv_{ds}^r\right) = r_s\left(i_{qs}^r - ji_{ds}^r\right) + L_{ss}\left(pi_{qs}^r - jpi_{ds}^r\right) + \omega_r L_{ss}\left(i_{ds}^r + ji_{qs}^r\right) + \omega_r\lambda_m' \tag{6.6-2}$$

which may be written as

$$\tilde{v}_{as} = (r_s + j\omega_r L_{ss})\tilde{i}_{as} + \tilde{e}_a + L_{ss}p\tilde{i}_{as} \tag{6.6-3}$$

where the functional notation has been dropped and we are assuming that

$$\tilde{e}_a = \omega_r \lambda'^{r}_{m} \underline{/0^\circ} \tag{6.6-4}$$

You should be wondering why all of the assumptions; first, we have assumed that ω_e and ω_r are always equal and we have used them interchangeably, now we are assuming that $\tilde{e}_a$ is at zero degrees. Well, please go along with us. We are setting the stage for the analysis of the brushless dc drive where these assumptions become real by the action of the drive inverter.

Now, for balanced steady-state operation the last term of (6.6-3) becomes zero and the steady-state phasor voltage equation becomes

$$\tilde{V}_{as} = (r_s + j\omega_r L_{ss})\tilde{I}_{as} + \tilde{E}_a \tag{6.6-5}$$

where

$$\tilde{E}_a = \frac{\omega_r \lambda'^{r}_{m}}{\sqrt{2}} \underline{/0^\circ} \tag{6.6-6}$$

For a three-phase machine, we will replace L_{ms}, which is part of L_{ss}, with $\frac{3}{2}L_{ms}$ in (6.6-3) and (6.6-5).

Let us take a moment to mention something that we should keep in mind. Back in Chapter 5 when we were dealing with steady-state calculations, we set the zero position of the q axis equal to zero; that is, $\theta(0) = 0$ and therefore $\theta_r(0) = 0$, in the case of the synchronous-type machine. This allowed us to express the instantaneous and steady-state phasor as given by (5.7-2) and (5.5-16), respectively. In Chapter 7, on the synchronous generator, we will have to deal with nonzero time-zero positions; however, we will keep $\theta(0)$ and $\theta_r(0)$ both equal to zero for the time being.

A convenient expression for the calculation of steady-state torque for balanced operation using phase variables can be obtained by expressing the instantaneous phase currents as

$$I_{as} = \sqrt{2}I_s \cos\left[\omega_e t + \theta_{esi}(0)\right] \tag{6.6-7}$$

$$I_{bs} = \sqrt{2}I_s \sin\left[\omega_e t + \theta_{esi}(0)\right] \tag{6.6-8}$$

Substituting (6.6-7) and (6.6-8) into (6.4-2) and since $\omega_e = \omega_r$, the steady-state torque may be expressed as

$$\begin{aligned} T_e &= \frac{P}{2}\lambda'^{r}_{m} I^{r}_{qs} \\ &= \frac{P}{2}\lambda'^{r}_{m}\sqrt{2}I_s \cos\left[\theta_{esi}(0) - \theta_r(0)\right] \end{aligned} \tag{6.6-9}$$

Equations (6.6-5), (6.6-6), and (6.6-9) are given again in Table 6.6-1.

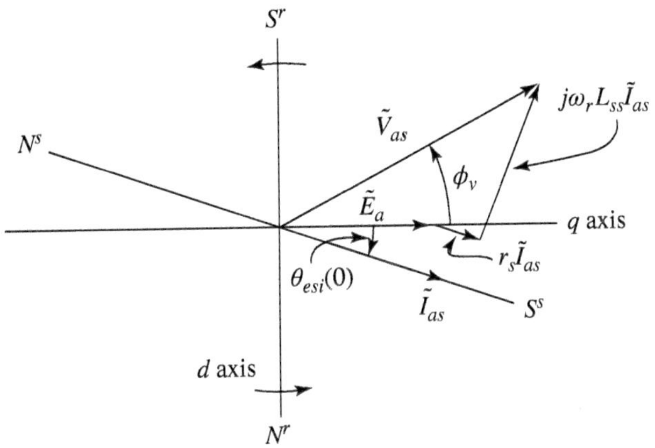

Figure 6.6-1 Phasor diagram showing voltages, currents, and rotor and stator poles of a permanent-magnetic ac machine with $\omega_r = \omega_e$ and $\widetilde{E}_a$ at zero degrees.

We can use the phasor diagram to advantage in portraying the steady-state operation of the permanent-magnet ac machine. During balanced steady-state operation and with the assumption of $\omega_e = \omega_r$, the rotor and phasor diagram are rotating in unison; therefore, they can be superimposed with $\widetilde{E}_a$ at zero degrees. Recall that $\widetilde{E}_a$ and the q axis are controlled to be at zero degrees. Moreover, since mmf$_r$ is also rotating at ω_r, N^r and S^r can be superimposed with the rotor and the phasor diagram. Now, since all are rotating in unison and the voltage and current phasors are constant, we can stop the rotation at any time or we can run at ω_e and observe the operation of the machine.

In Fig. 6.6-1, we have located the phasors associated with (6.6-5) and the stator poles relative to the stator current. We have assumed that the fundamental component of the north pole of the rotor is fixed in the d axis. Does Fig. 6.6-1 tell us anything about torque? Yes. We have motor action and we can see this since the poles established by the stator N^s and S^s, which are rotating counterclockwise at $\omega_e(\omega_r)$, are "trying" to pull the rotor poles N^r and S^r, into alignment with them. This "pulling" is being resisted by the load torque. When this "pulling torque" is equal to the load torque on the rotor shaft, steady-state operation occurs with torque in the counterclockwise direction; motor action. Is it clear that complete alignment of the poles occurs only if there is no load torque?

Example 6B Steady-State Operation of Permanent-Magnet ac Machine with $\phi_v = 0$

Consider the permanent-magnet ac machine with the parameters given in Example 6A. Assume that the applied stator voltages form a balanced two-phase set with $V_{as} = \sqrt{2}\,11.25 \cos \omega_r t$ ($\phi_v = 0$). The machine is operating as a brushless dc motor and the steady-state rotor speed is 600 r/min. Calculate T_e and $\widetilde{I}_{as}$, draw the phasor diagram, and locate the poles.

From (4.6-9),

$$\omega_r = \frac{P}{2}\omega_{rm} = \frac{4}{2}\frac{(600)(2\pi)}{60} = 40\pi \text{ rad/s} \tag{6B-1}$$

We can calculate $\widetilde{I}_{as}$ from (6.6-5),

$$\widetilde{V}_{as} = (r_s + j\omega_r L_{ss})\widetilde{I}_{as} + \widetilde{E}_a$$

$$11.25\underline{/0^\circ} = \left(3.4 + j(40\pi)\left(12.1 \times 10^{-3}\right)\right)\widetilde{I}_{as} + \frac{(40\pi)(0.0826)}{\sqrt{2}}\underline{/0^\circ} \tag{6B-2}$$

Solving for $\widetilde{I}_{as}$:

$$\widetilde{I}_{as} = \frac{11.25\underline{/0^\circ} - 7.34\underline{/0^\circ}}{3.4 + j1.52} = 1.05\underline{/-24.1^\circ} \tag{6B-3}$$

The torque is (6.6-9):

$$\begin{aligned} T_e &= \frac{P}{2}\lambda_m'^r \sqrt{2} I_s \cos\left[\theta_{esi}(0) - \theta_r(0)\right] \\ &= \left(\frac{4}{2}\right)(0.0826)\sqrt{2}1.05 \cos(-24.1^\circ) = 0.223 \text{ N}\cdot\text{m} \end{aligned} \tag{6B-4}$$

The phasor diagram is given in Fig. 6B-1.

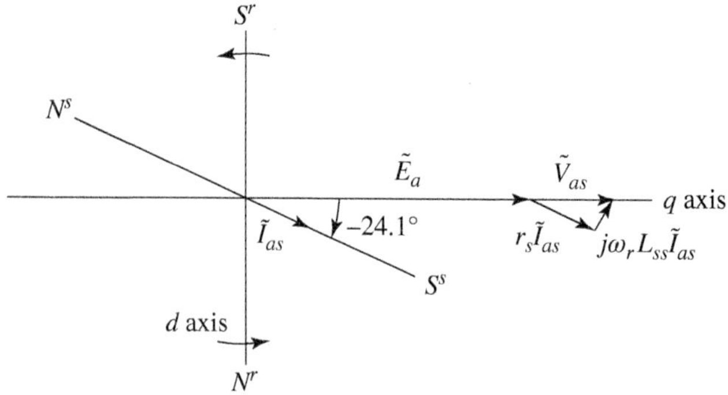

Figure 6B-1 Phasor diagram for $\phi_v = 0$.

SP6.6-1 What is ϕ_v in Fig. 6.6-1? [(6.5-17)]

SP6.6-2 What is the justification of assuming $\widetilde{E}_a$ is at zero degrees? [We will see later that it is controlled to zero]

SP6.6-3 What would be the phase angle of $\widetilde{I}_{as}$ in Fig. 6.6-1 for maximum motor torque for a given stator current? Why? [$\widetilde{I}_{as}$ at zero degrees. Stator and rotor poles would be orthogonal]

6.7 THREE-PHASE PERMANENT-MAGNET AC MACHINE

We will find that if we replace $L_{ss} = L_{ls} + L_{ms}$ in Table 6.6-1 with $L_{ls} + \frac{3}{2}L_{ms}$, the voltage equations are valid for a three-phase machine. This comes about due to the mutual coupling between stator phases. As a result of the $\frac{2}{3}$ in Park's transformation, we must also multiply the torque (power) equation for the two-phase machine by $\frac{3}{2}$. Otherwise, the equations given in Table 6.6-1 are valid for a three-phase machine. Some may wish to skip this section, or cover the first part up to the change of variables, (6.7-15).

A two-pole three-phase permanent-magnet ac machine is shown in Fig. 6.7-1. The stator windings are identical, displaced relative to each other by $\frac{2}{3}\pi$. The windings are sinusoidally distributed, each with N_s equivalent turns and resistance r_s. Electromagnetic torque is produced by the interaction of the poles of the permanent-magnet rotor and the poles resulting from the rotating air-gap mmf established by currents flowing in the stator windings. The rotating air-gap mmf established by symmetrical three-phase stator windings carrying balanced three-phase current is given by (4.4-18). It differs from that of the two-phase rotating air-gap mmf by the $\frac{3}{2}$ factor.

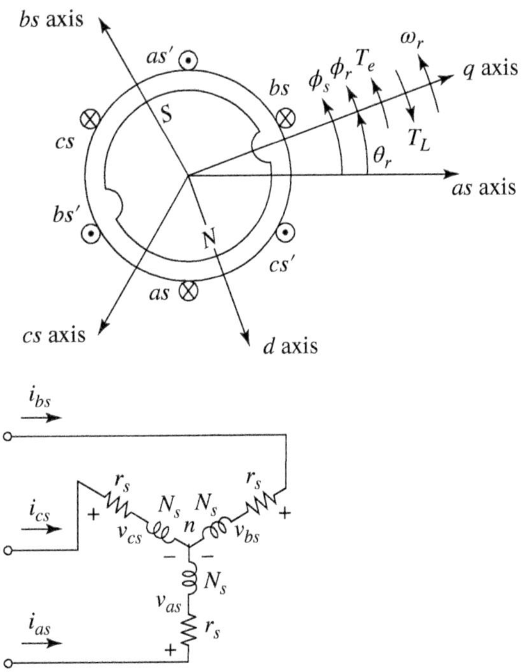

Figure 6.7-1 Two-pole three-phase permanent-magnet ac machine.

Voltage Equations and Winding Inductances

The voltage equations for the two-pole three-phase permanent-magnet ac machine shown in Fig. 6.7-1 may be expressed as

$$v_{as} = r_s i_{as} + \frac{d\lambda_{as}}{dt} \tag{6.7-1}$$

$$v_{bs} = r_s i_{bs} + \frac{d\lambda_{bs}}{dt} \tag{6.7-2}$$

$$v_{cs} = r_s i_{cs} + \frac{d\lambda_{cs}}{dt} \tag{6.7-3}$$

In matrix form,

$$\mathbf{v}_{abcs} = \mathbf{r}_s \mathbf{i}_{abcs} + p\boldsymbol{\lambda}_{abcs} \tag{6.7-4}$$

For voltages, currents, and flux linkages,

$$(\mathbf{f}_{abcs})^T = [f_{as} \;\; f_{bs} \;\; f_{cs}] \tag{6.7-5}$$

and

$$\mathbf{r}_s = \begin{bmatrix} r_s & 0 & 0 \\ 0 & r_s & 0 \\ 0 & 0 & r_s \end{bmatrix} \tag{6.7-6}$$

The flux linkage equations may be expressed as

$$\lambda_{as} = L_{asas} i_{as} + L_{asbs} i_{bs} + L_{ascs} i_{cs} + \lambda_{asm} \tag{6.7-7}$$

$$\lambda_{bs} = L_{bsas} i_{as} + L_{bsbs} i_{bs} + L_{bscs} i_{cs} + \lambda_{bsm} \tag{6.7-8}$$

$$\lambda_{cs} = L_{csas} i_{as} + L_{csbs} i_{bs} + L_{cscs} i_{cs} + \lambda_{csm} \tag{6.7-9}$$

In matrix form,

$$\boldsymbol{\lambda}_{abcs} = \mathbf{L}_s \mathbf{i}_{abcs} + \boldsymbol{\lambda}'_m \tag{6.7-10}$$

where $\boldsymbol{\lambda}'_m$ is the column vector:

$$\boldsymbol{\lambda}'_m = \begin{bmatrix} \lambda_{asm} \\ \lambda_{bsm} \\ \lambda_{csm} \end{bmatrix} = \lambda'_m \begin{bmatrix} \sin\theta_r \\ \sin\left(\theta_r - \frac{2}{3}\pi\right) \\ \sin\left(\theta_r + \frac{2}{3}\pi\right) \end{bmatrix} \tag{6.7-11}$$

In (6.7-11), λ'_m is the amplitude of the flux linkages established by the permanent magnet referred to the stator phase windings. In other words, the magnitude of $p\lambda_{asm}$, $p\lambda_{bsm}$, and $p\lambda_{csm}$ is proportional to the magnitude of the open-circuit voltage induced in the respective stator phase windings if the rotor were driven. The rotor position θ_r is defined by (6.3-10).

For our purposes, assume that the reluctance is the same in the q and d axis; whereupon, $\mathbf{L}_s$ may be written as

$$\mathbf{L}_s = \begin{bmatrix} L_{ls} + L_{ms} & -\frac{1}{2}L_{ms} & -\frac{1}{2}L_{ms} \\ -\frac{1}{2}L_{ms} & L_{ls} + L_{ms} & -\frac{1}{2}L_{ms} \\ -\frac{1}{2}L_{ms} & -\frac{1}{2}L_{ms} & L_{ls} + L_{ms} \end{bmatrix} \tag{6.7-12}$$

where L_{ls} represents the leakage inductance and L_{ms} the magnetizing inductance per phase. The off-diagonal terms are $-\frac{1}{2}L_{ms}$ since the mutual inductance between two stator windings displaced by $\frac{2}{3}\pi$ is $\cos\frac{2}{3}\pi$ times the mutual inductance when their magnetic axes coincide; that is, $L_{ms}\cos\frac{2}{3}\pi$.

Torque

Since the system is assumed to be magnetically linear, the field and coenergy are equal, thus,

$$\begin{aligned} W_c = {} & \frac{1}{2}L_{ss}\left(i_{as}^2 + i_{bs}^2 + i_{cs}^2\right) - \frac{1}{2}L_{ms}\left(i_{as}i_{bs} + i_{bs}i_{cs} + i_{cs}i_{as}\right) + \lambda_m' i_{as}\sin\theta_r \\ & + \lambda_m' i_{bs}\sin\left(\theta_r - \frac{2}{3}\pi\right) + \lambda_m' i_{cs}\sin\left(\theta_r + \frac{2}{3}\pi\right) + W_{pm} \end{aligned} \tag{6.7-13}$$

where W_{pm} is the energy associated with the permanent magnet, which is assumed constant. Taking the partial derivative with respect to θ_r and multiplying by $P/2$ yields

$$T_e = \frac{P}{2}\lambda_m'\left[i_{as}\cos\theta_r + i_{bs}\cos\left(\theta_r - \frac{2}{3}\pi\right) + i_{cs}\cos\left(\theta_r + \frac{2}{3}\pi\right)\right] \tag{6.7-14}$$

The above expression is positive for motor action. The torque and speed may be related as given by (6.4-3).

The following reference frame derivation may be de-emphasized and (6.6-5) and (6.6-6) may be used for the steady-state phasors and (6.6-9) for torque with the following modifications for the three-phase machine. Replace L_{ms} with $\frac{3}{2}L_{ms}$ and multiply T_e by $\frac{3}{2}$. This information is summarized in Table 6.6-1 for the two-phase machine. With these changes in mind, you can proceed to Section 6.8 or Section 6.9; however, it would be beneficial to work SP6.7-1 and SP6.7-3 on the way.

Since there are three stator variables (f_{as}, f_{bs}, and f_{cs}), recall from Chapter 5 that we must use three substitute variables in the transformation of the stator variables to the rotor reference frame. In particular,

$$\mathbf{f}^r_{qd0s} = \mathbf{K}^r_s \mathbf{f}_{abcs} \tag{6.7-15}$$

where

$$\left(\mathbf{f}^r_{qd0s}\right)^T = \left[f^r_{qs} \;\; f^r_{ds} \;\; f_{0s}\right] \tag{6.7-16}$$

$$\mathbf{K}^r_s = \frac{2}{3}\begin{bmatrix} \cos\theta_r & \cos\left(\theta_r - \frac{2}{3}\pi\right) & \cos\left(\theta_r + \frac{2}{3}\pi\right) \\ \sin\theta_r & \sin\left(\theta_r - \frac{2}{3}\pi\right) & \sin\left(\theta_r + \frac{2}{3}\pi\right) \\ \frac{1}{2} & \frac{1}{2} & \frac{1}{2} \end{bmatrix} \tag{6.7-17}$$

and where the rotor position θ_r is defined by (6.3-10). The inverse of $\mathbf{K}^r_s$ is

$$\left(\mathbf{K}^r_s\right)^{-1} = \begin{bmatrix} \cos\theta_r & \sin\theta_r & 1 \\ \cos\left(\theta_r - \frac{2}{3}\pi\right) & \sin\left(\theta_r - \frac{2}{3}\pi\right) & 1 \\ \cos\left(\theta_r + \frac{2}{3}\pi\right) & \sin\left(\theta_r + \frac{2}{3}\pi\right) & 1 \end{bmatrix} \tag{6.7-18}$$

It is important to recall that the same notation $\left(\mathbf{K}^r_s\right)$ was used for the two-phase transformation. Equation (6.7-17) is essentially Park's original transformation for the synchronous generator [2]. The zero variable, f_{0s}, is the third substitute variable and it is zero for balanced and even some unbalanced conditions. Also, f_{0s} is not a function of θ_r; therefore, the $0s$-variables are associated with stationary circuits and a raised index is not incorporated with the zero variables. Since we are dealing with a three-wire symmetrical stator as shown in Fig. 6.7-1, the f_{0s} variables will be zero.

Although we realize that the q^r_s and d^r_s voltage equations are going to be essentially those of a two-phase machine, we will still go through the derivation for the sake of completeness. Substituting the change of variables into the stator voltage equations given by (6.7-4) yields

$$\left(\mathbf{K}^r_s\right)^{-1}\mathbf{v}^r_{qd0s} = \mathbf{r}_s\left(\mathbf{K}^r_s\right)^{-1}\mathbf{i}^r_{qd0s} + p\left[\left(\mathbf{K}^r_s\right)^{-1}\boldsymbol{\lambda}^r_{qd0s}\right] \tag{6.7-19}$$

Premultiplying each side by $\mathbf{K}^r_s$ yields

$$\mathbf{v}^r_{qd0s} = \mathbf{r}_s\mathbf{i}^r_{qd0s} + \omega_r\boldsymbol{\lambda}^r_{dqs} + p\boldsymbol{\lambda}^r_{qd0s} \tag{6.7-20}$$

where

$$\left(\boldsymbol{\lambda}^r_{dqs}\right)^T = \left[\lambda^r_{ds} \;\; -\lambda^r_{qs} \;\; 0\right] \tag{6.7-21}$$

If one wishes to work through the steps necessary to go from (6.7-19) to (6.7-20), the trigonometric relations given in Appendix A will be very helpful.

For a magnetically linear system, the stator flux linkages are expressed by (6.7-10). Substituting the change of variables into (6.7-10) yields

$$(\mathbf{K}_s^r)^{-1}\boldsymbol{\lambda}_{qd0s}^r = \mathbf{L}_s(\mathbf{K}_s^r)^{-1}\mathbf{i}_{qd0s}^r + \boldsymbol{\lambda}_m' \tag{6.7-22}$$

Premultiplying by $\mathbf{K}_s^r$ yields

$$\boldsymbol{\lambda}_{qd0s}^r = \begin{bmatrix} L_{ls} + \frac{3}{2}L_{ms} & 0 & 0 \\ 0 & L_{ls} + \frac{3}{2}L_{ms} & 0 \\ 0 & 0 & L_{ls} \end{bmatrix} \begin{bmatrix} i_{qs}^r \\ i_{ds}^r \\ i_{0s} \end{bmatrix} + \lambda_m'^r \begin{bmatrix} 0 \\ 1 \\ 0 \end{bmatrix} \tag{6.7-23}$$

To be consistent with our previous notation, we have added the superscript r to λ_m'. In expanded form,

$$v_{qs}^r = r_s i_{qs}^r + \omega_r \lambda_{ds}^r + p\lambda_{qs}^r \tag{6.7-24}$$

$$v_{ds}^r = r_s i_{ds}^r - \omega_r \lambda_{qs}^r + p\lambda_{ds}^r \tag{6.7-25}$$

$$v_{0s} = r_s i_{0s} + p\lambda_{0s} \tag{6.7-26}$$

where

$$\lambda_{qs}^r = L_{ss} i_{qs}^r \tag{6.7-27}$$

$$\lambda_{ds}^r = L_{ss} i_{ds}^r + \lambda_m'^r \tag{6.7-28}$$

$$\lambda_{0s} = L_{ls} i_{0s} \tag{6.7-29}$$

and

$$L_{ss} = L_{ls} + \frac{3}{2}L_{ms} \tag{6.7-30}$$

We must be careful here. When dealing with the two-phase machine, we also used L_{ss} where it was $L_{ls} + L_{ms}$ rather than $L_{ls} + \frac{3}{2}L_{ms}$. The difference occurs due to the mutual coupling between the stator windings of a three-phase Y-connected stator. Although we should probably use a different notation for the two- and three-phase machines, we will not; instead, we will call attention to this difference as we go along.

Substituting (6.7-27) through (6.7-29) into (6.7-24) through (6.7-26) and since $\lambda_m'^r$ is constant, $p\lambda_m'^r = 0$, and we can write

$$v_{qs}^r = (r_s + pL_{ss})i_{qs}^r + \omega_r L_{ss} i_{ds}^r + \omega_r \lambda_m'^r \tag{6.7-31}$$

$$v_{ds}^r = (r_s + pL_{ss})i_{ds}^r - \omega_r L_{ss} i_{qs}^r \tag{6.7-32}$$

$$v_{0s} = (r_s + pL_{ls})i_{0s} \tag{6.7-33}$$

Note that v_{qs}^r and v_{ds}^r, (6.7-24) and (6.7-25) or (6.7-31) and (6.7-32), are identical in form to v_{qs}^r and v_{ds}^r for the two-phase machine given by (6.5-1) and (6.5-2) or (6.5-9) and (6.5-10), respectively. The above equations may be written in matrix form as

$$\begin{bmatrix} v_{qs}^r \\ v_{ds}^r \\ v_{0s} \end{bmatrix} = \begin{bmatrix} r_s + pL_{ss} & \omega_r L_{ss} & 0 \\ -\omega_r L_{ss} & r_s + pL_{ss} & 0 \\ 0 & 0 & r_s + pL_{ls} \end{bmatrix} \begin{bmatrix} i_{qs}^r \\ i_{ds}^r \\ i_{0s} \end{bmatrix} + \begin{bmatrix} \omega_r \lambda'^r_m \\ 0 \\ 0 \end{bmatrix} \tag{6.7-34}$$

The expression for electromagnetic torque is

$$T_e = \frac{3}{2}\frac{P}{2}\lambda'^r_m i_{qs}^r \tag{6.7-35}$$

which is positive for motor action. Please be aware that i_{qs}^r in (6.7-35) is Park's i_{qs}^r which is $\frac{2}{3}$ of the bracketed terms in (6.7-14). It is important to emphasize again that for the vast majority of operating conditions the rotor-reference frame voltage equations and torque equation for a three-phase machine are the same as the two-phase machine given in Table 6.6-1 except for the $\frac{3}{2}$ factor in L_{ss} and in T_e and P_e. Moreover, the steady-state phasor voltage equation given for $\tilde{V}_{as}$ is valid for a three-phase machine if we use $\frac{3}{2}L_{ms}$ when calculating L_{ss}. Clearly, the equivalent circuits given in Fig. 6.5-1b for the two-phase machine are valid for the three-phase if L_{ms} is replaced with $\frac{3}{2}L_{ms}$ in (6.5-6) through (6.5-10).

SP6.7-1 A four-pole three-phase permanent-magnet ac motor is operating with $\theta_r(0) = 0$, $V_{as} = 10\cos\omega_r t$, and $I_{as} = \cos(\omega_r t - 20°)$. Calculate λ'^r_m if $r_s = 4\ \Omega$, $L_{ss} = 0.01$ H, and $\omega_r = 146$ rad/s. $[\lambda'^r_m = 0.0393\ \text{V}\cdot\text{s/rad}]$

SP6.7-2 Repeat SP6.7-1 neglecting $\omega_r^2 L_{ss}^2$. $[\lambda'^r_m = 0.0427\ \text{V}\cdot\text{s/rad}]$

SP6.7-3 Calculate (a) the input power P_e, (b) T_e, and (c) efficiency for the operating conditions given in SP6.7-1. [(a) $P_e = 14.1$ W; (b) $T_e = 0.1108$ N · m; (c) eff = 57.4 %]

SP6.7-4 The expression for torque of a two-phase two-pole permanent-magnet ac machine is $T_e = \lambda'^r_m i_{qs}^r$. (a) What is the torque expression for a two-phase multipole permanent-magnet ac machine? (b) Multipole, three-phase machine? [(a) $T_e = \frac{P}{2}\lambda'^r_m i_{qs}^r$, (b) $\frac{3}{2}$ times (a) or (6.7-35)]

6.8 UNEQUAL DIRECT- AND QUADRATURE-AXIS INDUCTANCES

In our analysis of the permanent-magnet ac machine, we have assumed that the reluctance of the permanent magnet rotor is the same in the q and d axes. This assumption simplifies the analysis and allows us to portray the main operating modes of the brushless *dc* drive without significant error. There is, however, a difference in the q- and d-axis reluctances that produces a reluctance torque in addition to the main torque produced due to the interaction of the rotating magnetic

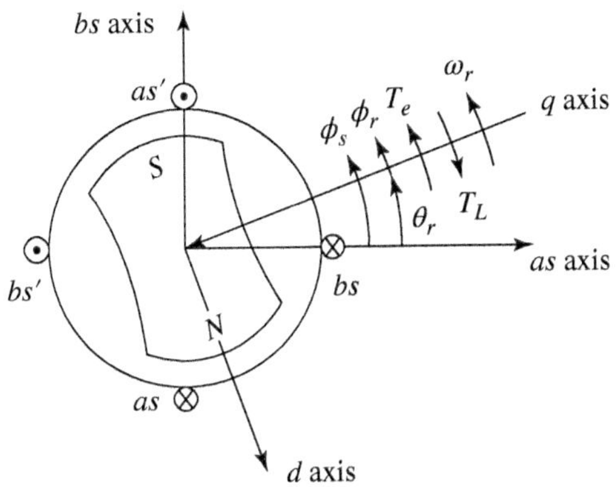

Figure 6.8-1 Two-phase permanent-magnet ac machine with unequal q- and d-axis reluctances.

field and the permanent-magnet rotor. This reluctance torque can be significant; therefore, the influence of the magnetically unsymmetrical rotor on the voltage and torque expressions warrants some discussion.

A two-phase permanent-magnet ac machine with unequal q- and d-axis inductances is shown in Fig. 6.8-1. Following the procedure set forth in Sections 1.7 and 2.7, we can express the stator inductance as

$$\mathbf{L}_s = \begin{bmatrix} L_{ls} + L_A - L_B \cos 2\theta_r & -L_B \sin 2\theta_r \\ -L_B \sin 2\theta_r & L_{ls} + L_A + L_B \cos 2\theta_r \end{bmatrix} \tag{6.8-1}$$

Those not interested in the reference frame derivation may skip to Table 6.8-2.

Now, in the rotor reference frame,

$$\begin{aligned} \begin{bmatrix} \lambda_{qs}^r \\ \lambda_{ds}^r \end{bmatrix} &= \mathbf{K}_s^r \mathbf{L}_s \left(\mathbf{K}_s^r\right)^{-1} \mathbf{i}_{qds}^r + \mathbf{K}_s^r \lambda_m'^r \begin{bmatrix} \sin\theta_r \\ -\cos\theta_r \end{bmatrix} \\ &= \begin{bmatrix} L_q & 0 \\ 0 & L_d \end{bmatrix} \begin{bmatrix} i_{qs}^r \\ i_{ds}^r \end{bmatrix} + \lambda_m'^r \begin{bmatrix} 0 \\ 1 \end{bmatrix} \end{aligned} \tag{6.8-2}$$

where

$$\begin{aligned} L_q &= L_{ls} + L_A - L_B \\ &= L_{ls} + L_{mq} \end{aligned} \tag{6.8-3}$$

$$\begin{aligned} L_d &= L_{ls} + L_A + L_B \\ &= L_{ls} + L_{md} \end{aligned} \tag{6.8-4}$$

In the standard reluctance machine $L_{mq} < L_{md}$; however, in case of some permanent-magnet ac machines $L_{md} < L_{mq}$ due to the permanent magnets in the d axis.

Substituting (6.8-2) into (6.5-1) and (6.5-2) yields

$$v_{qs}^r = (r_s + pL_q)i_{qs}^r + \omega_r L_d i_{ds}^r + \omega_r \lambda_m'^r \tag{6.8-5}$$

$$v_{ds}^r = (r_s + pL_d)i_{ds}^r - \omega_r L_q i_{ds}^r \tag{6.8-6}$$

Now, Power = $T_e \omega_r$ + losses; therefore, if we multiply v_{qs}^r by i_{qs}^r and v_{ds}^r by i_{ds}^r and add the result, then the coefficients of ω_r is T_e. Thus,

$$T_e = \frac{P}{2}\left[\lambda_m'^r i_{qs}^r + (L_d - L_q)i_{qs}^r i_{ds}^r\right] \tag{6.8-7}$$

Note that if i_{qs} and i_{ds} are greater than zero and if $L_q > L_d$, then the last term of (6.8-7) is negative.

For steady-state operation, lower-case variables are replaced with upper-case variables (for notational emphasis) and derivative terms are neglected. The resulting steady-state voltage and torque equations are summarized in Table 6.8-1.

Following a procedure similar to that used in (6.6-2) through (6.6-6), the *as*-phase steady-state voltage equation for a two-phase permanent-magnet ac machine may be expressed as

$$\tilde{V}_{as} = (r_s + j\omega_r L_q)\tilde{I}_{as} + \tilde{E}_a \tag{6.8-8}$$

where

$$\begin{aligned}\tilde{E}_a &= \frac{\omega_r}{\sqrt{2}}\left[(L_d - L_q)I_{ds}^r + \lambda_m'^r\right]e^{j0°}\\ &= \frac{\omega_r}{\sqrt{2}}\left\{-(L_d - L_q)\sqrt{2}I_s \sin\left[\theta_{esi}(0) - \theta_r(0)\right] + \lambda_m'^r\right\}e^{j0°}\end{aligned} \tag{6.8-9}$$

The torque expression becomes

$$\begin{aligned}T_e &= \frac{P}{2}\left[\lambda_m'^r I_{qs}^r + (L_d - L_q)I_{qs}^r I_{ds}^r\right]\\ &= \frac{P}{2}\{\lambda_m'^r \sqrt{2}I_s \cos\left[\theta_{esi}(0) - \theta_r(0)\right]\\ &\quad - \frac{P}{2}\left\{(L_d - L_q)\sqrt{2}I_s \cos\left[\theta_{esi}(0) - \theta_r(0)\right]\sqrt{2}I_s \sin\left[\theta_{esi}(0) - \theta_r(0)\right]\right\}\end{aligned} \tag{6.8-10}$$

The previous steady-state phasor equations are summarized in Table 6.8-2.

SP6.8-1 The inductive reactances in Table 6.8-1 are determined by multiplying by ω_r. Why? $[\omega_r = \omega_e]$

SP6.8-2 Determine the reluctance torque if $\tilde{I}_{as}$ is in phase with $\tilde{E}_a = |\tilde{E}_a|\underline{/0°}$. [Zero since $I_{ds}^r = 0$]

TABLE 6.8-1 Voltage and Torque Equations in q_s^r and d_s^r Variables for Steady-State Operation of an Unequal Quadrature- and Direct-Axis Inductance Permanent-Magnet AC Machine

Two Phase

$$\sqrt{2}\, V_s \cos\phi_v = r_s I_{qs}^r + \omega_r L_d I_{ds}^r + \omega_r \lambda_m'^r$$

$$-\sqrt{2}\, V_s \sin\phi_v = r_s I_{ds}^r - \omega_r L_q I_{qs}^r$$

$$T_e = \frac{P}{2}\left[\lambda_m'^r I_{qs}^r + \left(L_d - L_q\right) I_{qs}^r I_{ds}^r\right]$$

$$\omega_r = \omega_e$$

$$\phi_v = \theta_{esv} - \theta_r$$

$$= \theta_{esv}(0) \qquad \text{[in the steady state if } \theta_r(0) = 0\text{]}$$

$$L_q = L_{ls} + L_{mq}$$

$$L_d = L_{ls} + L_{md}$$

Three Phase

$$L_q = L_{ls} + \tfrac{3}{2} L_{mq}$$

$$L_d = L_{ls} + \tfrac{3}{2} L_{md}$$

Multiply T_e by $\frac{3}{2}$

TABLE 6.8-2 Voltage and Torque Equations for Steady-State Operation of an Unequal Direct- and Quadrature-Axis Inductance Permanent AC Machine, Expressed in Phasor Variables

Two Phase

$$\tilde{V}_{as} = \left(r_s + j\omega_r L_q\right)\tilde{I}_{as} + \tilde{E}_a$$

$$\tilde{E}_a = \frac{\omega_r}{\sqrt{2}}\left\{-\left(L_d - L_q\right)\sqrt{2} I_s \sin\left[\theta_{esi}(0) - \theta_r(0)\right] + \lambda_m'^r\right\} e^{j0°}$$

$$T_e = \frac{P}{2}\left\{\lambda_m'^r \sqrt{2} I_s \cos\left[\theta_{esi}(0) - \theta_r(0)\right]\right\}$$

$$-\frac{P}{2}\left\{\left(L_d - L_q\right)\sqrt{2}\, I_s \cos\left[\theta_{esi}(0) - \theta_r(0)\right]\sqrt{2} I_s \sin\left[\theta_{esi}(0) - \theta_r(0)\right]\right\}$$

Three Phase

$$L_q = L_{ls} + \tfrac{3}{2} L_{mq}$$

$$L_d = L_{ls} + \tfrac{3}{2} L_{md}$$

Multiply T_e by $\frac{3}{2}$

6.9 FIELD ORIENTATION OF A BRUSHLESS DC DRIVE

We are now ready to turn the permanent-magnet ac machine into what we called a brushless dc drive. To do this we must implement a method of sensing the position of the rotor poles and the rotor electrical angular velocity. Although this can be achieved in several ways, we will not become involved in these details in this introductory treatment of a brushless dc drive. Our purpose is to portray the possible operating characteristics of the machine–inverter combination and not the detailed design of the inverter control. With the rotor speed and the position of the rotor poles available, the dc-to-ac inverter is controlled so that the frequency of the fundamental component of the voltages applied to the stator windings is equal instantaneously to the electrical angular velocity of the rotor ω_r. Also, the switching of the inverter can be advanced or retarded thereby shifting the phase or time position of the phase voltages relative to the rotor poles.

In drive applications, the permanent-magnet ac machine is generally supplied from a voltage source inverter that is controlled to synchronize the frequency of the applied stator voltages with the electrical angular velocity of the rotor. Therefore, when the torque load on the shaft of the machine is increased, the machine slows and the drive inverter control decreases the frequency of the applied stator voltages which decreases the inductive reactances $(\omega_r L_{ss})$ and $\widetilde{E}_a$. Therefore, the decrease in rotor speed allows the current to increase, which, in turn, increases the strength of the stator rotating magnetic field to accommodate the increase in torque load. Although the purpose of the inverter is to control the frequency, it is also used to orient the rotating magnetic field of the stator relative to the permanent magnet of the rotor. This changes the relative position of the stator and rotor poles which changes the torque characteristics of the machine.

For purposes of the analysis of steady-state operation, it is generally assumed that $\widetilde{E}_a$ is at zero degrees and $\widetilde{V}_{as}$, the fundamental component of the as-phase voltage is at the phase angle $\theta_{esv}(0)$. Therefore, in the steady-state, from (6.5-17), $\phi_v = \theta_{esv}(0)$. If, for example, $\phi_v = 0$, then the inverter control ensures that the positive peak value of the fundamental component of v_{as} and the q axis are rotating in unison. That is, $\widetilde{E}_a$ and $\widetilde{V}_{as}$ would be at zero degrees if the phasor diagram is depicting the phasors each time the q axis is horizontal to the right; i.e., coinciding with the as axis or if we are running counterclockwise with the q axis, we would always see the peak value of v_{as}.

In this section, three control strategies are considered: $\phi_v = 0$, $\phi_v = \phi_{vMT/V}$, and $\phi_v = \phi_{vMT/A}$. When ϕ_v is controlled at $\phi_{vMT/V}$, the applied stator voltages are shifted relative to the q axis of the rotor to produce the maximum torque per volt that is possible at the instantaneous speed of the rotor (angular frequency of stator voltages). It is found that $\theta_{vMT/V}$ corresponds to the stator impedance angle which, of course, changes with frequency of the applied stator voltages (rotor speed). When ϕ_v is controlled at $\phi_{vMT/A}$, the applied stator voltages are shifted in phase relative to the q axis to produce the maximum torque per ampere possible at the instantaneous speed of the rotor. This occurs when $\widetilde{I}_{as}$ is in time phase with the q axis or in other words when $\widetilde{I}_{as}$ is orthogonal to the d axis; thus, the poles of the stator are orthogonal to the permanent magnet.

Before getting further into a discussion of the modes of operation it is helpful to talk a little more about ϕ_v. From (6.5-17),

$$\phi_v = \theta_{esv} - \theta_r \tag{6.9-1}$$

Assuming $\omega_r = \omega_e$, (6.9-1) becomes

$$\phi_v = \theta_{esv}(0) - \theta_r(0) \tag{6.9-2}$$

Also, for brushless dc machine operation, it is convenient to select $\theta_r(0)$ equal to zero whereupon

$$\phi_v = \theta_{esv}(0) \tag{6.9-3}$$

If steady-state operation is the main interest, the reference frame derivations of the different modes of operation may be de-emphasized and focus placed on the instantaneous and steady-state phasor voltage equations of the *as*-phase for each mode. Likewise, the q_s^r – and d_s^r – variables given in the plots of the operation of each mode can be de-emphasized. It is instructive to note the change in the instantaneous and steady-state phasor diagrams for the different modes of control.

Brushless DC Motor Operation with $\phi_v = 0$

The parameters of the fractional-horsepower four-pole three-phase permanent-magnet ac machine considered in this section are $r_s = 3.4\,\Omega$, $L_{ls} = 1.1$ mH, $L_{ms} = 11$ mH, and $\lambda_m' = 0.0826$ V·s/rad. Recall that L_{ms} must be multiplied by $\frac{3}{2}$ for a three-phase machine due to the mutual coupling of the stator windings when applying the transformation. For brushless dc drive operation, ω_e is made equal to ω_r and $\phi_v = 0$ is the common mode of operation, that is, the *as*-phase voltage phasor is controlled to be "in phase" with the q axis. Thus,

$$V_{as} = \sqrt{2}V_s \cos \omega_r t \tag{6.9-4}$$

with a balanced *abc* sequence and $V_s = 11.25$ V. With sinusoidal applied voltages and $\phi_v = 0$, the maximum positive value of V_{as} coincides with the q-axis and this unison is fixed by controlling the inverter. We will talk more about this in the following section.

The free-acceleration characteristics (starting from stall with $T_L = 0$) of brushless dc drive with $\phi_v = 0$ are shown in Fig. 6.9-1. The total inertia J of the rotor and mechanical load is 5×10^{-4} kg · m^2 and the damping coefficient B_m is neglected. The torque versus the rotor electrical angular velocity for Fig. 6.9-1 is shown in Fig. 6.9-2. The steady-state torque is also plotted in Fig. 6.9-2 for comparison purposes. The negative slope of the torque-speed characteristics ensures stable operation for motor operation. That is, for a given load torque, a slight slowing of the rotor from an operating point will cause T_e to increase, forcing the rotor back to the operating speed where $T_L = T_e$. A small increase in rotor speed causes T_e to decrease whereupon the load torque slows the rotor back to the original operating point; a stable operating condition.

Those not interested in reference frame derivations may skip to the free acceleration characteristics and focus on v_{as}, i_{as}, T_e, and ω_r in the plots. The voltage equation for a brushless dc motor during balanced steady-state operation may be obtained by setting V_{ds}^r, the imaginary part of $\widetilde{V}_{as}$, equal to zero (since $\phi_v = 0$, $V_{ds}^r = 0$). Solving (6.5-20) for I_{ds}^r yields

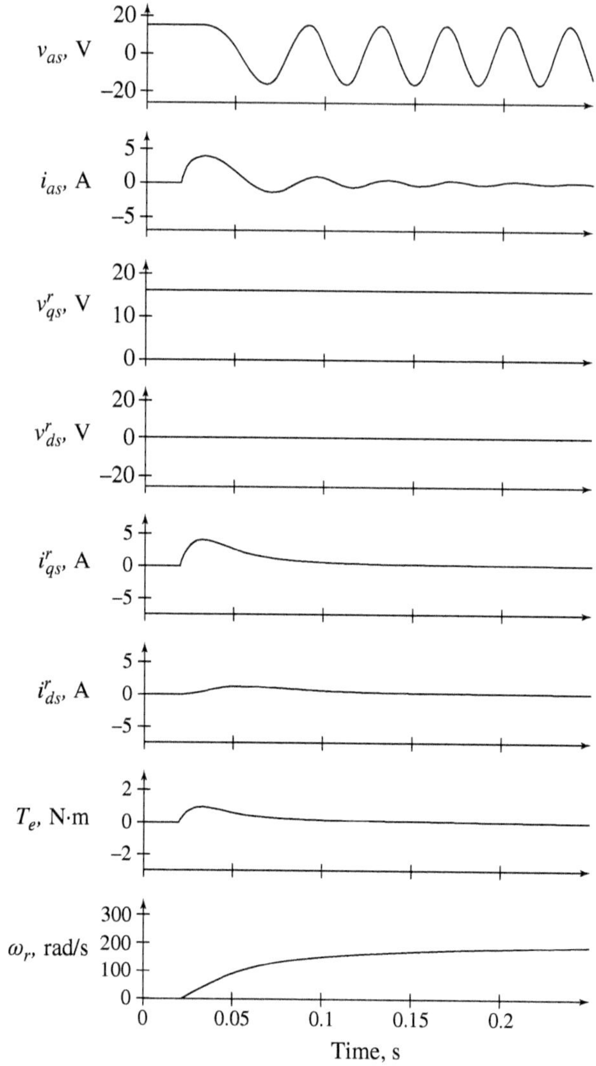

Figure 6.9-1 Free-acceleration characteristics of a brushless dc drive with $\phi_v = 0$ and a total inertia of 5×10^{-4} kg · m^2.

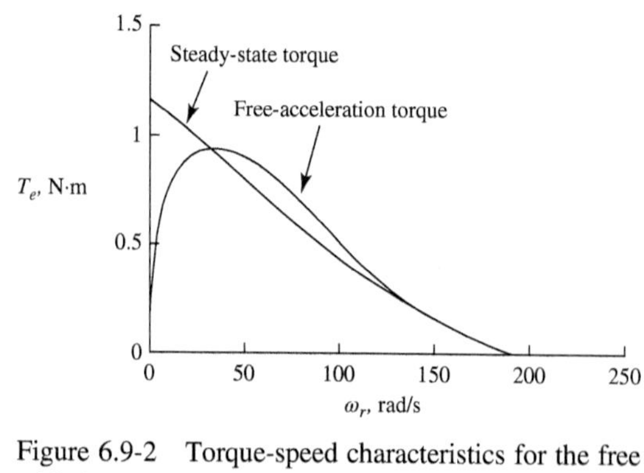

Figure 6.9-2 Torque-speed characteristics for the free acceleration shown in Fig. 6.9-1 with the steady-state torque shown for comparison purposes.

$$I_{ds}^{r} = \frac{\omega_r L_{ss}}{r_s} I_{qs}^{r} \tag{6.9-5}$$

Substituting (6.9-5) into (6.5-19) gives

$$V_{qs}^{r} = \frac{r_s^2 + \omega_r^2 L_{ss}^2}{r_s} I_{qs}^{r} + \omega_r \lambda_m'^r \tag{6.9-6}$$

During steady-state operation all quantities in (6.9-5) and (6.9-6) are constants.

If we choose to work with the *abc* variables, the phasor voltage equation is given by (6.6-5), that is

$$\widetilde{V}_{as} = (r_s + j\omega_r L_{ss})\widetilde{I}_{as} + \widetilde{E}_a \tag{6.9-7}$$

where from (6.6-6)

$$\widetilde{E}_a = \frac{1}{\sqrt{2}}\omega_r \lambda_m'^r \angle 0^\circ \tag{6.9-8}$$

We are dealing with a three-phase machine, thus $\frac{3}{2}L_{ms}$ is used in the evaluation of L_{ss}. For the three-phase brushless dc machine, the torque is given by (6.7-35) which is

$$\begin{aligned} T_e &= \frac{3}{2}\frac{P}{2}\lambda_m'^r i_{qs}^r \\ &= \frac{3}{2}\frac{P}{2}\lambda_m'^r \sqrt{2} I_s \cos\left[\theta_{esi}(0) - \theta_r(0)\right] \end{aligned} \tag{6.9-9}$$

where $\theta_r(0) = 0$.

In summary, the brushless dc drive is so named because the torque-versus-speed characteristics resemble those of a dc motor. Therefore, it seems logical that there is a reference frame where the voltage and current are dc rather than ac. This is very apparent from the traces in Fig. 6.9-1 of v_{qs}^r, i_{qs}^r, and T_e. Since the rotor and synchronous reference frames are the same, these variables are dc in the steady state and the steady-state torque-speed characteristics (Fig. 6.9-2) resemble those of a dc motor as shown in Fig. 3.4-1. It is also important to mention that the load torque determines the rotor speed, ω_r. The converter makes the frequency of the applied stator voltages equal to ω_r. For this reason ω_r is used in (6.9-7) and (6.9-8) to call attention to the fact.

It is instructive to consider the operating characteristics of the brushless dc motor during load torque changes as shown in Fig. 6.9-3. Initially, the machine is operating with $T_L = 0.1\,\text{N}\cdot\text{m}$. The load torque is suddenly stepped to $0.4\ \text{N}\cdot\text{m}$. The motor slows and once steady-state operation is established the load torque is stepped back to $0.1\ \text{N}\cdot\text{m}$. In this study, the inertia is $2 \times 10^{-4}\ \text{kg}\cdot\text{m}^2$. The decrease in the frequency of the electrical variables as the rotor slows and the increase in currents as the load is increased are evident.

Speed control can be achieved by varying the amplitude of the applied voltages. The steady-state torque-speed curves for various values of $|\widetilde{V}_{as}|$ are shown in Fig. 6.9-4.

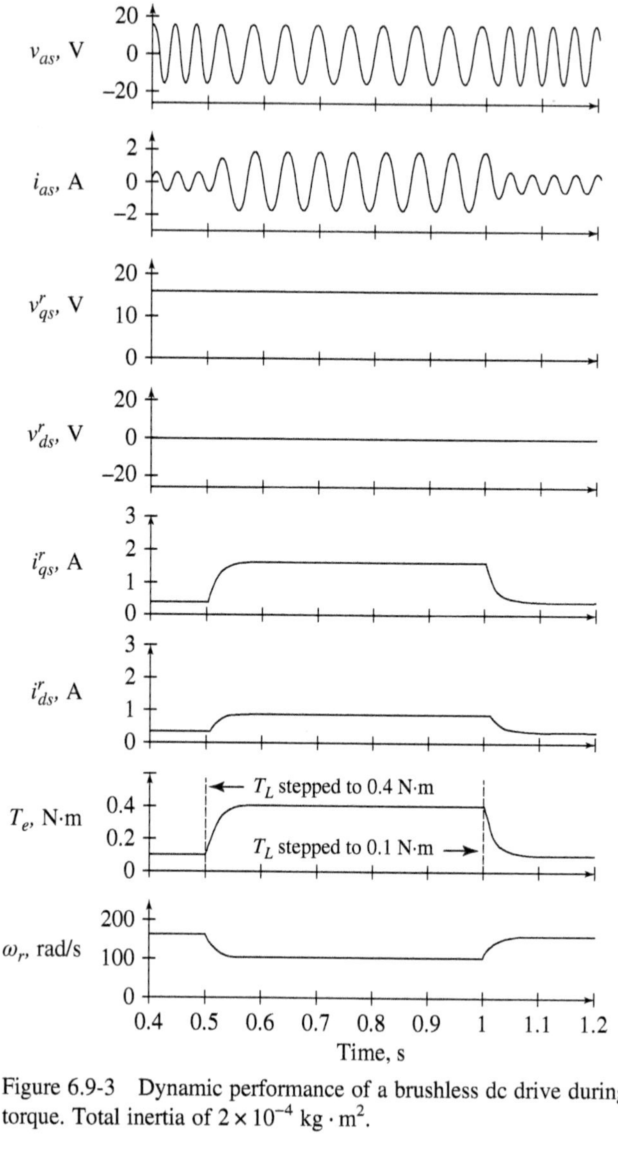

Figure 6.9-3 Dynamic performance of a brushless dc drive during step changes in load torque. Total inertia of 2×10^{-4} kg · m^2.

Example 6C Brushless DC Motor Operation with $\phi_v = 0$

A three-phase four-pole fractional-horsepower permanent-magnet ac machine has the following parameters: $r_s = 3.4\,\Omega$, $L_{ls} = 1.1$ mH, $L_{ms} = 11$ mH, and $\lambda'_m = 0.0826$ V·s/rad. The machine is wye-connected with balanced three-phase applied voltages. The as-phase voltage is $V_{as} = \sqrt{2}\,11.25\cos\omega_e t$. The actual steady-state rotor speed is 750 r/min. Calculate ω_e, T_e, and $\tilde{I}_{as}$ if $\phi_v = 0$ and draw a phasor diagram and show the stator and rotor poles. Also, calculate the power balance.

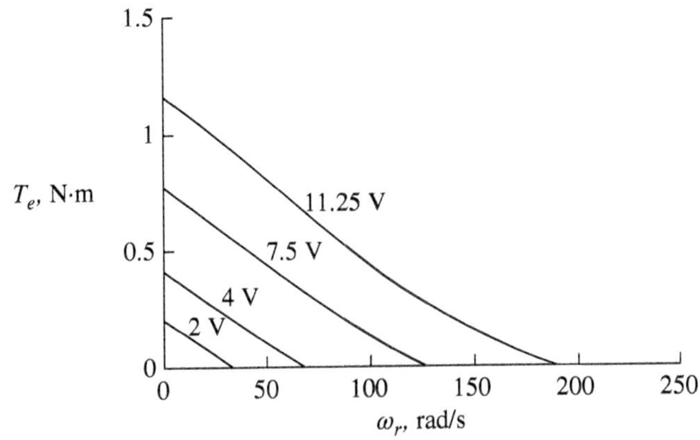

Figure 6.9-4 Torque-speed characteristics for $\phi_v = 0$ for different values of $|\widetilde{V}_{as}|$.

From (4.6-9) the electrical angular velocity of the rotor is

$$\begin{aligned} \omega_r &= \frac{P}{2}\omega_{rm} \\ &= \frac{P}{2}\frac{(\text{r/min})(\text{rad/r})}{\text{s/min}} \\ &= \frac{4}{2}\frac{(750)(2\pi)}{60} = 50\pi \text{ rad/s} \end{aligned} \tag{6C-1}$$

Using q_s^r and d_s^r Variables

The frequency of the voltages applied to the stator windings is equal to the electrical angular velocity of the rotor, therefore, $\omega_e = 50\pi$ rad/s. Now, since the q_s^r and d_s^r voltage equations are identical for the two- and three-phase machines except for the $\frac{3}{2}L_{ms}$ in L_{ss} and $\frac{3}{2}$ times the torque, we can use (6.5-21). But first, since $\phi_v = 0$ then from (5.5-16) with $\omega_r = \omega_e$,

$$\begin{aligned} \sqrt{2}\widetilde{V}_{as} &= V_{qs}^r - jV_{ds}^r \\ &= \sqrt{2}\, 11.25\angle 0^\circ \end{aligned} \tag{6C-2}$$

Therefore, $V_{ds}^r = 0$ and from (6.5-21) with the $\frac{3}{2}$ factor included

$$\begin{aligned} T_e &= \left(\frac{3}{2}\right)\left(\frac{P}{2}\right)\frac{r_s\lambda_m'^r}{r_s^2 + \omega_r^2 L_{ss}^2}\left(V_{qs}^r - \omega_r\lambda_m'^r\right) \\ &= \left(\frac{3}{2}\right)\left(\frac{4}{2}\right)\frac{(3.4)(0.0826)\left[\sqrt{2}11.25 - 50\pi(0.0826)\right]}{(3.4)^2 + (50\pi)^2\left[1.1 + \left(\frac{3}{2}\right)11\right]^2 \times 10^{-6}} \\ &= 0.129 \text{ N·m} \end{aligned} \tag{6C-3}$$

From (6.7-35),

$$\begin{aligned} I_{qs}^{r} &= \frac{2}{3}\frac{2}{P}\frac{T_e}{\lambda_m'^{r}} \\ &= \left(\frac{2}{3}\right)\left(\frac{2}{4}\right)\frac{(0.129)}{0.0826} = 0.521 \text{ A} \end{aligned} \tag{6C-4}$$

We can calculate I_{ds}^{r} from (6.9-5) since $V_{ds}^{r} = 0$:

$$\begin{aligned} I_{ds}^{r} &= \frac{\omega_r L_{ss}}{r_s} I_{qs}^{r} \\ &= \frac{50\pi\left(1.1 + \dfrac{3}{2}11\right) \times 10^{-3}(0.521)}{3.4} = 0.424 \text{ A} \end{aligned} \tag{6C-5}$$

From (5.5-16),

$$\sqrt{2}\tilde{I}_{as} = I_{qs}^{r} - jI_{ds}^{r}$$

$$\sqrt{2}\tilde{I}_{as} = 0.521 - j0.424 = 0.671\underline{/-39.1^\circ} \tag{6C-6}$$

Therefore,

$$\tilde{I}_{as} = \frac{1}{\sqrt{2}}(0.671)\underline{/-39.1^\circ} = 0.475\underline{/-39.1^\circ}\text{A} \tag{6C-7}$$

We need $\tilde{E}_a$, thus from (6.9-8),

$$\begin{aligned} \tilde{E}_a &= \frac{1}{\sqrt{2}}\omega_r \lambda_m'^{r} 0^\circ \\ &= \frac{1}{\sqrt{2}}(50\pi)(0.0826)\underline{/0^\circ} = 9.18\underline{/0^\circ}\text{V} \end{aligned} \tag{6C-8}$$

Using Phase Variables

Using phase variables we can solve for $\tilde{I}_{as}$ from (6.6-5) with $L_{ms} = L_{ls} + \frac{3}{2}L_{ms}$:

$$\begin{aligned} \tilde{I}_{as} &= \frac{\tilde{V}_{as} - \tilde{E}_a}{r_s + j\omega_r L_{ss}} \\ &= \frac{11.25\underline{/0^\circ} - 9.18\underline{/0^\circ}}{3.4 + j50\pi\left[1.1 + \left(\dfrac{3}{2}\right)11\right] \times 10^{-3}} \\ &= \frac{2.07\underline{/0^\circ}}{3.4 + j2.76} = \frac{2.07\underline{/0^\circ}}{4.38\underline{/39.1^\circ}} = 0.473\underline{/-39.1^\circ}\text{A} \end{aligned} \tag{6C-9}$$

The torque can be calculated using (6.6-9) multiplied by $\frac{3}{2}$, thus

$$\begin{aligned} T_e &= \left(\frac{3}{2}\right)\frac{P}{2}\lambda_m'\sqrt{2}I_s\cos\left[\theta_{esi}(0)-\theta_r(0)\right] \\ &= \frac{3}{2}\frac{4}{2}(0.0826)\sqrt{2}(0.473)\cos(-39.1^\circ-0) \\ &= 0.129\ \text{N·m} \end{aligned} \tag{6C-10}$$

The phasor diagram is given in Fig. 6C-1. We see from Fig. 6C-1 that the rotor poles can be considered as being "pulled" in the counterclockwise direction by the poles created by the stator currents; motor action.

The power balance for balanced conditions can be calculated starting with the power from the electrical system

$$P_e = \frac{3}{2}\left(v_{qs}^r i_{qs}^r + v_{ds}^r i_{ds}^r\right) \tag{6C-11}$$

or

$$P_e = 3\left|\tilde{V}_{as}\right|\left|\tilde{I}_{as}\right|\cos\left[\theta_{esv}(0)-\theta_{esi}(0)\right] \tag{6C-12}$$

Using (6C-11),

$$P_e = \frac{3}{2}\left(\sqrt{2}11.25\times 0.521\right) = 12.4\ \text{W} \tag{6C-13}$$

Using (6C-12),

$$\begin{aligned} P_e &= 3(11.25)(0.473)\cos\left[0-(-39.1^\circ)\right] \\ &= 12.4\ \text{W} \end{aligned} \tag{6C-14}$$

$$\begin{aligned} P_{\text{loss}} &= 3r_s\left|\tilde{I}_{as}\right|^2 \\ &= 3\times 3.4(0.473)^2 = 2.3\ \text{W} \end{aligned} \tag{6C-15}$$

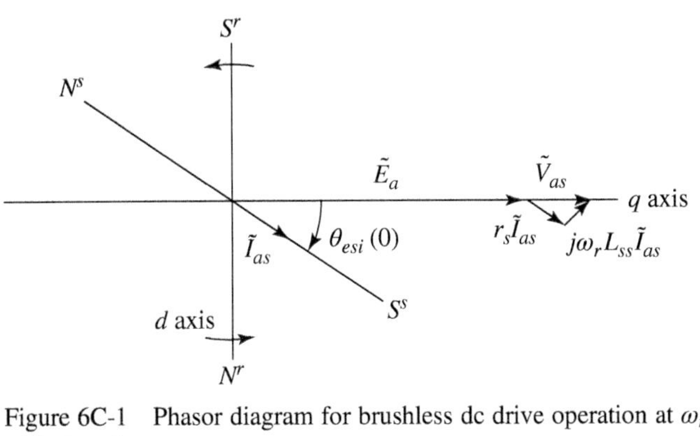

Figure 6C-1 Phasor diagram for brushless dc drive operation at $\omega_r = 50\pi$ rad/s with $\phi_v = 0$.

The power to the mechanical system is

$$P_m = T_e \omega_{rm}$$
$$= 0.129 \times \frac{2}{4} \times 50\pi = 10.13 \text{ W} \qquad (6C\text{-}16)$$

Note that $P_m = P_e - P_{loss}$ and the efficiency is

$$\text{eff} = \frac{P_m}{P_e} = \frac{10.13}{12.4} = 81.7\% \qquad (6C\text{-}17)$$

Maximum-Torque per Volt Operation of a Brushless DC Drive $\left(\phi_v = \phi_{vMT/V}\right)$

Although $\phi_v = 0$ is a common mode of operation of the brushless dc drive, researchers in [2, 3] discovered that advancing ϕ_v with respect to the q axis could increase the torque at rotor speeds greater than zero. This was shown analytically in [4] and illustrated by simulating the phase shifting (increasing ϕ_v) of the applied voltages to obtain maximum torque per volt $\left(\phi_v = \phi_{vMT/V}\right)$ at a given speed.

If the applied voltages and thus the stator poles are shifted relative to the magnetic field established by the permanent-magnet rotor, which is fixed in the d axis, the torque versus speed characteristics can be changed over a wide range by shifting ϕ_v from zero to 2π [4]. Here, we limit our discussion to shifting ϕ_v for the purpose of maximizing torque during motor operation. If desired, the reference frame derivation may be bypassed and the reader may proceed directly to (6.9-18).

Torque is proportional to i_{qs}^r and when ϕ_v is shifted from zero, v_{ds}^r is nonzero as expressed in (6.5-15). For the purpose of deriving an expression for the maximum torque per volt at a given rotor speed $\left(\phi_{vMT/V}\right)$, we will start with (6.9-10) and (6.9-11) for V_{qs}^r and V_{ds}^r, respectively. In particular, for steady-state operation,

$$V_{qs}^r = r_s I_{qs}^r + \omega_r L_{ss} I_{ds}^r + \omega_r \lambda_m'^r \qquad (6.9\text{-}10)$$

$$V_{ds}^r = r_s I_{ds}^r - \omega_r L_{ss} I_{qs}^r \qquad (6.9\text{-}11)$$

Please recall that these equations are valid for two- or three-phase devices for balanced steady-state operation if, for the three-phase device, $\frac{3}{2}L_{ms}$ is used when calculating L_{ss}. We also need the expressions for V_{qs}^r and V_{ds}^r as functions of ϕ_v; from (6.5-14) and (6.5-15), which are valid for transient and steady-state operation:

$$V_{qs}^r = \sqrt{2} V_s \cos \phi_v \qquad (6.9\text{-}12)$$

$$V_{ds}^r = -\sqrt{2}V_s \sin\phi_v \tag{6.9-13}$$

Since $\theta_r(0) = 0$, ϕ_v is the phase of $\widetilde{V}_{as}$.

Solving (6.9-11) for I_{ds}^r and substituting the result into (6.9-10) yields

$$V_{qs}^r = \frac{r_s^2 + \omega_r^2 L_{ss}^2}{r_s} I_{qs}^r + \frac{\omega_r L_{ss}}{r_s} V_{ds}^r + \omega_r \lambda_m'^r \tag{6.9-14}$$

Now, solving (6.9-14) for I_{qs}^r and substituting (6.9-12) and (6.9-13) for V_{qs}^r and V_{ds}^r, respectively, with $\theta_r(0) = 0$, we have

$$I_{qs}^r = \frac{r_s}{r_s^2 + \omega_r^2 L_{ss}^2}\left(\sqrt{2}V_s \cos\phi_v + \frac{\omega_r L_{ss}}{r_s}\sqrt{2}V_s \sin\phi_v - \omega_r \lambda_m'^r\right) \tag{6.9-15}$$

It is interesting to note from (6.9-15) that V_{ds}^r aids V_{qs}^r to increase I_{qs}^r for a given rotor speed. Since this results in a negative I_{ds}^r, it is often referred to as field weakening even though $\lambda_m'^r$ is not decreased in magnitude.

Since T_e is proportional to I_{qs}^r, (6.9-9), we can obtain the maximum torque for a given rotor speed by taking the derivative of I_{qs}^r with respect to ϕ_v and setting the result equal to zero and then solving for ϕ_v. Thus, from (6.9-15),

$$0 = -\sin\phi_v + \frac{\omega_r L_{ss}}{r_s}\cos\phi_v \tag{6.9-16}$$

whereupon

$$\frac{\sin\phi_v}{\cos\phi_v} = \frac{\omega_r L_{ss}}{r_s} \tag{6.9-17}$$

or

$$\phi_{vMT/V} = \tan^{-1}\frac{\omega_r L_{ss}}{r_s} \tag{6.9-18}$$

Equation (6.9-18) tells us that for a given positive rotor speed, $\phi_{vMT/V}$ will yield maximum possible torque per volt at that rotor speed.

The free-acceleration characteristics for $\phi_{vMT/V}$ are shown in Figs. 6.9-5 and 6.9-6. These characteristics may be compared to Figs. 6.9-1 and 6.9-2, respectively, where $\phi_v = 0$. Note the extended speed range with $\phi_{vMT/V}$ (Fig. 6.9-6) compared with $\phi_v = 0$ (Fig. 6.9-2). Also, note that i_{ds}^r is small positive in Fig. 6.9-1 but a larger negative value in Fig. 6.9-5. In other words, an increase in torque $\left(I_{qs}^r\right)$ and speed range occurs due to a decrease in I_{ds}^r (a larger negative value).

Figure 6.9-7 depicts the change in steady-state torque-speed characteristics for changes in $\left|\widetilde{V}_{as}\right|$.

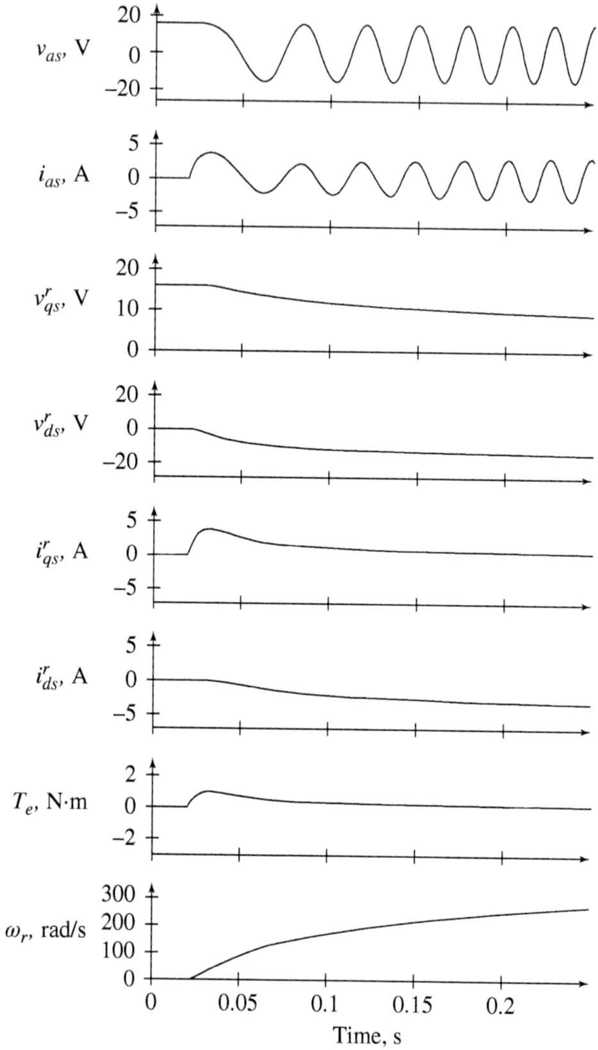

Figure 6.9-5 Free-acceleration characteristics of a brushless dc drive with $\phi_v = \phi_{vMT/V}$ and a total inertia of $5 \times 10^{-4} \mathrm{kg} \cdot \mathrm{m}^2$.

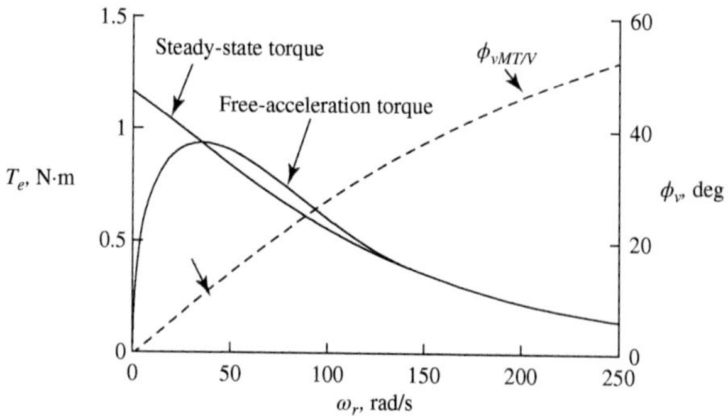

Figure 6.9-6 Torque-speed characteristics for free acceleration shown in Fig. 6.9-5 with the steady-state torque also shown. Compare to Fig. 6.9-2 where $\phi_v = 0$.

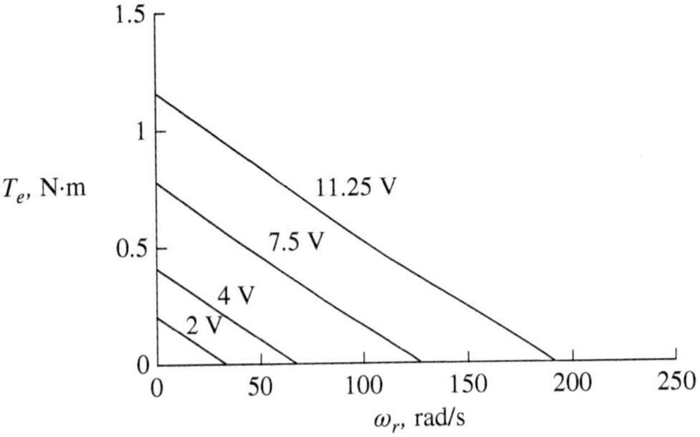

Figure 6.9-7 Torque-speed characteristics for $\phi_v = \phi_{vMT/V}$ for different values of $|\widetilde{V}_{as}|$.

Example 6D Brushless DC Drive Operation with $\phi_v = \phi_{vMT/V}$
In Example 6C, we determined the steady-state torque at 750 r/min ($\omega_r = 50\pi$ rad/s) with $\phi_v = 0$. In this example, we will calculate the torque at the same speed with $\phi_v = \phi_{vMT/V}$, draw the phasor diagram and locate the poles, and calculate the power balance. The parameters for the machine are repeated here for convenience: $r_s = 3.4\,\Omega$, $L_{ls} = 1.1$ mH, $L_{ms} = 11$ mH, and $\lambda'^r_m = 0.0826$ V · s/rad. From (6B-1), $\omega_r = 50\pi$ rad/s. Now, for a three-phase machine,

$$\begin{aligned} L_{ss} &= L_{ls} + \frac{3}{2}L_{ms} \\ &= \left(1.1 + \frac{3}{2}11\right) \times 10^{-3} = 17.6\,\text{mH} \end{aligned} \tag{6D-1}$$

From (6.9-18),

$$\begin{aligned} \phi_{vMT/V} &= \tan^{-1}\frac{\omega_r L_{ss}}{r_s} \\ &= \tan^{-1}\frac{50\pi(17.6)\times 10^{-3}}{3.4} = 39.1^\circ \end{aligned} \tag{6D-2}$$

Therefore,

$$V_{as} = \sqrt{2}11.25\cos(\omega_r t + 39.1^\circ) \tag{6D-3}$$

From (6.6-5),

$$\widetilde{V}_{as} = (r_s + j\omega_r L_{ss})\widetilde{I}_{as} + \widetilde{E}_a \tag{6D-4}$$

and from Example 6C, $\widetilde{E}_a = 9.18\angle 0^\circ$. Solving for $\widetilde{I}_{as}$:

$$\begin{aligned} \widetilde{I}_{as} &= \frac{\widetilde{V}_{as} - \widetilde{E}_a}{r_s + j\omega_r L_{ss}} \\ &= \frac{11.25\angle 39.1^\circ - 9.18\angle 0^\circ}{3.4 + j50\pi(17.6)\times 10^{-3}} = 1.62\angle 54.5^\circ\,\text{A} \end{aligned} \tag{6D-5}$$

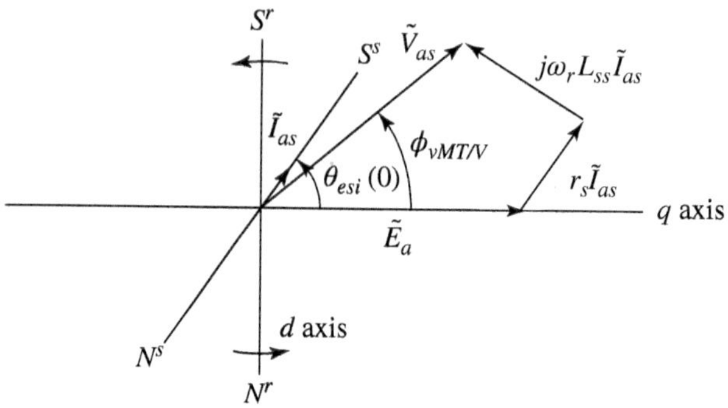

Figure 6D-1 Phasor diagram for brushless dc drive operation at $\omega_r = 50\pi$ rad/s with $\phi_v = \phi_{vMT/V}$.

From (6.9-9) the steady-state torque is

$$\begin{aligned} T_e &= \frac{3}{2}\frac{P}{2}\lambda_m'^r\sqrt{2}I_s\cos\theta_{esi}(0) \\ &= \left(\frac{3}{2}\right)\left(\frac{4}{2}\right)(0.0826)\left(\sqrt{2}\,1.62\cos 54.5^\circ\right) = 0.330\ \text{N}\cdot\text{m} \end{aligned} \tag{6D-6}$$

The phasor diagram is shown in Fig. 6D-1.

Calculating a power balance:

$$\begin{aligned} P_e &= 3\left|\tilde{V}_{as}\right|\left|\tilde{I}_{as}\right|\cos\left[\theta_{esv}(0) - \theta_{esi}(0)\right] \\ &= 3 \times 11.25 \times 1.62 \times \cos(39.1^\circ - 54.5^\circ) = 52.71\ \text{W} \end{aligned} \tag{6D-7}$$

$$\begin{aligned} P_{\text{loss}} &= 3r_s\left|\tilde{I}_{as}\right|^2 \\ &= 3 \times 3.4 \times (1.62)^2 = 26.77\ \text{W} \end{aligned} \tag{6D-8}$$

$$\begin{aligned} P_m &= T_e\omega_{rm} \\ &= 0.33 \times \frac{2}{4} \times 50\pi = 25.92\ \text{W} \end{aligned} \tag{6D-9}$$

The efficiency is

$$\text{eff} = \frac{P_m}{P_e} = \frac{25.92}{52.71} = 49.2\% \tag{6D-10}$$

From Fig. 6D-1, we see that the poles created by the stator can be considered as "pushing" the rotor poles in the counterclockwise direction. In Example 6C, T_e was calculated to be 0.129 N · m with $\phi_v = 0$. Here, with $\phi_{vMT/V}$, the torque is calculated to be 0.330 N · m; however, the magnitude of $\tilde{I}_{as}$ increased from 0.475 to 1.62 A. The torque is increased by a factor of 2.6 while the current is increased by a factor of 3.4. This is the maximum torque per volt that this device can produce at $\omega_r = 50\pi$ rad/s (750 r/min) with $V_s = 11.25$ V; however, when shifting the phase of the terminal voltage to achieve maximum torque, one must not exceed rated conditions for an extended period.

Maximum-Torque per Ampere Operation of a Brushless DC Drive $\left(\phi_v = \phi_{vMT/A}\right)$

Maximum-torque per ampere operation occurs when I^r_{ds} (imaginary part of $\widetilde{I}_{as}$) is made zero by controlling the position of $\widetilde{V}_{as}$ relative to the permanent magnet of the rotor $\left(\phi_v = \phi_{vMT/A}\right)$. The torque is directly related to the q-axis current (real part of $\widetilde{I}_{as}$). The d-axis current does not contribute to the torque, but decreases the efficiency of the machine as we have seen in Example 6D.

To derive an expression for $\phi_{vMT/A}$ for steady-state operation, we will substitute (6.9-12) and (6.9-13) into (6.9-10) and (6.9-11) for V^r_{qs} and V^r_{ds}, respectively, and solve for $\cos\phi_v$ and $\sin\phi_v$. If we set $I^r_{ds} = 0$ and perform several mathematical manipulations, we can express $\phi_{vMT/A}$, at a given rotor speed, as

$$\phi_{vMT/A} = \tan^{-1}\left[\omega_r\tau_s\left(\frac{-1 \pm \omega_r\tau_v\sqrt{1 + \omega_r^2\tau_v^2\left(1 - \omega_r^2\tau_v^2\right)}}{\omega_r^4\tau_s^2\tau_v^2 - 1}\right)\right] \tag{6.9-19}$$

where, for compactness,

$$\tau_s = \frac{L_{ss}}{r_s} \tag{6.9-20}$$

$$\tau_v = \frac{\lambda'^r_m}{\sqrt{2}V_s} \tag{6.9-21}$$

The free-acceleration characteristics are shown in Fig. 6.9-8. The torque-speed characteristics are shown in Fig. 6.9-9 including a plot of $\phi_{vMT/A}$.

The steady-state torque-speed characteristics for $\phi_{vMT/A}$ are shown in Fig. 6.9-10 and since (6.9-19) and (6.9-21) tell us that $\phi_{vMT/A}$ is a function of the magnitude of $\widetilde{V}_{as}$, Fig. 6.9-11 shows $\phi_{vMT/A}$ versus ω_r for each value of $\left|\widetilde{V}_{as}\right|$ shown in Fig. 6.9-10.

Example 6E Brushless DC Drive Operation with $\phi_v = \phi_{vMT/A}$

In this example, we will calculate the torque with $\phi_v = \phi_{vMT/A}$ and draw the phasor diagram and locate the stator and rotor poles. The parameters are repeated again; $r_s = 3.4\,\Omega$, $L_{ls} = 1.1$ mH, $L_{ms} = 11$ mH, and $\lambda'^r_m = 0.0826\ \text{V}\cdot\text{s/rad}$. In order to compare with Examples 6C and 6D, we will perform the calculations for $\omega_r = 50\pi$ rad/sec. From (6.9-19), $\phi_{vMT/A}$ is calculated to be 8.15° and 17° at $\omega_r = 50\pi$. The larger angle (17°) is an extraneous root. Also, $V_s = 11.25$ V, $L_{ss} = 17.6$ mH, and

$$V_{as} = \sqrt{2}\,11.25\cos\left(\omega_r t + 8.15°\right) \tag{6E-1}$$

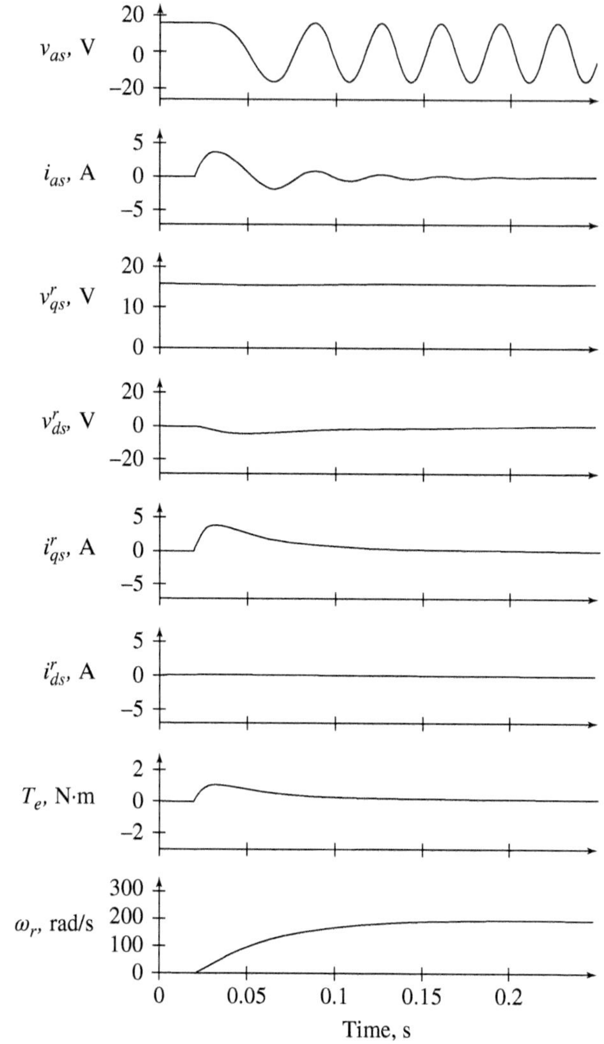

Figure 6.9-8 Free-acceleration characteristics of a brushless dc machine with $\phi_v = \phi_{vMT/A}$ $\left(I_{ds}^r = 0\right)$ and a total inertia of 5×10^{-4} kg · m^2.

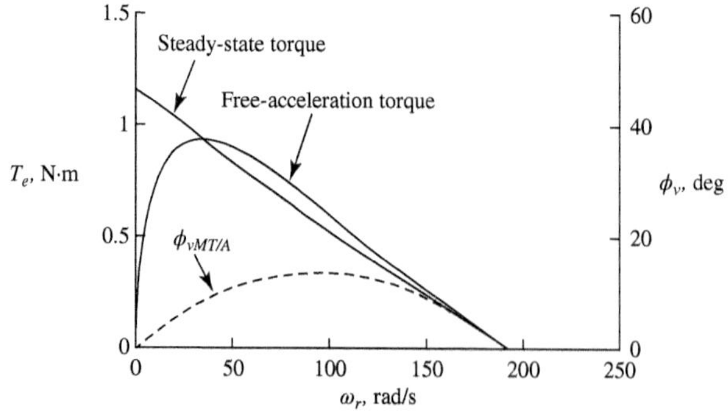

Figure 6.9-9 Torque-speed characteristics for free-acceleration shown in Fig. 6.9-8.

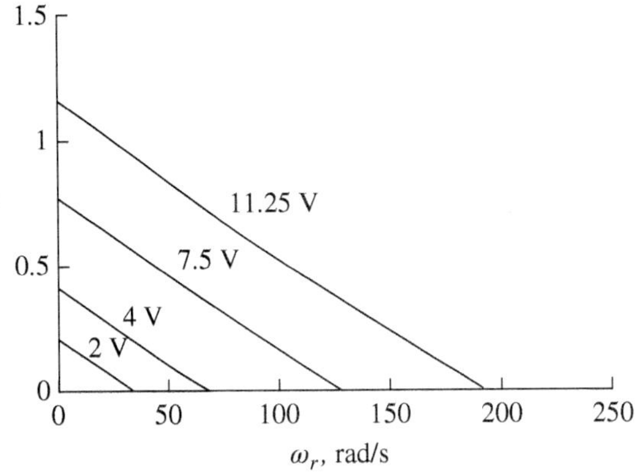

Figure 6.9-10 Torque-speed characteristics for $\phi_v = \phi_{vMT/A}$ for different values of $|\widetilde{V}_{as}|$.

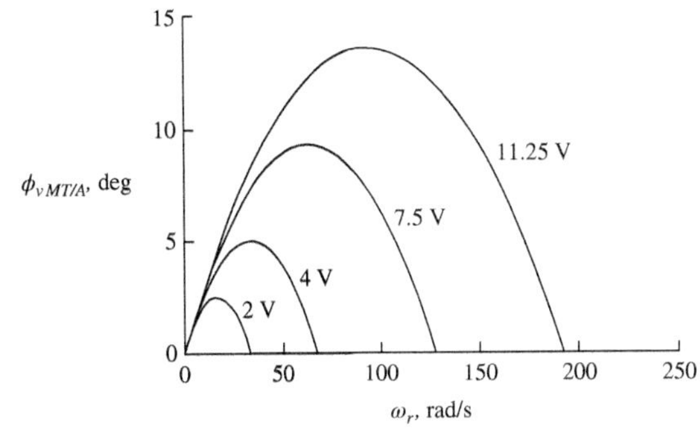

Figure 6.9-11 $\phi_{vMT/A}$ $(I_{ds}^r = 0)$ versus ω_r for different values of $|\widetilde{V}_{as}|$.

From (6.6-5),

$$\widetilde{V}_{as} = (r_s + j\omega_r L_{ss})\widetilde{I}_{as} + \widetilde{E}_a \tag{6E-2}$$

and again $\widetilde{E}_a = 9.18\angle 0°$. Solving for $\widetilde{I}_{as}$:

$$\begin{aligned}\widetilde{I}_{as} &= \frac{\widetilde{V}_{as} - \widetilde{E}_a}{r_s + j\omega_r L_{ss}} \\ &= \frac{11.25\underline{/8.15°} - 9.18\underline{/0°}}{3.4 + j50\pi(17.6)\times 10^{-3}} = 0.577\underline{/0°}\,\text{A}\end{aligned} \tag{6E-3}$$

From (6.9-9),

$$\begin{aligned}T_e &= \frac{3}{2}\frac{P}{2}\lambda_m'^r \sqrt{2} I_s \cos\theta_{esi}(0) \\ &= \left(\frac{3}{2}\right)\left(\frac{4}{2}\right)(0.0826)\left(\sqrt{2}\right)(0.577) = 0.202\ \text{N}\cdot\text{m}\end{aligned} \tag{6E-4}$$

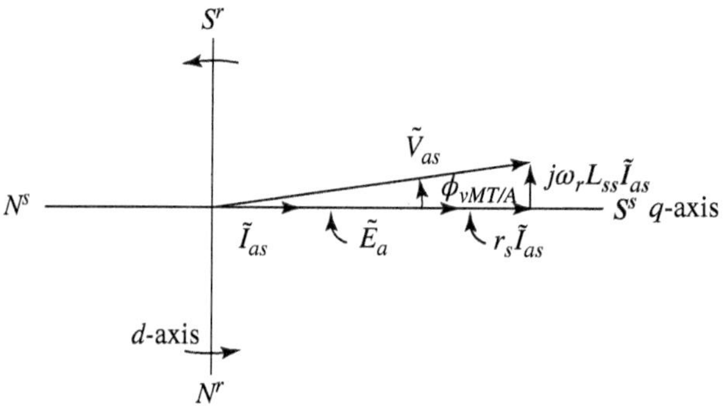

Figure 6E-1 Phasor diagram for brushless dc drive operation at $\omega_r = 50\pi$ rad/s with $\phi_v = \phi_{vMT/A}$.

The phasor diagram is shown in Fig. 6E-1. Note that the stator and rotor poles are orthogonal which yields the maximum torque per ampere for this device at $\omega_r = 50\pi$ rad/s.

The power balance becomes

$$\begin{aligned} P_e &= 3|\tilde{V}_{as}||\tilde{I}_{as}|\cos[\theta_{esv}(0) - \theta_{esi}(0)] \\ &= 3 \times 11.25 \times 0.577\cos(8.15^\circ - 0) = 19.3\ \text{W} \end{aligned} \tag{6E-5}$$

$$\begin{aligned} P_{loss} &= 3r_s|\tilde{I}_{as}|^2 \\ &= 3 \times 3.4(0.577)^2 = 3.4\ \text{W} \end{aligned} \tag{6E-6}$$

$$\begin{aligned} P_m &= T_e\omega_{rm} \\ &= 0.202 \times \frac{2}{4} \times 50\pi = 15.9\ \text{W} \end{aligned} \tag{6E-7}$$

The efficiency is

$$\text{eff} = \frac{P_m}{P_e} = \frac{15.9}{19.3} = 82.4\% \tag{6E-8}$$

The results of Examples 6C through 6E are summarized in Table 6.9-1. The three modes of operation $\phi_v = 0$, $\phi_v = \phi_{vMT/V}$, and $\phi_{vMT/A}$ were all performed at $\omega_r = 50\pi$ rad/s with the same parameters and $|\tilde{V}_{as}|$. Recall that $\sqrt{2}\,\tilde{I}_{as} = I_{qs}^r - jI_{ds}^r$.

A summary of expressions for the steady-state operation of the principle modes of operation of a brushless dc drive are given in Table 6.9-2. These expressions were obtained using only the fundamental component of the inverter output voltages and assuming ideal control action.

SP6.9-1 In Figs. 6.9-4, 6.9-7, and 6.9-10 it is noted that T_e at $\omega_r = 0$ are equal. Why? [ϕ_v is zero at all three]

SP6.9-2 Calculate the torque per volt, $T_{e/V}$, and torque per ampere, $T_{e/A}$, for the conditions given in Example 6C. Use rms values. [$T_{e/V} = 0.01147$ N·m/V; $T_{e/A} = 0.273$ N·m/A]

TABLE 6.9-1 Comparison of Steady-State Operating Modes Considered in Examples 6C Through 6E

ϕ_v (deg.)	T_e (N·m)	$\tilde{I}_{as}$ (A)	I^r_{qs} (A)	I^r_{ds} (A)	ω_r (rad/s)	eff (%)
0	0.129	$0.475\angle -39.1°$	0.521	0.424	50π	81.4
$\phi_{vMT/V}$ 39.1	0.330	$1.62\angle 54.5°$	1.33	−1.87	50π	49.2
$\phi_{vMT/A}$ 8.15	0.202	$0.577\angle 0°$	0.816	0	50π	82.4

TABLE 6.9-2 Voltage and Torque Expressions for Steady-State Operating Modes of a Brushless DC Drive – Phase Variables

$$\tilde{V}_{as} = (r_s + j\omega_r L_{ss})\tilde{I}_{as} + \tilde{E}_a$$

$$\tilde{E}_a = \frac{1}{\sqrt{2}}\omega_r \lambda'^r_m e^{j\theta_r(0)}$$

$$T_e = \frac{P}{2}\lambda'^r_m \sqrt{2} I_s \cos[\theta_{esi}(0) - \theta_r(0)]$$

$$\theta_r(0) = 0°$$

$$\phi_v = \theta_{esv}(0)$$

The brushless dc drives are generally three-phase; therefore,

- $L_{ss} = L_{ls} + \frac{3}{2}L_{ms}$
- T_e is multiplied by $\frac{3}{2}$

For $\phi_v = 0°$ and $\theta_r(0) = 0$

$$\theta_{esv}(0) = 0°$$

For $\phi_v = \phi_{vMT/V}$ and $\theta_r(0) = 0$

$$\phi_{vMT/V} = \tan^{-1}\frac{\omega_r L_{ss}}{r_s}$$

For $\phi_v = \phi_{vMT/A}$ and $\theta_r(0) = 0$

$$\phi_{vMT/A} = \tan^{-1}\left[\omega_r \tau_s \left(\frac{-1 \pm \omega_r \tau_v \sqrt{1 + \omega_r^2 \tau_v^2 (1 - \omega_r^2 \tau_v^2)}}{\omega_r^4 \tau_s^2 \tau_v^2 - 1}\right)\right]$$

where

$$\tau_s = \frac{L_{ss}}{r_s}$$

$$\tau_v = \frac{\lambda'^r_m}{\sqrt{2} V_s}$$

SP6.9-3 Repeat SP6.9-2 for Example 6D. $[T_{e/V} = 0.0293\ \text{N}\cdot\text{m/V}; T_{e/A} = 0.204\ \text{N}\cdot\text{m/A}]$

SP6.9-4 Repeat SP6.9-2 for Example 6E. $[T_{e/V} = 0.0179\ \text{N}\cdot\text{m/V};\ T_{e/A} = 0.350\ \text{N}\cdot\text{m/A}]$

6.10 INVERTER-SUPPLIED BRUSHLESS DC DRIVE

The brushless dc drive that we will consider consists of the three-phase six-step inverter shown in Fig. 6.10-1 and the three-phase permanent-magnet ac machine shown in Fig. 6.10-2. The three sensors shown in Fig. 6.10-2 are generally

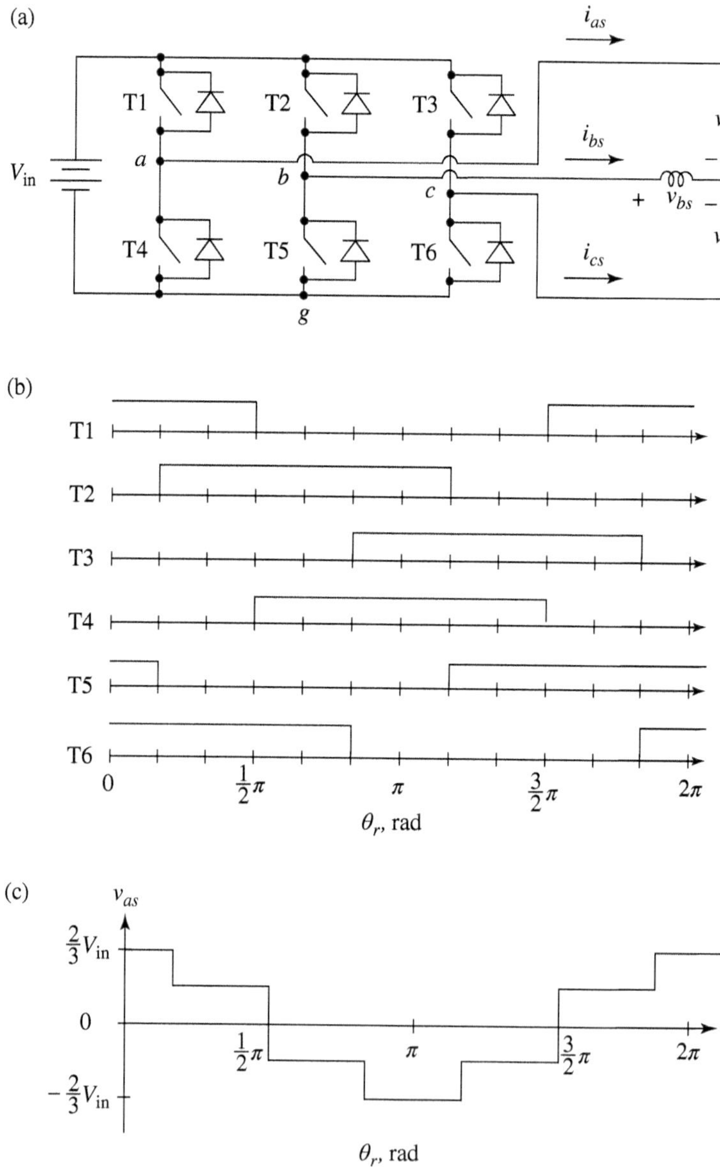

Figure 6.10-1 Inverter-machine drive, (a) Inverter configuration, (b) transistor switching logic, (c) plot of v_{as}.

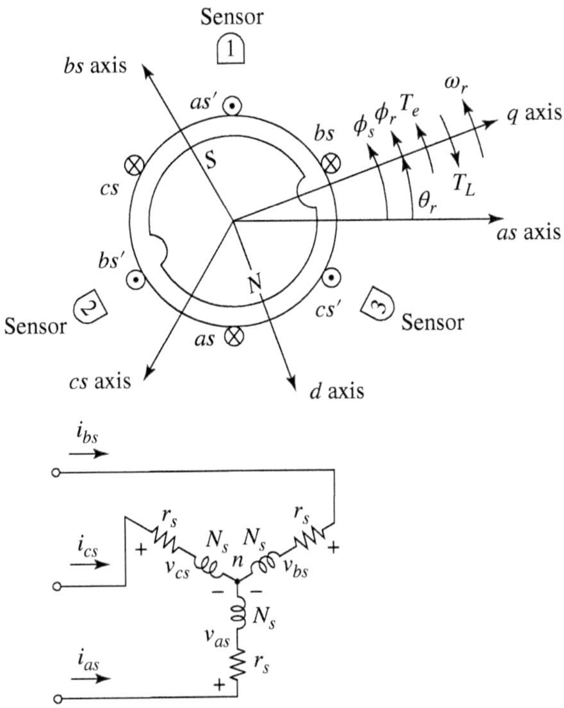

Figure 6.10-2 Two-pole three-phase permanent-magnet ac machine with sensors.

Hall-effect devices; when the south pole of the permanent-magnet rotor is under a sensor, its output is nonzero; with the rotor north pole under the sensor, its output is zero. The states of the sensors are used to determine the switching logic for the inverter, which, in turn, determines the output frequency of the inverter. In the actual machine, the sensors are not positioned over the rotor as shown in Fig. 6.10-2. Instead, they are placed over a ring that is mounted on the shaft external to the stator and magnetized the same as the rotor.

The inverter shown in Fig. 6.10-1a consists of six transistors each with an anti-parallel diode supplied from a dc-source V_{in}. The logic (switching) signals for the transistors are shown in Fig. 6.10-1b. It is assumed that the forward resistance of the diodes and transistors is negligibly small and the turn-on and turn-off times are neglected, whereupon the transistors and diodes are considered to be ideal switches. With these assumptions, the instant one of the "top-rail" transistors in a phase is turned off, the "bottom-rail" transistor of that phase can be turned on and conversely. In other words, each phase is either connected to the top or bottom rail; this is referred to as continuous-current or 180° mode of operation. The operation of this inverter is detailed in Chapter 10.

The inverter is supplying the wye-connected permanent-magnet ac machine, thus

$$v_{as} = r_s i_{as} + \frac{d\lambda_{as}}{dt} \tag{6.10-1}$$

$$v_{bs} = r_s i_{bs} + \frac{d\lambda_{bs}}{dt} \tag{6.10-2}$$

$$v_{cs} = r_s i_{cs} + \frac{d\lambda_{cs}}{dt} \tag{6.10-3}$$

Since the machine is three-wire and wye-connected, the sum of the currents is zero and since the load is symmetrical, the sum of the phase voltages, which is v_{0s}, is zero. Now, from Fig. 6.10-1, we can write

$$v_{ag} = v_{as} + v_{ng} \tag{6.10-4}$$

$$v_{bg} = v_{bs} + v_{ng} \tag{6.10-5}$$

$$v_{cg} = v_{cs} + v_{ng} \tag{6.10-6}$$

Adding (6.10-4) through (6.10-6) and since for a symmetrical wye-connected machine $v_{as} + v_{bs} + v_{cs} = 0$, we can write

$$v_{ng} = \frac{1}{3}\left(v_{ag} + v_{bg} + v_{cg}\right) \tag{6.10-7}$$

Substituting (6.10-7) into (6.10-4) through (6.10-6) and then solving for v_{as}, v_{bs}, and v_{cs} yields

$$v_{as} = \frac{2}{3}v_{ag} - \frac{1}{3}\left(v_{bg} + v_{cg}\right) \tag{6.10-8}$$

$$v_{bs} = \frac{2}{3}v_{bg} - \frac{1}{3}\left(v_{cg} + v_{ag}\right) \tag{6.10-9}$$

$$v_{cs} = \frac{2}{3}v_{cg} - \frac{1}{3}\left(v_{ag} + v_{bg}\right) \tag{6.10-10}$$

where v_{ag}, v_{bg}, and v_{cg} are either V_{in} or zero depending upon the state of the transistors. A plot of v_{as} is shown in Fig. 6.10-1c. The voltages v_{bs} and v_{cs} are of the same waveform as v_{as} but lag by 120° and 240°, respectively, for an *abc* sequence. The firing of the inverter is controlled so that the fundamental frequency of the applied voltages (v_{as}, v_{bs}, and v_{cs}) corresponds to the rotor speed ω_r of the permanent-magnet ac machine. That is, T1 is turned on by sensor 1, T2 by sensor 2, and T3 by sensor 3.

The plot of v_{as} given in Fig. 6.10-1c is readily established from (6.10-8) and the firing signals given in Fig. 6.10-1b. When T4, T5, and T6 are high terminals *a*, *b*, and *c*, respectively, they are connected to the lower rail and are zero. They allow a return path for the currents. The signals T1, T2, and T3 determine the plot of v_{as}, v_{bs}, and v_{cs}. In the case of v_{as}, (6.10-8) tells us that when T1 is high and T2 and T3 are low, v_{as} is $\frac{2}{3}V_{in}$. When T2 or T3 are high they each contribute $-\frac{1}{3}V_{in}$ to v_{as}. Clearly v_{as} is $-\frac{2}{3}V_{in}$ when T1 is low and both T2 and T3 are high.

In order to analytically "connect" the inverter to the machine equations, we need to transform v_{as}, v_{bs}, and v_{cs} to v_{qs}^r and v_{ds}^r; the voltage equations used to analyze the permanent-magnet ac machine. A Fourier expansion of v_{as} shown in Fig. 6.10-1c may be written as

$$v_{as} = \frac{2V_{in}}{\pi}\left(\cos\theta_r + \frac{1}{5}\cos 5\theta_r - \frac{1}{7}\cos 7\theta_r + \cdots\right) \tag{6.10-11}$$

The inverter controls the frequency of v_{as}, v_{bs}, and v_{cs} to correspond to ω_r, the electrical angular velocity of the rotor; hence, the voltages are expressed in terms of ω_r.

It should be noted that the maximum value of Fourier expansion of v_{as} given in Fig. 6.10-1c is $\frac{2}{3}V_{in}$; however, the peak value of the fundamental component of v_{as} given in (6.10-11) is $\frac{2}{\pi}V_{in}$. The voltages v_{bs} and v_{cs} may be expressed, for an *abc* sequence, by substituting $\theta_r - \frac{2}{3}\pi$ and $\theta_r + \frac{2}{3}\pi$, respectively, for θ_r in (6.10-11).

Before getting into the operation of the brushless dc drive, let us take a closer look at the voltage and current waveforms of the continuous-current inverter. For this purpose it is sufficient to focus on v_{as} and i_{as} shown in Fig. 6.10-3. We will work with Figs. 6.10-1a and 6.10-3. In Fig. 6.10-3, v_{as}, i_{as}, i_{aT1}, $-i_{ad1}$, $-i_{aT4}$, and i_{ad4} are plotted for a typical operating condition with a motor load, where $i_{ad1}(i_{ad4})$ is the current flowing in the anti-parallel diode of T1(T4). The diodes are not labeled in Fig. 6.10-1a. Let us start at the center of the peak value of v_{as}. We see from Fig. 6.10-1a that at this instant T1 is carrying positive i_{as} and it is being returned to the source through T5(i_{bs}) and T6(i_{cs}) (not shown in Fig. 6.10-3). At the first step T2 is turned on, T5 is turned off, and v_{as} drops to $\frac{1}{3}V_{in}$. Now, T1(i_{as}) and T2(i_{bs}) (not shown in Fig. 6.10-3) will share the positive current and T6(i_{cs}) will return the current to the source. At the next step, T1 is turned off and T4 is turned on. Now T2(i_{bs}) has all of positive current and T4(i_{as}) and T6(i_{cs}) will share the return current. However, since we have an inductive circuit, before T4 can share the return current there must be path for the positive current that was flowing in T1(i_{as}) at the time it was turned off. This is through i_{d4} (the anti-parallel diode of T4). When T4 is turned off, the anti-parallel diode of T1 springs into action. The pattern is clear.

The free-acceleration characteristics of this brushless dc motor-inverter drive with $\phi_v = 0$ are shown in Fig. 6.10-4. The three-phase machine parameters are those given in the previous section; in particular, $r_s = 3.4\ \Omega$, $L_{ls} = 1.1$ mH, $L_{ms} = 11$ mH, $L_{ss} = 17.6$ mH, and $\lambda_m'^r = 0.0826$ V · s/rad.

If we use (6.7-17) to transform v_{as}, v_{bs}, and v_{cs}, where v_{as} is given by (6.10-11), to the ω_r reference frame, we obtain

$$v_{qs}^r = \frac{2V_{in}}{\pi}\left(1 + \frac{2}{35}\cos 6\theta_r - \frac{2}{143}\cos 12\theta_r + \cdots\right) \tag{6.10-12}$$

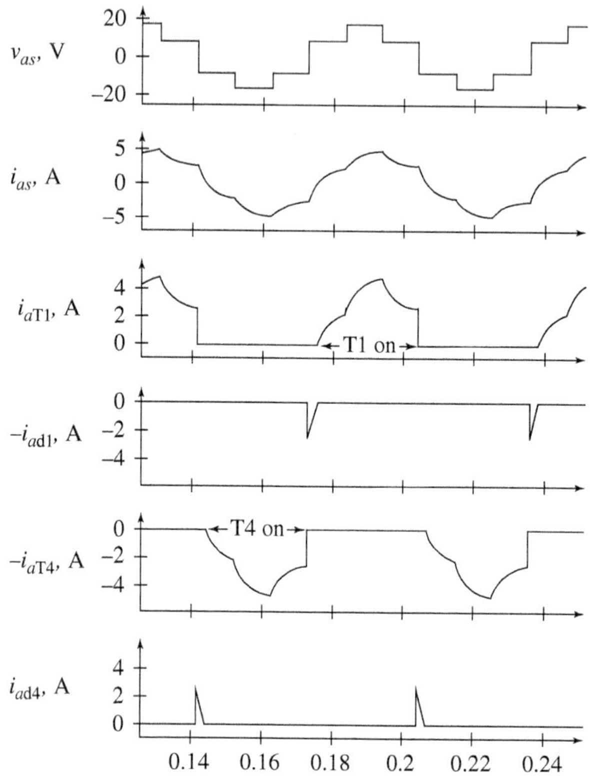

Figure 6.10-3 Plots of v_{as} and i_{as} and the components of i_{as}.

$$v_{ds}^{r} = \frac{2V_{\text{in}}}{\pi}\left(\frac{12}{35}\sin 6\theta_r - \frac{24}{143}\sin 12\theta_r + \cdots\right) \tag{6.10-13}$$

When we transformed the fundamental component of the applied voltages (two- or three-phase) with $v_{as} = \sqrt{2}V_s \cos\theta_r$, we obtained $\sqrt{2}V_s$ for v_{qs}^{r} and zero for v_{ds}^{r}. This is similar to (6.10-12) and (6.10-13), respectively, with the harmonics neglected. We are not surprised by this; however, it is interesting that the frequency of the harmonics has changed. The 5th and 7th of the three-phase voltages have become the 6th harmonic and the 11th and 13th have become the 12th in the q_s^r- and d_s^r-voltages.

For purposes of comparison with Fig. 6.9-1, the inverter voltage V_{in} is 25 V, where the constant component of v_{qs}^{r} from (6.10-12) with $\phi_v = 0$ is equal to the value used in the case of sinusoidal applied stator voltages. Free acceleration with $\phi_v = \phi_{vMT/V}$ is given in Fig. 6.10-5. Free acceleration with $\phi_v = \phi_{vMT/A}$ is given in Fig. 6.10-6.

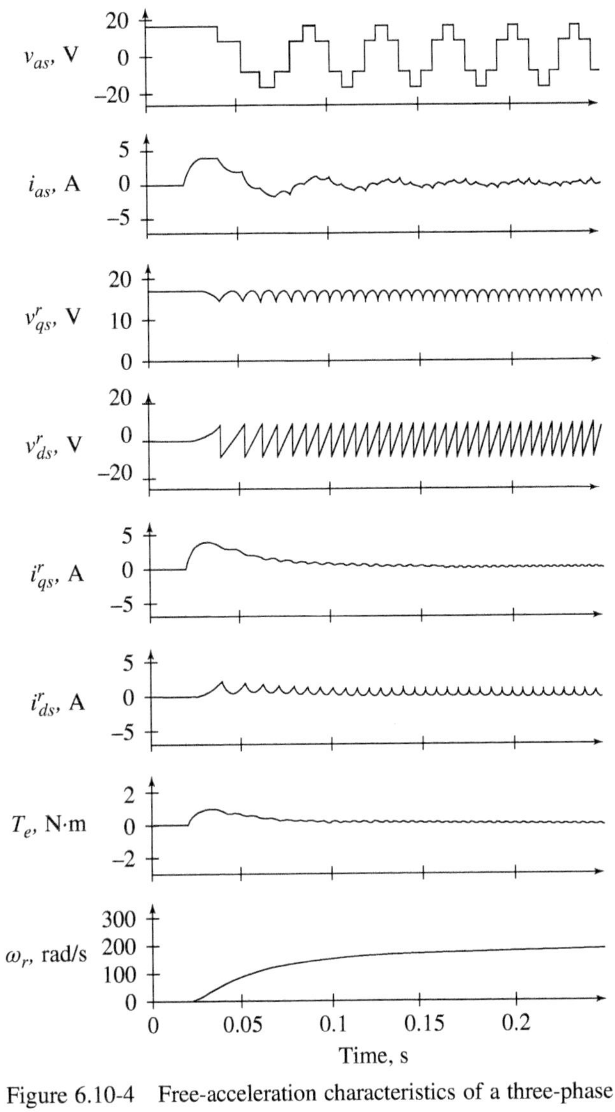

Figure 6.10-4 Free-acceleration characteristics of a three-phase brushless dc motor supplied by a six-step inverter with $\phi_v = 0$.

Modes of Control of a Brushless DC Drive

Field orientation by shifting of the applied voltage, $\widetilde{V}_{as}$, relative to the permanent-magnet rotor and/or the change in the magnitude of the applied voltages provide a wide range of control options. We are aware that phase shifting of the applied voltages relative to the position of the permanent-magnet rotor is possible by controlling the firing of the inverter based on the information retrieved from the Hall effect

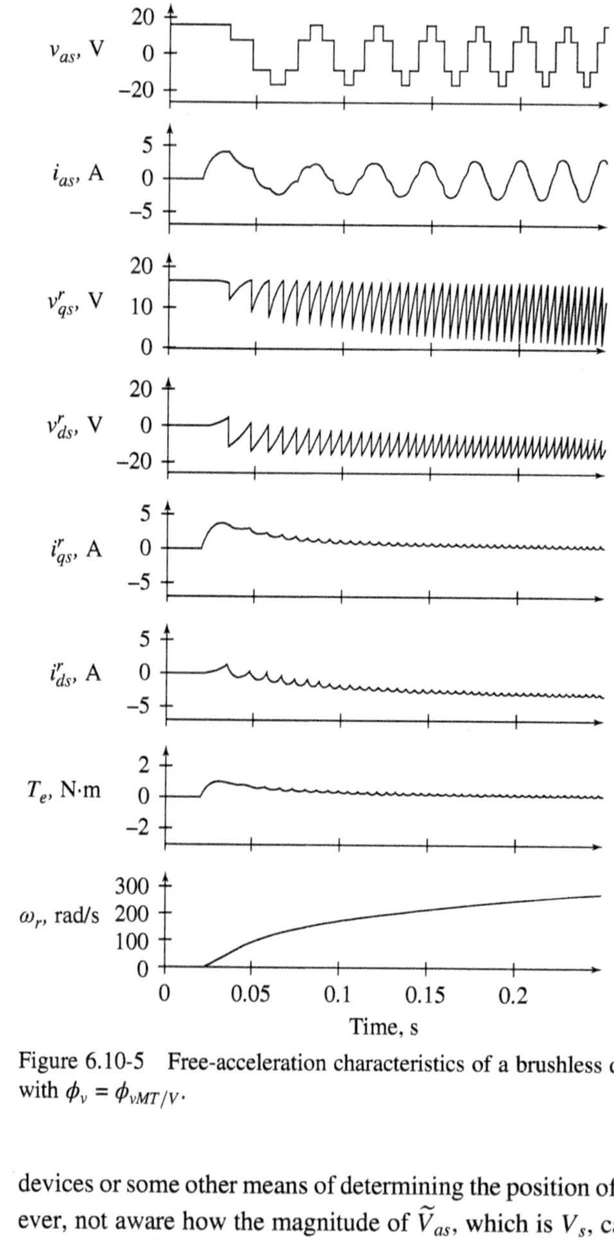

Figure 6.10-5 Free-acceleration characteristics of a brushless dc machine with $\phi_v = \phi_{vMT/V}$.

devices or some other means of determining the position of the rotor. We are, however, not aware how the magnitude of $\widetilde{V}_{as}$, which is V_s, can be changed. We will assume that the maximum rms voltage of the example machine is 11.25 V. We have not discussed how this voltage can be decreased. The output voltage of inverter can be changed by changing the input voltage V_{in}; however, this may not be an option, e.g., when the inverter input voltage is supplied from a constant source. In these cases, voltage control of the inverter output voltage can be achieved by pulse width modulation (PWM). For the purpose of discussion, let

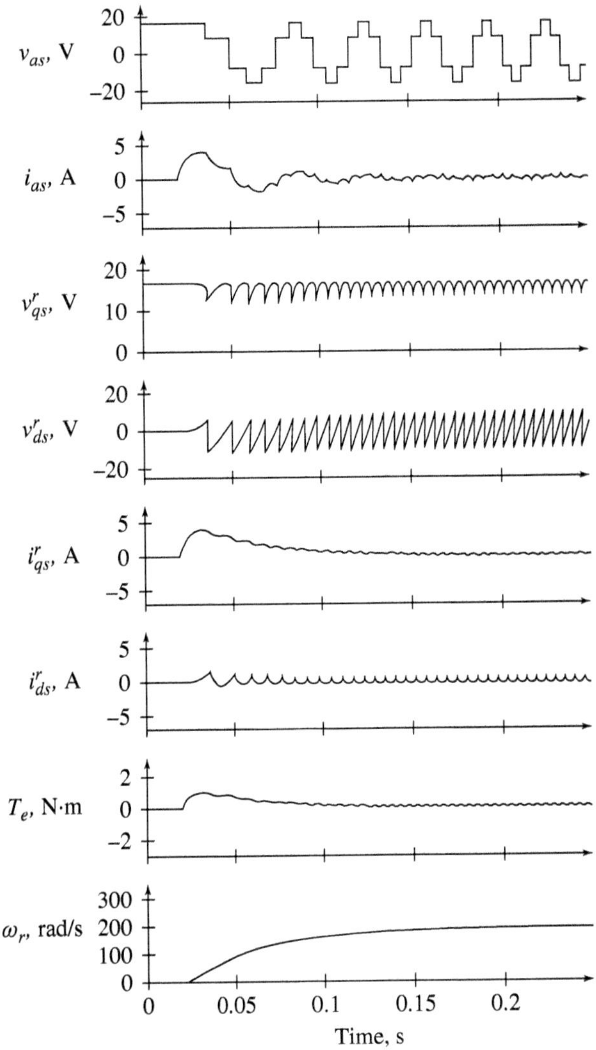

Figure 6.10-6 Free-acceleration characteristics of a brushless dc machine with $\phi_{vMT/A}$.

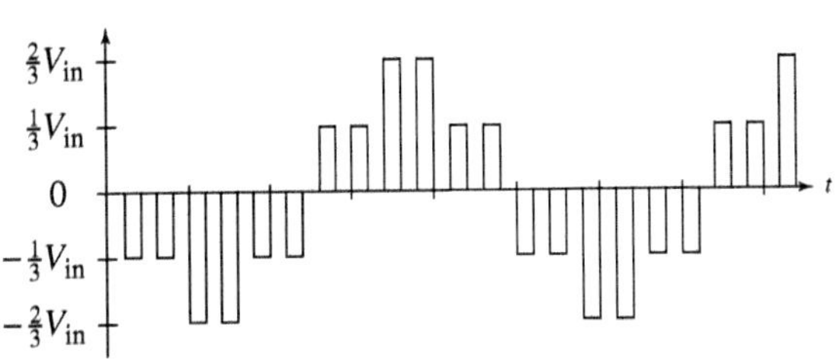

Figure 6.10-7 Pulse width modulated six-step inverter phase voltage.

us consider Fig. 6.10-7 wherein one of the output phase voltages of a six-step inverter with a PWM capability is shown. The "notching" of the six-step phase voltages is accomplished by instantaneously connecting all three of the motor phases to the same side of the dc source by appropriate switching of the transistors. Recall that during normal operation of the inverter, two phases are connected to one terminal of the dc source while the third is connected to the other terminal. During the time all three phases of the machine are connected to the same side of the inverter, the stator windings are short-circuited; thus, the machine is effectively disconnected from the inverter.

There are different PWM procedures, the type shown in Fig. 6.10-7 is one where the on and off times of the transistors are equal. In this case a change in the effective value of the stator phase voltages is achieved by controlling the time that the phase voltages are held at zero during one cycle of the voltages. Regardless which of the variations of PWM is used, the purpose is to change the magnitude of the effective sinusoidal approximation of the six-step phase voltages.

We now have some understanding of how the shifting of the phase and the change of the magnitude of the applied voltages can be achieved. Please be aware that (6.9-18) and (6.9-19) for $\phi_{vMT/V}$ and $\phi_{vMT/A}$, respectively, have been derived assuming an "idealized" machine, in that the assumptions include sinusoidal distribution of the windings, sinusoidal voltages, constant parameters at all operating temperatures, and that L_{ms} is the same in the q and d axes. Unfortunately, these assumptions are idealizations of the real world. Nevertheless, making these approximations for analysis purposes allows the salient operating characteristics to be readily portrayed. When controlling a "real world" drive, the ideal machine offers only guidelines. Nevertheless, an understanding of the operating characteristics with these simplifying assumptions is very helpful; however, in reality, it is necessary to implement a closed-loop control. For example, an on-line determination of I_{ds}^r implemented with a means of correcting any variation in I_{ds}^r from zero is necessary for $\phi_{vMT/A}$ operation. Regardless of all this, it is our purpose to establish an insight to the controlled potential of the brushless dc drive as simply and straightforward as possible and to lay the ground work for an advanced treatment of electric drives.

Example 6F Limiting $\tilde{I}_{as}$ to Rated and I_{ds}^r to Zero

As a closing example let us consider the control of $\tilde{I}_{as}$ from an ideal standpoint. Let us assume that we are dealing with the machine used in Examples 6C through 6E. In particular, $r_s = 3.4\ \Omega$, $L_{ls} = 1.1$ mH, $L_{ms} = 11$ mH, and $\lambda_m'^r = 0.0826\ \text{V}\cdot\text{s/rad}$. Also, let us assume that rated rms phase current is 0.9 A and rated rms phase voltage is 11.25 V. We want to maintain $I_{ds}^r = 0$ over the speed range from stall to no load. In addition, we want to limit the stator current to its rated value (T_e, rated) for as long as we can without exceeding rated voltage; whereupon, we hold voltage constant at rated while maintaining $\phi_{vMT/A}$. We can portray this as shown in Fig. 6F-1.

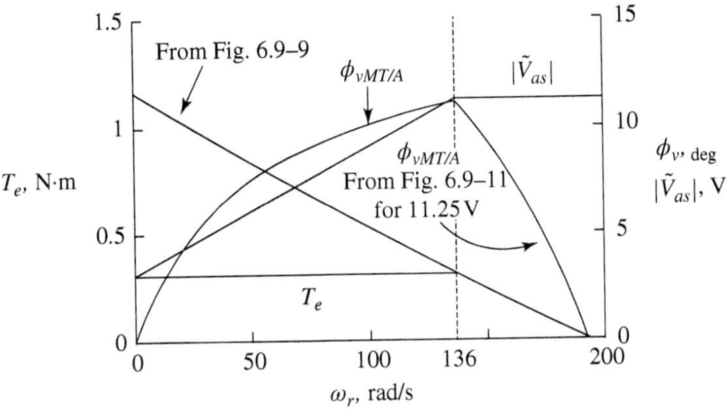

Figure 6F-1 Plot of T_e, $|\widetilde{V}_{as}|$, and $\phi_{vMT/A}$ for rated torque until rated voltage is reached.

Rated torque is

$$\begin{aligned} T_e &= \frac{3}{2}\frac{P}{2}\lambda'^r_m I^r_{qs} \\ &= \frac{3}{2}\frac{4}{2}0.0826\sqrt{2}(0.9) \\ &= 0.315\ \text{N}\cdot\text{m} \end{aligned} \tag{6F-1}$$

The voltage equation for rated current is

$$\widetilde{V}_{as} = (r_s + j\omega_r L_{ss})\tilde{I}_{as} + \tilde{E}_a \tag{6F-2}$$

where

$$\widetilde{V}_{as} = V_s\underline{/\phi_v} \tag{6F-3}$$

$$\tilde{I}_{as} = I_s\underline{/0^\circ} \tag{6F-4}$$

$$\tilde{E}_a = \frac{\omega_r \lambda'^r_m}{\sqrt{2}}\underline{/0^\circ} \tag{6F-5}$$

Let $\omega_r = 136$ rad/s and solve for $\widetilde{V}_{as}$:

$$\begin{aligned} \widetilde{V}_{as} &= \left[3.4 + j136\left(1.1 + \frac{3}{2}11\right) \times 10^{-3}\right]0.9\underline{/0^\circ} + \frac{136 \times 0.0826}{\sqrt{2}} \\ &= [3.4 + j(136)(17.6) \times 10^{-3}]0.9 + 7.94 \\ &= 11.00 + j2.15 \\ &= 11.21\underline{/11.06^\circ} \end{aligned} \tag{6F-6}$$

When the speed reaches 136 rad/s, the rated torque intersects the $\phi_{vMT/A}$ torque versus ω_r plot. Thus, we can no longer maintain rated current with the voltage rated at 11.25 V. Therefore, we must follow the $\phi_{vMT/A}$ plot to no-load operation; however, for $\omega_r < 136$ rad / s, we can control rated current. It is clear that any current less than rated can also be maintained.

Let us calculate $\widetilde{V}_{as}$ over the current controlled range using (6F-2). For stall:

$$V_{as} = 3.4 \times 0.9\underline{/0^\circ} = 3.06\underline{/0^\circ}\text{V} \tag{6F-7)}$$

For $\omega_r = 50$ rad/s:

$$\begin{aligned}\widetilde{V}_{as} &= \left[3.4 + j(50)\left(17.6 \times 10^{-3}\right)\right]0.9 + \frac{(50)(0.0826)}{\sqrt{2}} \\ &= 5.98 + j0.792 \\ &= 6.03\underline{/7.54^\circ}\text{V}\end{aligned} \tag{6F-8}$$

For $\omega_r = 100$ rad/s:

$$\begin{aligned}\widetilde{V}_{as} &= \left[3.4 + (100)\left(17.6 \times 10^{-3}\right)\right]0.9 + \frac{(100)(0.0826)}{\sqrt{2}} \\ &= 8.9 + j1.58 \\ &= 9.04\underline{/10.1^\circ}\text{ V}\end{aligned} \tag{6F-9}$$

Although we are not going to get into the details of the control to accomplish Fig. 6F-1, the ideal control that could be used is shown in Fig. 6F-2.

In Fig. 6F-2, the asterisked quantities are commanded values. In this case, $T_e^* = 0.315$ N · m for $\omega_r < 136$; however, $I_{ds}^{r*} = 0$ applies over the complete speed range. We must understand that the actual control would be much more involved since it would be necessary to ensure that I_{ds}^r is maintained at zero over the complete range.

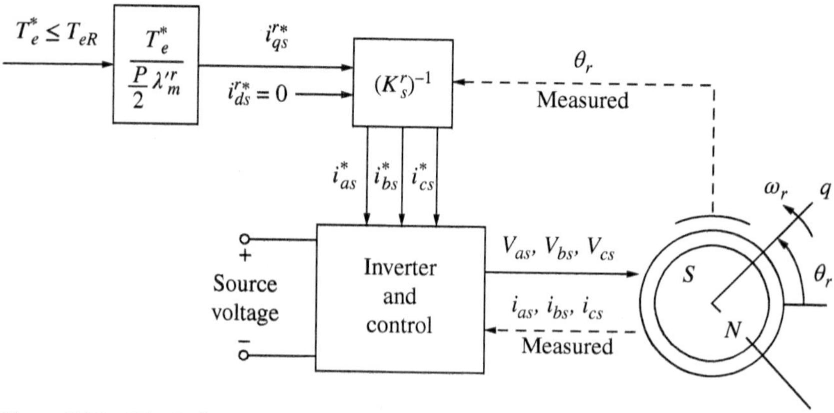

Figure 6F-2 Block diagram of constant-torque control of a three-phase permanent-magnet ac machine for ω_r less than rated value.

Example 6G Torque Control

The parameters of the four-pole three-phase permanent-magnet ac machine are $r_s = 3.4\,\Omega$, $L_{ls} = 1.1$ mH, $L_{ms} = \frac{3}{2}$ 11 mH, $\lambda_m'^{r} = 0.0826$ V · s/rad, and rated $V_s = 11.25$ V. The torque is controlled with $\phi_v = 0$ at 0.315 N · m until rated voltage is reached at $\omega_r = 120.1$ rad/s. The voltage, V_s, is then maintained at 11.25 V with $\phi_v = 0$ and the load line and Operating Point 1 are as shown in Fig. 6G-1. The load line is calculated from

$$T_L = K\omega_r^2 \tag{6G-1}$$

where $T_L = 0.315$ N·m and $\omega_r = 120.1$ rad/s. Substituting into (6G-1) and solving for K yields

$$K = \frac{0.315}{(120.1)^2} = 2.184 \times 10^{-5}\ \text{N} \cdot \text{m} \cdot \text{s} \tag{6G-2}$$

The commanded torque T_e^* is suddenly reduced to one-half the original. Thus,

$$\begin{aligned} T_e^* &= \left(\frac{1}{2}\right)(0.315) \\ &= 0.1575\ \text{N} \cdot \text{m} \end{aligned} \tag{6G-3}$$

The trajectory from Operating Point 1 to Operating Point 2 is shown in Fig. 6G-1. The rotor speed at Operating Point 2 can be calculated as

$$T_L = K\omega_r^2 \tag{6G-4}$$

where $T_L = 0.1575$ N · m and $K = 2.184 \times 10^{-5}$ N · m · s^2. This yields

$$\omega_r = \left(\frac{0.1575}{2.184 \times 10^{-5}}\right)^{\frac{1}{2}} = 84.9\ \text{rad/s} \tag{6G-5}$$

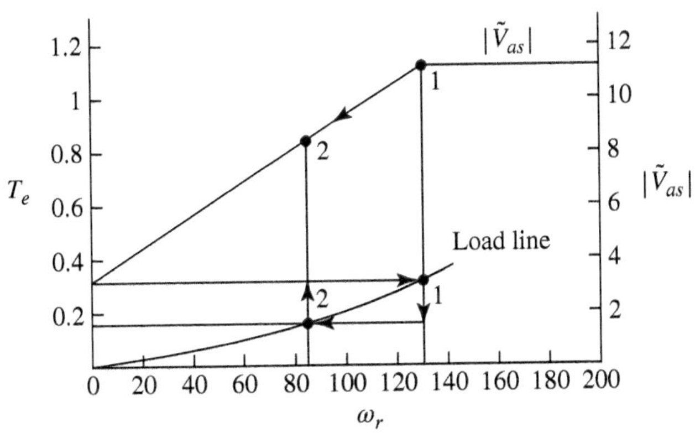

Figure 6G-1 Drive operation with T_e^* Switching.

Now, from (6.9-9),

$$T_e = \left(\frac{3}{2}\right)\left(\frac{P}{2}\right)\lambda'^r_m \sqrt{2} I_s \cos\theta_{esi}(0) \tag{6G-6}$$

We can write

$$\sqrt{2} I_s \cos\theta_{esi}(0) = \frac{T_e}{\left(\frac{3}{2}\right)\left(\frac{P}{2}\right)\lambda'^r_m} = \frac{0.1575}{\left(\frac{3}{2}\right)\left(\frac{4}{2}\right)0.0826} = 0.636 \text{ A} \tag{6G-7}$$

We can calculate the voltage from (6.6-5) and (6.6-6); however, we first need $\tilde{I}_{as}$. Since $\varphi_v = 0$, $\tilde{V}_{as}$, and $\tilde{E}_a$ are at zero degrees; therefore, the phase angle of $\tilde{I}_{as}$ is

$$\begin{aligned} \theta_{esi}(0) &= -\tan^{-1}\frac{\omega_r L_{ss}}{r_s} \\ &= -\tan^{-1}\frac{(8.49)\left(1.1 + \frac{3}{2}11\right) \times 10^{-3}}{r_s} \\ &= -\tan^{-1}0.439 = -23.7^\circ \end{aligned} \tag{6G-8}$$

From (6G-7),

$$I_s = \frac{0.636}{\sqrt{2}\cos 23.7} = 0.4912 \tag{6G-9}$$

Therefore,

$$\tilde{I}_{as} = 0.4912\underline{/-23.7^\circ} \tag{6G-10}$$

Substituting into (6.6-5) and (6.6-6) yields

$$\begin{aligned} \tilde{V}_{as} &= (r_s + j\omega_r L_{ss})\tilde{I}_{as} + \frac{1}{\sqrt{2}}\omega_r \lambda'^r_m \underline{/0^\circ} \\ &= (3.4 + j84.9)\left(1.1 + \frac{3}{2}11 \times 10^{-3}\right)(0.4912\underline{/-23.7^\circ}) + \frac{(84.9)(0.0826)}{\sqrt{2}} \\ &= (3.714\underline{/27.3^\circ})(0.4912\underline{/-23.7^\circ}) + 4.96 = 6.78 \text{ V} \end{aligned} \tag{6G-11}$$

After steady-state operation is reached at Operating Point 2, the commanded torque is increased back to 0.315 N · m. The trajectory from Operating Point 2 to Operating Point 1 is shown in Fig. 6G-1. The mechanical dynamics are determined by

$$T_e^* = J\frac{d\omega_r}{dt} + B_m\omega_r + T_L \tag{6G-12}$$

where $J = 5 \times 10^{-4}$ kg · m² and $B_m = 0$.

SP6.10-1 With the initial conditions assumed, the stall torque of an inverter-driven brushless dc machine is slightly larger than the sinusoidal approximation. Why? $\left[\frac{2}{3}V_{\text{in}} \text{ vs } \frac{2}{\pi}V_{\text{in}}\right]$

SP6.10-2 For a given inverter voltage what factors determine the starting torque. [rotor position relative to position of mmf_s]

6.11 RECAPPING

We have covered a considerable amount of material in this chapter; however, the material has been arranged so that some of the derivations could be bypassed if the steady-state performance is the focus. Also, the availability of animations and the tables with the important voltage and torque expressions for steady-state operation should help the student to focus on the important material. Although reference frame theory is convenient, if not necessary, to analyze and simulate the electric transients of an ac machine and to design and implement controls, the steady-state performance can be readily and sufficiently portrayed by a phasor diagram. We have attempted to enhance the use of the phasor diagram by including the location of the stator and rotor poles during dynamic and steady-state operation.

6.12 PROBLEMS

1. It is found that $\lambda'^r_m = 0.1$ V·s/rad for a permanent-magnet six-pole two-phase ac machine. Calculate the amplitude (peak value) of the open-circuit phase voltage measured when the rotor is turned at 60 revolutions per second (r/s).
2. Starting with (6.5-3), obtain (6.5-5).
3. In the analysis of the brushless dc motor, we have selected $\theta_r(0) = 0$, where $\sqrt{2}\tilde{F}_{as} = F^r_{qs} - jF^r_{ds}$. Express the relationship between these same variables if we had selected $\theta_r(0) = \frac{1}{2}\pi$.
4. Refer to Example 6B. Construct the phasor diagram showing $\tilde{V}_{as}, \tilde{E}_a, r_s\tilde{I}_{as}, j\omega_r L_{ss}\tilde{I}_{as}$, and $\tilde{I}_{as}$. Show the pole locations.
5. A four-pole two-phase permanent-magnet ac machine is driven by a mechanical source at $\omega_r = 3600$ r/min. The open-circuit voltage across one of the phases is 50 V (rms). (a) Calculate λ'^r_m. The mechanical source is removed and the following voltages are applied: $V_{as} = \sqrt{2}\,25\cos\theta_r$, $V_{bs} = \sqrt{2}\,25\sin\theta_r$, where $\theta_r = \omega_r t$. (b) Neglect friction ($B_m = 0$) and calculate the no-load rotor speed ω_r in rad/s.
6. The parameters of a two-pole three-phase permanent-magnet ac machine are $r_s = 4\ \Omega$, $\lambda'^r_m = 0.07$ V · s/rad, $L_{ls} = 1$ mH, and $L_{ms} = 9$ mH. The applied voltages are $V_{as} = \sqrt{2}\,14.1\cos\theta_r$ and $V_{bs} = \sqrt{2}\,14.1\sin\theta_r$, where $\theta_r = 200t$. Determine (a) $\tilde{I}_{as}$, (b) T_e, and (c) show the phasor diagram and locate the poles.
7. Repeat Problem 6 with $\phi_{vMT/V}$.
8. Repeat Problem 6 with $\phi_{vMT/A}$.

9. Consider the machine given in Example 6F. Determine the rotor speed where $|\tilde{I}_{as}| = 0.9$ A can no longer be maintained with $I^r_{ds} = 0$. Compare with Example 6F.

10. Plot v_{as}, v_{bs}, v_{cs}, v_{ab}, and v_{ng} for the 180° continuous-current inverter for $0 < \theta_r < 2\pi$ with the transistor switching shown in Fig. 6.10-1. Assume the inverter voltage V_{in} is constant.

11. Verify the rotor speed for Operating Point 1 of Example 6G. ■

6.13 REFERENCES

1. Paul Krause, Oleg Wasynczuk, Scott Sudhoff, and Steven Pekarek, *Analysis of Electric Machinery and Drive Systems*, 3rd Edition, Wiley, IEEE Press, 2013.
2. T. W. Nehl, F. A. Fouad, and N. A. Demerdash, "Digital Simulation of Power Conditioner-Machine Interaction for Electronically Commutated DC Permanent Magnet Machines," IEEE Trans. on Magnetics, Vol. 17, November 1981, pp. 3284–3286.
3. T. M. Jahns, "Torque Production in Permanent-Magnet Synchronous Motor Drives with Rectangular Current Excitations," IAS Conf. Rec., October 1983, pp. 476–487.
4. P. C. Krause, *Analysis of Electric Machinery*, McGraw-Hill Book Company, New York, 1986.

CHAPTER 7

SYNCHRONOUS MACHINES

7.1 INTRODUCTION

Nearly all electric power is generated by synchronous machines driven either by hydroturbines, steam turbines, or combustion engines. Just as the induction machine is the workhorse when it comes to converting energy from electric to mechanical, the synchronous generator is the principal means of converting energy from mechanical to electrical. Although nearly all electric power is generated with three-phase synchronous machines, their electrical and electromechanical behavior can be predicted from the equations which describe the two-phase salient-pole synchronous machine. In particular, with only slight modifications, these equations can be used to predict the performance of large hydroturbine and steam turbine synchronous generators, synchronous motors, and reluctance motors used in low-power drive systems. It is for this reason that we will focus our attention on the two-phase machine since the work involved is much less than with the three-phase device. However, those who wish to study the three-phase synchronous machine may do so since there is a section devoted to it near the end of the chapter and those interested in drives may wish to focus on the reluctance machine treated in Section 7.6.

The stator windings of a synchronous machine are symmetrical; however, the rotor of a synchronous machine is equipped with a field winding and one or more short-circuited windings that are referred to as damper windings. In general, the rotor windings each have different electrical characteristics. Moreover, the rotor of a salient-pole synchronous machine is magnetically asymmetrical. Owing to these rotor asymmetries, a change of variables does not offer an advantage in the case of the rotor variables. However, we will find it beneficial to define a change of variables or transformation for the voltages, currents, and flux linkages of the stator circuits. In effect, this transformation (Park's transformation [1]) replaces these stator variables with substitute variables associated with fictitious circuits rotating at the angular velocity of the rotor (rotor reference frame).

Electromechanical Motion Devices: Rotating Magnetic Field-Based Analysis and Online Animations, Third Edition. Paul Krause, Oleg Wasynczuk, Steven D. Pekarek, and Timothy O'Connell.

Published 2020 by John Wiley & Sons, Inc.

Since the field winding and the damper windings have different parameters and the salient-pole rotor is magnetically unsymmetrical, analyzing a practical three-phase synchronous generator is quite involved and not particularly appropriate for a first-time student of machine analysis, even from a two-phase standpoint. Therefore, this chapter is arranged so that the undergraduate has the opportunity to become familiar with the basis of synchronous machine operation without wading through reference frame derivations and a maze of three-phase trigonometry. In particular, the two-phase round-rotor machine with only a field winding is analyzed first, followed by a two-phase reluctance machine before considering the three-phase round-rotor machine with only a field winding. In each case, reference frame theory is used to derive the equations that describe dynamic and steady-state performance. If, however, only steady-state operation is the main interest, the material is arranged so that the reference frame derivations may be de-emphasized or bypassed and only the resulting voltage and torque equations describing steady-state operation considered. Moreover, for each machine these equations are arranged in tabular form for easy reference and to point out variations of these equations when the machines are used in drives or power system applications and the modifications when going from a two- to a three-phase machine. In other words, this chapter is arranged so that the student may be made aware of the voltage and torque equations that describe steady-state operation of the synchronous machine without becoming involved in reference frame theory. However, the reference frame derivations are presented for those who have studied Chapter 5 and want to dig a little deeper.

Although the rotor windings and rotor asymmetry are shown and discussed briefly in the following section, the analysis begins in Section 7.3 with a two-phase round-rotor synchronous machine with only a field winding. Whereupon, this machine is considered and compared to the permanent-magnet ac synchronous machine (Chapter 6) in construction and operation. The material is then arranged so that the beginning student can become aware of the basic operating mode of a synchronous machine used in a drive and/or a power system.

7.2 WINDINGS OF THE SYNCHRONOUS MACHINE

Before becoming involved in synchronous machine analysis, it is helpful to describe the function of the windings that are present in most synchronous machines that are used to generate electric power. For this purpose it is convenient to consider the windings of a two-pole three-phase salient-pole synchronous machine shown in Fig. 7.2-1. The stator windings are identical, sinusoidally distributed windings, as described in Chapter 4. For analysis purposes, the electrical characteristics of the rotor of a synchronous machine may be adequately represented with a field winding (*fd* winding) and short-circuited *damper* or *amortisseur* windings (*kq* and *kd* windings). Although the damper windings are shown with provisions to apply a voltage, they are, in fact, short-circuited windings that

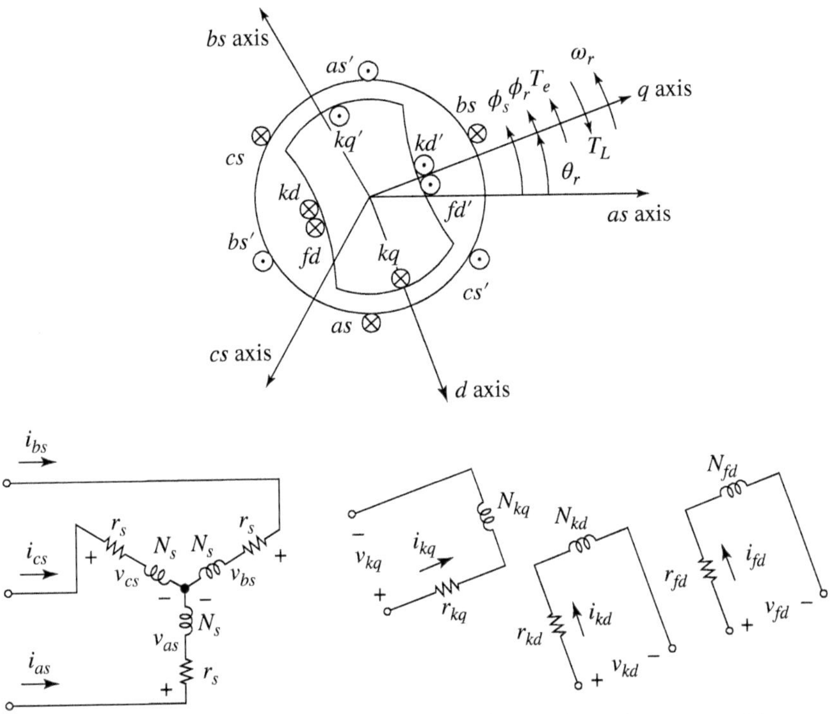

Figure 7.2-1 Salient-rotor two-pole three-phase salient-pole synchronous machine.

represent the paths for induced rotor currents. We will assume that the damper windings are approximated by two sinusoidally distributed windings displaced in space by 90°. The *kd* winding has the same magnetic axis as the *fd* winding, it has N_{kd} equivalent turns with resistance r_{kd}. The magnetic axis of *kq* winding is orthogonal with the magnetic axis of the *fd* and *kd* windings. It has N_{kq} equivalent turns and r_{kq} resistance. The rotor configuration shown in Fig. 7.2-1 for a three-phase machine is the same for any multiphase two-pole synchronous machine. The quadrature axis (*q* axis) and direct axis (*d* axis) are also shown in Fig. 7.2-1. The *q* axis is the magnetic axis of the *kq* winding, whereas the *d* axis is the magnetic axis of the *fd* and *kd* windings. In synchronous machine analysis, but not in general, the *q* and *d* axes are reserved to denote the rotor magnetic axes since, over the years, they have been associated with the physical structure of the synchronous machine rotor quite independent of any transformation. However, as we have seen in the previous chapters, these need not be associated with any physical axes when using them for general machine analysis.

With balanced steady-state stator currents, an air-gap mmf (mmf_s) is established that rotates about the air gap of a two-pole machine at ω_e, the angular velocity of the stator currents. A dc voltage can be applied to the *fd* winding by a brush and slip ring arrangement. The resulting field current i_{fd} establishes an air-gap mmf (mmf_r) which is fixed with respect to the rotor. We understand that the air-gap mmf (poles) established by the field winding must rotate at the same angular velocity as Tesla's rotating air-gap mmf established by the stator currents in order to produce a nonzero average electromagnetic torque during steady-state operation. Therefore, the rotor must rotate in synchronism with the air-gap mmf established by the stator windings ($\omega_r = \omega_e$); hence, the name synchronous machine. That is, the main torque production mechanism is this interaction of the air-gap mmf established by the stator currents mmf_s and the air-gap mmf due to the direct current flowing in the field winding mmf_r. Albeit small, an electromagnetic torque called reluctance torque is also developed at synchronous speed due to the nonuniform air gap as a result of the saliency of the rotor. The so-called salient-pole construction is common for slower speed machines (large number of poles) such as hydroturbine generators. In this type of rotor construction, the field winding is wound upon the rotor surface, as shown in Fig. 7.2-1, and the air gap is nonuniform to make room for the placement of the field winding. Therefore, the q-axis magnetic path has a higher reluctance than the d-axis magnetic path.

At this point it seems that other than the addition of the damper windings and the saliency of the rotor, the important difference between the permanent-magnetic ac machine considered in Chapter 6 and this machine is that the permanent magnet is replaced with a field winding. Why then are we considering these two synchronous machines separately? The answer is in the mode of operation. The primary use of a permanent-magnet machine is in a brushless dc drive, where the frequency of the applied stator voltages is controlled to be the electrical angular velocity of the rotor. The voltages applied to the stator of the synchronous machine used to generate power are fixed in frequency and magnitude by the power system. We will talk more about this as we go along.

We have yet to discuss the function of the damper windings. It was found early on that a synchronous machine with only a field winding would tend to oscillate about synchronous speed in a slowly damped manner following any slight disturbance. Adding damper windings (short-circuited rotor windings) provided the desired damping. To explain this damping action it is necessary to understand induction motor action (Chapter 8). Briefly, the damper windings are short-circuited as are the rotor windings of an induction motor and currents are induced in these rotor (damper) windings whenever the speed of the rotor differs from the angular velocity of the rotating air-gap mmf established by the stator currents (mmf_s).

The main torque of a synchronous machine is developed at synchronous speed because of the interaction of mmf_s and mmf_r. At synchronous speed, current is not induced in the damper windings and, hence, "induction motor" torque is not developed when $\omega_r = \omega_e$ (Chapter 8). However, if for any reason the speed of the rotor should vary from synchronous speed because of a disturbance, currents will be induced in the damper windings and the torque developed due to induction motor action, although small, will damp oscillations of the rotor speed. That is, a slight slowing down (speeding up) of the rotor will produce an induction motor torque to accelerate (decelerate) the rotor back to synchronous speed. The question arises as to why this unwanted oscillation does not occur in the brushless dc drive?

The answer is the fact that the frequency and amplitude control of the applied stator voltages inherently prevents this oscillation from occurring.

Torque is torque by whatever means it is developed and, perhaps, we should not emphasize the separation of torque into three types (interaction of mmf_s and mmf_r, reluctance, and induction) since the dynamic operation of the machine is described by nonlinear equations and superposition cannot be applied, in general. Nevertheless, this separation is helpful for our first look at the synchronous machine. In particular, we will first consider only the field winding with a round rotor. Next, we will analyze the two-phase reluctance machine and leave the detailed analysis of the induction torque until Chapter 8. Although we have discussed the characteristics of the practical synchronous machine from the separation standpoint, the analysis of the dynamic interaction of all windings shown in Fig. 7.2-1 is left to an advanced course.

A four-pole three-phase salient-pole synchronous machine is shown in Fig. 7.2-2. Note the dc machine connected to the shaft for purposes of supplying voltage to the field winding of the synchronous machine. Note also, the squirrel-cage damper windings embedded in the pole faces. Figure 7.2-3 shows the stator and rotor of a miniature two-pole three-phase alternator with an AlNiCO permanent-magnet rotor. This device produces 12 W at 4200 r/min to supply aircraft instruments. It mounts on an aircraft engine auxiliary drive pad where temperatures can be as high as 350 °F.

SP7.2-1 Express mmf_r for the two-pole three-phase synchronous machine shown in Fig. 7.2-1. $\left[\text{mmf}_r = -\left(N_{fd}/2\right) i_{fd} \sin\phi_r\right.$, where ϕ_r is the counterclockwise displacement from the *fd* axis]

Figure 7.2-2 Four-pole three-phase salient-pole synchronous machine.

Figure 7.2-3 Stator and rotor of a two-pole 12-W, 4200 r/min, permanent-magnet, synchronous machine. (Courtesy of Vickers Electromech.)

SP7.2-2 A dc voltage is applied to the *fd* winding of the machine shown in Fig. 7.2-1. The damper windings are short-circuited and the machine is driven at ω_r, counterclockwise. Assume that the stator currents are balanced 60-Hz currents with an acb-sequence. Determine the frequency of the currents flowing in the damper windings. $[\omega_r + \omega_e]$

7.3 TWO-PHASE ROUND-ROTOR SYNCHRONOUS MACHINE

An elementary two-phase synchronous machine is shown in Fig. 7.3-1. It consists of a two-phase symmetrical stator and a uniform air gap with a field winding on the rotor that is supplied from a dc source through slip rings. The angular displacement about the stator is denoted ϕ_s and it is referenced to the *as* axis. The angular displacement about the rotor is ϕ_r, which is referenced to the q axis. The electrical angular velocity of the rotor is ω_r and θ_r is the electrical angular position of the rotor measured from the *as* axis to the q axis. Thus, a given point on the rotor surface at the angular displacement ϕ_r may be related to an adjacent point on the inside stator surface with angular displacement ϕ_s as $\phi_s = \theta_r + \phi_r$. The electromechanical torque T_e and the load torque T_L are also shown in Fig. 7.3-1. We are aware from Chapter 2 that T_e is assumed positive in the positive direction of θ_r. The load torque is positive in the opposite direction, opposing rotation.

The voltage equations are

$$v_{as} = r_s i_{as} + \frac{d\lambda_{as}}{dt} \tag{7.3-1}$$

$$v_{bs} = r_s i_{bs} + \frac{d\lambda_{bs}}{dt} \tag{7.3-2}$$

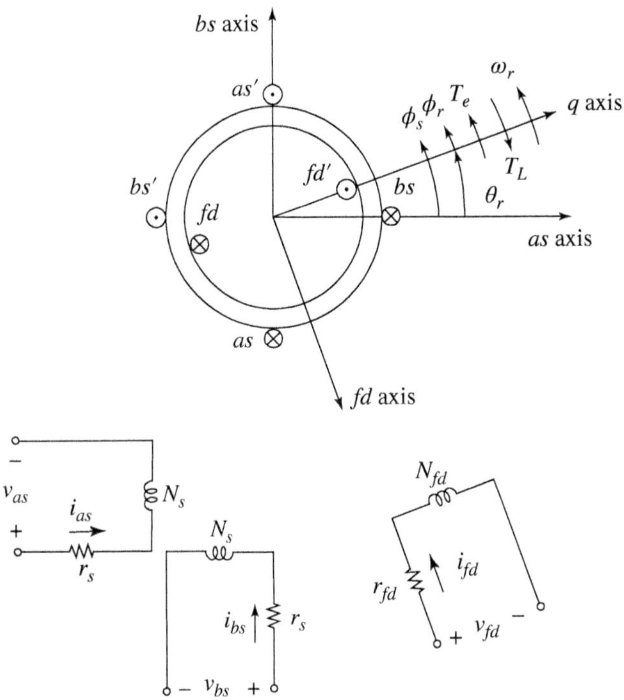

Figure 7.3-1 Two-phase round-rotor synchronous machine.

$$v_{fd} = r_{fd} i_{fd} + \frac{d\lambda_{fd}}{dt} \tag{7.3-3}$$

The rotor is round and the stator windings are orthogonal, thus, the mutual inductance between these windings (L_{asbs}) is zero; whereupon, the flux-linkage equations become

$$\lambda_{as} = L_{asas} i_{as} + L_{asfd} i_{fd} \tag{7.3-4}$$

$$\lambda_{bs} = L_{bsbs} i_{bs} + L_{bsfd} i_{fd} \tag{7.3-5}$$

Now,

$$L_{asas} = L_{bsbs} = L_{ls} + L_{ms} \tag{7.3-6}$$

where L_{ls} is the leakage inductance and the magnetizing inductance is

$$L_{ms} = \frac{N_s^2}{\mathcal{R}_m} \tag{7.3-7}$$

The mutual inductances are

$$L_{asfd} = L_{sr} \sin\theta_r \tag{7.3-8}$$

$$L_{bsfd} = -L_{sr} \cos\theta_r \tag{7.3-9}$$

where

$$\frac{d\theta_r}{dt} = \omega_r(t) \tag{7.3-10}$$

and

$$L_{sr} = \frac{N_s N_{fd}}{\mathcal{R}_m} \tag{7.3-11}$$

We can now write (7.3-4) and (7.3-5) as

$$\boldsymbol{\lambda}_{abs} = \mathbf{L}_s \mathbf{i}_{abs} + L_{sr} i_{fd} \begin{bmatrix} \sin\theta_r \\ -\cos\theta_r \end{bmatrix} \tag{7.3-12}$$

where

$$\mathbf{L}_s = \begin{bmatrix} L_{ss} & 0 \\ 0 & L_{ss} \end{bmatrix} \tag{7.3-13}$$

and

$$L_{ss} = L_{ls} + L_{ms} \tag{7.3-14}$$

It is helpful in developing an equivalent circuit to refer the *fd* variables to the N_s winding. Thus,

$$i'_{fd} = \frac{N_{fd}}{N_s} i_{fd} \tag{7.3-15}$$

Hence,

$$\begin{aligned} L_{sr} i_{fd} &= \left(\frac{N_{fd} N_s}{\mathcal{R}_m}\right)\left(\frac{N_s}{N_{fd}}\right) i'_{fd} \\ &= L_{ms} i'_{fd} \end{aligned} \tag{7.3-16}$$

Now the last term of (7.3-12) may be written as

$$L_{sr} i_{fd} \begin{bmatrix} \sin\theta_r \\ -\cos\theta_r \end{bmatrix} = L_{ms} i'_{fd} \begin{bmatrix} \sin\theta_r \\ -\cos\theta_r \end{bmatrix} \tag{7.3-17}$$

If you have studied Chapter 6, you will note that in (7.3-17) λ'_m has become $L_{ms} i'_{fd}$ because the synchronous machine is equipped with a field winding rather than a permanent magnet.

The torque may be expressed using the second entry in Table 2.5-1 where the coenergy may be expressed as

$$W_c = \frac{1}{2} L_{ss}\left(i_{as}^2 + i_{bs}^2\right) + L_{sr} i_{fd}\left(i_{as}\sin\theta - i_{bs}\cos\theta_r\right) + \frac{1}{2} L_{ffd} i_{fd}^2 \tag{7.3-18}$$

where L_{ffd} is the self-inductance of the *fd* winding. Now, from Table 2.5-1,

$$\begin{aligned} T_e &= \frac{P}{2}\frac{\partial W_c}{\partial \theta_r} \\ &= \frac{P}{2} L_{ms} i'_{fd} (i_{as} \cos \theta_r + i_{bs} \sin \theta_r) \end{aligned} \tag{7.3-19}$$

where we have substituted (7.3-16) for $L_{sr} i_{fd}$.

Reference Frame Derivation

Those who wish not to become involved with the reference frame derivation, but who are interested in the steady-state voltage equation, may proceed to Table 7.4-1.

The rotor of the synchronous machine is equipped with only a field winding with a dc current; therefore, the rotor is electrically unsymmetrical. Since we are unaware of a change of variables for a single circuit that offers an analytical advantage, it is necessary to transform the stator circuits and variables to fictitious circuits and substitute variables in the rotor reference frame as in the case of the permanent-magnet ac machine in Chapter 6. For this purpose, we will write (7.3-1) and (7.3-2) as

$$\mathbf{v}_{abs} = \mathbf{r}_s \mathbf{i}_{abs} + p\boldsymbol{\lambda}_{abs} \tag{7.3-20}$$

where

$$(\mathbf{f}_{abs})^T = [f_{as} \; f_{bs}] \tag{7.3-21}$$

and

$$\mathbf{r}_s = \begin{bmatrix} r_s & 0 \\ 0 & r_s \end{bmatrix} \tag{7.3-22}$$

The voltage equations in the rotor reference frame (v^r_{qs} and v^r_{ds}) may be expressed from (5.4-6) and (5.4-15) with $\omega = \omega_r$. In particular,

$$v^r_{qs} = r_s i^r_{qs} + \omega_r \lambda^r_{ds} + p\lambda^r_{qs} \tag{7.3-23}$$

$$v^r_{ds} = r_s i^r_{ds} - \omega_r \lambda^r_{qs} + p\lambda^r_{ds} \tag{7.3-24}$$

where the raised r is used to denoted the variables in the rotor reference frame.

The *fd* voltage equation referred to the N_s winding becomes

$$v'^r_{fd} = r'_{fd} i'^r_{fd} + p\lambda'^r_{fd} \tag{7.3-25}$$

where i'^r_{fd} is (7.3-15) with the raised r added and similarly

$$v'^r_{fd} = \frac{N_s}{N_{fd}} v^r_{fd} \tag{7.3-26}$$

$$\lambda'^{r}_{fd} = \frac{N_s}{N_{fd}} \lambda^{r}_{fd} \tag{7.3-27}$$

$$r'_{fd} = \left(\frac{N_s}{N_{fd}}\right)^2 r_{fd} \tag{7.3-28}$$

Our goal is to express the flux-linkage terms λ^r_{qs}, λ^r_{ds}, and λ'^{r}_{fd} in (7.3-23) through (7.3-25) in terms of currents. This will allow us to draw equivalent circuits to represent the synchronous machine in the rotor reference frame. From (5.3-9),

$$\boldsymbol{\lambda}^r_{qds} = \mathbf{K}^r_s \boldsymbol{\lambda}_{abs} \tag{7.3-29}$$

where $\boldsymbol{\lambda}_{abs}$ is given by (7.3-12) and (7.3-17). The transformation $\mathbf{K}^r_s$ is given by (5.3-12) with $\theta = \theta_r$. Thus,

$$\begin{aligned}\begin{bmatrix} \lambda^r_{qs} \\ \lambda^r_{ds} \end{bmatrix} &= \mathbf{K}^r_s \mathbf{L}_s \left(\mathbf{K}^r_s\right)^{-1} \begin{bmatrix} i^r_{qs} \\ i^r_{ds} \end{bmatrix} + L_{ms} i'^{r}_{fd} \mathbf{K}^r_s \begin{bmatrix} \sin\theta_r \\ -\cos\theta_r \end{bmatrix} \\ &= \mathbf{L}_s \mathbf{i}^r_{qds} + L_{ms} i'^{r}_{fd} \begin{bmatrix} 0 \\ 1 \end{bmatrix}\end{aligned} \tag{7.3-30}$$

Equation (7.3-30) may be written as

$$\lambda^r_{qs} = L_{ss} i^r_{qs} \tag{7.3-31}$$

$$\lambda^r_{ds} = L_{ss} i^r_{ds} + L_{ms} i'^{r}_{fd} \tag{7.3-32}$$

We have taken care of the flux linkages for the stator equations (7.3-23) and (7.3-24). The flux linkage λ^r_{fd} may be written as

$$\begin{aligned}\lambda^r_{fd} &= L_{ffd} i^r_{fd} + L_{sr}[\sin\theta_r - \cos\theta_r]\left(\mathbf{K}^r_s\right)^{-1} \mathbf{i}^r_{qds} \\ &= L_{ffd} i^r_{fd} + L_{sr} i^r_{ds}\end{aligned} \tag{7.3-33}$$

We have to refer to the stator circuits. From (7.3-15) and (7.3-27),

$$\begin{aligned}\lambda'^{r}_{fd} &= \frac{N_s}{N_{fd}}\left(L_{ffd}\frac{N_s}{N_{fd}} i'^{r}_{fd} + L_{sr} i^r_{ds}\right) \\ &= L'_{ffd} i'^{r}_{fd} + L_{ms} i^r_{ds}\end{aligned} \tag{7.3-34}$$

Now,

$$\begin{aligned}L'_{ffd} &= \left(\frac{N_s}{N_{fd}}\right)^2 L_{lfd} + \frac{N_s}{N_{fd}} \frac{N_s N_{fd}}{\mathcal{R}_m} \\ &= L'_{lfd} + L_{ms}\end{aligned} \tag{7.3-35}$$

Equations (7.3-31), (7.3-32), and (7.3-34) may now be written as

$$\lambda^r_{qs} = L_{ls} i^r_{qs} + L_{ms} i^r_{qs} \tag{7.3-36}$$

$$\lambda_{ds}^r = L_{ls}i_{ds}^r + L_{ms}\left(i_{ds}^r + i_{fd}'^r\right) \tag{7.3-37}$$

$$\lambda_{fd}'^r = L_{lfd}'i_{fd}'^r + L_{ms}\left(i_{ds}^r + i_{fd}'^r\right) \tag{7.3-38}$$

Equations (7.3-23) through (7.3-25) and (7.3-36) through (7.3-38) suggest the equivalent circuits shown in Fig. 7.3-2.

The torque may be expressed from (7.3-19) and the expression for i_{qs}^r from $\mathbf{K}_s^r\,\mathbf{i}_{abs}$.

$$T_e = \frac{P}{2}L_{ms}i_{fd}'^r i_{qs}^r \tag{7.3-39}$$

Here, we see the similarity with the expression for torque given by (6.5-18) for the permanent-magnet ac machine. That is, $L_{ms}i_{fd}'^r$ in (7.3-17) and (7.3-39) is $\lambda_m'^r$ in (6.5-18).

The torque and rotor speed are related by

$$T_e = J\left(\frac{2}{P}\right)\frac{d\omega_r}{dt} + B_m\left(\frac{2}{P}\right)\omega_r + T_L \tag{7.3-40}$$

where J is the inertia expressed in kilogram · meter² (kg · m²) or joule · second² (J · s²). The load torque T_L is positive for a torque load. The constant B_m is a damping coefficient associated with the rotational system of the machine and mechanical load. It has the units of N · m · s/rad.

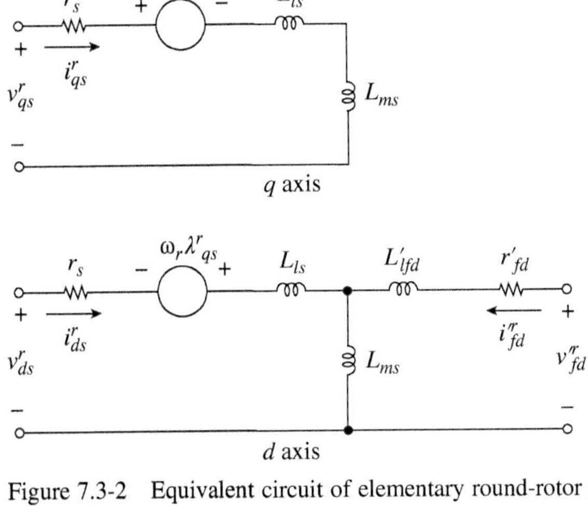

Figure 7.3-2 Equivalent circuit of elementary round-rotor synchronous machine in rotor reference frame.

SP7.3-1 Express v_{qs}^r and v_{ds}^r for open-circuit stator phases. $\left[v_{qs}^r = \omega_r L_{ms} i_{fd}^{\prime r}, v_{ds}^r = 0\right]$

SP7.3-2 Repeat SP7.3-1 for v_{as} and v_{bs}. $\left[v_{as} = \omega_r L_{ms} i_{fd}^{\prime r} \cos\theta_r,\ v_{bs} = \omega_r L_{ms} i_{fd}^{\prime r} \sin\theta_r\right]$

SP7.3-3 In (7.3-18) we used $\frac{1}{2}L_{ffd}\left(i_{fd}^r\right)^2$ rather than $\frac{1}{2}L_{ffd}^{\prime}\left(i_{fd}^{\prime r}\right)^2$. Why does this not make a difference when evaluating T_e? [Neither is a function of θ_r]

7.4 ANALYSIS OF STEADY-STATE OPERATION

In the case of the synchronous machine, we have transformed the stator variables to the rotor reference frame. During steady-state operation, the electrical angular velocity of the rotor, ω_r, is equal to ω_e. Hence, the circuits that exist in the rotor reference frame are the field winding and the fictitious windings ($\overset{r}{q}s$ and $\overset{r}{d}s$ windings). These windings rotate at ω_r and therefore do not experience a change of flux linkages as a result of relative motion between windings. Moreover, since $i_{fd}^{\prime r}$ is constant, mmf$_r^r$ is a constant amplitude sinusoidal function of ϕ_r and, for balanced steady-state conditions, mmf$_s^e$ is also a constant amplitude sinusoidal function of ϕ_e. Therefore, there can be no induced voltage due to transformer action in any of the circuits in the rotor reference frame. One would then guess that the substitute currents and voltages associated with all windings in the rotor reference frame, actual or fictitious, would be constant (zero in the case of the damper windings) during balanced steady-state operation. This seems logical, but what has happened to the balanced, sinusoidal stator variables? Remember the balanced, steady-state sinusoidal stator currents give rise to a constant amplitude mmf$_s^e$ rotating at ω_e (i.e., Tesla's rotating magnetic field). But, if this constant amplitude mmf$_s^e$ is now to be produced by substitute currents flowing in fictitious windings (i_{qs}^r and i_{ds}^r) that are mathematically fixed in the rotor reference frame, which rotates at ω_e during steady-state operation, these substitute currents must be constant.

It should be pointed out that during steady-state operation we can relate F_{qs}^r and F_{ds}^r to $\widetilde{F}_{as}$; however, when ω_r differs from ω_e, variables are induced due to the unsymmetrical rotor that are of $\omega_e - \omega_r$ angular velocity. The synchronous reference frame variables will then contain multiple frequencies and the instantaneous phasor concept is not valid for sustained rotor speed different from ω_e. Although we will not set forth the derivation, we will use the instantaneous phasor later for small, temporary changes in rotor speed from synchronous.

Now that we know what to expect, let us proceed. During balanced steady-state operation, the stator variables may be expressed as

$$F_{as} = \sqrt{2}F_s \cos\left[\omega_e t + \theta_{esf}(0)\right] \tag{7.4-1}$$

$$F_{bs} = \sqrt{2}F_s \sin\left[\omega_e t + \theta_{esf}(0)\right] \tag{7.4-2}$$

Substituting F_{as}, and F_{bs}, into the equations of transformation, (5.3-8), with $\theta = \theta_r$; wherein the steady state with $\omega_r = \omega_e$:

$$\theta_r = \omega_e t + \theta_r(0) \tag{7.4-3}$$

Thus,

$$F^r_{qs} = \sqrt{2}F_s \cos\left[\theta_{esf}(0) - \theta_r(0)\right] \tag{7.4-4}$$

$$F^r_{ds} = -\sqrt{2}F_s \sin\left[\theta_{esf}(0) - \theta_r(0)\right] \tag{7.4-5}$$

Since, $\theta_{esf}(0)$ and $\theta_r(0)$ are constants, F^r_{qs} and F^r_{ds} are constants.

Now, since all variables in the rotor reference frame are constant during steady-state operation, all variables multiplied by the operator p are zero. Hence, (7.3-23) through (7.3-25) may be written as

$$V^r_{qs} = r_s I^r_{qs} + \omega_r \lambda^r_{ds} \tag{7.4-6}$$

$$V^r_{ds} = r_s I^r_{ds} - \omega_r \lambda^r_{qs} \tag{7.4-7}$$

$$V'^r_{fd} = r'_{fd} I'^r_{fd} \tag{7.4-8}$$

where capital letters are used to denote constant, steady-state voltages, and currents and λ^r_{qs} and λ^r_{ds} are (7.3-36) and (7.3-37), respectively. Appropriate substitution of (7.3-36) and (7.3-37) into (7.4-6) and (7.4-7), wherein ω_r is set equal to ω_e yields

$$V^r_{qs} = r_s I^r_{qs} + X_{ss} I^r_{ds} + X_{ms} I'^r_{fd} \tag{7.4-9}$$

$$V^r_{ds} = r_s I^r_{ds} - X_{ss} I^r_{qs} \tag{7.4-10}$$

Note these are dc voltage equations and, yet, we have reactances, X_{ss} and X_{ms}, where $X_{ss} = \omega_e L_{ss}$ and $X_{ms} = \omega_e L_{ms}$. We are not dealing with phasors, yet X times I is a voltage, nevertheless. The above voltage equations can be used in their present form to analyze the synchronous machine; however, it is convenient and customary to relate the F^r_{qs} and F^r_{ds} quantities, which are constants, to $\tilde{F}_{as}$, which is a phasor representing a sinusoidal voltage.

In Chapter 5 we considered the situation where $\theta(0)$ was nonzero and mentioned that this would occur when analyzing steady-state operation of synchronous machines. The situation is present here. From (7.4-1),

$$\tilde{F}_{as} = F_s e^{j\theta_{esf}(0)} \tag{7.4-11}$$

If each side of (7.4-11) is multiplied by $\sqrt{2}e^{-j\theta_r(0)}$ we will obtain

$$\sqrt{2}\tilde{F}_{as}e^{-j\theta_r(0)} = \sqrt{2}F_s \cos\left[\theta_{esf}(0) - \theta_r(0)\right] + j\sqrt{2}F_s \sin\left[\theta_{esf}(0) - \theta_r(0)\right] \tag{7.4-12}$$

Comparing (7.4-4) and (7.4-5) to the terms of (7.4-12), we can write

$$\sqrt{2}\tilde{F}_{as}e^{-j\theta_r(0)} = F^r_{qs} - jF^r_{ds} \tag{7.4-13}$$

We have only a few more steps. From (7.4-13) we can write

$$\sqrt{2}\tilde{V}_{as}e^{-j\theta_r(0)} = V^r_{qs} - jV^r_{ds} \tag{7.4-14}$$

Substituting (7.4-9) and (7.4-10) into (7.4-14) yields

$$\sqrt{2}\widetilde{V}_{as}e^{-j\theta_r(0)} = r_s I^r_{qs} + X_{ss}I^r_{ds} + X_{ms}I'^r_{fd} + j\left(-r_s I^r_{ds} + X_{ss}I^r_{qs}\right) \tag{7.4-15}$$

Now, from (7.4-13),

$$j\sqrt{2}\widetilde{I}_{as}e^{-j\theta_r(0)} = I^r_{ds} + jI^r_{qs} \tag{7.4-16}$$

then (7.4-15) may be written as

$$\widetilde{V}_{as} = (r_s + jX_{ss})\widetilde{I}_{as} + \frac{1}{\sqrt{2}}\omega_e L_{ms} I'^r_{fd} e^{j\theta_r(0)} \tag{7.4-17}$$

where $X_{ss} = \omega_e L_{ss}$. It is convenient to define the last term of (7.4-17) as

$$\widetilde{E}_a = \frac{1}{\sqrt{2}}\omega_e L_{ms} I'^r_{fd} e^{j\theta_r(0)} \tag{7.4-18}$$

Thus, (7.4-17) becomes

$$\widetilde{V}_{as} = (r_s + jX_{ss})\widetilde{I}_{as} + \widetilde{E}_a \tag{7.4-19}$$

We can express the steady-state electromagnetic torque for a magnetically linear system from (7.3-19) for balanced steady-state operation

$$\begin{aligned} T_e &= \frac{P}{2}L_{ms}I'^r_{fd}\sqrt{2}I_s\{\cos\left[\omega_e t + \theta_{esi}(0)\right]\cos\left[\omega_e t + \theta_r(0)\right] \\ &\quad + \sin\left[\omega_e t + \theta_{esi}(0)\right]\sin\left[\omega_e t + \theta_r(0)\right]\} \\ &= \frac{P}{2}L_{ms}I'^r_{fd}\sqrt{2}I_s\cos\left[\theta_{esi}(0) - \theta_r(0)\right] \end{aligned} \tag{7.4-20}$$

Please note from (7.4-4) that (7.4-20) can be expressed as

$$T_e = \frac{P}{2}L_{ms}I'^r_{fd}I^r_{qs} \tag{7.4-21}$$

It should be mentioned that the power system engineer commonly uses a slightly different expression to calculate steady-state torque. This expression is given in the next section by (7.5-4) wherein, the stator resistance, r_s, is neglected.

Before leaving this section to focus on the analysis of balanced steady-state operation in a large power system, it is interesting and instructive to compare the permanent-magnet ac machine covered in Chapter 6 to the round-rotor synchronous machine. Both are synchronous machines and we have assumed both to be round-rotor machines. Thus, for analysis purpose they differ only in the structure of the rotor; that is, the rotor magnetic field, mmf$_r$, is created by a permanent magnet in one and by a field winding in the other. Therefore, we would expect the equations used to predict their performance to be very similar. However, in Chapter 6 we used the permanent-magnet ac machine in a variable frequency drive system application and in this chapter we are assuming the synchronous machine is connected to a constant frequency system. These would be the modes of operation common to these devices. Let us explain this by a comparison of the balanced steady-state equations given for each machine in Table 7.4-1.

TABLE 7.4-1 Comparison of Two-Phase Synchronous-Type Machines Used in Drive and Constant Frequency Systems During Balanced, Steady-State Operation

Permanent-Magnet ac Machine in a Variable-Frequency Drive System with $\theta_r(0) = 0°$	Round-Rotor Synchronous Machine in a Constant Frequency System with $\theta_{esv}(0) = 0°$
$\tilde{V}_{as} = (r_s + j\omega_r L_{ss})\tilde{I}_{as} + \tilde{E}_a$	$\tilde{V}_{as} = (r_s + j\omega_e L_{ss})\tilde{I}_{as} + \tilde{E}_a$
$\tilde{E}_a = \frac{1}{\sqrt{2}}\omega_r \lambda'^r_m e^{j0°}$	$\tilde{E}_a = \frac{1}{\sqrt{2}}\omega_e L_{ms} I'^r_{fd} e^{j\theta_r(0)}$
$T_e = \frac{P}{2}\lambda'^r_m I^r_{qs}$	$V'^r_{fd} = r'^r_{fd} I'^r_{fd}$
$T_e = \frac{P}{2}\lambda'^r_m \sqrt{2} I_s \cos\theta_{esi}(0)$	$T_e = \frac{P}{2} L_{ms} I'^r_{fd} I^r_{qs}$
	$= \frac{P}{2} L_{ms} I'^r_{fd} \sqrt{2} I_s \cos[\theta_{esi}(0) - \theta_r(0)]^*$
	*See (7.5-4) for torque expression with r_s neglected

We have previously noted that λ'^r_m for the permanent-magnet ac machine is analogous to $L_{ms}I'^r_{fd}$ for the round-rotor synchronous machine with a field winding. In the case of the permanent-magnet ac machine, the frequency of the stator voltages is made equal to the angular velocity of the rotor, ω_r, by inverter control. Therefore, $\omega_r L_{ss}$ in $\tilde{V}_{as}$ (6.6-5) and ω_r in $\tilde{E}_a$ (6.6-6) change as the rotor slows or speeds up. In the case of the synchronous machine connected to a constant frequency system, ω_e is fixed equal to the angular velocity (synchronous speed) of the system and ω_r must equal ω_e in the steady state.

In the case of the permanent-magnet ac machine, $\tilde{E}_a$ is fixed in phase by the inverter control and load torque changes can be satisfied by changes in rotor speed (ω_r) which alters $\tilde{E}_a$ and the stator reactance and/or the magnitude and perhaps the phase of $\tilde{V}_{as}$. In the case of the synchronous machine in a constant frequency system, ω_e is constant and $\tilde{V}_{as}$ is fixed in magnitude and phase by the system. Therefore, sudden changes in load or input torque are satisfied by a change in the phase of $\tilde{E}_a$ due to a transient change in ω_r since I'^r_{fd} is slow to change due to the large field inductance. Thus, the temporary change in the rotor speed from synchronous changes the relative position of the rotor poles established by mmf_r and the poles of mmf_s which, in effect, changes the steady-state operating point, $\theta_r(0)$.

SP7.4-1 Briefly describe the principle mode of operation of the synchronous machine used in an inverter drive and in a power system. [drive is variable frequency; power system is constant frequency]

SP7.4-2 Indicate the reference phasor for each mode of operation in SP7.4-1. [drive, $\tilde{E}_a$; power system, $\tilde{V}_{as}$]

SP7.4-3 Describe the principle method of satisfying changes in load torque for each mode of operation in SP7.4-1. [drive, change in steady-state speed; power system, change in relative position of stator and rotor poles]

7.5 ANALYSIS OF STEADY-STATE OPERATION IN POWER SYSTEMS

There are features of the analysis of synchronous machines in power system applications that warrant mentioning. Perhaps one of the most engrained notation is the *rotor angle*, δ. In its broadest definition it is defined as $\delta = \theta_r - \theta_{esv}$, where θ_r is the position of the q axis of the rotor and θ_{esv} is generally the position of the maximum positive value of the as phase voltage. However, since $\theta_{esv}(0)$ is generally selected as zero and since $\omega_e = \omega_r$ in the steady state, it is sufficient for the analysis of steady-state operation to define δ as

$$\delta = \theta_r(0) \tag{7.5-1}$$

Therefore, the power system engineer would replace the $\theta_r(0)$ in all of the equations for steady-state operation in Section 7.4 with δ. The as-phase voltage phasor, $\widetilde{V}_{as}$, is generally selected to be at zero degrees and is the reference phasor, thus, the rotor angle is the position of the q axis relative to $\widetilde{V}_{as}$. Now, we are aware that this is "mixing" mechanical and electrical coordinates. We are allowed to do this regardless of the state of the system; however, we are interested only in the steady state at this time.

A common approximation is to neglect the stator resistance in the steady-state calculations. This is a reasonable approximation since the stator winding resistance is much smaller than the inductive reactances at synchronous frequency. This, however, is not the case in variable frequency drive applications such as the brushless dc drive considered in Chapter 6, since the resistance becomes a significant part of the input impedance at low operating frequencies.

The power system engineer is interested primarily in generator operation of the synchronous machine and current is generally assumed to be positive out of the machine rather than into the machine, as we have done. Although this is convenient in that it makes T_e positive for generator action, we will not take the time to play power system engineer to this extent even though it is a straightforward reversal of the polarity or direction of stator variables.

A large power system is often considered to be an infinite bus, since, regardless of a connected machine's mode of operation, it cannot change the system's voltage, neither fixed in phase nor amplitude, or frequency. Since the frequency remains at ω_e (60 or 50 Hz), X_{ss} in (7.4-19) is much larger than r_s. Therefore, it is common to neglect r_s and, with some work, (7.4-20) becomes

$$T_e = -\frac{P}{2} L_{ms} I_{fd}^{\prime r} \sqrt{2} \frac{V_s}{X_{ss}} \sin \delta \tag{7.5-2}$$

Also, some prefer to define an excitation voltage as

$$E_{xfd}^{\prime r} = X_{ms} I_{fd}^{\prime r} \tag{7.5-3}$$

where the superscripts r were introduced in the derivation to denote variables in the rotor frame of reference. Substituting (7.5-3) into (7.5-2) yields an expression very familiar to power system engineers:

$$T_e = -\frac{P}{2}\left(\frac{1}{\omega_e}\right)\left(\frac{1}{X_{ss}}\right)\sqrt{2}E'^r_{xfd}V_s \sin\delta \tag{7.5-4}$$

It should be pointed out that (7.5-4) may be obtained quite directly by using the reference frame approach. In particular, with r_s neglected I^r_{qs} may be obtained from (7.4-10) and substituted into (7.3-39) and since $V^r_{ds} = -\sqrt{2}V_s \sin\delta$, (7.5-4) results.

A plot of (7.5-4) is shown in Fig. 7.5-1. The rotor angle δ, which is $\theta_r(0)$, is the angle between the phasor $\widetilde{E}_a$ which lays along the q axis and the phasor $\widetilde{V}_{as}$ which is assumed to be at zero degrees. Also, for a given load or input torque, the rotor angle is constant during steady-state operation and, since T_e versus δ is a sinusoidal function, we need only consider the plot for $-\pi < \delta < \pi$. Now, the load torque and the electromagnetic torque are related by

$$T_e = J\left(\frac{2}{P}\right)\frac{d\omega_r}{dt} + B_m\left(\frac{2}{P}\right)\omega_r + T_L \tag{7.5-5}$$

which is a repeat of (7.3-40). It is clear that during steady-state operation $T_e = T_L$ if B_m is neglected. For illustrative purposes, let the synchronous machine be connected to an electric system and let the load torque be negative (generator action). That is, torque is applied to the shaft by some external means, a steam or hydro-turbine or a combustion engine, or, perhaps a wind turbine. Regardless of how the applied torque is developed, it is an input torque to the shaft of the machine and, if the ohmic losses along with the friction and windage losses are neglected, the steady-state T_e must equal T_L, thus T_e is negative also. Let us think about this for a minute. The machine is connected to an electric system. If a torque is applied to the shaft and since torque times rotor speed is equal to power, the synchronous machine must deliver an equal amount of power (neglecting friction, windage, and

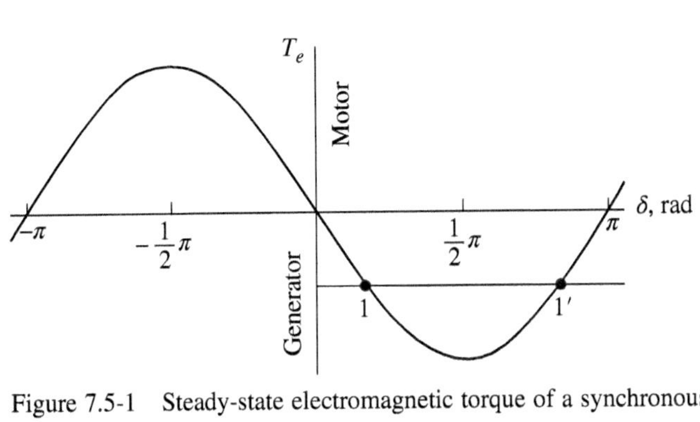

Figure 7.5-1 Steady-state electromagnetic torque of a synchronous machine (stator resistance neglected).

ohmic losses) to the electric system. Otherwise, there would be a torque or power imbalance and, if T_e is not equal to T_L, the synchronous machine would either accelerate for $T_e - T_L > 0$ or decelerate for $T_e - T_L < 0$.

In a later section, we will discuss briefly the dynamic performance of the synchronous machine. Our purpose now is to continue our example of steady-state operation to determine the stable operating region. We can do as we did in the case of the relay back in Chapter 2. With T_L negative, there are two possible operating points between $-\pi < \delta < \pi$. They are denoted 1 and 1′ in Fig. 7.5-1. Now, we must remember that T_L is constant and negative. First, assume that steady-state operation occurs at point 1. This is a valid operating point if the system will return to this point when distributed from it. To test this, let δ decrease ever so slightly. In this case $T_e - T_L > 0$ and the rotor will accelerate, thereby increasing δ; hence, a torque is developed to move the machine back to operating point 1. If now δ increases ever so slightly, $T_e - T_L < 0$, and the system will again move back to point 1. Hence, point 1 is a stable operating point.

Although we all suspect what is going to happen at point 1′, let us go through the exercise anyway. If we assume that the system is operating stably at point 1′, then if it is a stable operating point, a displacement in δ from this operating point should cause a torque to restore the system to this operating point. If δ decreases slightly from point 1′, $T_e - T_L < 0$ and the machine would decelerate which would further decrease δ. The system would move away from point 1′ and, after all transients have subsided, the system would operate stably at point 1. If, instead, δ increases from point 1′, $T_e - T_L > 0$, whereupon the machine will accelerate, moving away from point 1′. We can conclude that, although point l′ satisfies the "torque balance equation," it is not a stable operating condition. Let us go back a step. If δ increases from point 1′ and the rotor accelerates, where will it end up? At point $\delta_1 + 2\pi$ if it does not go unstable dynamically. We really cannot appreciate the meaning of "unstable dynamically" until we discuss the dynamic characteristics in a later section. It is sufficient here to say that we are dealing with steady-state characteristics, and one can get into trouble using steady-state characteristics to explain the large excursion, dynamic (transient) characteristics of a system.

It is apparent that the only way that the steady-state electric power output or input (or T_e) can be changed is to change the torque T_L. However, the electrical characteristics of a synchronous machine connected to a system can be changed by changing the field voltage $V_{fd}'^r$. Although the following is, perhaps, of most interest to power system engineers, it is worth a passing consideration by all.

To set the stage for this discussion, it is understood that the majority of power system loads are predominately inductive, and when these increase in magnitude, the system voltage may become depressed (i.e., sag) as a result of the need to supply the reactive power to the inductive loads. This reactive power may be supplied by switching in banks of capacitors at the load centers. The synchronous generator may also supply reactive power by controlling $E_{xfd}'^r \left(V_{fd}'^r\right)$. To explain this we will assume that the load torque T_L is zero and if we neglect friction and windage losses along with the stator resistance, then T_e and δ are also zero and the machine will run at synchronous speed without absorbing energy from either the electric or mechanical systems. Although this mode of operation is not feasible in practice since the machine will actually absorb a small amount of electrical power to satisfy the ohmic, friction, and windage losses, a small δ would exist at no load. It is, however, convenient, for purposes of explanation to assume δ is zero. With the machine "floating on the line" the field voltage can be adjusted to establish the desired terminal conditions. Three situations may exist (remember we are assuming $\delta = 0°$):

1. $|\tilde{E}_a| = |\tilde{V}_{as}|$, whereupon $\tilde{I}_{as} = 0$.
2. $|\tilde{E}_a| > |\tilde{V}_{as}|$, whereupon $\tilde{I}_{as}$ leads $\tilde{V}_{as}$ by 90° and the synchronous machine appears as a capacitor supplying reactive power to the system.
3. $|\tilde{E}_a| < |\tilde{V}_{as}|$, whereupon $\tilde{I}_{as}$ lags $\tilde{V}_{as}$ by 90°, and the machine appears as an inductor absorbing reactive power from the system.

Recall that reactive power is generally denoted as Q and expressed per phase as

$$\begin{aligned} Q &= |\tilde{V}_{as}||\tilde{I}_{as}| \sin[\theta_{esv}(0) - \theta_{esi}(0)] \\ &= |\tilde{V}_{as}||\tilde{I}_{as}| \sin\phi_{pf} \end{aligned} \tag{7.5-6}$$

where ϕ_{pf} is the power factor angle and the units of Q are var (volt ampere reactive). An inductance is said to absorb reactive power and thus, by definition, Q is positive for an inductor and negative for a capacitor. Actually, Q is a measure of the exchange of energy stored in the electric (capacitor) field and magnetic (inductance) field; however, real power is not interchanged between these energy storage devices.

Now, to maintain the voltage in a power system at rated value, the synchronous generators are normally operated in the overexcited mode, $|\tilde{E}_a| > |\tilde{V}_{as}|$, since the generators are a source of reactive power for the inductive loads. In the past, some synchronous machines have been placed in the power system for the sole purpose of supplying reactive power without any provision to absorb or provide real power. During peak load conditions when the system voltage is depressed, these so-called synchronous condensers are brought on line and the field voltage is adjusted to help increase the system voltage. In this mode of operation, the synchronous machine behaves like an adjustable capacitor. On the other hand, it may be necessary for a generator to absorb reactive power in order to regulate voltage in a high-voltage transmission system during light load conditions. This mode of operation is, however, not desirable and should be avoided since machine oscillations become less damped as the reactive power required is decreased [2]. The influence of the field voltage during motor operation is illustrated in Example 7A.

As a finale to the analysis of steady-state operation of synchronous machines, let us consider the procedure by which generator action is established and then look at the phasor diagram for this mode of operation. A prime mover is mechanically connected to the shaft of the synchronous generator. As mentioned, this prime mover can be either a steam or wind turbine, a hydroturbine, or a combustion engine. We will assume that the machine is synchronized with the power system and the torque input to the shaft from the prime mover is very small. Therefore, the synchronous machine is essentially floating on the line. If now the input torque is increased incrementally to some value (T_L negative), for example, by supplying steam to the turbine blades, a torque imbalance occurs. Briefly, T_e must remain at its original value (near zero) until ω_r temporarily advances above ω_e, thereby increasing δ. In other words, the rotor will temporarily accelerate slightly above synchronous speed in accordance with (7.5-5). Thus, T_e increases negatively and a new operating point (e.g., point 1 in Fig. 7.5-1) will be established where T_e is equal to T_L. The rotor will again rotate at synchronous speed with a new $\theta_r(0)$ or δ. The stepwise loading would then be repeated and the operating point would move to the right. If, during generator operation, the torque input from the prime mover is increased (T_L negative) to a value greater than the maximum possible value of T_e, the machine will be unable to maintain steady-state operation since it cannot electrically transmit the power supplied to the shaft. In this case, the device will accelerate above synchronous speed theoretically without bound. However, protection is normally provided in power systems which disconnects the machine from the system and reduces the input torque to zero by closing the steam valves of the steam turbine, for example, when it exceeds synchronous speed by 3–5 percent.

Normally, steady-state generator operation is depicted by a phasor diagram similar to that shown in Fig. 7.5-2. The phase angle of $\widetilde{V}_{as}$ is $\theta_{esv}(0)$ and if it is selected to be zero, $\theta_{esi}(0)$ is then the angle between the voltage and the current. In the steady state the phasor diagram and the q and d axes of the machine may be superimposed along with the stator and rotor poles.

The relationship between the rotor reference frame variables and the phasors is also shown in Fig. 7.5-2. Care must be taken when interpreting this diagram since $\widetilde{V}_{as}$, $\widetilde{I}_{as}$, and $\widetilde{E}_a$ are phasors representing sinusoidal quantities; however, all rotor reference frame quantities are constants during balanced steady-state operation. They are not phasors even though they are displayed on a phasor diagram

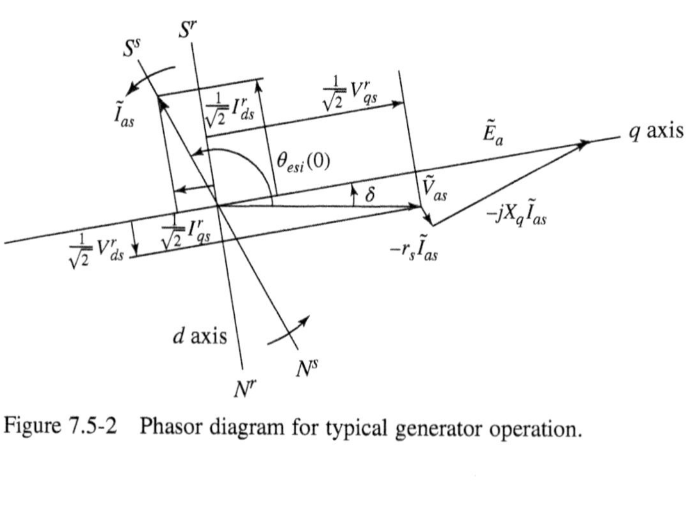

Figure 7.5-2 Phasor diagram for typical generator operation.

TABLE 7.5-1 Two-Phase Round-Rotor Synchronous Machine During Steady-State Power System Operation

Conditions: r_s Neglected, $\theta_{esv}(0) = 0°$, $\theta_r(0) = \delta$

$V^r_{qs} = X_{ss}I^r_{ds} + X_{ms}I'^r_{fd}$	$\tilde{V}_{as} = j\omega_e L_{ss}\tilde{I}_{as} + \tilde{E}_a$
$V^r_{ds} = -X_{ss}I^r_{qs}$	$\tilde{E}_a = \frac{1}{\sqrt{2}}\omega_e L_{ms}I'^r_{fd}e^{j\delta}$
$T_e = \frac{P}{2}L_{ms}I'^r_{fd}I^r_{qs}$	$T_e = -\frac{P}{2}\left(\frac{1}{\omega_e}\right)\left(\frac{1}{X_{ss}}\right)\sqrt{2}E'^r_{xfd}V_s \sin\delta$

$$E'^r_{xfd} = X_{ms}I'^r_{fd}$$
$$V'^r_{fd} = r'_{fd}I'^r_{fd}$$
$$P = |\tilde{V}_{as}||\tilde{I}_{as}|\cos\phi_{pf}, \text{ per phase}$$
$$Q = |\tilde{V}_{as}||\tilde{I}_{as}|\sin\phi_{pf}, \text{ per phase}$$
$$\phi_{pf} = \theta_{esv}(0) - \theta_{esi}(0)$$

Note: See Table 7.8-1 for changes for a three-phase machine.

which actually rotates at ω_e, counterclockwise. We see the rotor poles pushing the stator poles in the counterclockwise direction; generator action. The analytical information involved in this section is summarized in Table 7.5-1.

Example 7A Analysis of Synchronous Motor Operation with Different Power Factors

A six-pole two-phase round-rotor synchronous machine is supplied from a 440-V (rms) 60-Hz source. The machine is operated as a motor with a total power input of 40 kW at the terminals. The parameters of the machine are $r_s = 0.3\ \Omega$, $L_{ls} = 0.001$ H, $L_{ms} = 0.015$ H, $r'_{fd} = 0.03\ \Omega$, and $L'_{lfd} = 0.001$ H. (a) The excitation (i.e., the field current) is adjusted so that $\tilde{I}_{as}$ lags $\tilde{V}_{as}$ by 30°. Calculate $\tilde{E}_a$. (b) Repeat (a) with the excitation adjusted so that $\tilde{I}_{as}$ is in phase with $\tilde{V}_{as}$. (c) Repeat (a) with the excitation adjusted so that $\tilde{I}_{as}$ leads $\tilde{V}_{as}$ by 30°.

For part (a). The phase current may be calculated from the power as

$$|\tilde{I}_{as}| = \frac{\frac{1}{2}(40 \times 10^3)}{440 \cos 30°} = 52.5 \text{ A} \tag{7A-1}$$

Note that the 40×10^3 is the total power which is the sum of the two phases. With

$$\tilde{V}_{as} = 440\angle 0°$$
$$\tilde{I}_{as} = 52.5\angle -30° \text{A} \tag{7A-2}$$

From (7.4-19),

$$\begin{aligned}\widetilde{E}_a &= \widetilde{V}_{as} - (r_s + jX_{ss})\widetilde{I}_{as} \\ &= 440\underline{/0^\circ} - [0.3 + j377(0.001 + 0.015)]52.5\underline{/-30^\circ} \\ &= 377.8\underline{/-44.8^\circ}\ \text{V}\end{aligned} \tag{7A-3}$$

For part (b). The phase current is

$$|\widetilde{I}_{as}| = \frac{20 \times 10^3}{440} = 45.4\ \text{A} \tag{7A-4}$$

From (7.4-19),

$$\begin{aligned}\widetilde{E}_a &= 440\underline{/0^\circ} - (0.3 + j6.032)45.4\underline{/0^\circ} \\ &= 506.7\underline{/-32.7^\circ}\ \text{V}\end{aligned} \tag{7A-5}$$

For part (c). The phase current is

$$|\widetilde{I}_{as}| = \frac{20 \times 10^3}{440\cos(-30^\circ)} = 52.5\ \text{A} \tag{7A-6}$$

From (7.4-19),

$$\begin{aligned}\widetilde{E}_a &= 440\underline{/0^\circ} - (0.3 + j6.032)52.5\underline{/30^\circ} \\ &= 649.4\underline{/-25.8^\circ}\ \text{V}\end{aligned} \tag{7A-7}$$

Example 7B Phasor Diagrams for Synchronous Motor Operation
It is instructive to construct the phasor diagram and to calculate the torque for each operating condition in Example 7A. Let us first repeat the parameters of the 440-V 60-Hz six-pole machine, $r_s = 0.3\ \Omega$, $L_{ls} = 0.001$ H, $L_{ms} = 0.015$ H, $r'_{fd} = 0.03\ \Omega$, and $L'_{lfd} = 0.001$ H. The power input is 40×10^3 W at rated voltage.

The power input for each operating mode is the same. Therefore, the power output at the shaft is

$$P_{\text{out}} = P_{\text{in}} - P_{\text{loss}} \tag{7B-1}$$

where we will assume that the P_{loss} is only the ohmic loss, thus,

$$P_{\text{loss}} = 2|\widetilde{I}_{as}|^2 r_s \tag{7B-2}$$

From part (a) of Example 7A,

$$\widetilde{I}_{as} = 52.5\underline{/-30^\circ}\text{A} \tag{7B-3}$$

$$\widetilde{E}_a = 377.8\underline{/-44.8^\circ}\ \text{V} \tag{7B-4}$$

The phasor diagram is shown in Fig. 7B-1. We see that the stator poles are "pushing" the rotor in the counterclockwise direction; motor action.

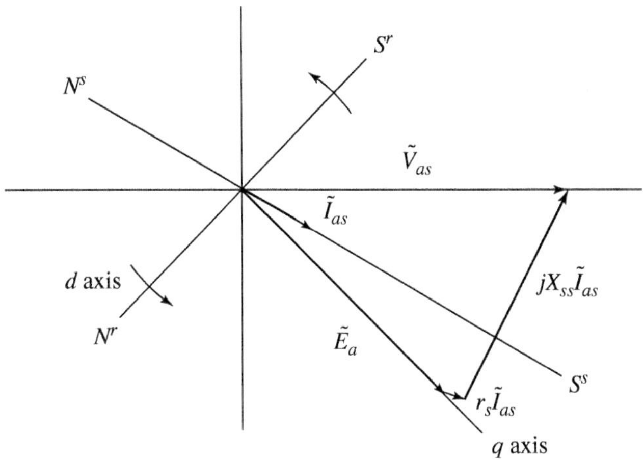

Figure 7B-1 Phasor diagram for part (a).

Now,

$$\begin{aligned} P_{\text{loss}} &= 2r_s\left|\widetilde{I}_{as}\right|^2 \\ &= 2(0.3)(52.5)^2 = 1.654 \text{ kW} \end{aligned} \tag{7B-5}$$

The P_{out} becomes

$$\begin{aligned} P_{\text{out}} &= P_{\text{in}} - P_{\text{loss}} \\ &= 40 - 1.654 = 38.346 \text{ kW} \end{aligned} \tag{7B-6}$$

Now, during steady-state operation

$$\begin{aligned} T_e &= \frac{1}{\omega_{rm}} P_{\text{out}} \\ &= \frac{P}{2\omega_e} P_{\text{out}} \\ &= \frac{6}{(2)(377)} 38.346 \times 10^3 = 305.1 \text{ N·m} \end{aligned} \tag{7B-7}$$

Before continuing to part (b), let us calculate the torque using (7.4-20) which is repeated here as

$$T_e = \frac{P}{2} L_{ms} I_{fd}^{\prime r} \sqrt{2} I_s \cos\left[\theta_{esi}(0) - \theta_r(0)\right] \tag{7B-8}$$

The value of $I_{fd}^{\prime r}$ can be obtained from (7.4-18), which is

$$\widetilde{E}_a = \frac{1}{\sqrt{2}} \omega_e L_{ms} I_{fd}^{\prime r} e^{j\delta} \tag{7B-9}$$

Solving (7B-9) for I'^r_{fd} yields

$$
\begin{aligned}
I'^r_{fd} &= \frac{\sqrt{2}}{\omega_e L_{ms}} |\tilde{E}_a| \\
&= \frac{\sqrt{2}(377.8)}{(377)(0.015)} = 94.5\ \text{A}
\end{aligned}
\tag{7B-10}
$$

Substituting (7B-10) for $\tilde{I}'^r_{fd}$ into (7B-8) yields

$$
T_e = \frac{6}{2}(0.015)(94.5)\sqrt{2}(52.5)\cos(-30^\circ + 44.8^\circ) = 305.33\ \text{N·m} \tag{7B-11}
$$

From part (b) of Example 7A,

$$
\tilde{I}_{as} = 45.4\angle 0^\circ\text{A} \tag{7B-12}
$$

$$
\tilde{E}_a = 506.7\angle -32.7^\circ\text{V} \tag{7B-13}
$$

The resulting phasor diagram is shown in Fig. 7B-2. Now,

$$
\begin{aligned}
P_{\text{loss}} &= 2r_s |\tilde{I}_{as}|^2 \\
&= 2(0.3)(45.4)^2 = 1.237\ \text{kW}
\end{aligned}
\tag{7B-14}
$$

The P_{out} becomes

$$
40 - 1.237 = 38.763 \tag{7B-15}
$$

The steady-state torque is

$$
\begin{aligned}
T_e &= \frac{P}{2\omega_e} P_{\text{out}} \\
&= \frac{6}{(2)(377)} 38.763 \times 10^3 = 308.5\ \text{N·m}
\end{aligned}
\tag{7B-16}
$$

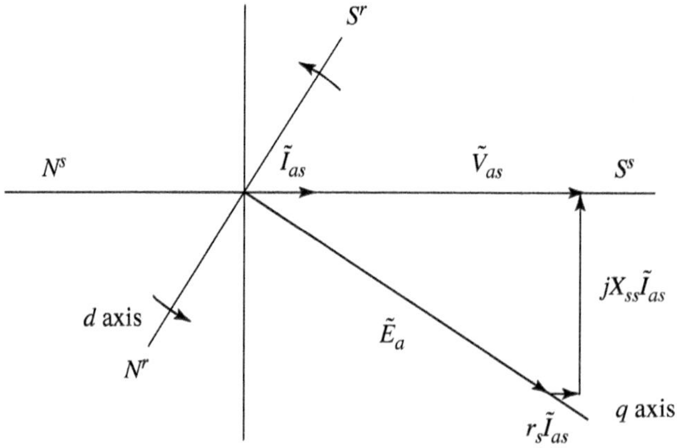

Figure 7B-2 Phasor diagram for part (b).

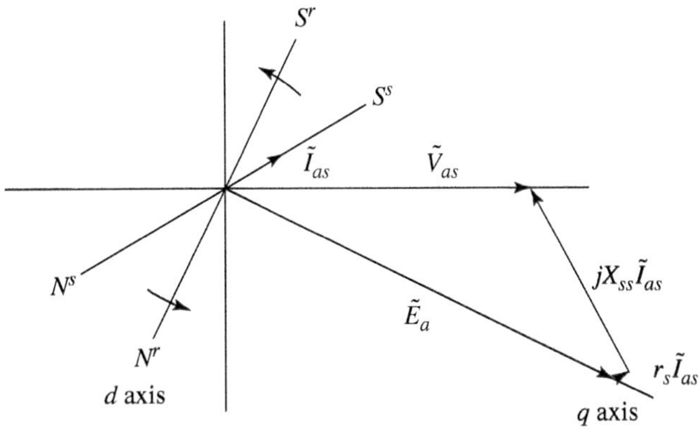

Figure 7B-3 Phasor diagram for part (c).

From part (c) of Example 7A,

$$\tilde{I}_{as} = 52.5\angle 30°\text{A} \tag{7B-17}$$

$$\tilde{E}_a = 649.4\angle -25.8°\text{V} \tag{7B-18}$$

The torque is 305.1 the same as in part (a) which is given by (7B-7). The resulting phasor diagram is shown in Fig. 7B-3.

SP7.5-1 A two-pole two-phase synchronous machine is operated as a generator with $\tilde{V}_{as} = 110\underline{/0°}$ V and $\tilde{I}_{as} = 5\underline{/150°}$ A. Calculate (a) total real power and (b) total reactive power. [(a) $P = -952.6$ W; (b) $Q = -550$ var]

SP7.5-2 The machine in SP7.5-1 is a round-rotor device with $\omega_r = 377$ rad/s. $L_{ls} = 4$ mH, $L_{ms} = 50$ mH, and $r_s \cong 0$. Calculate δ. [$\delta = 28.7°$]

SP7.5-3 Calculate $I_{fd}^{\prime r}$ for SP7.5-2. [$I_{fd}^{\prime r} = 13.76$ A]

SP7.5-4 Neglect r_s and with $\tilde{V}_{as} = V_s\underline{/0°}$ draw the phasor diagram for (a) $\tilde{I}_{as} = 0$, $|\tilde{E}_a| = |\tilde{V}_{as}|$, (b) $\tilde{I}_{as}$ leads $\tilde{V}_{as}$ by 90° with $|\tilde{E}_a| > |\tilde{V}_{as}|$, and (c) $\tilde{I}_{as}$ lagging $\tilde{V}_{as}$ by 90° with $|\tilde{E}_a| < |\tilde{V}_{as}|$.

7.6 TWO-PHASE RELUCTANCE MACHINE

An elementary two-phase reluctance machine is shown in Fig. 7.6-1. The voltage equations are

$$v_{as} = r_s i_{as} + \frac{d\lambda_{as}}{dt} \tag{7.6-1}$$

$$v_{bs} = r_s i_{bs} + \frac{d\lambda_{bs}}{dt} \tag{7.6-2}$$

The self-inductances of the stator windings and the mutual inductances between stator windings are functions of θ_r. Although this is a review of the material in Section 1.7, let us consider L_{asas}. With $\theta_r = 0$, we see from Fig. 7.6-1 that the magnetizing inductance of L_{asas} is less than it would be when $\theta_r = \frac{1}{2}\pi$. Let the magnetizing inductance of the *as* winding be denoted L_{mq} when $\theta_r = 0$, since in this position the q axis (high-reluctance path) is aligned with the magnetic axis of the *as* winding. Thus,

$$L_{asas} = L_{ls} + L_{mq}, \text{ when } \theta_r = 0 \tag{7.6-3}$$

where L_{ls} is the leakage inductance of the stator windings and

$$L_{mq} = \frac{N_s^2}{\mathcal{R}_{mq}} \tag{7.6-4}$$

where $\mathcal{R}_{mq}$ is an equivalent reluctance of the magnetic path in the q axis. We called this $\mathcal{R}_m(0)$ in (1.7-24). Now, at $\theta_r = \frac{1}{2}\pi$ the d axis (lower-reluctance path) is aligned with the magnetic axis of the *as* winding. Hence, denoting this magnetizing inductance as L_{md}, we can write

$$L_{asas} = L_{ls} + L_{md}, \quad \text{when } \theta_r = \frac{1}{2}\pi \tag{7.6-5}$$

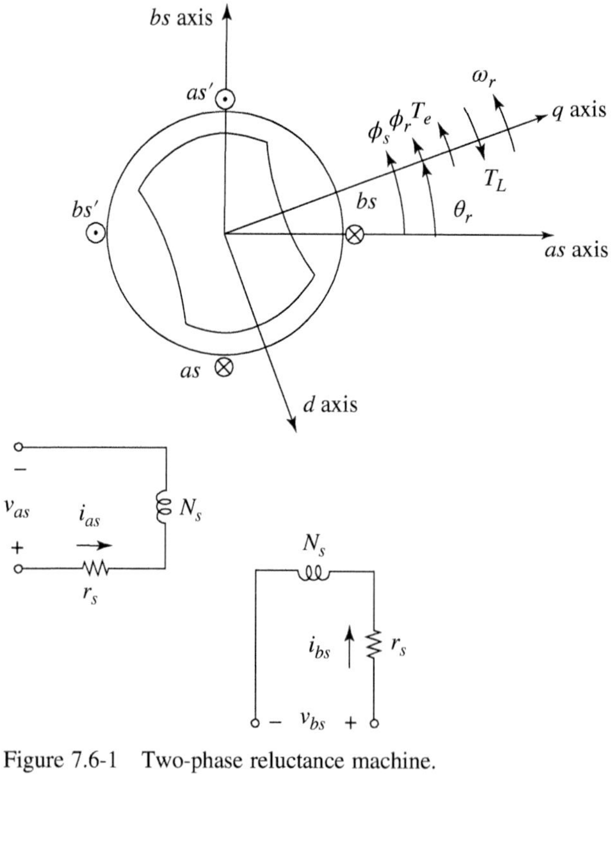

Figure 7.6-1 Two-phase reluctance machine.

where

$$L_{md} = \frac{N_s^2}{\mathcal{R}_{md}} \tag{7.6-6}$$

where $\mathcal{R}_{md}$ is an equivalent reluctance of the magnetic path in the d axis. This is $\mathcal{R}_m\left(\frac{1}{2}\pi\right)$ in (1.7-25).

Since $\mathcal{R}_{mq} > \mathcal{R}_{md}$, $L_{mq} < L_{md}$, and we see that a minimum L_{asas} occurs at $\theta_r = 0$ and also again at $\theta_r = \pi$. Therefore, (7.6-3) is valid for $\theta_r = 0$ and π. Similarly, maximum L_{asas} occurs at $\theta_r = \frac{1}{2}\pi$; and again at $\theta_r = \frac{3}{2}\pi$; hence (7.6-5) applies for $\theta_r = \frac{1}{2}\pi$ and $\frac{3}{2}\pi$. The magnetizing inductance varies about an average value (which must be positive) and if we assume this variation to be sinusoidal, it would vary as a function of $2\theta_r$ (Fig. 1.7-3). Let L_A be the average value and L_B the amplitude of the sinusoidal variation about this average value. In this case,

$$L_{mq} = L_A - L_B \tag{7.6-7}$$

$$L_{md} = L_A + L_B \tag{7.6-8}$$

Substituting (7.6-4) and (7.6-6) for L_{mq} and L_{md}, respectively, into (7.6-7) and (7.6-8) and solving for L_A and L_B yields

$$L_A = \frac{N_s^2}{2}\left(\frac{1}{\mathcal{R}_{md}} + \frac{1}{\mathcal{R}_{mq}}\right) \tag{7.6-9}$$

$$L_B = \frac{N_s^2}{2}\left(\frac{1}{\mathcal{R}_{md}} - \frac{1}{\mathcal{R}_{mq}}\right) \tag{7.6-10}$$

Assuming a sinusoidal variation, we can write (Fig. 1.7-3)

$$L_{asas} = L_{ls} + L_A - L_B \cos 2\theta_r \tag{7.6-11}$$

If the air gaps were uniform as is the case in a round-rotor synchronous machine, $\mathcal{R}_{mq} = \mathcal{R}_{md}$ and, hence, from (7.6-10), $L_B = 0$.

By a similar procedure, it follows that, for the salient-pole device,

$$L_{bsbs} = L_{ls} + L_A + L_B \cos 2\theta_r \tag{7.6-12}$$

Note that, when $\theta_r = 0$, L_{asas} is a minimum according to (7.6-11) and, according to (7.6-12), L_{bsbs} is a maximum. This, of course, corresponds to that which is portrayed in Fig. 7.6-1.

The mutual inductance $L_{asbs}(L_{bsas})$ is next. One would think that since the windings are orthogonal, the mutual coupling would always be zero; however, this is not the case due to the nonuniform air gap. Let us consider Fig. 7.6-2 where various rotor positions are shown with only the flux paths of the *as* winding depicted. Coupling occurs when flux produced by one winding links the other winding; for example, when the flux of the *as* winding links the *bs* winding. This will give us L_{bsas} and we know that $L_{asbs} = L_{bsas}$.

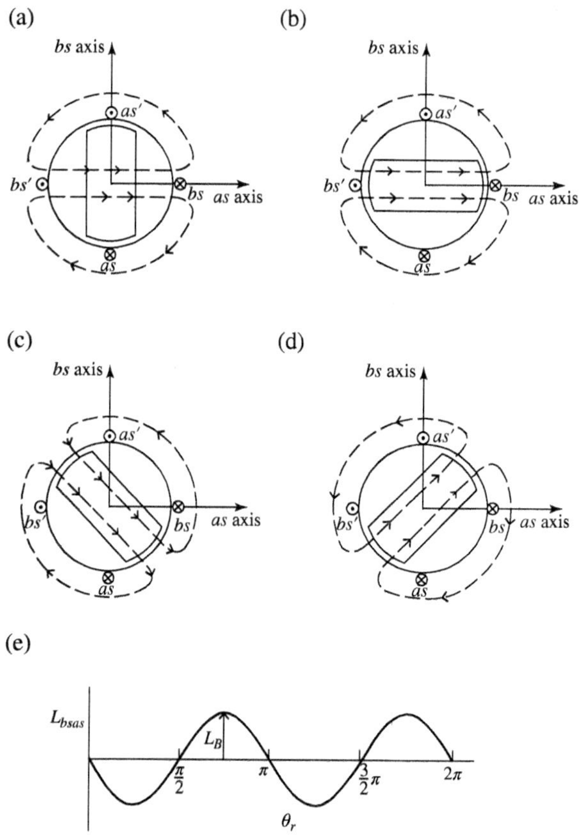

Figure 7.6-2 Flux path of *as* winding illustrating the mutual coupling between stator winding to determine L_{bsas} and L_{asbs}, (a) $\theta_r = 0$, π, and 2π; (b) $\theta_r = \frac{1}{2}\pi$ and $\frac{3}{2}\pi$; (c) $\theta_r = \frac{1}{4}\pi$ and $\frac{5}{4}\pi$; (d) $\theta_r = \frac{3}{4}\pi$ and $\frac{7}{4}\pi$; and (e) approximation of L_{bsas} and L_{asbs}.

Note that, when $\theta_r = 0$, π, and 2π as shown in Fig. 7.6-2a or when $\theta_r = \frac{1}{2}\pi$ and $\frac{3}{2}\pi$ as shown in Fig. 7.6-2a, L_{bsas} is zero. In these positions, there is no channeling of the flux of one winding through the other. However, let the rotor start to turn counterclockwise from zero toward $\frac{1}{2}\pi$ and consider the flux produced by the *as* winding. As the rotor turns, the configuration of the rotor provides a low-reluctance path to the flux produced by the *as* winding and the flux is channeled through the *bs* winding with maximum coupling occurring at $\theta_r = \frac{1}{4}\pi$, as illustrated in Fig. 7.6-2b. We see that this same rotor position relative to the windings occurs also at $\theta_r = \frac{5}{4}\pi$. Maximum coupling will again occur at $\theta_r = \frac{3}{4}\pi$ and $\frac{7}{4}\pi$, as illustrated in Fig. 7.6-2d. Now, what is the sign of the mutual inductance? With the assumed direction of positive currents, the right-hand rule tells us that L_{bsas} (or L_{asbs}) is negative at $\theta_r = \frac{1}{4}\pi, \frac{5}{4}\pi, \ldots$ (the fluxes of the windings oppose each other for positive currents) and positive for $\theta_r = \frac{3}{4}\pi, \frac{7}{4}\pi, \ldots$ (the fluxes aid each other). If we sketch L_{bsas} versus θ_r using the above information, we see, from Fig. 7.6-2e, that, as a first approximation, L_{bsas} or L_{asbs} may be expressed as

$$L_{bsas} = L_{asbs} = -L_B \sin 2\theta_r \tag{7.6-13}$$

In order for us to prove that the coefficient is L_B, it would be necessary to become quite involved [2]. We will accept this without proving it.

Reference Frame Derivation

For those only interested in the steady-state characteristics, the following reference frame derivation may be de-emphasized and study resumed with (7.6-39) and/or Table 7.6-1. Now, (7.6-1) and (7.6-2) may be written as

$$\mathbf{v}_{abs} = \mathbf{r}_s \mathbf{i}_{abs} + p\boldsymbol{\lambda}_{abs} \tag{7.6-14}$$

where

$$(\mathbf{f}_{abs})^T = [f_{as} \ f_{bs}] \tag{7.6-15}$$

$$\mathbf{r}_s = r_s \mathbf{I} \tag{7.6-16}$$

$$\boldsymbol{\lambda}_{abs} = \mathbf{L}_s \mathbf{i}_{abs} \tag{7.6-17}$$

where we can now write

$$\mathbf{L}_s = \begin{bmatrix} L_{asas} & L_{asbs} \\ L_{bsas} & L_{bsbs} \end{bmatrix} = \begin{bmatrix} L_{ls} + L_A - L_B \cos 2\theta_r & -L_B \sin 2\theta_r \\ -L_B \sin 2\theta_r & L_{ls} + L_A + L_B \cos 2\theta_r \end{bmatrix} \tag{7.6-18}$$

You have seen (7.6-18) before. It is (5.4-18) wherein it was used to demonstrate the facility of the change of variables in eliminating the rotor position-dependence of the inductances. We will show this again here. From (5.4-6) and (5.4-15), the rotor reference frame voltage equations are

$$v_{qs}^r = r_s i_{qs}^r + \omega_r \lambda_{ds}^r + p\lambda_{qs}^r \tag{7.6-19}$$

$$v_{ds}^r = r_s i_{ds}^r - \omega_r \lambda_{qs}^r + p\lambda_{ds}^r \tag{7.6-20}$$

Now, from (7.6-17),

$$\begin{bmatrix} \lambda_{qs}^r \\ \lambda_{ds}^r \end{bmatrix} = \mathbf{K}_s^r \mathbf{L}_s (\mathbf{K}_s^r)^{-1} \mathbf{i}_{qds}^r = \begin{bmatrix} L_q & 0 \\ 0 & L_d \end{bmatrix} \begin{bmatrix} i_{qs}^r \\ i_{ds}^r \end{bmatrix} \tag{7.6-21}$$

where the q-axis and d-axis inductances are

$$\begin{aligned} L_q &= L_{ls} + L_A - L_B \\ &= L_{ls} + L_{mq} \end{aligned} \tag{7.6-22}$$

$$\begin{aligned} L_d &= L_{ls} + L_A + L_B \\ &= L_{ls} + L_{md} \end{aligned} \tag{7.6-23}$$

The voltage equations (7.6-19) and (7.6-20) may now be written as

$$v_{qs}^r = r_s i_{qs}^r + \omega_r L_d i_{ds}^r + pL_q i_{qs}^r \tag{7.6-24}$$

$$v_{ds}^r = r_s i_{ds}^r - \omega_r L_q i_{qs}^r + pL_d i_{ds}^r \tag{7.6-25}$$

We can obtain an expression for torque through a power balance. The power entering the coupling field from the electrical system may be expressed as

$$pW_e = i_{qs}^r p\lambda_{qs}^r + i_{ds}^r p\lambda_{ds}^r + \left(\lambda_{ds}^r i_{qs}^r - \lambda_{qs}^r i_{ds}^r\right) p\theta_r \tag{7.6-26}$$

We have moved r_s external to the coupling field to be consistent with our work in Chapter 2. Now,

$$pW_e = pW_f - pW_m \tag{7.6-27}$$

where

$$pW_m = -T_e\left(\frac{2}{P}\right)p\theta_r \tag{7.6-28}$$

Now, λ_{qs}^r and λ_{ds}^r are not functions of θ_r, therefore, from (7.6-28) and (7.6-26) we can equate the coefficients of $p\theta_r$, whereupon,

$$T_e = \frac{P}{2}\left(\lambda_{ds}^r i_{qs}^r - \lambda_{qs}^r i_{ds}^r\right) \tag{7.6-29}$$

The first two terms of (7.6-26) must be pW_f. Substituting (7.6-21) for λ_{qs}^r and λ_{ds}^r in (7.6-29) yields

$$T_e = \frac{P}{2}\left(L_d - L_q\right) i_{ds}^r i_{qs}^r \tag{7.6-30}$$

In constant frequency applications where the reactance component of the stator impedance is much larger than r_s, such as a power system, the resistance is often neglected. Thus, from (7.6-24) and (7.6-25) for balanced steady-state operation

$$I_{ds}^r = \frac{V_{qs}^r}{X_d} \tag{7.6-31}$$

$$I_{qs}^r = -\frac{V_{ds}^r}{X_q} \tag{7.6-32}$$

Now,

$$V_{qs}^r = \sqrt{2}V_s \cos\left[\theta_{esv}(0) - \theta_r(0)\right] \tag{7.6-33}$$

$$V_{ds}^r = -\sqrt{2}V_s \sin\left[\theta_{esv}(0) - \theta_r(0)\right] \tag{7.6-34}$$

Substituting (7.6-33) and (7.6-34) into (7.6-31) and (7.6-32) and then I_{qs}^r and I_{ds}^r into (7.6-30) yields a familiar expression for the steady-state torque of a reluctance machine during balanced operation with the stator resistance neglected:

$$T_e = \left(\frac{P}{2}\right)\left(\frac{1}{\omega_e}\right)(V_s)^2\left(\frac{1}{X_q} - \frac{1}{X_d}\right)\sin 2\left[\theta_{esv}(0) - \theta_r(0)\right] \tag{7.6-35}$$

Equation (7.6-35) may be expressed in terms of the rotor angle δ. In the steady state $\theta_r(0) - \theta_{esv}(0)$ and since $\theta_{esv}(0)$ is often selected to be zero, δ becomes $\theta_r(0)$. This was discussed in the first paragraph of Section 7.5.

From (7.6-24) and (7.6-25) for balanced steady-state operation

$$V_{qs}^r = r_s I_{qs}^r + X_d I_{ds}^r \tag{7.6-36}$$

$$V_{ds}^r = r_s I_{ds}^r - X_q I_{qs}^r \tag{7.6-37}$$

Substituting (7.6-36) and (7.6-37) into (7.4-14) yields

$$\sqrt{2}\widetilde{V}_{as}e^{-j\theta_r(0)} = r_s I_{qs}^r + X_d I_{ds}^r - j\left(r_s I_{ds}^r - X_q I_{qs}^r\right) \tag{7.6-38}$$

If we now add and subtract $X_q I_{ds}^r$ from the right-hand side of (7.6-38) and then use (7.4-16) to obtain $jX_q\widetilde{I}_{as}$, $\widetilde{V}_{as}$ may be expressed as

$$\widetilde{V}_{as} = \left(r_s + j\omega_e L_q\right)\widetilde{I}_{as} + \widetilde{E}_a \tag{7.6-39}$$

where

$$\widetilde{E}_a = \omega_e\left(L_d - L_q\right)\frac{1}{\sqrt{2}}I_{ds}^r e^{j\theta_r(0)} \tag{7.6-40}$$

and L_q and L_d are defined by (7.6-22) and (7.6-23), respectively. We can express $\widetilde{E}_a$ in terms of phase variables by substituting (7.4-5) for I_{ds}^r, in particular,

$$\widetilde{E}_a = -\omega_e\left(L_d - L_q\right)I_s \sin\left[\theta_{esi}(0) - \theta_r(0)\right]e^{j\theta_r(0)} \tag{7.6-41}$$

If we substitute (7.4-4) for I_{qs}^r and (7.4-5) for I_{ds}^r into the steady-state version of (7.6-30), the torque may be expressed in phase variables as

$$\begin{aligned} T_e &= -\frac{P}{2}\left(L_d - L_q\right)\sqrt{2}I_s \cos\left[\theta_{esi}(0) - \theta_r(0)\right]\sqrt{2}I_s \sin\left[\theta_{esi}(0) - \theta_r(0)\right] \\ &= -\frac{P}{2}\left(L_d - L_q\right)I_s^2 \sin 2\left[\theta_{esi}(0) - \theta_r(0)\right] \end{aligned} \tag{7.6-42}$$

A summary of the information provided in this section is given in Table 7.6-1.

TABLE 7.6-1 Equations for the Analysis of the Steady-State Performance of a Two-Phase Reluctance Machine

$$\widetilde{V}_{as} = \left(r_s + j\omega_e L_q\right)\widetilde{I}_{as} + \widetilde{E}_a$$
$$\widetilde{E}_a = -\omega_e\left(L_d - L_q\right)I_s \sin\left[\theta_{esi}(0) - \theta_r(0)\right]e^{j\theta_r(0)}$$
$$L_q = L_{ls} + L_{mq}$$
$$L_d = L_{ls} + L_{md}$$
$$T_e = -\frac{P}{2}\left(L_d - L_q\right)\sqrt{2}I_s \cos\left[\theta_{esi}(0) - \theta_r(0)\right]\sqrt{2}I_s \sin\left[\theta_{esi}(0) - \theta_r(0)\right]$$
$$= -\frac{P}{2}\left(L_d - L_q\right)I_s^2 \sin 2\left[\theta_{esi}(0) - \theta_r(0)\right]$$

Notes: Neglect r_s if $\omega_e L_q \gg r_s$. See Table 7.8-1 for changes for three-phase machine.

Before leaving this section it is instructive to consider briefly the steady-state torque-angle characteristics of a salient-pole synchronous machine. To establish this would require a detailed analysis; however, we can get some idea of these characteristics by "superimposing" the information from this section with that from the two previous sections. Recall that the steady-state torque-angle characteristics of a round-rotor synchronous machine with a field winding and with the stator resistance neglected were depicted in Fig. 7.5-1. This plot is given again in Fig. 7.6-3a. The torque-angle plot with rotor saliency given by (7.6-35) is shown in Fig. 7.6-3b and the sum of Fig. 7.6-3a and b is shown in Fig. 7.6-3c. Although this approach is not rigorous, one is able to realize the main features of the large synchronous generator with rotor saliency. Moreover, we are aware that saliency occurs as a convenience for winding the field winding on the rotor and the torque produced due to saliency is small and is essentially inconsequential to power system operation; however, the salient torque is the only torque produced in the case of a low-power reluctance motor used in drives. If the drive is variable frequency, the stator resistance may not be neglected for purposes of analysis accuracy.

SP7.6-1 The parameters of a two-pole two-phase 60-Hz reluctance machine are $r_s = 10\ \Omega$, $L_{ls} = 26.5$ mH, $L_{mq} = 132.6$ mH, and $L_{md} = 318.3$ mH. It is operating as a motor with $\theta_r(0) = -30°$. Calculate T_e. Neglect r_s and $\widetilde{V}_{as} = 115\underline{/0°}$. [$T_e = 0.271$ N·m]

SP7.6-2 Repeat SP7.6-1 for maximum T_e. [$T_e = 0.312$ N·m]

7.7 DYNAMIC AND STEADY-STATE PERFORMANCE

It is instructive to observe the variables of the synchronous machine during dynamic and steady-state operation. In this section, the operation of a two-phase synchronous generator and a two-phase reluctance machine are illustrated by computer traces. Two-phase machines are not often used in practice; instead, three-phase machines are more common. Nevertheless, our purpose is to understand the theory and principles of operation of a synchronous machine. A two-phase machine is just as applicable in this regard as a three-phase machine. Information necessary to make the straightforward "transformation" from a two-phase to a three-phase machine is provided in a later section.

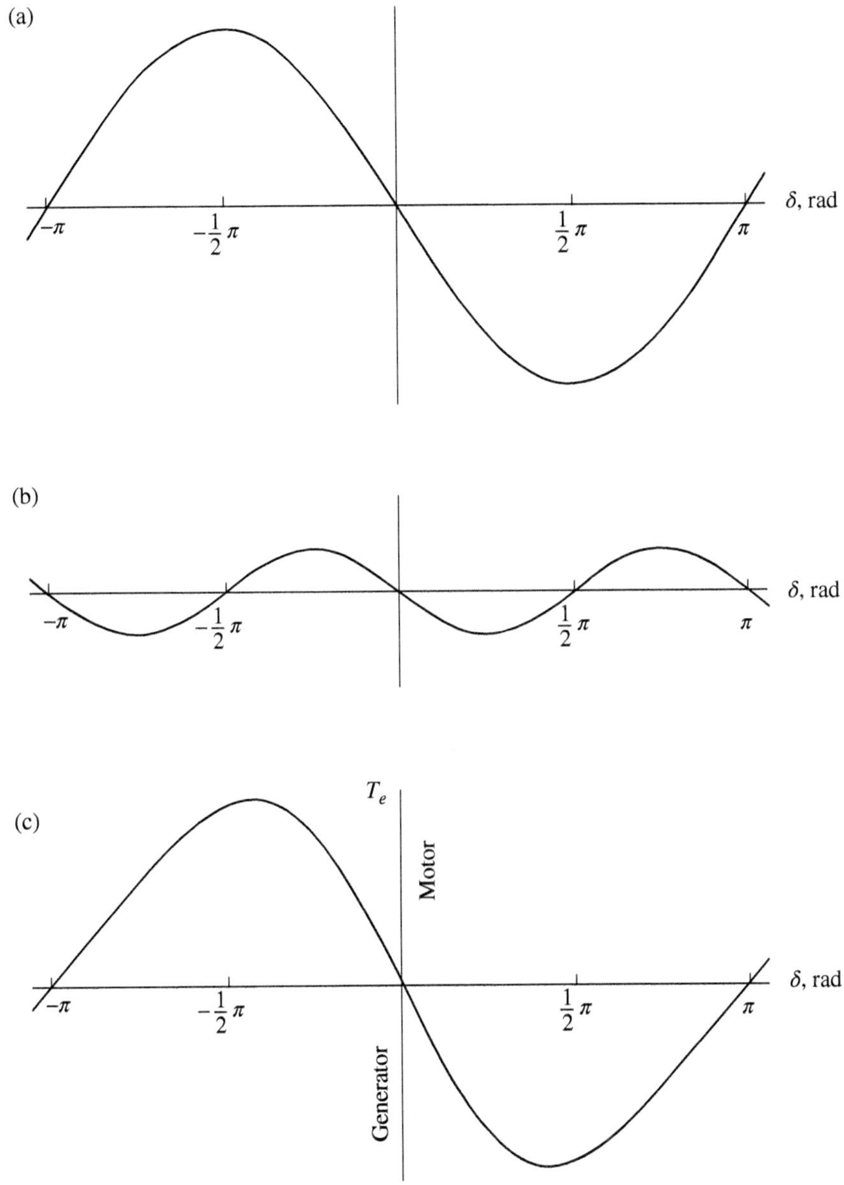

Figure 7.6-3 Steady-state electromagnetic torque of a synchronous machine with a salient rotor [$\delta = \theta_r(0)$ and stator resistances neglected]: (a) plot of (7.5-4), torque due to field winding alone, (b) plot of (7.6-35), torque due to rotor saliency alone, and (c) combination of (a) and (b) (not to scale).

Two-Phase Synchronous Machine

The two-phase synchronous machine which we will consider is a four-pole 150-hp 440-V (rms) 60-Hz machine with the following parameters: $r_s = 0.2\ \Omega$, $L_{ls} = 1.14$ mH, $L_{mq} = 11$ mH, $L_{md} = 13.7$ mH, $r'_{fd} = 0.013\ \Omega$, and $L'_{lfd} = 2.1$ mH. The inertia of the rotor and connected mechanical load is $J = 16.6\ \text{kg}\cdot\text{m}^2$ and B_m is assumed to be zero. The machine is a salient-pole device and equipped with damper windings.

The dynamic performance of this synchronous machine during a step decrease in load torque from zero to $-400\ \text{N}\cdot\text{m}$ is illustrated in Fig. 7.7-1. Since this is generator operation, perhaps it is more appropriate to consider this as a step increase in input torque from zero to $400\ \text{N}\cdot\text{m}$. In any event, the machine is initially operating at synchronous speed with the field voltage adjusted so that the open-circuit voltage of the stator windings is equal to the rated voltage of the

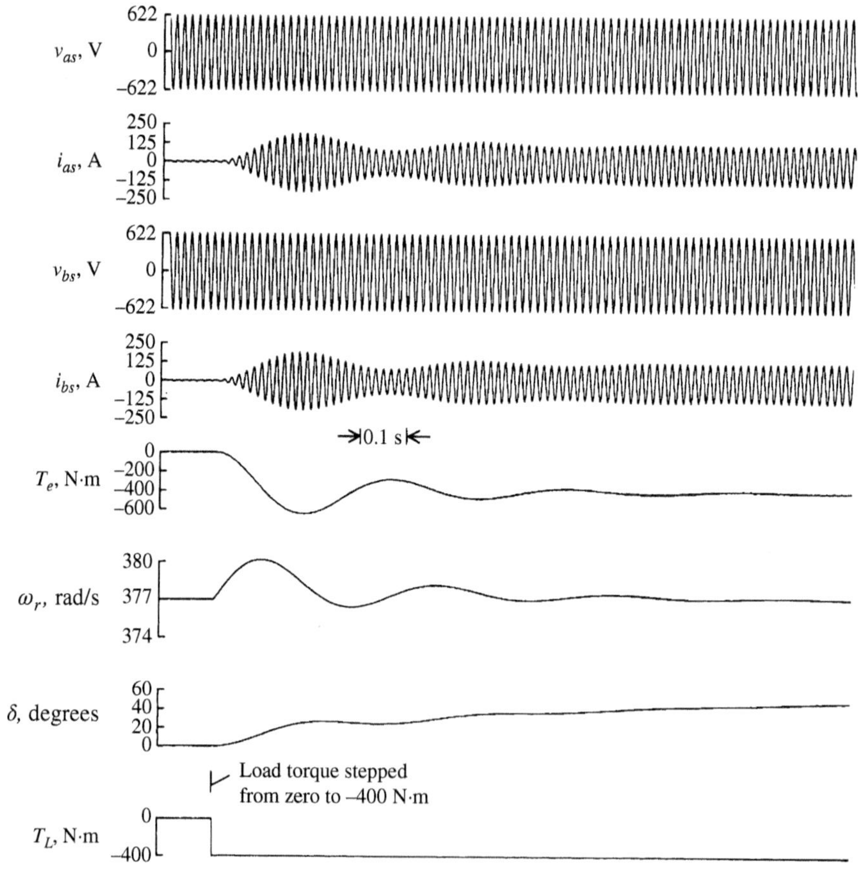

Figure 7.7-1 Dynamic performance of a two-phase synchronous generator during a step decrease in load torque (step increase in input torque).

machine (440 V). Therefore, the stator currents are very small since $T_L = 0$. Plotted are v_{as}, i_{as}, v_{bs}, i_{bs}, T_e, ω_r (electrical angular velocity), δ, and T_L.

Immediately upon the application of the input torque $(-T_L)$, the machine accelerates above synchronous speed as predicted by (7.5-5) and the rotor angle, δ, increases since $\delta = \theta_r - \theta_{esv}$. The rotor continues to speed up until the accelerating torque on the rotor is zero. This occurs when T_e is equal in magnitude to the input torque. As noted in Fig. 7.7-1, this occurs when the speed has increased to approximately 380 rad/s (electrical angular velocity). Even though the accelerating torque on the rotor is zero at this time, the rotor is still running above synchronous speed. Hence, δ will continue to increase and, consequently, T_e will continue to decrease (increase negatively) causing the rotor to decelerate and the speed of the rotor decreases toward synchronous speed. When ω_r becomes equal to ω_e, the rotor angle is approximately 28 electrical degrees and T_e is approximately -600 N · m. Now, there is a decelerating torque of approximately 200 N · m causing the rotor speed to decrease below synchronous speed, whereupon the rotor angle will begin to decrease. Damped oscillations of the rotor about synchronous speed continue due to the damper windings until the new steady-state operating point is reached. We can think of the instantaneous electromagnetic torque during this disturbance resulting from interaction between (a) the stator and field currents, (b) the stator currents and saliency of the rotor, and (c) the stator and damper winding currents. Although this line of thinking may be helpful in visualizing what is going on, we must be careful since dynamically the device is an interactive system. The dynamic torque versus rotor angle characteristics during and following this step change in input torque are shown in Fig. 7.7-2. It is interesting to note that

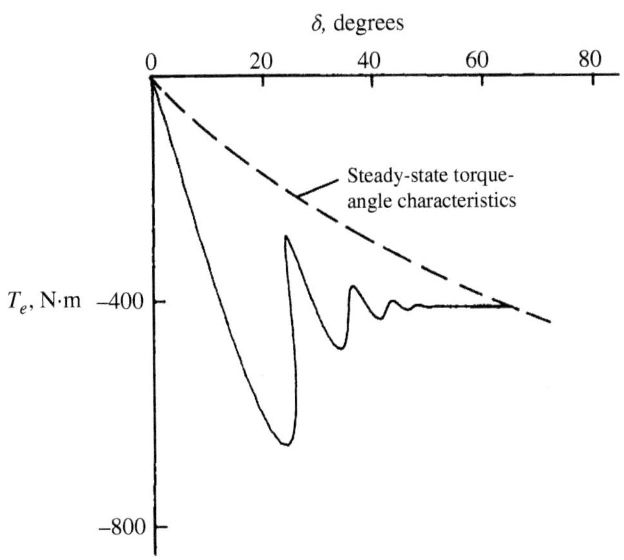

Figure 7.7-2 Dynamic torque versus rotor-angle characteristic for Fig. 7.7-1. The steady-state characteristics are also shown.

it requires considerable time before the machine establishes steady-state operation at $T_L = -400$N·m. The steady-state torque-angle curve which is also shown, in part, in Fig. 7.7-2 will pass through $T_e = 0$ at $\delta \cong 0$ and $T_L = -400$N·m at $\delta = 68°$; however, it is important to note that the steady-state T_e versus δ characteristic is much different than the transient T_e versus δ characteristic.

Recall that, if we slowly increase the input torque in small increments, theoretically we could reach the maximum value of T_e shown in Fig. 7.6-3c before the machine would fall out of synchronism. It is interesting to mention in passing that the maximum value of input torque (or load torque) that can be applied (step input), with T_L initially zero and with the machine still being able to return to synchronous speed, is referred to as the *transient stability limit.* Since the dynamic characteristics are predicted by nonlinear differential equations, it is necessary to employ a computer to predict the dynamic torque-angle characteristics and to determine the transient stability limit. We shall leave this to the power system engineer because it is, indeed, a topic which should be studied by one working in the area of power system stability [3]. However, our purpose here is to make the first-time reader aware that the steady-state and dynamic torque versus rotor angle characteristics are different, sometimes markedly different as illustrated here.

Figure 7.7-3 is a repeat of Fig. 7.7-1 with the rotor reference frame variables plotted rather than the stator or machine variables and the plot of the load torque omitted. Also plotted is the field current i'^r_{fd}. Although, for this machine, the field current changes only slightly owing to a change in flux linkages, this is not typical of all machines. In some cases, depending upon the parameters and the type of the disturbance, a considerable voltage may be induced in the field winding resulting in a marked change in field current during the transient period [2].

Two-Phase Reluctance Machine

It is instructive to observe the steady-state and dynamic performance of a low-power two-phase reluctance motor. The parameters of a 115-V (rms) 60-Hz two-pole two-phase $\frac{1}{10}$ – hp reluctance motor are $r_s = 10\,\Omega$, $L_{ls} = 26.5$ mH, $L_{mq} = 132.6$ mH, $L_{md} = 318.3$ mH, $B_m = 0$, and $J = 1 \times 10^{-3}$ kg·m^2. The machine is also equipped with damper windings.

The dynamic performance when T_L is stepped from zero to 0.2 N · m is shown in Fig. 7.7-4. The following variables are plotted: v_{as}, i_{as}, v_{bs}, i_{bs}, T_e, ω_r, δ, and T_L. The steady-state torque versus rotor angle characteristic is shown in Fig. 7.7-5. The stator resistance of this small reluctance motor is relatively large; thus, the characteristics shown in Figs. 7.7-4 and 7.7-5 are determined without neglecting r_s. Note the initial value of δ with $T_e = 0$; therein $\theta_r(0)$ appears to be between 4° and 5°. It can be shown that the zero torque angle is

$$\theta_r(0) = \frac{\pi}{2} + \tan^{-1}\left(-\frac{\omega_e L_d}{r_s}\right) \tag{7.7-1}$$

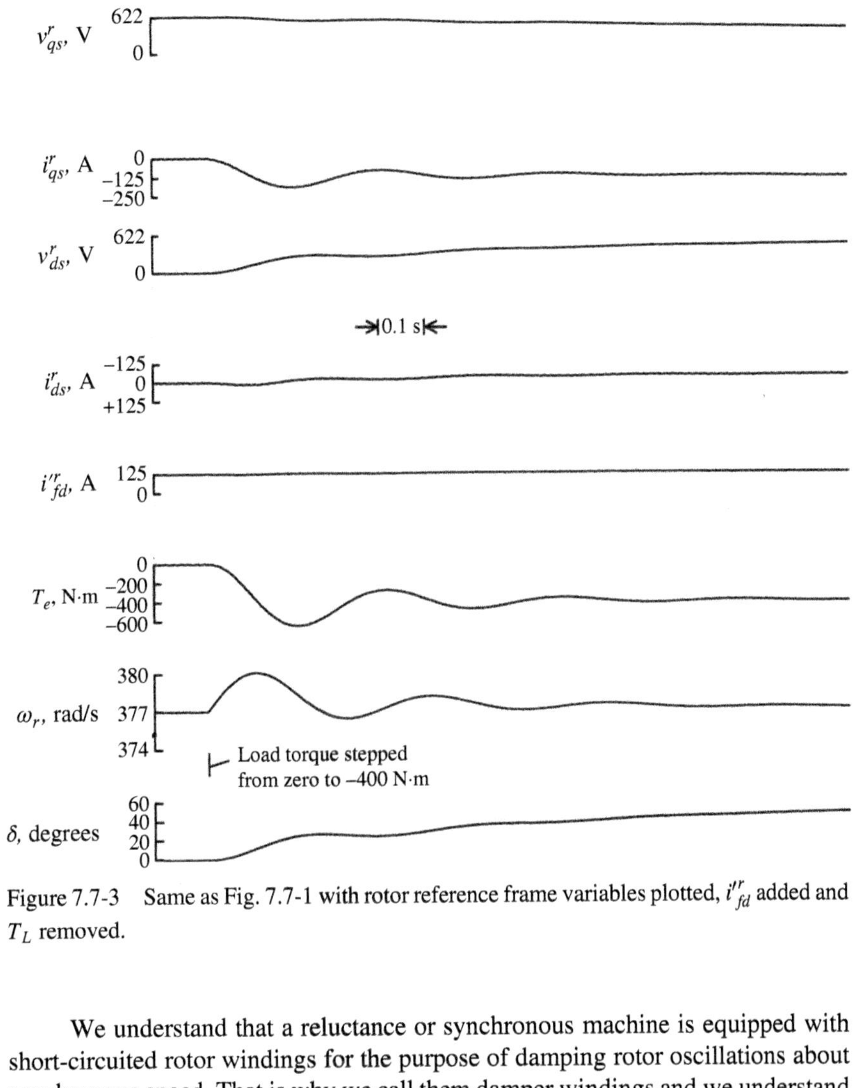

Figure 7.7-3 Same as Fig. 7.7-1 with rotor reference frame variables plotted, i'^r_{fd} added and T_L removed.

We understand that a reluctance or synchronous machine is equipped with short-circuited rotor windings for the purpose of damping rotor oscillations about synchronous speed. That is why we call them damper windings and we understand that the damping torque occurs because of currents induced in the rotor circuits whenever $\omega_r \neq \omega_e$. Although important, the damper winding currents are small and not plotted. As we have mentioned, this is often called induction motor action since, with the short-circuited rotor windings, the reluctance machine produces an average torque-speed characteristic due to induction-motor action. If the rotor should slow down ever so slightly from synchronous speed, the induction motor torque would become positive which would tend to accelerate the rotor back to synchronous speed. Likewise, if the rotor should speed up from synchronous speed, the induction-motor torque would be negative, which would tend to slow the rotor speed. Induction motor action is covered in Chapter 8.

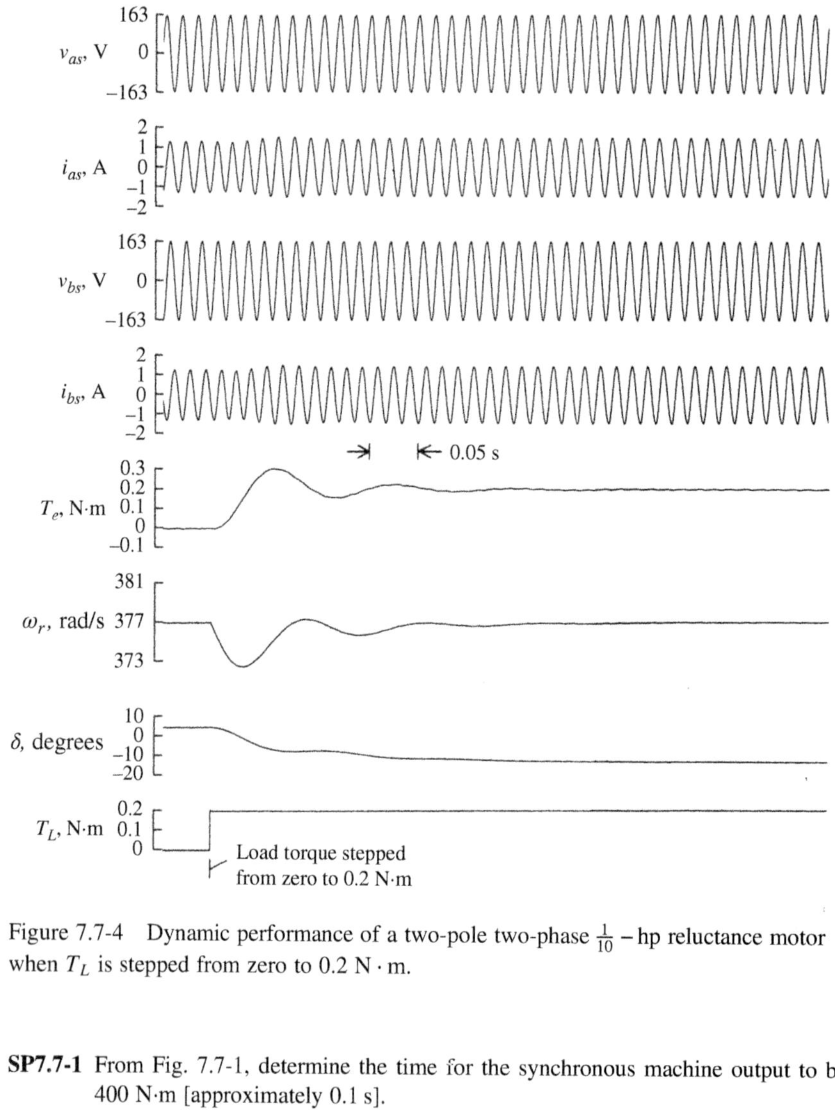

Figure 7.7-4 Dynamic performance of a two-pole two-phase $\frac{1}{10}$ − hp reluctance motor when T_L is stepped from zero to 0.2 N · m.

SP7.7-1 From Fig. 7.7-1, determine the time for the synchronous machine output to be 400 N·m [approximately 0.1 s].

7.8 THREE-PHASE ROUND-ROTOR SYNCHRONOUS MACHINE

A two-pole three-phase round-rotor synchronous machine is shown in Fig. 7.8-1. The stator windings are identical and sinusoidally distributed with their magnetic axes displaced 120° from each other. The extension from the analysis of a two-phase machine to a three-phase machine is straightforward; however, it is worthwhile to note the expressions of the mutual inductances of the stator windings and the derivation of the equivalent circuit. Also, the addition of a third substitute

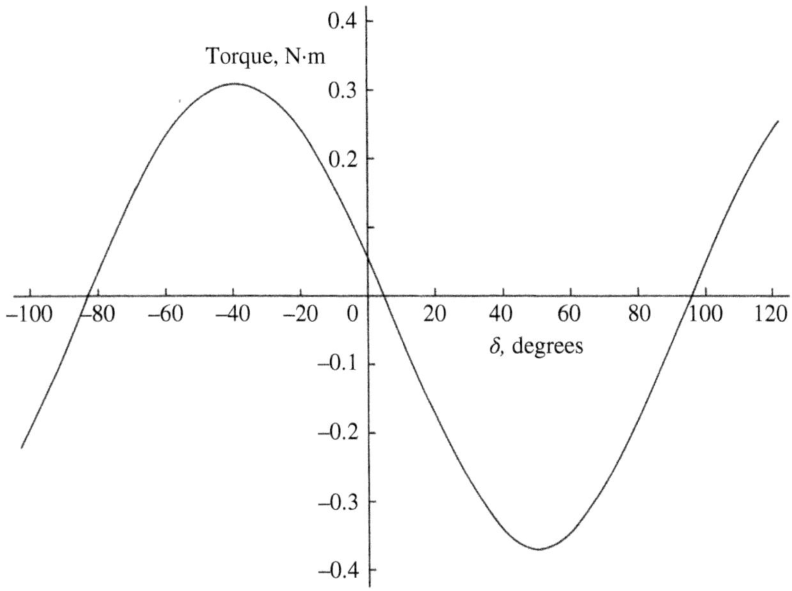

Figure 7.7-5 Steady-state torque versus rotor angle for a two-pole, two-phase, $\frac{1}{10}$ – hp reluctance machine.

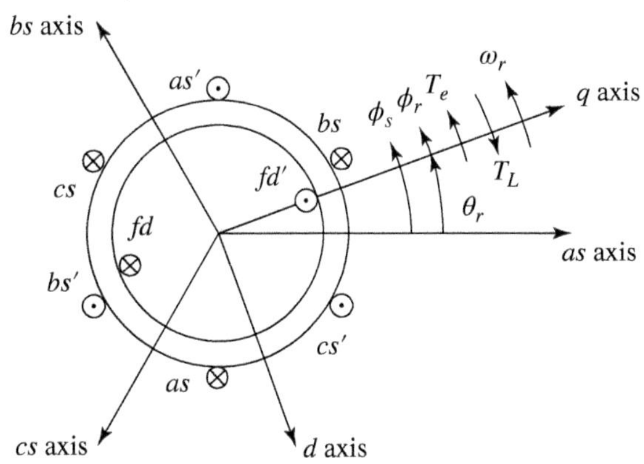

Figure 7.8-1 Three-phase round-rotor synchronous machine.

variable, the zero variable, is necessary since we have three stator variables *as*, *bs*, and *cs*.

The voltage equations are

$$v_{as} = r_s i_{as} + \frac{d\lambda_{as}}{dt} \tag{7.8-1}$$

$$v_{bs} = r_s i_{bs} + \frac{d\lambda_{bs}}{dt} \tag{7.8-2}$$

$$v_{cs} = r_s i_{cs} + \frac{d\lambda_{cs}}{dt} \tag{7.8-3}$$

$$v_{fd} = r_{fd} i_{fd} + \frac{d\lambda_{fd}}{dt} \tag{7.8-4}$$

Equations (7.8-1) through (7.8-3) may be written as

$$\mathbf{v}_{abcs} = \mathbf{r}_s \mathbf{i}_{abcs} + p\boldsymbol{\lambda}_{abcs} \tag{7.8-5}$$

where

$$(\mathbf{f}_{abcs})^T = (f_{as}\ f_{bs}\ f_{cs}) \tag{7.8-6}$$

and

$$\mathbf{r}_s = r_s \mathbf{I} \tag{7.8-7}$$

where $\mathbf{I}$ is a 3×3 identity matrix.

The flux-linkage equations may be written as

$$\boldsymbol{\lambda}_{abcs} = \mathbf{L}_s \mathbf{i}_{abcs} + L_{sr} i_{fd} \begin{bmatrix} \sin\theta_r \\ \sin\left(\theta_r - \dfrac{2\pi}{3}\right) \\ \sin\left(\theta_r + \dfrac{2\pi}{3}\right) \end{bmatrix} \tag{7.8-8}$$

where L_{sr} is given by (7.3-11) and

$$\mathbf{L}_s = \begin{bmatrix} L_{ls} + L_{ms} & -\dfrac{1}{2}L_{ms} & -\dfrac{1}{2}L_{ms} \\ -\dfrac{1}{2}L_{ms} & L_{ls} + L_{ms} & -\dfrac{1}{2}L_{ms} \\ -\dfrac{1}{2}L_{ms} & -\dfrac{1}{2}L_{ms} & L_{ls} + L_{ms} \end{bmatrix} \tag{7.8-9}$$

where L_{ms} is given by (7.3-7) and $L_{sr} i_{fd}$ in (7.8-8) will be replaced by $L_{ms} i'_{fd}$ which is (7.3-16).

If balanced, steady-state operation is the focus, the following reference frame derivation may be de-emphasized and study resumed with Table 7.8-1. The transformation of the stator variables to substitute variables in the rotor reference frame can be written from (5.8-11) through (5.8-13). In particular,

$$v_{qs}^r = r_s i_{qs}^r + \omega_r \lambda_{ds}^r + p\lambda_{qs}^r \tag{7.8-10}$$

$$v_{ds}^r = r_s i_{ds}^r - \omega_r \lambda_{qs}^r + p\lambda_{ds}^r \tag{7.8-11}$$

$$v_{0s} = r_s i_{0s} + p\lambda_{0s} \tag{7.8-12}$$

The *fd*-voltage equation may be written directly from (7.3-25),

$$v_{fd}'^r = r_{fd}' i_{fd}'^r + p\lambda_{fd}'^r \tag{7.8-13}$$

where we have added the raised r to emphasize that the field winding is physically in the rotor reference frame. The flux-linkage equations in the $qd0s$ variables may be obtained from (7.8-8). Thus,

$$\left(\mathbf{K}_s^r\right)^{-1}\boldsymbol{\lambda}_{qd0s}^r = \mathbf{L}_s\left(\mathbf{K}_s^r\right)^{-1}\mathbf{i}_{qd0s} + L_{sr}i_{fd}^r\begin{bmatrix}\sin\theta_r \\ \sin\left(\theta_r - \dfrac{2\pi}{3}\right) \\ \sin\left(\theta_r + \dfrac{2\pi}{3}\right)\end{bmatrix} \tag{7.8-14}$$

Solving for $\boldsymbol{\lambda}_{qd0s}^r$ yields

$$\boldsymbol{\lambda}_{qd0s}^r = \mathbf{K}_s^r\mathbf{L}_s\left(\mathbf{K}_s^r\right)^{-1}\mathbf{i}_{qd0s}^r + \mathbf{K}_s^r\begin{bmatrix}\sin\theta_r \\ \sin\left(\theta_r - \dfrac{2\pi}{3}\right) \\ \sin\left(\theta_r + \dfrac{2\pi}{3}\right)\end{bmatrix}L_{sr}i_{fd}^r \tag{7.8-15}$$

The transformations $\mathbf{K}_s^r$ and $\left(\mathbf{K}_s^r\right)^{-1}$ are (5.8-5) and (5.8-7), respectively, with θ replaced with θ_r. Now, $\mathbf{L}_s$ is (7.8-9); whereupon the first term on the right-hand side of (7.8-15) becomes

$$\mathbf{K}_s^r\mathbf{L}_s\left(\mathbf{K}_s^r\right)^{-1}\mathbf{i}_{qd0s} = \begin{bmatrix}L_{ls} + \dfrac{3}{2}L_{ms} & 0 & 0 \\ 0 & L_{ls} + \dfrac{3}{2}L_{ms} & 0 \\ 0 & 0 & L_{ls}\end{bmatrix}\mathbf{i}_{qd0s}^r \tag{7.8-16}$$

The second term on the right-hand side of (7.8-15) can be written as

$$\mathbf{K}_s^r\begin{bmatrix}\sin\theta_r \\ \sin\left(\theta_r - \dfrac{2\pi}{3}\right) \\ \sin\left(\theta_r + \dfrac{2\pi}{3}\right)\end{bmatrix}L_{sr}i_{fd}^r = L_{sr}i_{fd}^r\begin{bmatrix}0 \\ 1 \\ 0\end{bmatrix} \tag{7.8-17}$$

Combining (7.8-16) and (7.8-17), we can express $\boldsymbol{\lambda}_{qd0s}^r$ in expanded form as

$$\lambda_{qs}^r = \left(L_{ls} + \frac{3}{2}L_{ms}\right)i_{qs}^r \tag{7.8-18}$$

$$\lambda_{ds}^r = \left(L_{ls} + \frac{3}{2}L_{ms}\right)i_{ds}^r + L_{sr}i_{fd}^r \tag{7.8-19}$$

Since we are assuming symmetrical three-wire stator windings $i_{0s} = 0$, thus, $\lambda_{0s} = 0$. It is convenient to set forth equivalent circuits as we did in Fig. 7.3-2 for the two-phase device. Unfortunately, there is a little problem. In Fig. 7.3-2 the mutual inductance between the d axis and the field winding is L_{ms}; however, this is not the form of (7.8-19). In order to accommodate an equivalent circuit, the last term of (7.8-19) must be $\frac{3}{2}L_{ms}i'^{r}_{fd}$. That is,

$$L_{sr}i^{r}_{fd} = \frac{3}{2}L_{ms}i'^{r}_{fd} \tag{7.8-20}$$

The only way (7.8-20) can be, is if we redefine i'^{r}_{fd}. Solving for i'^{r}_{fd} from (7.8-20) yields

$$\begin{aligned} i'^{r}_{fd} &= \frac{2}{3}\frac{L_{sr}}{L_{ms}}i^{r}_{fd} \\ &= \frac{2}{3}\frac{N_{fd}N_s}{N_s^2}i^{r}_{fd} = \frac{2}{3}\frac{N_{fd}}{N_s}i^{r}_{fd} \end{aligned} \tag{7.8-21}$$

We can now write (7.8-19) as

$$\lambda^{r}_{ds} = L_{ls}i^{r}_{ds} + \frac{3}{2}L_{ms}\left(i^{r}_{ds} + i'^{r}_{fd}\right) \tag{7.8-22}$$

Although it is instructive to show that the equivalent circuit for the three-phase device is very similar to Fig. 7.3-2 for the two-phase, the algebraic manipulations are a bit boring. One may wish to skip to the paragraph following (7.8-29).

Recall that the $\frac{3}{2}$ factor came about due to the coupling between the three-phase stator windings. Therefore, we are not quite ready to present an equivalent circuit. It is necessary to refer the field winding voltage equation to the stator windings. The field voltage equation may be written as

$$\begin{aligned} v^{r}_{fd} &= \left(r_{fd} + pL_{lfd}\right)i^{r}_{fd} + L_{mf}pi^{r}_{fd} \\ &\quad + pL_{sr}\left[\sin\theta_r i_{as} + \sin\left(\theta_r - \frac{2\pi}{3}\right)i_{bs} + \sin\left(\theta_r + \frac{2\pi}{3}\right)i_{cs}\right] \end{aligned} \tag{7.8-23}$$

The terms in [ ] are $\frac{3}{2}i^{r}_{ds}$. Therefore, (7.8-23) may be written as

$$v^{r}_{fd} = \left(r_{fd} + pL_{lfd}\right)i^{r}_{fd} + L_{mfd}pi^{r}_{fd} + pL_{sr}\frac{3}{2}i^{r}_{ds} \tag{7.8-24}$$

We will leave the turns-ratio on v'^{r}_{fd} and λ'^{r}_{fd} as given for the two-phase machine, (7.3-26) and (7.3-27), respectively. Although with this choice field power will not be preserved; that is, $v'^{r}_{fd}i'^{r}_{fd} \neq v^{r}_{fd}i^{r}_{fd}$. Nevertheless, this allows derivation of the desired equivalent circuit. Substituting (7.8-21) for i^{r}_{fd}, (7.3-26) for v^{r}_{fd}, and expressing the inductances in terms of turns-ratios, we can write (7.8-24) as

$$\frac{N_{fd}}{N_s}v'^{r}_{fd} = \left(r_{fd} + pL_{lfd}\right)\left(\frac{3}{2}\right)\frac{N_s}{N_{fd}}i'_{fd} + \frac{N_{fd}^2}{\mathcal{R}_m}\left(\frac{3}{2}\right)\frac{N_s}{N_{fd}}pi'^{r}_{fd} + \frac{3}{2}\frac{N_sN_{fd}}{\mathcal{R}_m}pi^{r}_{ds} \tag{7.8-25}$$

where L_{ffd} has been written in terms of leakage inductance L_{lfd} and magnetizing inductance L_{mfd}. Solving for v'^{r}_{fd} yields

$$v'^{r}_{fd} = r'_{fd}i'^{r}_{fd} + L'_{lfd}pi'^{r}_{fd} + \frac{N_s^2}{\mathcal{R}_m}\left(\frac{3}{2}\right)pi'^{r}_{fd} + \frac{N_s^2}{\mathcal{R}_m}\left(\frac{3}{2}\right)pi^{r}_{ds} \tag{7.8-26}$$

Finally

$$v_{fd}'^r = r_{fd}' i_{fd}'^r + L_{lfd}' p i_{fd}'^r + \frac{3}{2} L_{ms} \left(i_{ds}^r + i_{fd}'^r \right) \tag{7.8-27}$$

where

$$r_{fd}' = \frac{3}{2} \left(\frac{N_s}{N_{fd}} \right)^2 r_{fd} \tag{7.8-28}$$

$$L_{lfd}' = \frac{3}{2} \left(\frac{N_s}{N_{fd}} \right)^2 L_{lfd} \tag{7.8-29}$$

The equivalent circuit becomes Fig. 7.3-2 with $\frac{3}{2}L_{ms}$ rather than L_{ms}, $i_{fd}'^r$ given by (7.8-21), and r_{fd}' and L_{lfd}' given by (7.8-28) and (7.8-29), respectively. The voltage equation for the zero variables is

$$v_{0s} = r_s i_{0s} + pL_{ls} i_{0s} \tag{7.8-30}$$

However, since we are considering only balanced conditions, zero variables are zero. Also, the steady-state voltage equation for $\widetilde{V}_{as}$ for the three-phase machine is given by (7.4-19) with X_{ms} replaced with $\frac{3}{2}X_{ms}$ in X_{ss} and in $\widetilde{E}_{as}$, the last term on the right-side of (7.4-19), and $I_{fd}'^r$ equal to (7.8-21).

TABLE 7.8-1 Conversion of Two-Phase Machine Equations to Three Phase

For Table 7.4-1

Permanent-Magnet Machine

- L_{ss} is replaced by $L_{ls} + \dfrac{3}{2}L_{ms}$
- T_e is multiplied by $\frac{3}{2}$

Round-Rotor Machine

- L_{ss} is replaced by $L_{ls} + \frac{3}{2}L_{ms}$
- L_{ms} in $\widetilde{E}_a$ and T_e replaced by $\frac{3}{2}L_{ms}$ if (7.8-21) is used for $I_{fd}'^r$
- T_e is multiplied by $\frac{3}{2}$

Note: $I_{fd}'^r$ was expressed by (7.8-21) to allow an equivalent circuit; a field current was not present in the permanent-magnet machine.

For Table 7.5-1

- L_{ss} is replaced by $L_{ls} + \dfrac{3}{2}L_{ms}$ in $\widetilde{V}_{as}$ and in X_{ss} and in the expression for T_e
- L_{ms} is replaced with $\frac{3}{2}L_{ms}$ in $\widetilde{E}_a$
- If (7.8-21) is used for $I_{fd}'^r$, then X_{ms} in $E_{xfd}'^r$ is replaced by $\frac{3}{2}X_{ms}$ and r_{fd}' becomes (7.8-28)
- T_e is multiplied by $\frac{3}{2}$

For Table 7.6-1

- L_q is replaced by $L_{ls} + \frac{3}{2}L_{mq}$ in $\widetilde{V}_{as}$, $\widetilde{E}_a$, and T_e
- L_d is replaced by $L_{ls} + \frac{3}{2}L_{md}$ in $\widetilde{V}_{as}$, $\widetilde{E}_a$, and T_e
- T_e is multiplied by $\frac{3}{2}$

SP7.8-1 The synchronous machines in Examples 7A and 7B are for two-phase devices. What changes would be necessary in the calculations if these devices were three-phase machines? [Only X_{ms} in X_{ss}; it would be multiplied by $\frac{3}{2}$]

SP7.8-2 The stator zero variables do not exist if the voltages are balanced. When do the stator zero variables not exist for unbalanced voltages? [For three-wire wye-connected symmetrical stator windings since $i_{as} + i_{bs} + i_{cs} = 0$.]

7.9 RECAPPING

Since the rotor of the synchronous machine is equipped with a field winding and damper windings, it is magnetically unsymmetrical, and the analysis becomes rather involved. Moreover, because the rotor windings are not symmetrical it becomes a chore to deal with turns-ratios of these windings, especially when deriving an equivalent circuit. Unfortunately, sorting this all out tends to impede the understanding of the basic concepts of synchronous machine operation. Although this is not to say the power system engineer should not be aware of the details of the analysis of the synchronous machine, it certainly is unnecessary to subject the first-time student of machine analysis to details that are of minor importance in understanding basic synchronous machine operation.

In Chapter 4 we became aware of Tesla's rotating magnetic field and its role in portraying the operation of the electric machine, starting with a two-phase device and establishing the principals of machine operation through pole interactions and reference frame theory. We have attempted to follow Tesla's lead in presenting the synchronous machine to the junior-level undergraduate. If the student successfully completes the material through Section 7.4 and has grasped the differences and similarities between the two-phase permanent-magnet synchronous machine and the two-phase round-rotor synchronous machine with a field winding, we have accomplished 90% of our mission. We have made an effort to present the material by building on the work in Chapters 4 and 5 and doing it as concisely as possible.

The later sections of this chapter are certainly of importance. For example, the reluctance machine gives a first look at the influence, albeit generally minor, of the reluctance torque upon the operation of synchronous machines with a field winding. Also, since reluctance machines are used in low-power electric drives, the material is helpful to the student interested in electric drives. It is important to note that the form of the balanced, steady-state voltage and torque equations are similar in form for all variations of the synchronous machine considered in this chapter. The summaries given in Tables 7.4-1, 7.5-1, and 7.6-1 provide convenient access to the steady-state voltage and torque expressions.

Unfortunately, in the section on the three-phase synchronous machine it is necessary to become rather involved when developing the equivalent circuit. This material is reserved for the student interested in power systems. However, Table 7.8-1 which follows the three-phase derivation provides the necessary changes to convert the equations given in the previously mentioned tables for two-phase machines to equations for three-phase machines.

7.10 PROBLEMS

1. A four-pole 2-hp two-phase round-rotor synchronous machine is connected to a 110-V, 60-Hz source. The machine is operating as a generator with a total steady-state power output of 1 kW at the terminals. The phase current lags the phase voltage by 160°. The parameters are $r_s = 0.5\ \Omega$, $L_{ls} = 0.005$ H, and $L_{ms} = 0.05$ H. Calculate $\widetilde{E}_a$ and draw the phasor diagram showing $\widetilde{V}_{as}$, $\widetilde{I}_{as}$, $\widetilde{E}_a$, and $\left(r_s + jX_q\right)\widetilde{I}_{as}$ and locate the poles.
2. Repeat Problem 1 if the machine is operating as a motor with $\widetilde{I}_{as}$ in phase with $\widetilde{V}_{as}$. The input power is 1 kW.
3. Calculate the torque for Problem 2.
4. Refer to Example 7A. Calculate $E_{xfd}'^{r}$ for each $\widetilde{E}_a$ given in (a) through (c).
5. Calculate Q for each value of $\widetilde{I}_{as}$ in Example 7B.
6. The field of a 60-Hz three-phase round-rotor synchronous machine is adjusted so that the open-circuited phase voltage is 14 kV. The parameters are $r_s = 0.003\ \Omega$, $X_{ls} = 0.1\ \Omega$, and $X_{ms} = 1.1\ \Omega$. Calculate $i_{fd}'^{r}$ and $E_{xfd}'^{r}$.
7. A three-phase synchronous generator is a 64-pole hydroturbine generator rated at 325 MVA with 11.56 kV phase voltage. The machine delivers real power with a power factor of −0.85. The machine parameters in ohms at 60-Hz are $r_s = 0.003\ \Omega$, $X_{ls} = 0.1\ \Omega$, and $X_{ms} = 1.2\ \Omega$. Calculate $\widetilde{E}_a$, $E_{xfd}'^{r}$, T_e and draw the phasor diagram and locate the poles. ■

7.11 REFERENCES

1. R. H. Park, "Two Reaction Theory of Synchronous Machines-Generalized Method of Analysis-Part I," *AIEE Trans.*, Vol. 48, July 1929, pp. 716–727.
2. P. C. Krause, O. Wasynczuk, and S. Sudhoff, *Analysis of Electric Machinery*, 2nd Edition. Wiley, IEEE Press, 1994.
3. P. C. Krause, O. Wasynczuk, S. Sudhoff, and S. Pekarek, *Analysis of Electric Machinery and Drive Systems*, 3rd Edition, Wiley, IEEE Press, 2013.

CHAPTER 8

SYMMETRICAL INDUCTION MACHINES AND FIELD ORIENTATION

8.1 INTRODUCTION

Although the induction machine is generally used as a means to convert electric power to mechanical work, it can operate as either a motor or a generator. Three-phase induction motors are commonly used in large-horsepower applications, for example, pump drives, steel mill drives, hoist drives, vehicle drives, and as a generator in wind turbine and low-head hydro applications. Additionally, they are used extensively as variable-speed drive motors. Single-phase induction motors, which develop torque in a manner similar to the multiphase induction motor, are used in many household appliances.

In order to conduct a rigorous analysis of an induction machine, it is convenient to perform a change of variables (transformation) that eliminates the position-varying mutual inductances that occur due to windings in relative motion as witnessed in Section 1.7. Fortunately, in the case of the symmetrical induction machine, the equivalent circuit that portrays balanced steady-state operation is similar to that of a two-winding transformer.

The analysis of symmetrical two- and three-phase induction machines is essentially the same. Therefore, we will focus our attention on the two-phase machine, since this enables us to become familiar with the theory and performance of induction machines without becoming inundated with three-phase trigonometric manipulations. Once the theory has been established, the dynamic and steady-state performances of a 5-hp and a 0.1-hp two-phase induction motor during balanced operation are illustrated by instantaneous and steady-state phasors and by computer traces. This is followed with an analysis of the three-phase induction

Electromechanical Motion Devices: Rotating Magnetic Field-Based Analysis and Online Animations, Third Edition. Paul Krause, Oleg Wasynczuk, Steven D. Pekarek, and Timothy O'Connell.

Published 2020 by John Wiley & Sons, Inc.

machine, where it is shown that for balanced conditions the equations that describe the performance of a three-phase machine are identical in form to the equations for a two-phase machine. We need only to multiply an inductance and the torque by a factor of $\frac{3}{2}$.

In the last section of the chapter, the principles of field orientation are presented. The implementation of the control is not detailed; instead, the machine equations are modified assuming the control is functioning properly. This approach enables a clear and concise derivation and a straightforward description of field orientation methodology. The steady-state phasor voltage equation is valid if the control is functioning correctly and it is not only appropriate to portray steady-state performance but also the dynamic electrical response since the stator currents are commanded variables and can be changed nearly instantaneously, essentially without electrical transients. Moreover, the phasor voltage equation is similar in form to the phasor voltage equation of the permanent-magnet ac machine (brushless dc drive) given in Chapter 6 and the synchronous machine given in Chapter 7. Animation F of Appendix B deals with an induction machine.

8.2 TWO-PHASE INDUCTION MACHINE

A two-pole two-phase induction machine is shown in Fig. 8.2-1. It is assumed that the stator windings may be portrayed by orthogonal, sinusoidally distributed windings as described in Chapter 4. It is convenient to assume that the rotor of the two-pole two-phase induction machine may also be portrayed electrically by two sinusoidally distributed windings displaced 90°. We will talk about squirrel cage rotors of induction motors later. Hence, for our present purposes, we will consider that the *ar* and *br* windings are sinusoidally distributed, each with the same total number of turns and winding resistance. Thus, both the stator and rotor are symmetrical. For this reason, the device is often referred to as a *symmetrical induction machine*.

Note that the air gap distance between the stator and rotor is uniform. The rotor windings of an induction machine are generally short-circuited ($v_{ar} = v_{br} = 0$) with only the stator windings connected to a source. In this case, the induction machine is said to be single-fed. In some special applications, such as wind turbines, both the rotor and stator windings are connected to sources. In particular, the rotor windings are connected to a stationary multiphase source by a brush and slip-ring arrangement. In this case, the induction machine is double-fed. We will not deal with a double-fed machine.

As established in Chapter 4, the angular displacement about the stator is denoted ϕ_s, and it is referenced to the *as* axis. We see from Fig. 8.2-1 that the angular displacement about the rotor is denoted ϕ_r and it is referenced to the *ar* axis. The angular velocity of the rotor is ω_r and θ_r is its angular position. In particular, θ_r is the angular position between the *ar* and *as* axes. The electromechanical torque T_e and the load torque T_L are also indicated in Fig. 8.2-1. The torque T_e is assumed to be positive in the direction of increasing θ_r whereas the load torque is positive in the opposite direction (opposing rotation).

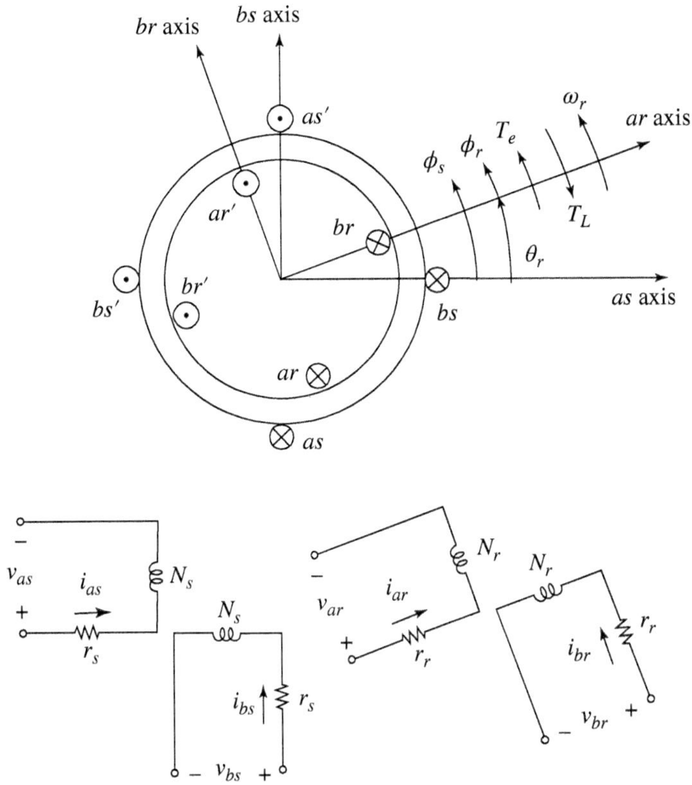

Figure 8.2-1 A two-pole two-phase symmetrical induction machine.

Let us take a few minutes to review the things we have already learned about an induction machine. In our analysis, we will assume that the stator and equivalent windings of the rotor are sinusoidally distributed and symmetrical. Since the rotor currents of frequency $\omega_e - \omega_r$ are flowing in circuits that are rotating at ω_r, Tesla's rotating magnetic field of the stator (mmf$_s$) and the rotor (mmf$_r$) rotate about the air gap at ω_e. Therefore, mmf$_s$ and mmf$_r$ are stationary relative to each other and a constant torque is produced due to the force between these magnetic systems.

All this sounds pretty exciting; however, all depends on the rotor currents being balanced with an angular frequency of $\omega_e - \omega_r$. Well, the rotor windings are short-circuited, so Tesla's rotating magnetic field established by the balanced stator currents rotates about the air gap at ω_e. The rotor is revolving at ω_r. Therefore, the time rate of change of the flux linking the rotor windings due to the stator rotating magnetic field is $\omega_e - \omega_r$. Therefore, the angular frequency of the currents induced in the rotor windings would be $\omega_e - \omega_r$. Moreover, since the stator rotating magnetic field is produced by balanced stator currents, it is not too much of a stretch to accept that the currents induced in the short-circuited rotor windings due to this rotating magnetic field would also be a balanced two-phase set.

Now, before we get into the analysis there are still several things that we can anticipate from the things we have learned from previous chapters. Since the currents induced in the rotor circuits are due to a difference in the speed of the rotor and the stator rotating magnetic field, rotor currents would not be induced when the angular velocity of the rotor is equal to the angular velocity of the stator rotating magnetic field ($\omega_r = \omega_e$). Thus, the induction machine will not develop a torque at synchronous speed; only when $\omega_r \neq \omega_e$ will rotor currents be induced and a torque produced. This, of course, is completely opposite to a synchronous machine which produces an average torque only when $\omega_r = \omega_e$. By now you have probably guessed that the induction machine is so named from the fact that torque is produced as a result of the rotor currents being "induced" by the stator rotating magnetic field. It is interesting that the induction machine was first referred to as a rotating transformer. This is understandable, but we will see this more clearly when we derive the steady-state equivalent circuit for the induction machine and find that it is similar to that of a transformer. Also, we should mention that generator action occurs if the stator windings are connected to a source, thus establishing a rotating magnetic field, and the rotor, with its short-circuited windings, is driven above synchronous speed.

One more thing. Recall that in the case of the synchronous machine the rotor is unsymmetrical since there is only one field winding or a set of permanent magnets on the rotor. Therefore, a rotor rotating magnetic field cannot be produced. That is, the rotor magnetic field is fixed in the d axis of the rotor by the field winding or permanent magnets. Therefore, the rotor must rotate in synchronism with the stator rotating magnetic field in order to produce an average torque or power. Remember that when an asymmetry exists the only reference frame where position-dependent inductances are eliminated is the reference frame where the asymmetry exists. However, in the case of the symmetrical induction machine, the rotating magnetic field due to rotor currents (mmf_r) and the stator rotating magnetic field (mmf_s) always rotate in unison during balanced conditions regardless of the speed of the rotor (except when $\omega_r = \omega_e$). Therefore, we would anticipate that rotor position-dependent inductances would not be present in any reference frame view of mmf_s and mmf_r. What in the world are we trying to say? Well, in Chapter 5 we found that reference frame theory establishes the equations for viewing the variables in a given frame of reference that produce the view of Tesla's rotating magnetic field in that reference frame. The frequency of those reference frame variables is, of course, a function of the angular velocity of the reference frame; however, the variables are independent of rotor position during steady-state operation. Thus, for balanced steady-state operation, Tesla's rotating magnetic field due to the stator currents (mmf_s) and Tesla's rotating magnetic field due to the induced rotor currents (mmf_r) are fixed in amplitude and relative position. In other words, the frequency of mmf_s and mmf_r depends on the reference frame from which they are being viewed; however, their amplitude and relative phase will remain unchanged. We are more than ready to start the analysis of an induction machine. In fact, without writing an equation we have a good idea what is ahead and we are correctly anticipating that there will be no surprises.

A cutaway of a four-pole three-phase 6.5-hp 460-V squirrel-cage induction motor is shown in Fig. 8.2-2. It is an enclosed fan-cooled severe-duty motor for use in the chemical, paper, cement, and mining industries. A disassembled four-pole two-phase $\frac{1}{10}$ – hp 115-V induction motor, which is used in low-power control applications, is shown in Fig. 8.2-3. Also shown in Fig. 8.2-3 is the case that houses the speed-reduction gears. The rotors of the induction motors shown in Figs. 8.2-2 and 8.2-3 are referred to as "squirrel-cage" rotors. Although it is difficult to discern, the squirrel cage rotors are made up of laminated punched steel with aluminum bars die casted in the openings of the laminated rotor and the bars terminated at each end of the rotor in an aluminum ring, which is visible in Figs. 8.2-2 and 8.2-3. The protrusions from the aluminum rings are for cooling purposes. If we remove the steel laminations, the remaining aluminum bars and end rings resemble the rotor (blades) of a "squirrel-cage fan." The question often arises as to the validity of considering these short-circuited bars as producing a rotating magnetic field. It would require considerable work to show this analytically; however, let us try to talk our way through it. Recall that the rotating magnetic field produced by the stator comes about due to the sinusoidally distributed windings and the balanced stator currents.

Figure 8.2-2 Four-pole three-phase 6.5-hp 460-V severe-duty, squirrel-cage induction motor. (Courtesy of General Electric.)

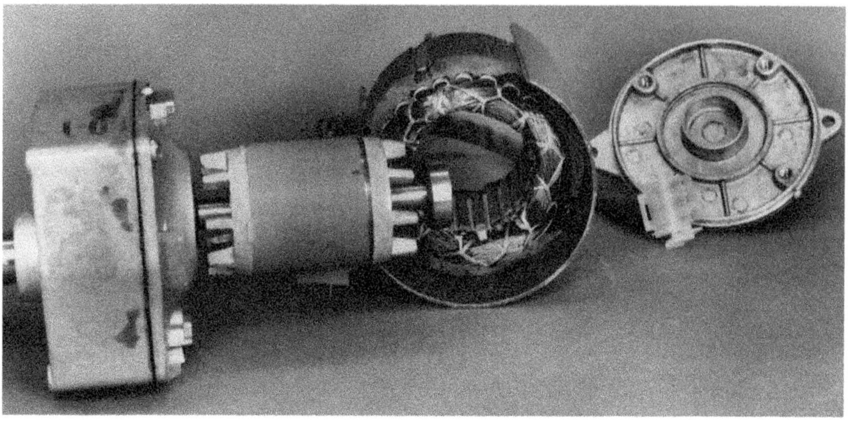

Figure 8.2-3 Four-pole two-phase 1/10-hp 115-V induction motor with reduction gear.

The currents induced in the short-circuited rotor bars are unrestricted and one can argue that the stator mmf will induce rotor currents such that the resulting rotor rotating magnetic field (mmf_r) would be sinusoidal in waveform like the inducing mmf (mmf_s), with an angular velocity equal to the difference in rotor speed from synchronous speed. This is essentially what would occur if the rotor were equipped with symmetrical sinusoidally distributed windings with currents of an angular frequency of $\omega_e - \omega_r$. Clearly, this is a first-order explanation or approximation, but it is commonly accepted, so we will follow suit.

SP8.2-1 Assume sinusoidally distributed windings on the stator and rotor of the machine shown in Fig. 8.2-1. Express (a) mmf_{as} in terms of θ_r and ϕ_r and (b) mmf_{ar} in terms of θ_r and ϕ_s. [(a) $\text{mmf}_{as} = (N_s/2)i_{as}\cos(\phi_r + \theta_r)$; (b) $\text{mmf}_{ar} = (N_r/2)i_{ar}\cos(\phi_s - \theta_r)$]

SP8.2-2 The frequency of the balanced stator currents of an induction machine is 60 Hz and mmf_s rotates counterclockwise. The device is operating as a motor, and the rotor of the two-pole machine is rotating counterclockwise at $0.9\omega_e$. (a) Determine the frequency of the balanced rotor currents. Determine the angular velocity of mmf_s and mmf_r relative to an observer sitting (b) on the rotor and (c) on the stator. [(a) 6 Hz; (b) 37.7 rad/s, ccw; (c) 377 rad/s, ccw]

SP8.2-3 Repeat SP8.2-2 for a six-pole induction machine operating as a generator which is being driven at $\omega_r = 1.1\omega_e$. [(a) 6 Hz; (b) 37.7/3 rad/s, cw; (c) 377/3 rad/s, ccw]

SP8.2-4 Give the phase relationship of the rotor currents for (a) SP8.2-2 and (b) SP8.2-3. [(a) $\tilde{I}_{ar} = j\tilde{I}_{br}$; (b) $\tilde{I}_{ar} = -j\tilde{I}_{br}$]

8.3 VOLTAGE EQUATIONS AND WINDING INDUCTANCES

The voltage equations for the induction machine depicted in Fig. 8.2-1 may be expressed as

$$v_{as} = r_s i_{as} + \frac{d\lambda_{as}}{dt} \tag{8.3-1}$$

$$v_{bs} = r_s i_{bs} + \frac{d\lambda_{bs}}{dt} \tag{8.3-2}$$

$$v_{ar} = r_r i_{ar} + \frac{d\lambda_{ar}}{dt} \tag{8.3-3}$$

$$v_{br} = r_r i_{br} + \frac{d\lambda_{br}}{dt} \tag{8.3-4}$$

where r_s is the resistance of each of the stator windings and r_r is the resistance of each of the rotor windings. It is convenient, for future derivations, to write (8.3-1) through (8.3-4) in matrix form as

$$\mathbf{v}_{abs} = \mathbf{r}_s \mathbf{i}_{abs} + p\boldsymbol{\lambda}_{abs} \tag{8.3-5}$$

$$\mathbf{v}_{abr} = \mathbf{r}_r \mathbf{i}_{abr} + p\boldsymbol{\lambda}_{abr} \tag{8.3-6}$$

where

$$(\mathbf{f}_{abs})^T = [f_{as} \;\; f_{bs}] \tag{8.3-7}$$

$$(\mathbf{f}_{abr})^T = [f_{ar} \;\; f_{br}] \tag{8.3-8}$$

In (8.3-7) and (8.3-8), f can represent voltage, current, or flux linkage and the superscript T denotes the transpose. In (8.3-5) and (8.3-6), p is the shorthand notation for the operator d/dt. Also,

$$\mathbf{r}_s = \begin{bmatrix} r_s & 0 \\ 0 & r_s \end{bmatrix} = r_s\mathbf{I} \tag{8.3-9}$$

and

$$\mathbf{r}_r = \begin{bmatrix} r_r & 0 \\ 0 & r_r \end{bmatrix} = r_r\mathbf{I} \tag{8.3-10}$$

where **I** is the identity matrix.

Although (8.3-5) and (8.3-6) are valid for a nonlinear magnetic system, we generally assume that the system is magnetically linear, whereupon the flux linkages may be expressed as linear functions of inductances and currents. In particular, we can write

$$\lambda_{as} = L_{asas} i_{as} + L_{asbs} i_{bs} + L_{asar} i_{ar} + L_{asbr} i_{br} \tag{8.3-11}$$

$$\lambda_{bs} = L_{bsas} i_{as} + L_{bsbs} i_{bs} + L_{bsar} i_{ar} + L_{bsbr} i_{br} \tag{8.3-12}$$

$$\lambda_{ar} = L_{aras} i_{as} + L_{arbs} i_{bs} + L_{arar} i_{ar} + L_{arbr} i_{br} \tag{8.3-13}$$

$$\lambda_{br} = L_{bras} i_{as} + L_{brbs} i_{bs} + L_{brar} i_{ar} + L_{brbr} i_{br} \tag{8.3-14}$$

The self- and mutual inductances given in (8.3-11) through (8.3-14) are defined by their subscripts. Reciprocity applies, thus $L_{asbs} = L_{bsas}$, $L_{asar} = L_{aras}$, etc. For future derivations, it is convenient to write (8.3-11) through (8.3-14) in matrix form as

$$\boldsymbol{\lambda}_{abs} = \mathbf{L}_s \mathbf{i}_{abs} + \mathbf{L}_{sr} \mathbf{i}_{abr} \tag{8.3-15}$$

$$\boldsymbol{\lambda}_{abr} = (\mathbf{L}_{sr})^T \mathbf{i}_{abs} + \mathbf{L}_r \mathbf{i}_{abr} \tag{8.3-16}$$

which may also be written as

$$\begin{bmatrix} \boldsymbol{\lambda}_{abs} \\ \boldsymbol{\lambda}_{abr} \end{bmatrix} = \begin{bmatrix} \mathbf{L}_s & \mathbf{L}_{sr} \\ (\mathbf{L}_{sr})^T & \mathbf{L}_r \end{bmatrix} \begin{bmatrix} \mathbf{i}_{abs} \\ \mathbf{i}_{abr} \end{bmatrix} \tag{8.3-17}$$

The inductance matrices are defined from (8.3-11) through (8.3-14) and $\boldsymbol{\lambda}$ and $\mathbf{i}$ from (8.3-7) and (8.3-8).

Our job now is to express the self- and mutual inductances of all windings. As in the case of the transformer, the self-inductance of each winding is made up of a leakage inductance caused by the flux that fails to cross the air gap and a magnetizing inductance due to the flux that traverses the air gap and circulates through the stator and rotor steel. For symmetrical stator windings, the self-inductances L_{asas} and L_{bsbs} are equal and will be denoted L_{ss}, where

$$L_{ss} = L_{ls} + L_{ms} \tag{8.3-18}$$

In (8.3-18), L_{ls} is the leakage inductance and L_{ms} the magnetizing inductance. The machine is designed to minimize the leakage inductance; it generally makes up approximately 10 percent of the self-inductance. The self-inductance of symmetrical rotor windings may be expressed similarly,

$$L_{rr} = L_{lr} + L_{mr} \tag{8.3-19}$$

The magnetizing inductances L_{ms} and L_{mr} may be expressed in terms of turns and reluctance. In particular,

$$L_{ms} = \frac{N_s^2}{\mathcal{R}_m} \tag{8.3-20}$$

$$L_{mr} = \frac{N_r^2}{\mathcal{R}_m} \tag{8.3-21}$$

The magnetizing reluctance $\mathcal{R}_m$ is total non-leakage reluctance seen by the winding. Expressions for L_{ms} and L_{mr}, as defined in (8.3-20) and (8.3-21), may be derived. In particular, it can be shown that [1]

$$L_{ms} = N_s^2 \frac{\pi \mu_0 r l}{4g} \tag{8.3-22}$$

$$L_{mr} = N_r^2 \frac{\pi \mu_0 r l}{4g} \tag{8.3-23}$$

where μ_0 is the permeability of free space, r is the mean radius of the air gap, l is the axial length of the air gap, and g is the radial length of the air gap. One must perform a rather involved and lengthy derivation to obtain (8.3-22) and (8.3-23). We will not do this derivation.

Since the stator (rotor) windings are orthogonal as depicted in Fig. 8.2-1, coupling does not exist between the *as* and *bs* windings (L_{asas} or L_{bsbs}) or between the *ar* and *br* windings (L_{arbr} or L_{brar}). Recall that the equivalent, sinusoidally distributed windings are depicted by one coil placed at the maximum turn density. The mutual inductances L_{asbs}, L_{bsas}, L_{arbr}, and L_{brar} are all zero in the case of a two-phase stator. In a three-phase machine, where the stator (rotor) windings are displaced by 120° magnetically in space, a coupling exists between the stator (rotor) windings. Nevertheless, for the two-phase machine we can write

$$\mathbf{L}_s = \begin{bmatrix} L_{ss} & 0 \\ 0 & L_{ss} \end{bmatrix} = L_{ss}\mathbf{I} \tag{8.3-24}$$

$$\mathbf{L}_r = \begin{bmatrix} L_{rr} & 0 \\ 0 & L_{rr} \end{bmatrix} = L_{rr}\mathbf{I} \tag{8.3-25}$$

The stator and rotor windings are in relative motion. Therefore, coupling will occur between the stator and rotor windings and this coupling will vary with the position (θ_r) of the rotor windings relative to the stator windings. For example, when the *as* and *ar* windings are aligned, $\theta_r = 0$, the magnitude of coupling between these windings is maximum and, with the assumed direction of positive i_{as} and i_{ar}, the right-hand rule tells us that the mutual fluxes are aiding. Hence, the mutual inductance at $\theta_r = 0$ is a positive maximum and can be expressed in terms of turns and $\mathcal{R}_m$ as

$$L_{asar} = \frac{N_s N_r}{\mathcal{R}_m} \quad \text{for } \theta_r = 0 \tag{8.3-26}$$

Now, when $\theta_r = \frac{1}{2}\pi$, the *as* and *ar* windings are orthogonal and

$$L_{asar} = 0 \quad \text{for } \theta_r = \frac{1}{2}\pi \tag{8.3-27}$$

For $\theta_r = \pi$, the windings are again aligned but now, with the assumed direction of positive i_{as} and i_{ar}, they oppose, thus,

$$L_{asar} = -\frac{N_s N_r}{\mathcal{R}_m} \quad \text{for } \theta_r = \pi \tag{8.3-28}$$

at $\theta_r = \frac{3}{2}\pi$, the windings are again orthogonal and

$$L_{asar} = 0 \quad \text{for } \theta_r = \frac{3}{2}\pi \tag{8.3-29}$$

From (8.3-26) through (8.3-29), we see that mutual inductances might be approximated as a cosine function of θ_r. In particular, if we define L_{sr} as

$$L_{sr} = \frac{N_s N_r}{\mathcal{R}_m} \tag{8.3-30}$$

we can approximate L_{asar} or L_{aras} as

$$L_{asar} = L_{sr} \cos \theta_r \tag{8.3-31}$$

If we were to carry out the derivation as in [1], we would find that (8.3-31) is, indeed, a valid expression for the mutual inductance between the *as* and *ar* windings. It follows by inspection of Fig. 8.2-1 that

$$L_{asbr} = -L_{sr} \sin \theta_r \tag{8.3-32}$$

$$L_{bsar} = L_{sr} \sin \theta_r \tag{8.3-33}$$

$$L_{bsbr} = L_{sr} \cos \theta_r \tag{8.3-34}$$

Hence,

$$\mathbf{L}_{sr} = L_{sr} \begin{bmatrix} \cos \theta_r & -\sin \theta_r \\ \sin \theta_r & \cos \theta_r \end{bmatrix} \tag{8.3-35}$$

One should now be able to write the mutual inductances by inspection. For practice, express the stator and rotor mutual inductances if the positive direction of i_{bs} is reversed. In this case, (8.3-31) and (8.3-32) remain unchanged; however, the sign of (8.3-33) and (8.3-34) would be changed.

Once the expressions for the mutual inductances are known, we begin to understand the complexities involved in the analysis of electric machines. The stator-to-rotor mutual inductances are sinusoidal functions of θ_r because of their relative motion. Hence, when in the voltage equations we take the derivative of the flux linkages with respect to time, we no longer obtain only the familiar $L(di/dt)$. Instead, two terms result; one due to the derivative of the mutual inductance, since θ_r is a function of time, and one due to the derivative of the current. For example,

$$\frac{d(L_{aras}i_{as})}{dt} = \frac{\partial(L_{aras}i_{as})}{\partial\theta_r}\frac{d\theta_r}{dt} + \frac{\partial(L_{aras}i_{as})}{\partial i_{as}}\frac{di_{as}}{dt} \tag{8.3-36}$$

Although we are going to have to deal with this problem, we shall not do it now. However, before leaving this work, let us incorporate a turns ratio into the equations as we did for the transformer. We may not see the purpose of this at this point since it cannot yield a constant-parameter equivalent *T* circuit, as in the case of the transformer, because of the variation of the mutual inductances. In Section 8.5, we will incorporate a change of variables which will allow us to treat the induction machine from the standpoint of an equivalent *T* circuit with constant inductances. In preparation for this event, we will incorporate a turns ratio at this time. In particular, we will refer the rotor variables to a winding with N_s turns by the following turns ratios:

$$\mathbf{i}'_{abr} = \frac{N_r}{N_s}\mathbf{i}_{abr} \tag{8.3-37}$$

$$\mathbf{v}'_{abr} = \frac{N_s}{N_r}\mathbf{v}_{abr} \tag{8.3-38}$$

$$\boldsymbol{\lambda}'_{abr} = \frac{N_s}{N_r}\boldsymbol{\lambda}_{abr} \tag{8.3-39}$$

Thus, (8.3-5) and (8.3-6) may be written as

$$\mathbf{v}_{abs} = \mathbf{r}_s\mathbf{i}_{abs} + p\boldsymbol{\lambda}_{abs} \tag{8.3-40}$$

$$\mathbf{v}'_{abr} = \mathbf{r}'_r\mathbf{i}'_{abr} + p\boldsymbol{\lambda}'_{abr} \tag{8.3-41}$$

where

$$\mathbf{r}'_r = \left(\frac{N_s}{N_r}\right)^2 \mathbf{r}_r \tag{8.3-42}$$

Substitution of (8.3-37) and (8.3-39) into (8.3-17) yields

$$\begin{bmatrix} \boldsymbol{\lambda}_{abs} \\ \boldsymbol{\lambda}'_{abr} \end{bmatrix} = \begin{bmatrix} \mathbf{L}_s & \dfrac{N_s}{N_r}\mathbf{L}_{sr} \\ \dfrac{N_s}{N_r}(\mathbf{L}_{sr})^T & \mathbf{L}'_r \end{bmatrix} \begin{bmatrix} \mathbf{i}_{abs} \\ \mathbf{i}'_{abr} \end{bmatrix} \tag{8.3-43}$$

where

$$\mathbf{L}'_r = \left(\frac{N_s}{N_r}\right)^2 \mathbf{L}_r = \begin{bmatrix} L'_{rr} & 0 \\ 0 & L'_{rr} \end{bmatrix} \tag{8.3-44}$$

Since L_{mr} and L_{ms} may be related from (8.3-20) and (8.3-21) and $L'_{lr} = \left(\frac{N_s}{N_r}\right)^2 L_{lr}$,

$$\begin{aligned} L'_{rr} &= L'_{lr} + \left(\frac{N_s}{N_r}\right)^2 L_{mr} \\ &= L'_{lr} + L_{ms} \end{aligned} \tag{8.3-45}$$

Note that

$$\frac{N_s}{N_r}\mathbf{L}_{sr} = \frac{N_s}{N_r}L_{sr}\begin{bmatrix} \cos\theta_r & -\sin\theta_r \\ \sin\theta_r & \cos\theta_r \end{bmatrix} \tag{8.3-46}$$

Comparing L_{ms}, (8.3-20), and L_{sr}, (8.3-30), we see that

$$\frac{N_s}{N_r}L_{sr} = L_{ms} \tag{8.3-47}$$

Hence, (8.3-46) may be expressed in terms of L_{ms} and, for compactness, we will define L'_{sr} as

$$\begin{aligned} \mathbf{L}'_{sr} &= \frac{N_s}{N_r}\mathbf{L}_{sr} \\ &= L_{ms}\begin{bmatrix} \cos\theta_r & -\sin\theta_r \\ \sin\theta_r & \cos\theta_r \end{bmatrix} \end{aligned} \tag{8.3-48}$$

Thus, (8.3-43) becomes

$$\begin{bmatrix} \boldsymbol{\lambda}_{abs} \\ \boldsymbol{\lambda}'_{abr} \end{bmatrix} = \begin{bmatrix} \mathbf{L}_s & \mathbf{L}'_{sr} \\ \left(\mathbf{L}'_{sr}\right)^T & \mathbf{L}'_r \end{bmatrix} \begin{bmatrix} \mathbf{i}_{abs} \\ \mathbf{i}'_{abr} \end{bmatrix} \tag{8.3-49}$$

SP8.3-1 Assume that θ_r is positive in the clockwise direction in Fig. 8.2-1 rather than in the counterclockwise direction. Express all inductances. $[L_{asbr} = L_{sr} \sin\theta_r;\ L_{bsar} = -L_{sr}\sin\theta_r$; all others unchanged$]$

SP8.3-2 The as and bs windings in Fig. 8.2-1 are rotated $\frac{1}{4}\pi$ clockwise from the position shown. Express L_{asar} and L_{asbr}. $\left[L_{asar} = L_{sr}\cos\left(\theta_r + \frac{1}{4}\pi\right);\ L_{asbr} = -L_{sr} \sin\left(\theta_r + \frac{1}{4}\pi\right)\right]$

SP8.3-3 Consider the device shown in Fig. 8.2-1. All windings are open-circuited except the as winding. $I_{as} = \sin t$, $L_{ms} = 0.1\text{H}$, $L_{rr} = \frac{1}{4}L'_{rr}$, $\omega_r = 0$, and $\theta_r = \frac{1}{3}\pi$. Determine V_{ar}. $[V_{ar} = 0.025 \cos t]$

8.4 TORQUE

From Table 2.5-1, for a P-pole machine:

$$T_e(\mathbf{i}, \theta_r) = \frac{P}{2}\frac{\partial W_c(\mathbf{i}, \theta_r)}{\partial \theta_r} \tag{8.4-1}$$

In a magnetically linear system, the energy in the coupling field W_f and the coenergy W_c are equal. The field energy can be expressed as

$$\begin{aligned} W_f(\mathbf{i}, \theta_r) = {} & \frac{1}{2}L_{ss}i_{as}^2 + \frac{1}{2}L_{ss}i_{bs}^2 + \frac{1}{2}L'_{rr}i'^2_{ar} + \frac{1}{2}L'_{rr}i'^2_{br} + L_{ms}i_{as}i'_{ar}\cos\theta_r - L_{ms}i_{as}i'_{br}\sin\theta_r \\ & + L_{ms}i_{bs}i'_{ar}\sin\theta_r + L_{ms}i_{bs}i'_{br}\cos\theta_r \end{aligned} \tag{8.4-2}$$

Since $W_f = W_c$, substituting (8.4-2) into (8.4-1) yields the electromagnetic torque for a magnetically linear P-pole two-phase symmetrical induction machine. In particular,

$$T_e = -\frac{P}{2}L_{ms}\left[\left(i_{as}i'_{ar} + i_{bs}i'_{br}\right)\sin\theta_r + \left(i_{as}i'_{br} - i_{bs}i'_{ar}\right)\cos\theta_r\right] \tag{8.4-3}$$

The torque and rotor speed are related by

$$T_e = J\frac{d\omega_{rm}}{dt} + B_m\omega_{rm} + T_L \tag{8.4-4}$$

where ω_{rm} is the actual rotor speed. For a P-pole machine, since $\omega_{rm} = (2/P)\omega_r$,

$$T_e = J\frac{2}{P}\frac{d\omega_r}{dt} + B_m\frac{2}{P}\omega_r + T_L \tag{8.4-5}$$

where J is the inertia of the rotor and, in some cases, the shaft-connected load. The first term on the right-hand side is the inertial torque. The units of J are kilogram-meter^2 (kg · m^2) or joules.second^2 (J · s^2). Often, the inertia is given as a quantity called WR^2 expressed in units of pound mass · feet^2 (lbm · ft^2). The load torque T_L is positive for a torque load on the shaft of the induction machine, as shown in Fig. 8.2-1. The constant B_m is a damping coefficient associated with the rotational system of the machine and mechanical load. It has the units of N · m · s/rad of mechanical rotation and it is generally small and often neglected.

SP8.4-1 Why is the sixth term on the right-hand side of (8.4-2) negative? [(8.3-32)]

8.5 VOLTAGE EQUATIONS IN THE ARBITRARY REFERENCE FRAME

Although the following detailed reference frame-based derivation is straightforward, you may wish not to go through this derivation but instead focus on steady-state performance. If so, you can de-emphasize the material up to Section 8.8 or (8.8-5) and (8.8-6) and Fig. 8.8-1 where the focus on steady-state operation begins.

In Chapter 5, we introduced the concept of reference frame theory wherein we set forth a change of variables that related the variables associated with symmetrical stationary and rotating circuits to substitute variables associated with fictitious circuits rotating at an arbitrary angular velocity. From (5.4-6) and (5.4-15) or (5B-5) and (5B-6) the stator variables in the arbitrary reference frame become

$$v_{qs} = r_s i_{qs} + \omega \lambda_{ds} + p\lambda_{qs} \tag{8.5-1}$$

$$v_{ds} = r_s i_{ds} - \omega \lambda_{qs} + p\lambda_{ds} \tag{8.5-2}$$

From (5.9-13) and (5.9-14) and incorporating the turns ratio, the rotor variables in the arbitrary reference frame become

$$v'_{qr} = r'_r i'_{qr} + (\omega - \omega_r)\lambda'_{dr} + p\lambda'_{qr} \tag{8.5-3}$$

$$v'_{dr} = r'_r i'_{dr} - (\omega - \omega_r)\lambda'_{qr} + p\lambda'_{dr} \tag{8.5-4}$$

There are several things we should talk about. Equations (8.5-1) through (8.5-4) are the voltage equations for a symmetrical, two-phase induction machine in the arbitrary reference frame; symmetrical, in that the two stator (rotor) windings are orthogonal magnetically with the same winding resistance and the same number of equivalent turns. It is important to realize that the machine need not be linear magnetically. The above voltage equations are valid for either a magnetically linear or a magnetically nonlinear symmetrical induction machine. Recall that we restrict the analysis to a linear magnetic system when we express the flux linkages as linear functions of inductances.

Example 8A Commonly Used Reference Frames

Depending on the application, there are three reference frames that are commonly used in the analysis and control of induction machines: the stationary, the rotor, and the synchronously rotating. Let us write the voltage equations in each of these reference frames.

For the stationary reference frame $\omega = 0$; whereupon, (8.5-1) through (8.5-4) become

$$v_{qs}^{s} = r_s i_{qs}^{s} + p\lambda_{qs}^{s} \tag{8A-1}$$

$$v_{ds}^{s} = r_s i_{ds}^{s} + p\lambda_{ds}^{s} \tag{8A-2}$$

$$v_{qr}'^{s} = r_r' i_{qr}'^{s} - \omega_r \lambda_{dr}'^{s} + p\lambda_{qr}'^{s} \tag{8A-3}$$

$$v_{dr}'^{s} = r_r' i_{dr}'^{s} + \omega_r \lambda_{qr}'^{s} + p\lambda_{dr}'^{s} \tag{8A-4}$$

In the above equations the raised s is used to denote variables in the stationary reference frame.

For the rotor reference frame $\omega = \omega_r$, (8.5-1) through (8.5-4) become

$$v_{qs}^{r} = r_s i_{qs}^{r} + \omega_r \lambda_{ds}^{r} + p\lambda_{qs}^{r} \tag{8A-5}$$

$$v_{ds}^{r} = r_s i_{ds}^{r} - \omega_r \lambda_{qs}^{r} + p\lambda_{ds}^{r} \tag{8A-6}$$

$$v_{qr}'^{r} = r_r' i_{qr}'^{r} + p\lambda_{qr}'^{r} \tag{8A-7}$$

$$v_{dr}'^{r} = r_r' i_{dr}'^{r} + p\lambda_{dr}'^{r} \tag{8A-8}$$

Here, we have used the raised r to denote variables in the rotor reference frame.

For the synchronously rotating reference frame $\omega = \omega_e$, (8.5-1) through (8.5-4) become

$$v_{qs}^{e} = r_s i_{qs}^{e} + \omega_e \lambda_{ds}^{e} + p\lambda_{qs}^{e} \tag{8A-9}$$

$$v_{ds}^{e} = r_s i_{ds}^{e} - \omega_e \lambda_{qs}^{e} + p\lambda_{ds}^{e} \tag{8A-10}$$

$$v_{qr}'^{e} = r_r' i_{qr}'^{e} + (\omega_e - \omega_r)\lambda_{dr}'^{e} + p\lambda_{qr}'^{e} \tag{8A-11}$$

$$v_{dr}'^{e} = r_r' i_{dr}'^{e} + (\omega_e - \omega_r)\lambda_{qr}'^{e} + p\lambda_{dr}'^{e} \tag{8A-12}$$

The raised e denotes the variables in the synchronously rotating reference frame. It should be clear that all of the above equations are valid for linear or nonlinear magnetic systems. When we substitute inductances in the flux-linkage equations, the equations are then valid only for magnetically linear systems.

Before leaving this example, let us note an important property of reference frame theory; the speed voltages disappear from the q- and d-voltage equations if the reference frame is fixed where the windings physically exist. This occurs in (8A-1) and (8A-2) or the reference frame fixed in the stator and in (8A-7) and (8A-8) for the rotor reference frame. Unfortunately, the $\omega\lambda$ terms became known as the speed voltages probably as a result of Park's original work. He was working on the rotor reference frame and the voltage equations for the transformed stator circuits of the synchronous machine were essentially (8A-5) and (8A-6), wherein the $\omega_r\lambda$ terms appeared. These terms were referred to as the speed voltages since rotor speed, ω_r, was present in these terms. We now know that the speed voltages are a function of the reference frame speed and appear whenever the reference frame angular velocity is different than that of the circuits being transformed.

Example 8B Reference Frame Applied Voltages
Assume a two-phase *ab*-sequence balanced set with

$$v_{as} = \sqrt{2}v_s \cos\theta_{esv} \tag{8B-1}$$

where

$$\frac{d\theta_{esv}}{dt} = \omega_e \tag{8B-2}$$

Express v_{qs} and v_{ds} in the arbitrary reference frame where

$$\frac{d\theta}{dt} = \omega \tag{8B-3}$$

Then, express the applied voltages in the reference frames considered in Example 8A.

From (5.5-7) and (5.5-8),

$$v_{qs} = \sqrt{2}v_s \cos(\theta_{esv} - \theta) \tag{8B-4}$$

$$v_{ds} = -\sqrt{2}v_s \sin(\theta_{esv} - \theta) \tag{8B-5}$$

In the stationary reference frame, $\omega = 0$,

$$v_{qs}^s = \sqrt{2}v_s \cos[\theta_{esv} - \theta(0)] \tag{8B-6}$$

$$v_{ds}^s = -\sqrt{2}v_s \sin[\theta_{esv} - \theta(0)] \tag{8B-7}$$

where $\theta(0)$ is the time-zero position of the reference frame. In the rotor reference frame, $\omega = \omega_r$:

$$v_{qs}^r = \sqrt{2}v_s \cos(\theta_{esv} - \theta_r) \tag{8B-8}$$

$$v_{ds}^r = -\sqrt{2}v_s \sin(\theta_{esv} - \theta_r) \tag{8B-9}$$

In the synchronous reference frame, $\omega = \omega_e$,

$$v_{qs}^e = \sqrt{2}v_s \cos(\theta_{esv} - \theta_e) \tag{8B-10}$$

$$v_{ds}^e = -\sqrt{2}v_s \sin(\theta_{esv} - \theta_e) \tag{8B-11}$$

SP8.5-1 Are the voltage equations given in this section valid for a nonlinear magnetic system? Why? [yes; λ is not restricted to be a linear function of i]

SP8.5-2 When are (8A-5) through (8A-8) identical in form to (a) (8A-1) through (8A-4)? (b) to (8A-9) through (8A-12)? [(a) $\omega_r = 0$; (b) $\omega_r = \omega_e$]

SP8.5-3 You are given a set of four voltage equations in substitute variables; however, the raised indexes are missing. How can you tell if the reference frame is in either the stator or rotor? [speed voltages are absent in the qs and ds or qr and dr voltage equations]

SP8.5-4 When are (8B-6) and (8B-7) v_{as} and $-v_{bs}$, respectively? [$\theta(0) = 0$]

8.6 MAGNETICALLY LINEAR FLUX-LINKAGE EQUATIONS AND EQUIVALENT CIRCUITS

The flux-linkage equations for a magnetically linear system given by (8.3-49) are in terms of as, bs, ar', and br' variables. In the arbitrary reference frame, (8.3-49) becomes

$$\begin{bmatrix} \boldsymbol{\lambda}_{qds} \\ \boldsymbol{\lambda}'_{qdr} \end{bmatrix} = \begin{bmatrix} \mathbf{K}_s\mathbf{L}_s(\mathbf{K}_s)^{-1} & \mathbf{K}_s\mathbf{L}'_{sr}(\mathbf{K}_r)^{-1} \\ \mathbf{K}_r(\mathbf{L}'_{sr})^T(\mathbf{K}_s)^{-1} & \mathbf{K}_r\mathbf{L}'_r(\mathbf{K}_r)^{-1} \end{bmatrix} \begin{bmatrix} \mathbf{i}_{qds} \\ \mathbf{i}'_{qdr} \end{bmatrix} \tag{8.6-1}$$

Since, from (8.3-24) and (8.3-44), respectively, $\mathbf{L}_s = L_{ss}\mathbf{I}$ and $\mathbf{L}'_r = L'_{rr}\mathbf{I}$,

$$\mathbf{K}_s\mathbf{L}_s(\mathbf{K}_s)^{-1} = \begin{bmatrix} L_{ss} & 0 \\ 0 & L_{ss} \end{bmatrix} = \mathbf{L}_s \tag{8.6-2}$$

$$\mathbf{K}_r\mathbf{L}'_r(\mathbf{K}_r)^{-1} = \begin{bmatrix} L'_{rr} & 0 \\ 0 & L'_{rr} \end{bmatrix} = \mathbf{L}'_r \tag{8.6-3}$$

Now, let us work on the upper-right element of the 2×2 matrix in (8.6-1):

$$\mathbf{K}_s\mathbf{L}'_{sr}(\mathbf{K}_r)^{-1} = \begin{bmatrix} \cos\theta & \sin\theta \\ \sin\theta & -\cos\theta \end{bmatrix} L_{ms} \begin{bmatrix} \cos\theta & -\sin\theta_r \\ \sin\theta_r & \cos\theta_r \end{bmatrix} \begin{bmatrix} \cos(\theta-\theta_r) & \sin(\theta-\theta_r) \\ \sin(\theta-\theta_r) & -\cos(\theta-\theta_r) \end{bmatrix} \tag{8.6-4}$$

It is left to the reader to show that (8.6-4) reduces to

$$\mathbf{K}_s\mathbf{L}'_{sr}(\mathbf{K}_r)^{-1} = \begin{bmatrix} L_{ms} & 0 \\ 0 & L_{ms} \end{bmatrix} = \mathbf{L}_{ms} \tag{8.6-5}$$

and that

$$\mathbf{K}_s\mathbf{L}'_{sr}(\mathbf{K}_r)^{-1} = \mathbf{K}_r(\mathbf{L}'_{sr})^T(\mathbf{K}_s)^{-1} \tag{8.6-6}$$

Therefore,

$$\begin{bmatrix} \boldsymbol{\lambda}_{qds} \\ \boldsymbol{\lambda}'_{qdr} \end{bmatrix} = \begin{bmatrix} \mathbf{L}_s & \mathbf{L}_{ms} \\ \mathbf{L}_{ms} & \mathbf{L}'_r \end{bmatrix} \begin{bmatrix} \mathbf{i}_{qds} \\ \mathbf{i}'_{qdr} \end{bmatrix} \tag{8.6-7}$$

Since $L_{ss} = L_{ls} + L_{ms}$ and $L'_{rr} = L'_{lr} + L_{ms}$, (8.6-7) may be expressed as

$$\lambda_{qs} = L_{ls} i_{qs} + L_{ms}\left(i_{qs} + i'_{qr}\right) \tag{8.6-8}$$

$$\lambda_{ds} = L_{ls} i_{ds} + L_{ms}\left(i_{ds} + i'_{dr}\right) \tag{8.6-9}$$

$$\lambda'_{qr} = L'_{lr} i'_{qr} + L_{ms}\left(i_{qs} + i'_{qr}\right) \tag{8.6-10}$$

$$\lambda'_{dr} = L'_{lr} i'_{dr} + L_{ms}\left(i_{ds} + i'_{dr}\right) \tag{8.6-11}$$

Equations (8.5-1) through (8.5-4) and (8.6-8) through (8.6-11) suggest the equivalent circuits given in Fig. 8.6-1. Equations (8.6-8) through (8.6-11) and Fig. 8.6-1 are important in that for a symmetrical induction machine the rotor position-dependent terms of the mutual inductances are eliminated in all frames of reference. We would expect this since both the stator and rotor windings are assumed to be symmetrical and we are able to transform the variables to a common frame of reference; the arbitrary reference frame.

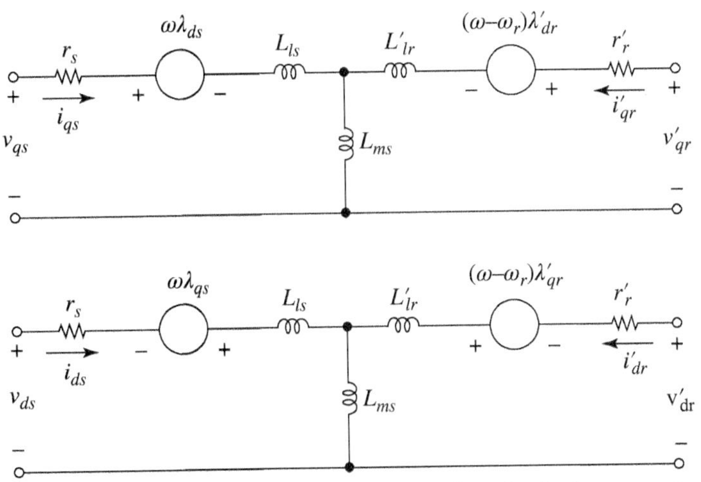

Figure 8.6-1 Arbitrary reference frame equivalent circuits for a two-phase, symmetrical induction machine.

SP8.6-1 If in the equivalent circuits given in Fig. 8.6-1 there are speed voltages only in the q'_r and d'_r part of the T equivalent circuits, what reference frame is being used? [stationary]

SP8.6-2 Repeat SP8.6-1 if speed voltages are only in the qs and ds part. [rotor]

8.7 TORQUE EQUATIONS IN ARBITRARY REFERENCE FRAME VARIABLES

Equation (8.4-3) is an expression for the electromagnetic torque in machine currents i_{as}, i_{bs}, i'_{ar}, and i'_{br} for a magnetically linear system. If we express the torque in terms of the arbitrary reference frame currents, the electromagnetic torque for a magnetically linear system may be expressed as

$$T_e = \frac{P}{2} L_{ms}\left(i_{qs}i'_{dr} - i_{ds}i'_{qr}\right) \tag{8.7-1}$$

where T_e is positive for motor action. Equivalent expressions for torque are

$$T_e = \frac{P}{2}\left(\lambda'_{qr}i'_{dr} - \lambda'_{dr}i'_{qr}\right) \tag{8.7-2}$$

$$T_e = \frac{P}{2}\left(\lambda_{ds}i_{qs} - \lambda_{qs}i_{ds}\right) \tag{8.7-3}$$

When T_e is expressed in terms of flux linkages as in (8.7-2) and (8.7-3), it appears as if the leakage inductances play a role in the evaluation of torque. However, the product of leakage inductances times the currents cancel when the flux linkages are expressed in terms of inductances. It should also be mentioned that (8.7-2) and (8.7-3) are valid for magnetically linear and nonlinear systems [3]. For a three-phase machine, (8.7-1) through (8.7-3) must be multiplied by $\frac{3}{2}$. Also, for a magnetically linear assumption, L_{ms} in the flux-linkage equations must also be multiplied by $\frac{3}{2}$.

SP8.7-1 Show that (8.7-2) and (8.7-3) are equal for a linear magnetic system.

8.8 PHASORS AND STEADY-STATE OPERATING MODES

Phasors and Steady-State Equivalent Circuit

In this section we will set forth the instantaneous and steady-state phasors and the steady-state equivalent circuit. We will then use this circuit to demonstrate the steady-state characteristics of several common modes of operation of an induction machine.

The instantaneous phasor equations may be obtained by a procedure used in Chapter 5. In particular, we will substitute the voltage equations in the synchronously rotating reference frame given by (8A-9) through (8A-12) into

$$\tilde{f}_{as} = f^e_{qs} - jf^e_{ds} \tag{8.8-1}$$

$$\tilde{f}'_{ar} = f'^e_{qr} - jf'^e_{dr} \tag{8.8-2}$$

to obtain the instantaneous phasor voltage equations. Then, we will express (8.6-8) through (8.6-11) in terms of synchronous reference frame variables and substitute into the instantaneous phasor voltage equations to obtain

$$\tilde{v}_{as} = r_s \tilde{i}_{as} + j\omega_e L_{ss}\tilde{i}_{as} + j\omega_e L_{ms}\tilde{i}'_{ar} + pL_{ss}\tilde{i}_{as} + pL_{ms}\tilde{i}'_{ar} \tag{8.8-3}$$

$$\tilde{v}'_{ar} = r'_r \tilde{i}'_{ar} + j(\omega_e - \omega_r)L'_{rr}\tilde{i}'_{ar} + j(\omega_e - \omega_r)L_{ms}\tilde{i}_{as} + pL'_{rr}\tilde{i}'_{ar} + pL_{ms}\tilde{i}_{as} \tag{8.8-4}$$

For steady-state conditions the last two terms of (8.8-3) and (8.8-4) become zero and we obtain the following steady-state voltage equations:

$$\tilde{V}_{as} = r_s \tilde{I}_{as} + j\omega_e (L_{ls} + L_{ms})\tilde{I}_{as} + j\omega_e L_{ms}\tilde{I}'_{ar} \tag{8.8-5}$$

$$\tilde{V}'_{ar} = r'_r \tilde{I}'_{ar} + j(\omega_e - \omega_r)\left(L'_{lr} + L_{ms}\right)\tilde{I}'_{ar} + j(\omega_e - \omega_r)L_{ms}\tilde{I}_{as} \tag{8.8-6}$$

The so-called *slip* is

$$s = \frac{\omega_e - \omega_r}{\omega_e} \tag{8.8-7}$$

If we divide (8.8-6) by the slip, it becomes

$$\frac{\tilde{V}'_{ar}}{s} = \frac{r'_r}{s}\tilde{I}'_{ar} + j\omega_e\left(L'_{lr} + L_{ms}\right)\tilde{I}'_{ar} + j\omega_e L_{ms}\tilde{I}_{as} \tag{8.8-8}$$

Equations (8.8-5) and (8.8-8) suggest the single-phase equivalent T circuit of a two-phase symmetrical induction machine during steady-state balanced operation shown in Fig. 8.8-1. Please note that the inductive reactances are calculated as $X = \omega_e L$. One tends to calculate the inductive reactances of the rotor circuit as $X = (\omega_e - \omega_r)L$ when using (8.8-8). It is clear that (8.8-6) is of the form we would expect; however, we have divided (8.8-6) by $(\omega_r - \omega_e)$ and multiplied by ω_e to arrive at (8.8-8). We understand that current is not induced in the rotor windings when $\omega_r = \omega_e$. Since the rotor windings are generally short-circuited and from (8.8-7) the slip is zero and $\frac{r'_r}{s}$ is infinite at synchronous speed, the rotor circuit appears to be open-circuited, thus correctly portraying synchronous speed "operation." We will find that, with a slight modification (X_{ms} becomes $\frac{3}{2}X_{ms}$), this circuit may also be used to calculate the steady-state performance of a three-phase symmetrical induction machine.

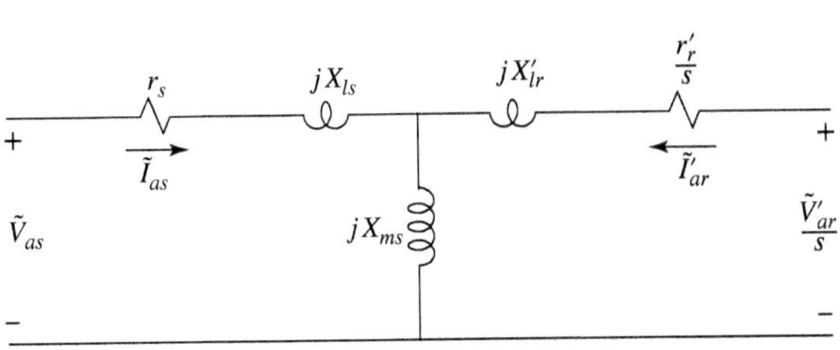

Figure 8.8-1 Equivalent circuit for a two-phase symmetrical induction machine for balanced steady-state operation.

Steady-State Torque

An expression for the steady-state electromagnetic torque may be obtained by first writing (8.7-1) in terms of I_{qs}^e, I_{ds}^e, $I_{qs}^{\prime e}$, and $I_{dr}^{\prime e}$, and then using (8.8-1) and (8.8-2) to express $\widetilde{I}_{as}$ and $\widetilde{I}_{ar}^{\prime}$. The expression may be reduced to

$$T_e = 2\left(\frac{P}{2}\right) L_{ms} \,\mathrm{Re}\left[j\widetilde{I}_{as}^{*}\widetilde{I}_{ar}^{\prime}\right] \tag{8.8-9}$$

where $\widetilde{I}_{as}^{*}$ is the complex conjugate of $\widetilde{I}_{as}$. The phasor currents may be calculated from the equivalent circuit given in Fig. 8.8-1.

Let us take a moment to tie this back to the work in Chapter 4 and derive an expression for torque using the air-gap mmfs. For this purpose, (4.5-16) for mmf_s^e and (4.5-17) for mmf_r^e are repeated here:

$$\mathrm{mmf}_s^e = \frac{N_s}{2}\sqrt{2}I_s \cos\left[\theta_{esi}(0) - \phi_e\right] \tag{8.8-10}$$

$$\mathrm{mmf}_r^e = \frac{N_r}{2}\sqrt{2}I_r \cos\left[\theta_{eri}(0) - \phi_e\right] \tag{8.8-11}$$

Now, torque is the cross product of the maximum value of mmf_s^e, which occurs at $\phi_e = \theta_{esi}(0)$, and the maximum value of rotor flux, which occurs at $\phi_e = \theta_{eri}(0)$ and is mmf_r^e divided by $\mathcal{R}_m$, the total reluctance of the uniform air gap. Thus, from (8.8-10) and (8.8-11) and since we must use the total mmf drop and change I_r to $I_r^{\prime}$,

$$T_e = \left(\frac{P}{2}\right)\frac{N_s^{\,2}}{\mathcal{R}_m}\sqrt{2}I_s\sqrt{2}I_r^{\prime} \sin\left[\theta_{esi}(0) - \theta_{eri}(0)\right] \tag{8.8-12}$$

where the $\frac{P}{2}$ accounts for more than two poles. We see that (8.8-12) is (8.8-9) since

$$L_{ms} = \frac{N_s^2}{\mathcal{R}_m} \tag{8.8-13}$$

The phasor currents may be expressed as

$$\widetilde{I}_{as} = I_s[\cos\theta_{esi}(0) + j\sin\theta_{esi}(0)] \tag{8.8-14}$$

$$\widetilde{I}_{ar}^{\prime} = I_r^{\prime}[\cos\theta_{eri}(0) + j\sin\theta_{eri}(0)] \tag{8.8-15}$$

If (8.8-14) and (8.8-15) are substituted in (8.8-9), it becomes (8.8-12). We will find use for (8.8-9) and (8.8-12) when calculating balanced steady-state torque.

The balanced steady-state torque-speed or torque-slip characteristic of a single-fed induction machine warrants discussion. The vast majority of induction machines in use today are single-fed, wherein electric power is transferred to or from the induction machine via the stator circuits with the rotor windings short-circuited. In this case, $\widetilde{V}'_{ar}$ is zero, whereupon (8.8-8) may be written as

$$\widetilde{I}'_{ar} = -\frac{jX_{ms}}{r'_r/s + jX'_{rr}}\widetilde{I}_{as} \tag{8.8-16}$$

Substituting (8.8-16) into (8.8-9) yields the following expression for electromagnetic torque of a single-fed two-phase symmetrical induction machine during balanced steady-state operation:

$$T_e = \frac{2(P/2)\left(X_{ms}^2/\omega_e\right)\left(r'_r/s\right)\left|\widetilde{I}_{as}\right|^2}{\left(r'_r/s\right)^2 + X'^2_{rr}} \tag{8.8-17}$$

It is important to note from (8.8-17) that torque is positive (motor action) when slip is positive which occurs when $\omega_r < \omega_e$, and negative (generator action) when the slip is negative which occurs when the rotor is being driven above synchronous speed $\omega_r > \omega_e$, and zero when the slip is zero $(\omega_r = \omega_e)$. In other words, the single-fed induction machine develops torque at all speeds except at synchronous speed.

With the rotor windings short-circuited, the input impedance of the equivalent circuit shown in Fig. 8.8-1 is

$$Z = \frac{\left(r_s r'_r/s\right) + \left(X_{ms}^2 - X_{ss}X'_{rr}\right) + j\left[\left(r'_r/s\right)X_{ss} + r_s X'_{rr}\right]}{\left(r'_r/s\right) + jX'_{rr}} \tag{8.8-18}$$

Now, $\left|\widetilde{I}_{as}\right|^2$ is I_s^2 and

$$I_s = \frac{\left|\widetilde{V}_{as}\right|}{|Z|} \tag{8.8-19}$$

Hence, the expression for the steady-state electromagnetic torque for a single-fed two-phase symmetrical induction machine becomes

$$T_e = \frac{2(P/2)\left(X_{ms}^2/\omega_e\right)r'_r s\left|\widetilde{V}_{as}\right|^2}{\left[r_s r'_r + s\left(X_{ms}^2 - X_{ss}X'_{rr}\right)\right]^2 + \left(r'_r X_{ss} + s r_s X'_{rr}\right)^2} \tag{8.8-20}$$

Thus, for a given set of parameters and source frequency ω_e, the steady-state torque varies as the square of the magnitude of the applied voltages. The steady-state torque-speed characteristics typical of many single-fed two-phase induction machines are shown in Fig. 8.8-2. The parameters of the machine are often selected so that, for rated frequency operation, the maximum torque occurs between 80 and 90 percent of synchronous speed. Generally, the maximum torque is two or three times the rated torque of the machine. Although we are considering a two-phase machine, the general shape of the torque-speed characteristic is similar for multiphase induction machines. However, when extending the two-phase analysis to a three-phase induction machine, we must multiply all expressions for torque by $\frac{3}{2}$ due to the $\frac{2}{3}$ factor used in Park's three-phase transformation and we must also

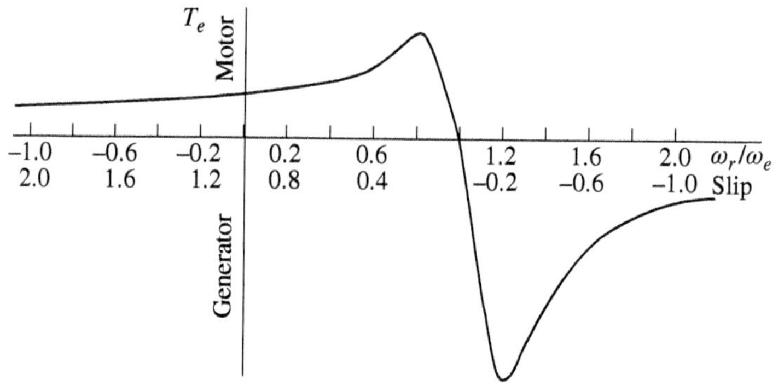

Figure 8.8-2 Steady-state torque-speed characteristic of a symmetrical induction machine.

replace $L_{ms}(X_{ms})$ with $\frac{3}{2}L_{ms}\left(\frac{3}{2}X_{ms}\right)$ in the voltage and torque expressions because of the mutual coupling between the three-phase stator windings. Please recall that $L_{ms}(X_{ms})$ is contained in $L_{ss}(X_{ss})$ and $L'_{rr}(X'_{rr})$. We will point all this out again when we consider the three-phase induction machine later in this chapter.

Example 8C Phasor Diagram for Steady-State Operation

Consider a single-fed two-pole two-phase 5-hp 110-V (rms) 60-Hz induction machine with the following parameters: $r_s = 0.295\ \Omega$, $L_{ls} = 0.944$ mH, $L_{ms} = 35.15$ mH, $r'_r = 0.201\ \Omega$, and $L'_{lr} = 0.944$ mH. Calculate $\tilde{I}_{as}$, $\tilde{I}'_{ar}$, and T_e, draw the phasor diagram, and show the rotor and stator poles for (a) $s = 0.05$ and (b) $s = -0.05$. The equivalent circuit for (a) and (b) is shown in Fig. 8C-1. The inductive reactances are calculated as ω_e times the inductance in henrys.

$$r_s = 0.295\ \Omega \quad X_{ms} = 13.252\ \Omega \quad r'_r = 0.201\ \Omega$$
$$X_{ls} = 0.356\ \Omega \qquad\qquad X'_{lr} = 0.356\ \Omega$$

For part (a):

$$\frac{r'_r}{s} = \frac{0.201}{0.05} = 4.02\ \Omega \tag{8C-1}$$

$$\begin{aligned} Z_{\text{in}} &= r_s + jX_{ls} + \frac{\left(\frac{r'_r}{s} + jX'_{lr}\right)jX_{ms}}{\frac{r'_r}{s} + j(X'_{lr} + X_{ms})} \\ &= 0.295 + j0.356 + \frac{(4.02 + j0.356)j13.252}{4.02 + j(0.356 + 13.252)} \\ &= 4.18\angle 25.58^\circ\Omega \end{aligned} \tag{8C-2}$$

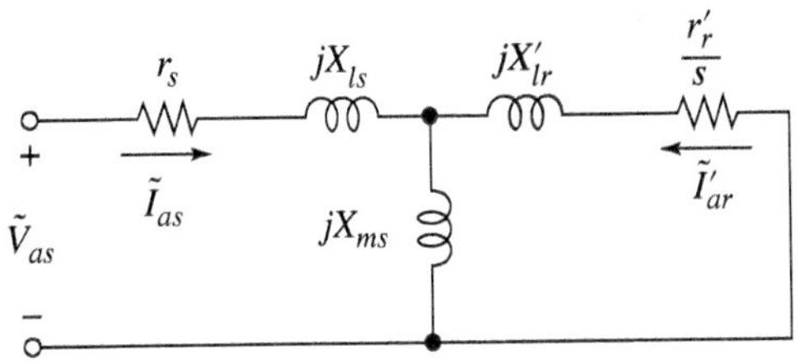

$r_s = 0.295\ \Omega$ $X_{ms} = 13.252\ \Omega$ $r'_r = 0.201\ \Omega$
$X_{ls} = 0.356\ \Omega$ $X'_{lr} = 0.356\ \Omega$

Figure 8C-1 Equivalent circuit for steady-state operation of a single-fed induction machine.

$$\begin{aligned}\tilde{I}_{as} &= \frac{\tilde{V}_{as}}{Z_{\text{in}}} \\ &= \frac{110\angle 0^\circ}{4.18\angle 25.58^\circ} = 26.32\angle -25.58^\circ \text{A}\end{aligned} \tag{8C-3}$$

From (8.8-16),

$$\begin{aligned}\tilde{I}'_{ar} &= -\frac{jX_{ms}}{\frac{r'_r}{s} + jX'_{rr}}\tilde{I}_{as} \\ &= -\frac{j13.252}{4.02 + j(0.356 + 13.252)}26.32\angle -25.58^\circ \\ &= -24.58\angle 9.12^\circ = 24.58\angle 170.9^\circ \text{A}\end{aligned} \tag{8C-4}$$

From (8.8-12),

$$\begin{aligned}T_e &= 2\left(\frac{P}{2}\right)L_{ms}I_sI'_r \sin\left[\theta_{esi}(0) - \theta_{eri}(0)\right] \\ &= 2\left(\frac{2}{2}\right)(35.15\times 10^{-3})(26.32)(24.58)\sin(-25.58^\circ - 170.9^\circ) \\ &= 12.9\ \text{N}\cdot\text{m}\end{aligned} \tag{8C-5}$$

The torque is positive in the direction of rotation; motor action. The phasor diagram for part (a) is shown in Fig. 8C-2. The stator poles are "pushing" the rotor poles counterclockwise; motor action.

For part (b):

$$\frac{r'_r}{-0.05} = -4.02\ \Omega \tag{8C-6}$$

In the calculation we need to replace $\frac{r'_r}{s} = 0.402\ \Omega$ in part (a) with $\frac{r'_r}{s} = -0.402\ \Omega$. Thus,

$$Z_{\text{in}} = 3.648\angle 151.5^\circ \tag{8C-7}$$

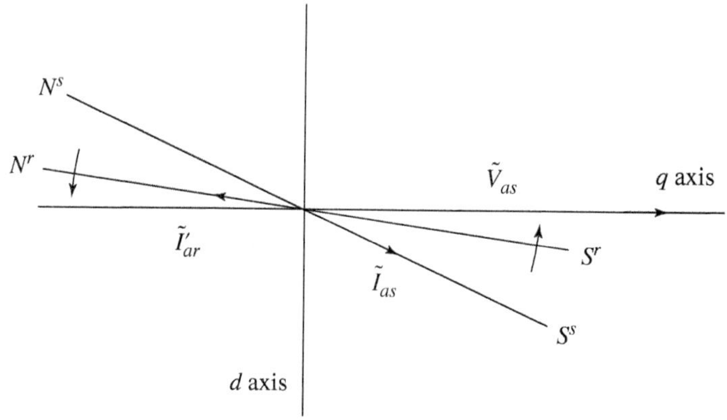

Figure 8C-2 Phasor diagram; motor action.

$$\tilde{I}_{as} = 30.154\angle -151.5°\text{A} \tag{8C-8}$$

$$\tilde{I}'_{ar} = 28.2\angle 12°\text{A} \tag{8C-9}$$

From (8C-5),

$$\begin{aligned} T_e &= 2\left(\frac{2}{2}\right)(35.15 \times 10^{-3})(30.154)(28.2)\sin(-151.5° - 12°) \\ &= -17.0\ \text{N·m} \end{aligned} \tag{8C-10}$$

The torque is negative in the direction of rotation; generator action. The phasor diagram is shown in Fig. 8C-3. Therein, the rotor poles are "pushing" the stator poles counterclockwise; generator actions.

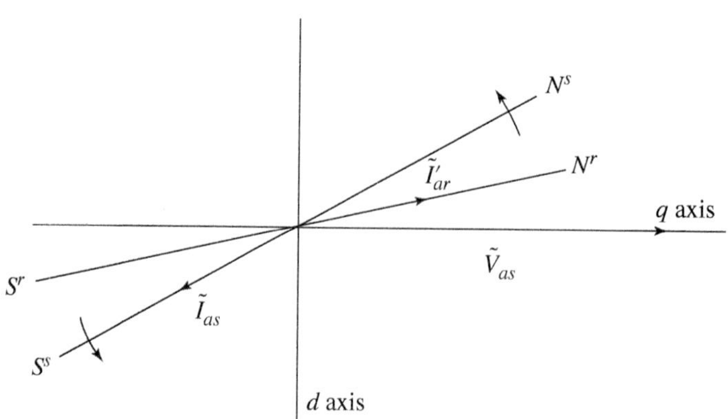

Figure 8C-3 Phasor diagram; generator action.

Additional Operating Characteristics

Let us take a moment to consider stable, steady-state operation of an induction machine. For steady-state operation, $d\omega_r/dt$ is zero in (8.4-5) since the speed is constant and B_m is generally neglected; thus, $T_e = T_L$. From Fig. 8.8-2, it would appear that, for a given load torque T_L there could be two operating points on the T_e versus ω_r plot. When we calculate the steady-state torque we are assuming, in the calculation, that the speed is constant without regard to whether or not it is a stable operating point. Let us consider the T_e versus ω_r plot shown in Fig. 8.8-3. Since induction machines are used primarily as motors, the characteristic shown in Fig. 8.8-3 is the range of speeds over which motors operate. For the load torque T_L, shown in Fig. 8.8-3, there appear to be two operating Points, 1 and $1'$. If stable steady-state operation occurs at either Point 1 or $1'$, then, if the system is displaced from this operating point, there will be a torque established to return the system to this operating point. This, of course, is the same procedure used to determine the stable operating points of the relay in Chapter 2.

Consider operation at Point 1. If the speed ω_r were to increase ever so slightly, T_e would become less than T_L and the load torque (T_L) requirements would exceed the electromagnetic torque (T_e) developed by the machine. Consequently, the machine would slow down and return to operation at Point 1. If, instead, the speed were to decrease from operating Point 1, T_e would exceed T_L and the machine would accelerate back to Point 1. It follows that Point 1 is a stable steady-state operating point.

Now, consider operation at Point $1'$. An increase in ω_r would cause T_e to exceed T_L and the machine to accelerate and move away from Point $1'$. In fact, the machine would accelerate to Point 1 where it would operate stably. If ω_r decreased slightly from Point $1'$, T_e would become less than T_L. The machine would decelerate to stall ($\omega_r = 0$) and, if T_L remained applied to the shaft, the rotor would reverse direction and accelerate in the opposite direction. Point $1'$ is an unstable operating point. It follows that for the assumed direction of positive current, stable operation, for either motor or generation operation, occurs on the negatively sloped portion of the T_e versus ω_r characteristic.

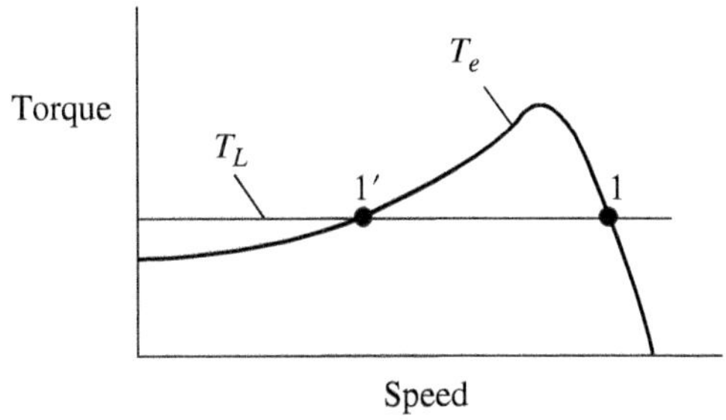

Figure 8.8-3 Stable operation of an induction motor.

Regarding Fig. 8.8-3, one may wonder how operation could be established at Point 1 to begin with, since T_L is larger than T_e at stall. Thus, the machine could not develop a large enough starting torque to accelerate the machine up to operation at Point 1. In most cases, the load torque is a function of ω_r, say, $T_L = K\omega_r^2$, for example. In these cases, the machine can develop sufficient starting torque and, if T_L and T_e match on the negative slope portion, stable operation will occur. If, on the other hand, T_L is constant and greater than T_e at $\omega_r = 0$, we have at least three choices: (a) increase the stator voltage; (b) increase the rotor resistance; or (c) use a different machine. Increasing the rotor resistance to increase the starting torque is something that we have not yet discussed. We will now.

An expression for the slip at maximum torque may be obtained by taking the derivative of (8.8-20) with respect to slip and setting the result equal to zero. In particular,

$$s_m = r_r' G \tag{8.8-21}$$

where

$$G = \pm\sqrt{\frac{r_s^2 + X_{ss}^2}{\left(X_{ms}^2 - X_{ss}X_{rr}'\right)^2 + r_s^2 X_{rr}'^2}} \tag{8.8-22}$$

Two values of slip at maximum torque, s_m, are obtained, one for motor action and one for generator action. It is important to note that G is not a function of r_r'; thus, the slip at maximum torque, (8.8-21), is directly proportional to r_r'. Consequently, with all other machine parameters constant, the speed at which maximum steady-state torque occurs may be varied by inserting external rotor resistance. This feature is often used when starting large motors which have coil-wound rotor windings with slip rings. In this application, balanced external rotor resistances are placed across the terminals of the rotor windings so that maximum torque occurs near stall. As the machine speeds up, the external resistors are decreased in value. On the other hand, some induction machines are designed with high resistance rotor windings so that maximum torque is produced at or near stall to provide fast response.

It may at first appear that the magnitude of the maximum torque would be influenced by r_r'. However, if (8.8-21) is substituted into (8.8-20), the maximum torque may be expressed as

$$T_{e,\,\max} = \frac{2(P/2)\left(X_{ms}^2/\omega_e\right)G\left|\tilde{V}_{as}\right|^2}{\left[r_s + G\left(X_{ms}^2 - X_{ss}X_{rr}'\right)\right]^2 + \left(X_{ss} + G r_s X_{rr}'\right)^2} \tag{8.8-23}$$

Equation (8.8-23) is independent of r_r'. Thus, the maximum torque remains constant if only r_r' is varied; however, the slip at which maximum torque is produced varies in accordance with (8.8-21). Figure 8.8-4 illustrates the effect of changing r_r'. Therein $r_{r3}' > r_{r2}' > r_{r1}'$.

In variable-frequency drive systems, the operating speed of the electromechanical device (reluctance, synchronous, or induction machine) is controlled by changing the frequency of the applied voltages by either an inverter (solid-state

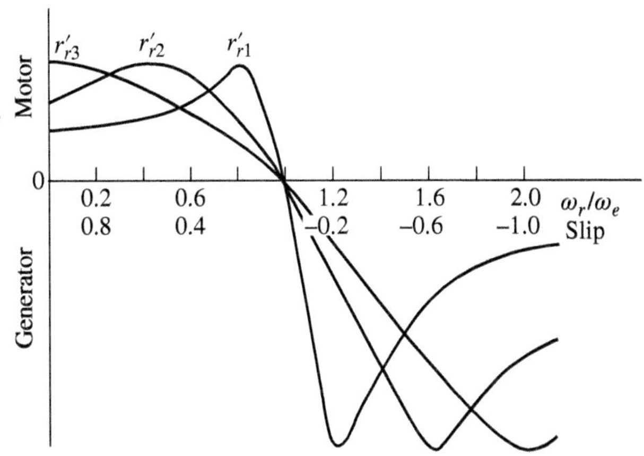

Figure 8.8-4 Steady-state torque-speed characteristics of a symmetrical induction machine for different values of r'_r.

dc-to-ac converter) or a cycloconverter (ac frequency changer). The phasor voltage equations are applicable regardless of the frequency of operation. It is only necessary to keep in mind that the reactances given in the steady-state equivalent circuit, Fig. 8.8-1, are defined as the product of ω_e and the inductances. As the frequency is decreased, the time rate of change of the steady-state variables is decreased proportionally. Thus, the inductive reactances decrease linearly with frequency. If the amplitude of the applied voltages is maintained at the rated value, the currents will become excessive. To prevent large currents, the magnitude of the stator voltages is decreased as the frequency is decreased. In many applications, the voltage magnitude is reduced linearly with frequency until a low frequency is reached, whereupon the decrease in voltage is programmed in a manner to compensate for the effects of the stator resistance.

The influence of frequency upon the steady-state torque-speed characteristics is illustrated in Fig. 8.8-5. These characteristics are for a linear relationship between the magnitude of the applied voltages and frequency. This machine is designed to operate at $\omega_e = \omega_b$, where ω_b corresponds to the rated frequency. Rated voltage is applied at rated frequency, that is, when $\omega_e = \omega_b$, $|\tilde{V}_{as}| = V_B$, where V_B is the base or rated voltage. Since the reactances $(\omega_e L)$ decrease with frequency, the voltage is reduced as frequency is reduced to avoid large stator currents. The maximum torque is reduced markedly at $\omega_e/\omega_b = 0.1$. At this frequency, the voltage would probably be increased somewhat so as to obtain a higher torque. Perhaps a voltage of, say, $0.15V_B$ or $0.2V_B$ would be used rather than $0.1V_B$. Saturation of the stator or rotor steel may, however, cause the stator currents to be excessive at this higher voltage. These practical considerations of variable-frequency drives are of major importance but well beyond the scope of this text. However, we will encounter variable frequency operation later in this chapter when we deal with field orientation of an induction machine.

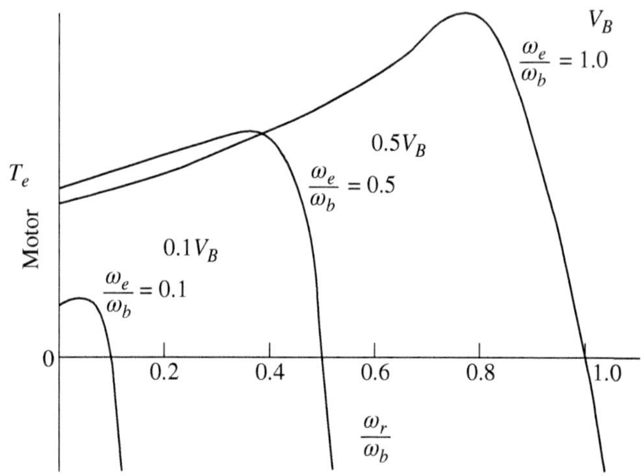

Figure 8.8-5 Steady-state torque-speed characteristics of a symmetrical induction machine for different operating frequencies.

Example 8D No-Load and Blocked-Rotor Tests

The parameters for the equivalent circuit shown in Fig. 8.8-1 may be calculated by using electromagnetic field theory or determined from tests. The tests generally performed are a dc test, a no-load test, and a blocked-rotor test. Test data are given for a 3-hp four-pole 110-V (rms) two-phase 60-Hz induction machine in Table 8D-1, where all ac voltages and currents are in rms values.

During the dc test, a dc voltage is applied to one phase while the machine is at standstill. Thus,

$$r_s = \frac{V_{dc}}{I_{dc}} = \frac{6.9}{13} = 0.531\ \Omega \tag{8D-1}$$

The no-load test, which is analogous to the transformer open-circuit test, is performed with balanced 60-Hz voltages applied to the stator windings without mechanical load on the rotor (no load). The total power input during this test is the sum of the stator ohmic losses per phase, the core losses due to hysteresis and eddy currents, and rotational losses due to friction and windage. The total stator ohmic losses (two phases) are

$$P_r = 2I_{\text{nl}}^2 r_s = 2(3.86)^2(0.531) = 15.8\ \text{W} \tag{8D-2}$$

Therefore, the power loss due to friction, windage, and core losses is

$$P_{\text{fWC}} = P_{\text{nl}} - P_r = 134 - 15.8 = 118.2\ \text{W} \tag{8D-3}$$

In the equivalent circuit shown in Fig. 8.8-1, the core loss is neglected. It is generally small and, in most cases, little error is introduced by neglecting it. It can be approximated by placing a resistor in shunt with the magnetizing reactance X_{ms}. The friction and windage losses may be approximated with B_m in (8.4-5).

TABLE 8D-1 Test Data to Determine Machine Parameters

DC Test	No-Load Test	Blocked-Rotor Test
$V_{dc} = 6.9$ V	$V_{\text{nl}} = 110$ V	$V_{\text{br}} = 23.5$ V
$I_{dc} = 13.0$ A	$I_{\text{nl}} = 3.86$ A	$I_{\text{br}} = 16.1$ A
	$P_{\text{nl}} = 134$ W	$P_{\text{br}} = 469$ W
	$f = 60$ Hz	$f = 15$ Hz

It is noted from the no-load test data that the power factor is very small since the total apparent two-phase power input to the motor is

$$|S_{\text{nl}}| = 2V_{\text{nl}}I_{\text{nl}} = 2(110)(3.86) = 849.2 \text{ VA} \tag{8D-4}$$

Therefore, the no-load impedance is highly inductive, and its magnitude is assumed to be the sum of the stator leakage reactance and the magnetizing reactance since the rotor speed is essentially synchronous, $(s \approx 0)$, whereupon r'_r/s is much larger than X_{ms} in Fig. 8.8-1. Thus,

$$X_{ls} + X_{ms} \cong \frac{V_{\text{nl}}}{I_{\text{nl}}} = \frac{110}{3.86} = 28.5\ \Omega \tag{8D-5}$$

During the blocked-rotor test, which is analogous to the transformer short-circuit test, the rotor is locked by some mechanical means and balanced two-phase stator voltages are applied. The frequency of the applied voltages is often less than rated in order to obtain a representative value of r'_r, since during normal operation the frequency of the rotor currents is low and the rotor resistance of some induction machines varies considerably with frequency. During stall $(s = 1)$, the rotor impedance $r'_r/s + jX'_{lr}$ is much smaller in magnitude than X_{ms}, whereupon the current flowing in the magnetizing reactance may be neglected in these calculations. Hence, the total two-phase power input to the motor during the blocked-rotor test is

$$P_{\text{br}} = 2I_{\text{br}}^2\left(r_s + r'_r\right) \tag{8D-6}$$

From which

$$r'_r = \frac{P_{\text{br}}}{2I_{\text{br}}^2} - r_s = \frac{469}{(2)(16.1)^2} - 0.531 = 0.374\ \Omega \tag{8D-7}$$

The magnitude of the blocked-rotor input impedance is

$$|Z_{\text{br}}| = \frac{V_{\text{br}}}{I_{\text{br}}} = \frac{23.5}{16.1} = 1.46\ \Omega \tag{8D-8}$$

Thus,

$$\left|\left(r_s + r'_r\right) + j\frac{15}{60}\left(X_{ls} + X'_{lr}\right)\right| = 1.46\ \Omega \tag{8D-9}$$

From which

$$\left[\frac{15}{60}\left(X_{ls}+X'_{lr}\right)\right]^2 = (1.46)^2-\left(r_s+r'_r\right)^2 \tag{8D-10}$$
$$= (1.46)^2-(0.531+0.374)^2 = 1.31\ \Omega$$

Thus,

$$X_{ls}+X'_{lr} = 4.58\ \Omega \tag{8D-11}$$

Generally, X_{ls} and X'_{lr} are assumed equal; however, in some types of induction machines a different ratio is suggested. We will assume $X_{ls} = X'_{lr}$, whereupon we have determined the machine parameters. In particular, for $\omega_e = 377$ rad/s, the parameters are $r_s = 0.531\ \Omega$, $X_{ls} = 2.29\ \Omega$, $X_{ms} = 26.2\ \Omega$, $r'_r = 0.374\ \Omega$, and $X'_{lr} = 2.29\ \Omega$.

Example 8E Starting Torque and Current Calculations

A four-pole 110-V (rms) 28-A 7.5-hp two-phase induction motor has the following parameters: $r_s = 0.3\ \Omega$, $L_{ls} = 0.0015$ H, $L_{ms} = 0.035$ H, $r'_r = 0.15\ \Omega$, and $L'_{lr} = 0.0007$ H. The machine is supplied from a 110-V 60-Hz source. Calculate the starting torque and starting current.

It would be necessary to use a computer to solve for the starting current and torque if the electrical and mechanical transients are to be considered. However, an approximation of the actual starting characteristics may be obtained from a constant-speed steady-state analysis. For this purpose, it is assumed that the speed is fixed at zero and the electric system has established steady-state operation.

$$X_{ss} = \omega_e(L_{ls}+L_{ms}) = 377(0.0015+0.035) = 13.76\ \Omega \tag{8E-1}$$

$$X'_{rr} = \omega_e\left(L'_{lr}+L_{ms}\right) = 377(0.0007+0.035) = 13.46\ \Omega \tag{8E-2}$$

$$X_{ms} = \omega_e L_{ms} = 377(0.035) = 13.2\ \Omega \tag{8E-3}$$

The steady-state torque with $\omega_r = 0$ ($s = 1$) may be calculated from (8.8-20).

$$\begin{aligned} T_e &= \frac{2(P/2)\left(X_{ms}^2/\omega_e\right)r'_r s\left|\tilde{V}_{as}\right|^2}{\left[r_s r'_r + s\left(X_{ms}^2 - X_{ss}X'_{rr}\right)\right]^2 + \left(r'_r X_{ss} + s r_s X'_{rr}\right)^2} \\ &= \frac{2\left(\frac{4}{2}\right)\left(13.2^2/377\right)(0.15)(1)(110)^2}{\left\{(0.3)(0.15)+(1)\left[13.2^2-(13.76)(13.46)\right]\right\}^2 + [(0.15)(13.76)+(1)(0.3)(13.46)]^2} \\ &= 21.4\ \text{N·m} \end{aligned} \tag{8E-4}$$

Since $s = 1$, the rotor impedance in parallel with X_{ms} is much smaller than X_{ms}. Thus, for this mode of operation the input impedance is approximately

$$\begin{aligned} Z &= (r_s + r_r') + j(X_{ls} + X_{lr}') \\ &= (0.3 + 0.15) + j377(0.0015 + 0.0007) \\ &= 0.45 + j0.83\ \Omega = 0.944\angle 61.5°\Omega \end{aligned} \tag{8E-5}$$

With $\widetilde{V}_{as}$ as the reference phasor,

$$\widetilde{I}_{as} = \frac{\widetilde{V}_{as}}{Z} = \frac{110\angle 0°}{0.944\angle 61.5°} = 117\angle -61.5°\text{A} \tag{8E-6}$$

The stall or starting current is over four times larger than the rated current. In some large machines, the starting current with rated voltage applied may be ten times the rated current. This high value of current may cause overheating and damage to the windings. Consequently, reduced voltage is applied to many large machines during the starting period, and rated voltage is not applied until the machine has accelerated to near-rated speed.

SP8.8-1 The rotor speed ω_{rm} of a six-pole two-phase induction motor is $0.3\omega_e$. Express (a) I'_{ar}, (b) I'^{s}_{qr}, and (c) $\widetilde{I}'_{ar}$ for balanced steady-state operation with $\widetilde{I}^{s}_{qr} = I'_r\underline{/30°}$. [(a) $I'_{ar} = \sqrt{2}\,I'_r \cos(0.1\omega_e t + 30°)$; (b) $I'^{s}_{qr} = \sqrt{2}\,I'_r \cos(\omega_e t + 30°)$; (c) $\widetilde{I}'_{ar} = \widetilde{I}'^{s}_{qr}$]

SP8.8-2 Neglecting the current flowing in X_{ms} is generally an acceptable approximation when calculating the machine parameters from the blocked-rotor test (Example 8D). However, this approximation is not valid when calculating the blocked-rotor torque. Use (8.8-9) and let $\widetilde{I}_{as} = a + jb$ to show that T_e is zero regardless of the rotor speed if the current flowing in X_{ms} is assumed to be negligibly small. Use (8.8-12) and repeat. [Real parts cancel; $\theta_{esi}(0) = \theta_{eri}(0) + \pi; T_e = 0$]

SP8.8-3 Neglect the core losses and assume that the friction and windage losses P_{fWC}, calculated in Example 8D, are to be represented by $B_m\omega_{rm}$, with B_m selected so that a load equivalent to 118.2 W occurs at $\omega_r = 0.9\omega_e$. Determine B_m. $[B_m = 4.11 \times 10^{-3}\ \text{N} \cdot \text{m} \cdot \text{s/rad}]$

SP8.8-4 The parameters of a 60-Hz four-pole two-phase induction motor are $r_s = r_r' = 20\ \Omega$, $L_{ls} = L_{lr}' = 25$ mH, and $L_{ms} = 0.3$ H. Calculate $\widetilde{I}_{as}$ for (a) no-load conditions and (b) approximate the blocked-rotor current $(\omega_r = 0)$, with $\widetilde{V}_{as} = 115\underline{/0°}$. [(a) $\widetilde{I}_{as} = 0.927\underline{/-80.75°}$ A; (b) $\widetilde{I}_{as} = 2.6\underline{/-25.2°}$ A]

8.9 DYNAMIC AND STEADY-STATE PERFORMANCE – MACHINE VARIABLES

It is instructive to observe the machine variables during transient and steady-state operation. For this purpose, the nonlinear differential equations which describe the induction machine have been programmed on a computer. Two induction machines are considered; a 5-hp general-purpose machine with characteristics typical of many two- and three-phase squirrel-cage industrial-type induction motors and a $\frac{1}{10}$ – hp two-phase induction motor with high-resistance rotor windings to produce a relatively large starting torque. The majority of the performance characteristics given in this and the following section are for the single-fed two-pole two-phase 5-hp 110-V (rms) 60-Hz induction machine with the following parameters: $r_s = 0.295\ \Omega$, $L_{ls} = 0.944$ mH, $L_{ms} = 35.15$ mH, $r'_r = 0.201\ \Omega$, and $L'_{lr} = 0.944$ mH. The inertia of the rotor and connected mechanical load is $J = 0.026$ kg · m^2. The four-pole two-phase $\frac{1}{10}$ – hp 115-V(rms) 60-Hz induction motor has the following parameters: $r_s = 24.5\ \Omega$, $L_{ls} = 27.06$ mH, $L_{ms} =$ 273.7 mH, $r'_r = 23\ \Omega$, and $L'_{lr} = 27.06$ mH. Friction and windage losses are neglected. The inertia of the rotor and connected mechanical load is $J = 1 \times 10^{-3}$ kg·m^2. This device is shown in Fig. 8.2-3.

Free Acceleration from Stall

The free-acceleration characteristics are depicted in Figs. 8.9-1 and 8.9-2 for the 5-hp machine and in Figs. 8.9-3 and 8.9-4 for the $\frac{1}{10}$ – hp motor. At $t = 0$, rated voltages are applied of the form $v_{as} = \sqrt{2}\, V_s \cos 377t$ and $v_{bs} = \sqrt{2} V_s \sin 377t$, where $V_s = 110$ V for the 5-hp machine and $V_s = 115$ V for the $\frac{1}{10}$ – hp motor. The machines accelerate from stall with zero load torque and, since friction and windage losses are not taken into account, the simulated machine accelerates to synchronous speed. In practice, friction and windage losses would exist and the machines would not reach synchronous speed. Instead, they would operate at a speed slightly less than synchronous, developing an electromagnetic torque T_e sufficient to satisfy the small torque load due to friction and windage. For the two-pole 5-hp machine, synchronous speed is 3600 r/min, where $\omega_{rm} = 377$ rad/s. For the four-pole $\frac{1}{10}$ – hp motor, synchronous speed is 1800 r/min, where $\omega_{rm} = 188.5$ rad/s. In both cases, the electrical angular velocity of the rotors, ω_r, is 377 rad/s at synchronous speed.

We will find that rated current (rms) for the 5-hp and the $\frac{1}{10}$ – hp machines, respectively, are on the order of 20 and 1.5 A. The starting current for the 5-hp machine (Fig. 8.9-1) is approximately 120 A (rms), or, on the order of six times the rated current. This ratio is typical of industrial-type induction motors which, depending upon the load to be accelerated from stall, may require reduced-voltage starting, as mentioned in Example 8E. The starting current for the $\frac{1}{10}$ – hp motor (Fig. 8.9-3) is about 2 A (rms), or less than twice the rated current. These devices are designed to withstand full-voltage starting since they are often used in systems requiring rapid, short-duration acceleration from stall and, depending upon the application, may seldom reach rated speed.

The transient dc offset of the stator and rotor currents, characteristic of rL circuits, is evident in Figs. 8.9-1 and 8.9-3. This offset in the stator and rotor currents gives rise to the transient pulsation in the electromagnetic torque. Note that the pulsation in torque, which is at the frequency of the stator source voltages (60 Hz), disappears when the transient offsets in the stator and rotor currents disappear. Also apparent in the stator and rotor currents, especially in the case of the 5-hp machine, is the variation in the envelope of the currents during the

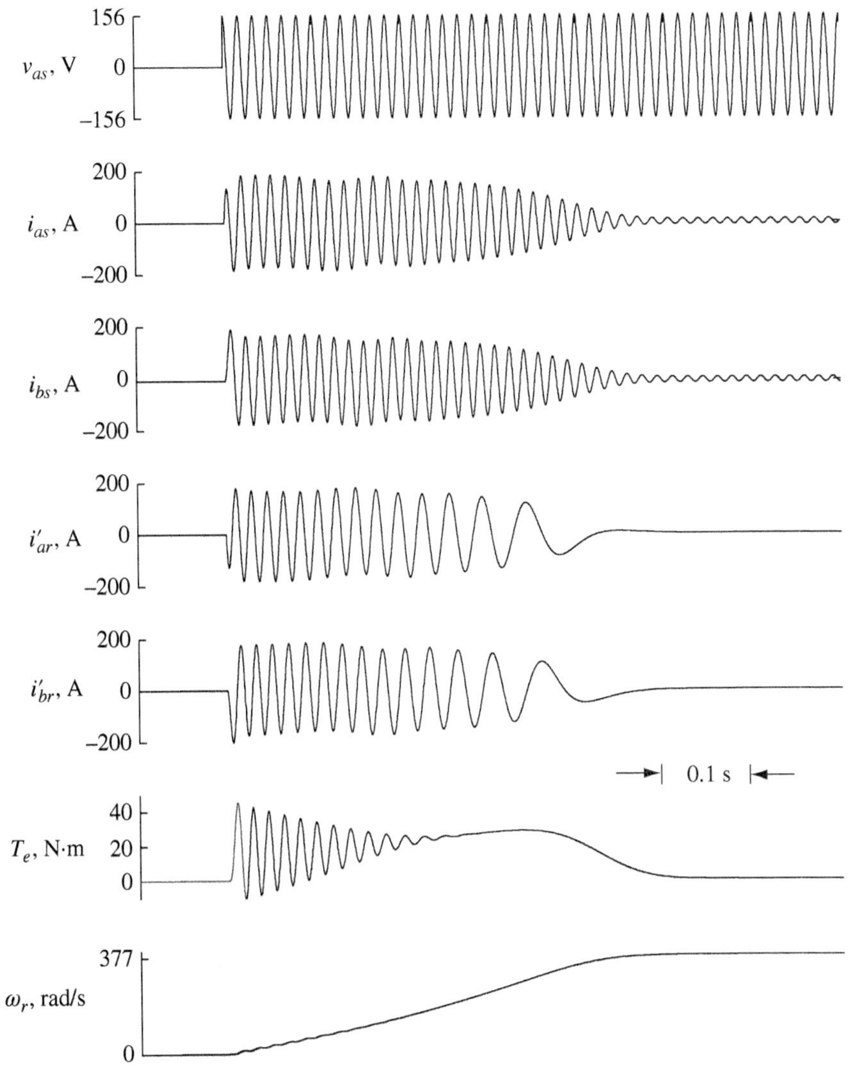

Figure 8.9-1 Free-acceleration characteristics of a two-pole two-phase 5-hp induction motor – machine variables.

transient period. This is caused by the interaction of the transient offset in the stator currents with the transient offset in the rotor currents and the fact that the stator and rotor circuits are in relative motion [1].

For small-horsepower induction machines, the steady-state torque-speed characteristic is essentially the average of the transient torque–speed curve during the time the transient pulsating torque exists. The inertia of the rotor and connected load is generally large enough to prevent this pulsating torque from causing

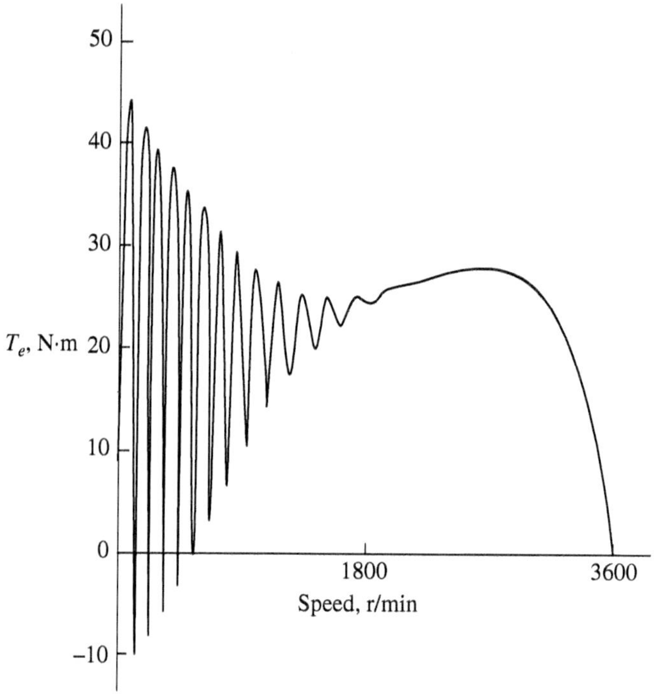

Figure 8.9-2 Torque versus speed during the free-acceleration shown in Fig. 8.9-1.

significant variations in rotor speed. This is not the case, however, if the induction machine is operated with stator voltages of low frequencies as may occur in variable-frequency drive systems.

Acceleration from Stall with a Load Torque

The acceleration characteristics of the 5-hp motor with a load torque are shown in Figs. 8.9-5 and 8.9-6. Here, the load torque is $T_L = K\omega_{rm}^2$ where K is selected as $10(377)^{-2}$ N · m · s^2/rad^2. A load torque characteristic of this type is typical of a fan load. Rated torque load is approximately 10 N · m. The constant K is selected so that a load torque of 10 N · m occurs at synchronous speed. In Fig. 8.9-6, the difference between the average value of the electromagnetic torque T_e and the load torque T_L is the torque that accelerates the rotor, commonly referred to as the accelerating torque. Acceleration occurs until $T_e = T_L$.

Step Changes in Load Torque

The dynamic performance of the 5-hp motor during load torque changes is shown in Fig. 8.9-7. Initially, the machine is operating with a constant load torque

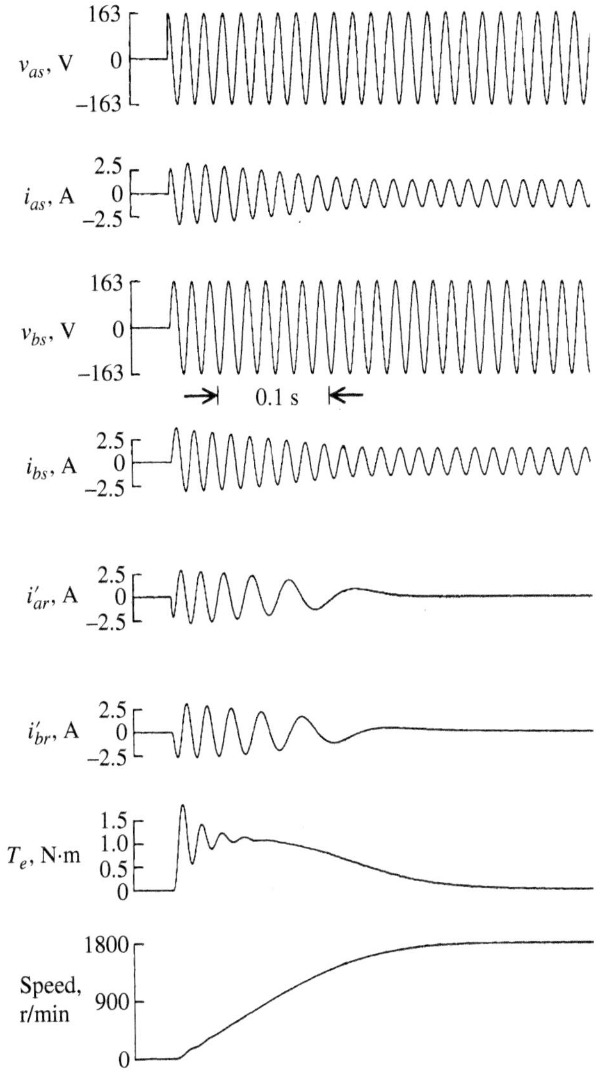

Figure 8.9-3 Free-acceleration characteristics of a four-pole two-phase 1/10-hp induction motor – machine variables.

of 5 N · m. The load torque is stepped to 10 N · m and the machine is allowed to reach steady-state operation, whereupon the load torque is stepped back to 5 N·m. You should be able to calculate the rotor speed for $T_L = 5$ N·m and $T_L = 10$ N·m from the frequency of the rotor currents shown in Fig. 8.9-7. For $T_L = 5$ N·m, $\omega_r \cong 370$ rad/s; for $T_L = 10$ N·m, $\omega_r \cong 362$ rad/s.

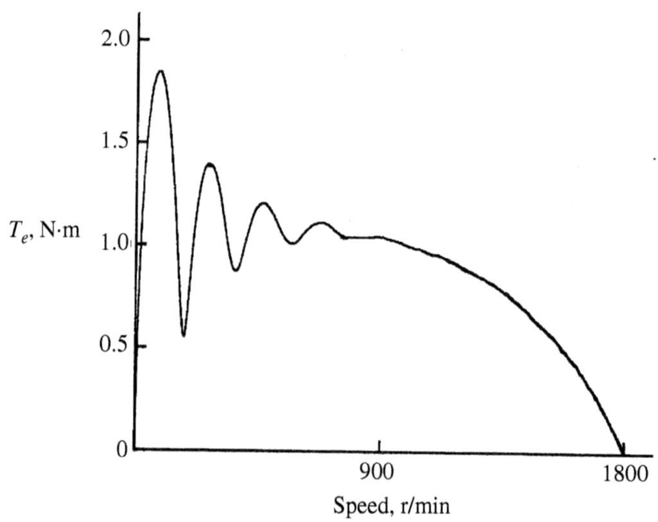

Figure 8.9-4 Torque versus speed during the free acceleration shown in Fig. 8.9-3.

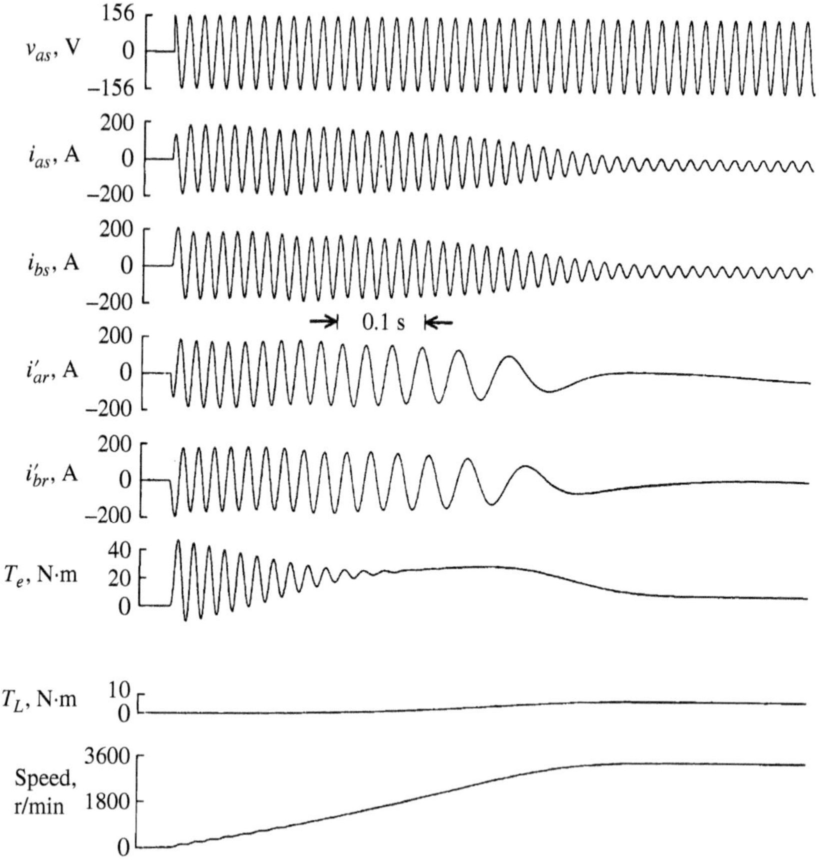

Figure 8.9-5 Acceleration from stall of a 5-hp induction motor with $T_L = K\omega_{rm}^2$, where $K = 10(377)^{-2}$ N·m·s^2/rad^2; 3600 r/min = 377 rad/s.

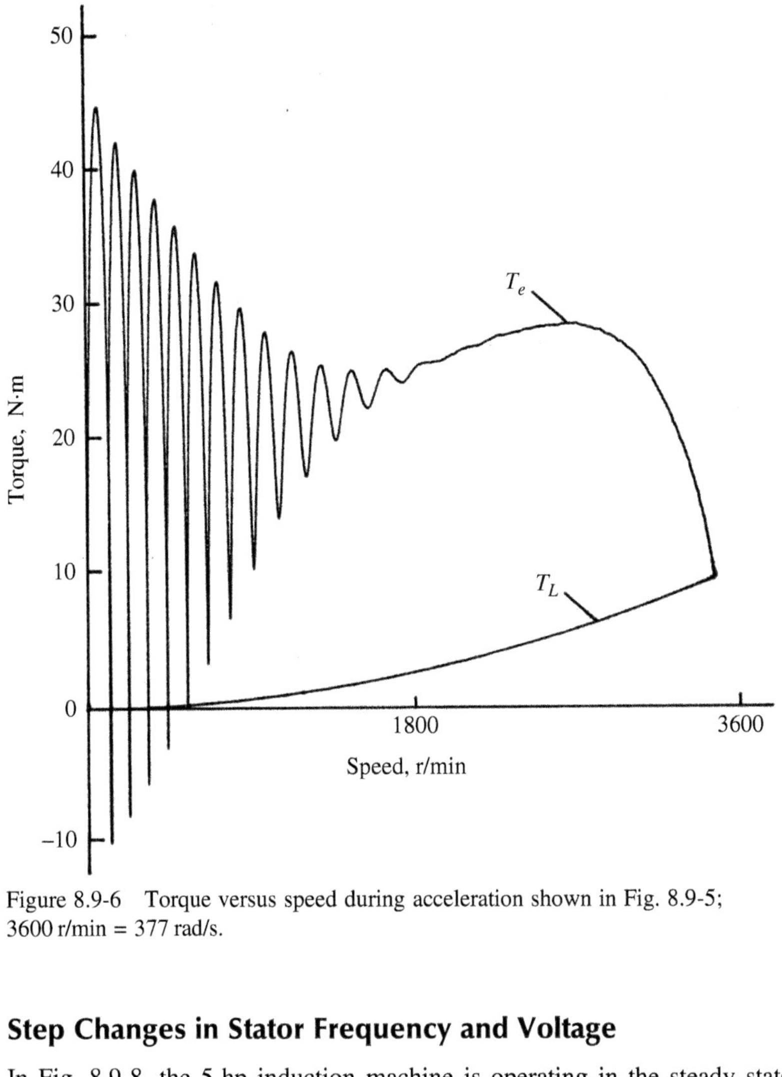

Figure 8.9-6 Torque versus speed during acceleration shown in Fig. 8.9-5; 3600 r/min = 377 rad/s.

Step Changes in Stator Frequency and Voltage

In Fig. 8.9-8, the 5-hp induction machine is operating in the steady state with $T_L = 10\,\mathrm{N\cdot m}$. The frequency of the stator voltages is stepped down to 50 Hz, and, at the same time, the amplitude of the stator voltages is decreased to $\left(\frac{5}{6}\right)(110)$ V. Once the system has reached steady-state operation, the frequency and amplitude of the stator voltages are stepped back to rated (60 Hz and 110 V). The load torque is held fixed at $10\,\mathrm{N\cdot m}$ during this change in stator frequency. This type of operation is possible when the induction machine is supplied from a variable-frequency inverter where the frequency and amplitude of the fundamental component of the stator voltages may be rapidly changed by controlling the switching of the inverter, which converts a dc voltage to a variable-frequency ac voltage.

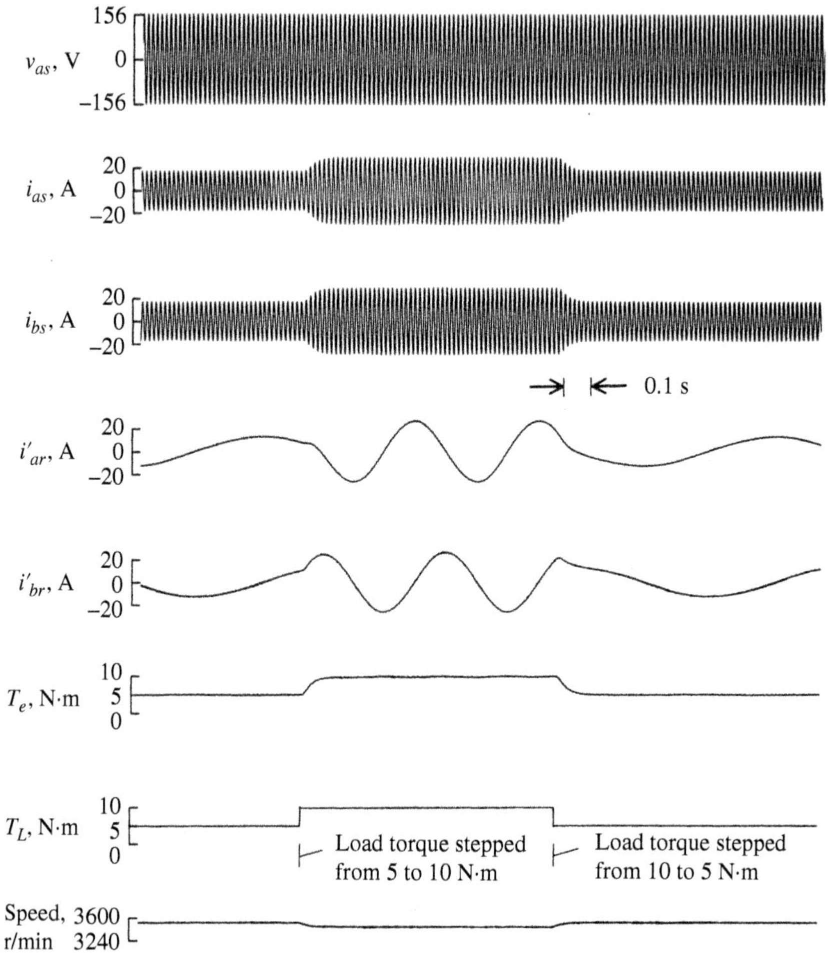

Figure 8.9-7 Step changes in load torque of a 5-hp induction motor.

The dynamic torque-speed characteristics during these step changes in frequency and voltage are shown in Fig. 8.9-9. Therein, the steady-state torque-speed characteristics for 50- and 60-Hz operations are also shown. Initially, operation is at Point 1. When the frequency and voltage amplitude are changed, the instantaneous torque decreases, whereupon the rotor slows down and steady-state operation finally occurs at Point 2. When the stator voltages are stepped back to rated conditions, the instantaneous torque increases, the rotor accelerates, and the original operating condition is re-established.

SP8.9-1 Approximate the accelerating torque from Fig. 8.9-6 when the rotor speed is 280 rad/s. [≅23 N·m]

SP8.9-2 In going from Point 1 to Point 2 in Fig. 8.9-9, the induction machine acts as a generator supplying energy to the electric system. What is the source of this energy? [change in rotor speed and applied voltage]

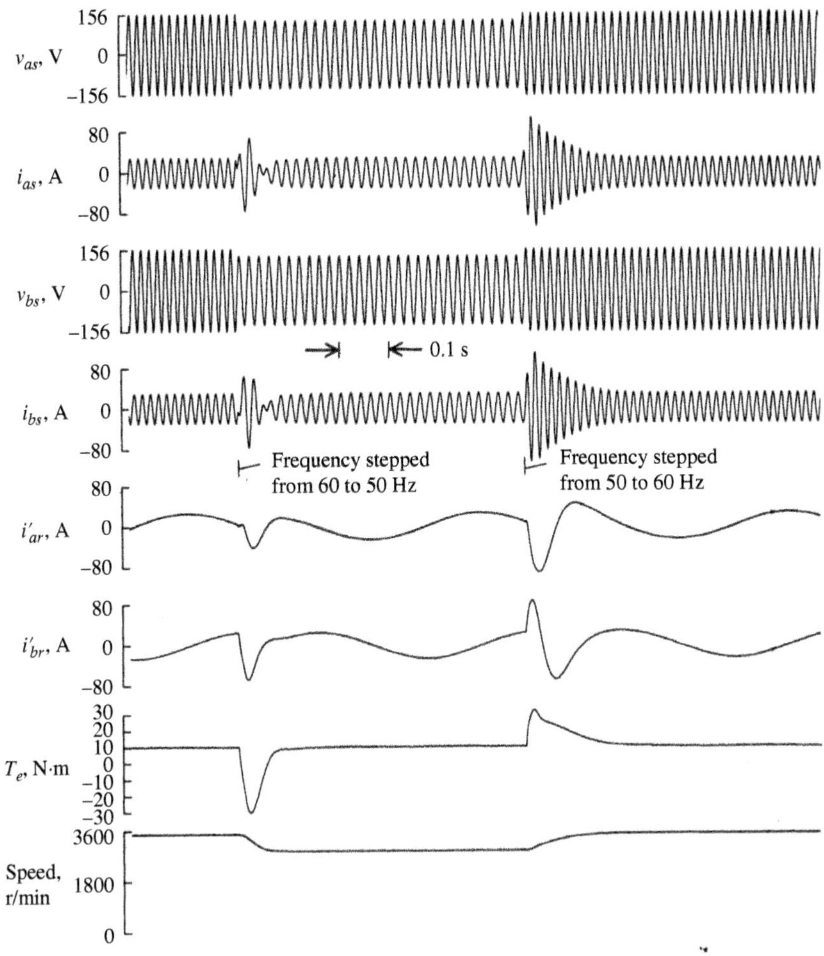

Figure 8.9-8 Step changes in the frequency and magnitude of the stator voltages of the 5-hp induction motor.

8.10 FREE ACCELERATION VIEWED FROM STATIONARY, ROTOR, AND SYNCHRONOUSLY ROTATING REFERENCE FRAMES

It is instructive to view the free acceleration of the 5-hp two-phase induction motor in the reference frames commonly used for induction machine analysis and control; however, if you have not studied reference frame theory, you will likely omit this section.

Free acceleration portrayed in stationary reference frame variables is shown in Fig. 8.10-1; rotor reference frame variables in Fig. 8.10-2; and synchronously rotating reference frame variables in Fig. 8.10-3. Each of these traces should be

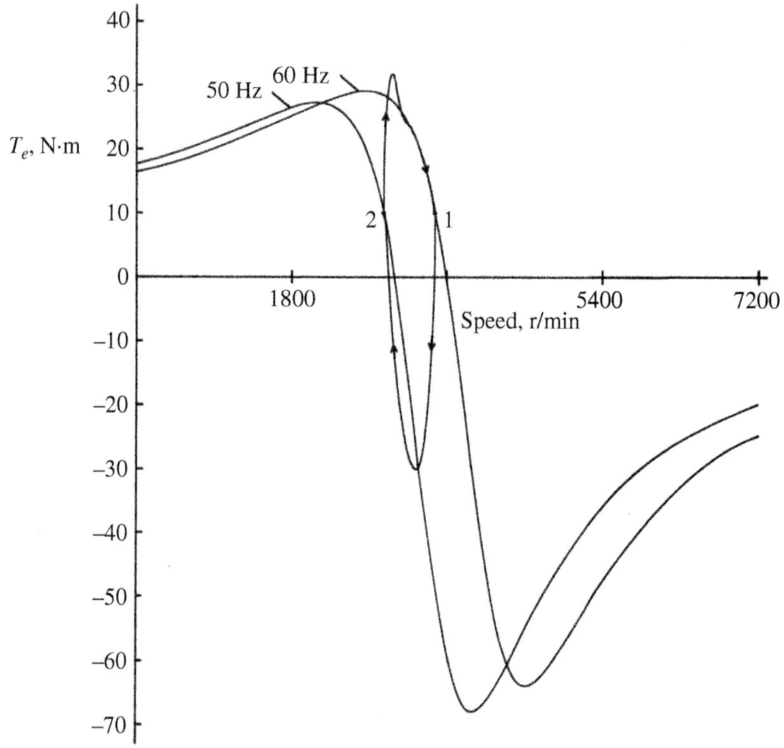

Figure 8.9-9 Torque versus speed for step changes in the stator frequency and voltage shown in Fig. 8.9-8.

compared to Fig. 8.9-1, which depicts the same free acceleration in machine (actual) variables.

It is very important to note that all steady-state voltages and currents are of the same frequency for a specific reference frame; ω_e for the stationary, $\omega_e - \omega_r$ for the rotor, and constant for the synchronously rotating. We established this in Chapter 4 with equations (4.5-12) through (4.5-17). This makes sense since as we pointed out in Chapter 5, the change of variables provides us with substitute variables that portray mmf$_s$ and mmf$_r$ as viewed from a specific reference frame. In particular, the change of variables transforms all variables not physically in the selected reference to that reference frame. The rotor variables to the stationary in Fig. 8.10-1, the stator to the rotor in Fig. 8.10-2, and both the stator and the rotor to the synchronously rotating in Fig. 8.10-3; thereby providing a correct view of the mmfs. Clearly, reference frame theory is a result of Tesla's rotating magnetic field.

We see that mmf$_s$ and mmf$_r$ rotate in unison regardless of the speed of the rotor, except when $\omega_r = \omega_e$, in which case no currents are induced in the rotor

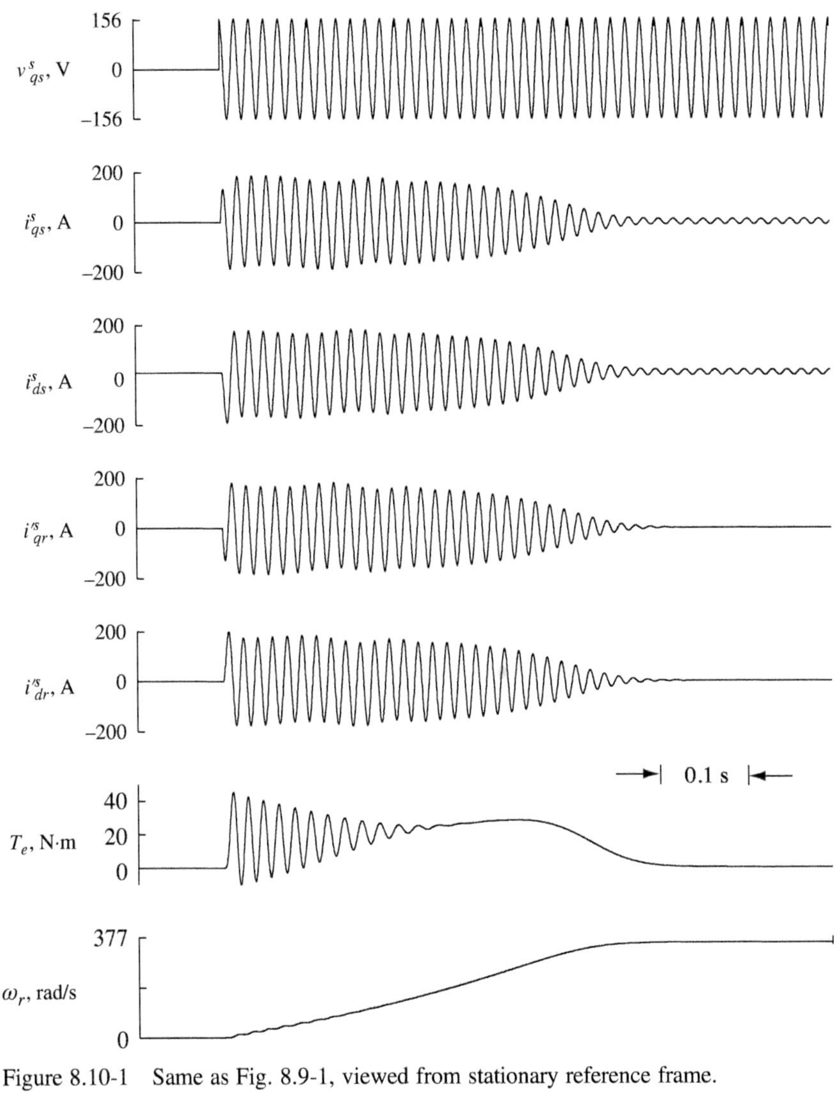

Figure 8.10-1 Same as Fig. 8.9-1, viewed from stationary reference frame.

circuits. However, it is important to realize that the reference frame from which we view operation is not in any way dependent on the speed of the rotor.

It is interesting to take a moment to consider Fig. 8.10-3. The torque pulsates negatively in the early part of the acceleration from stall. This would indicate brief periods of generator action. Recall from Example 8C and, in particular, Fig. 8C-3 where steady-state generator action is depicted. Therein, $\tilde{I}_{as}$ is in the third quadrant

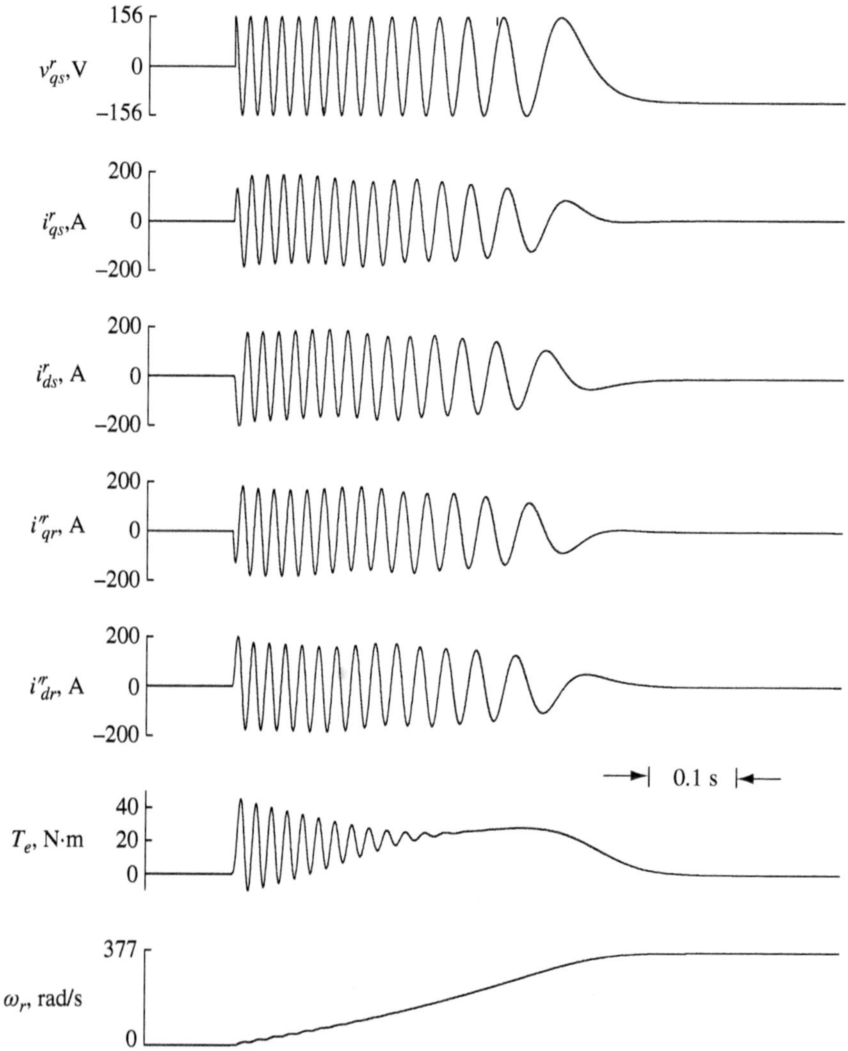

Figure 8.10-2 Same as Fig. 8.9-1, viewed from rotor reference frame.

with a negative real part (negative I_{qs}^e). In the case of the free acceleration depicted in Fig. 8.10-3, i_{qs}^e and i_{ds}^e are positive and i'^e_{qr} and i'^e_{dr} are negative during all of the acceleration period. Now, from (8.7-1),

$$T_e = \frac{P}{2} L_{ms} \left(i_{qs} i'_{dr} - i_{ds} i'_{qr} \right) \tag{8.10-1}$$

which is valid in all reference frames. We see that since i_{ds} is always positive and i'_{qr} is always negative, the product of i_{ds} and i'_{qr} will always be negative, thus

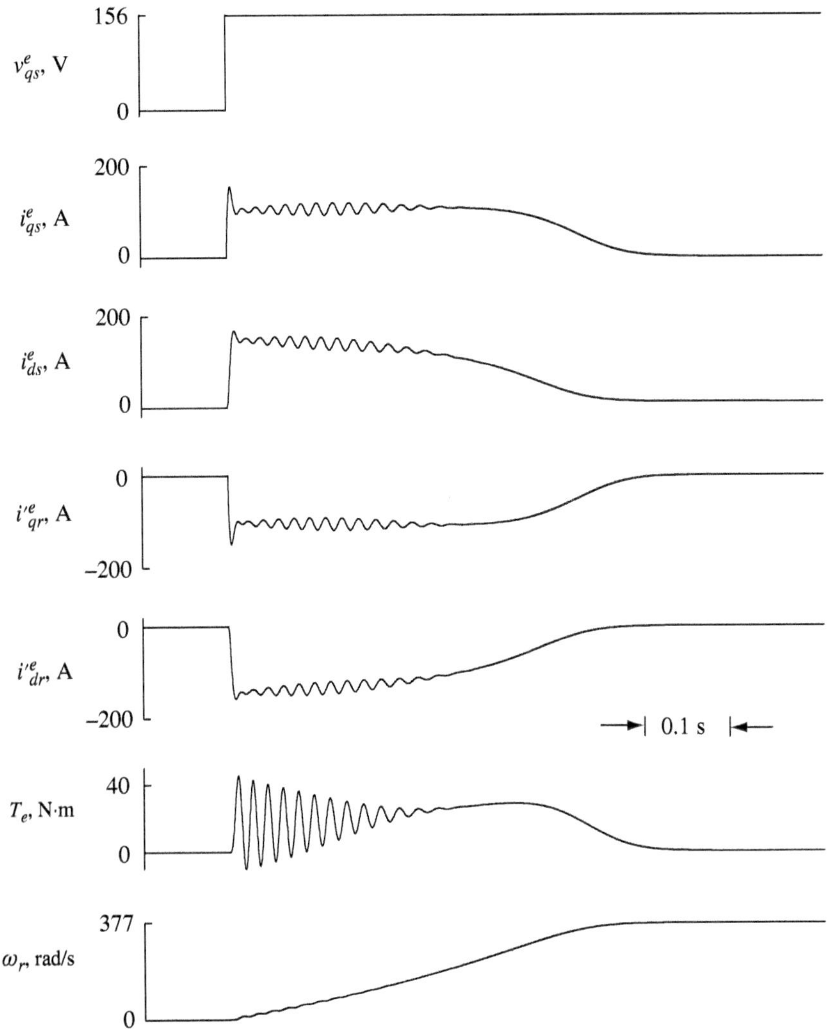

Figure 8.10-3 Same as Fig. 8.9-1, viewed from synchronously rotating reference frame.

producing a positive component of motor torque. Therefore, for generator action $(-T_e)$, the magnitude of $i_{qs}i'_{dr}$, which is negative, must be larger than the magnitude of $i_{ds}i'_{qr}$.

SP8.10-1 Determine the steady-state frequency of (a) i'_{ar} and i'_{br} in Fig. 8.9-1, (b) i'^s_{qr} and i'^s_{dr} in Fig. 8.10-1, (c) i'^r_{qr} and i'^r_{dr} in Fig. 8.10-2, and (d) i'^e_{qr} and i'^e_{dr} in Fig. 8.10-3. [(a) $\omega_e - \omega_r$, (b) ω_e, (c) $\omega_e - \omega_r$, (d) zero]

SP8.10-2 Repeat SP8.10-1 for the stator-related variables. [(a) ω_e, (b) ω_e, (c) $\omega_e - \omega_r$, (d) zero]

8.11 THREE-PHASE INDUCTION MACHINE

A two-pole three-phase symmetrical induction machine is shown in Fig. 8.11-1. The three-phase machine has three identical, sinusoidally distributed stator (rotor) windings with the magnetic axes displaced 120° from each other. For the most part, the analysis of a three-phase machine is a direct extension of that of a two-phase machine. In fact, once we are acquainted with the two-phase symmetrical induction machine, the analysis of the three-phase symmetrical induction machine is straightforward and is more involved only because of the trigonometry necessary to deal with three-phase variables. Perhaps the only aspect which may cause some study is the addition of the third substitute variable in the transformation; however, these so-called zero variables or zero quantities play no role in the analysis of balanced

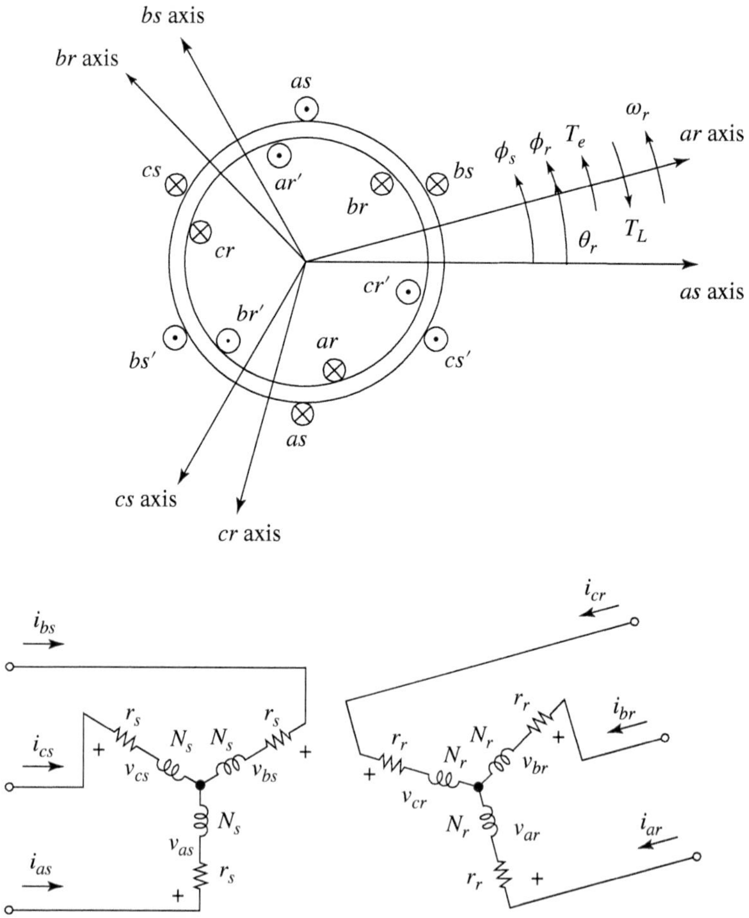

Figure 8.11-1 A two-pole three-phase symmetrical induction machine.

operation of three-phase symmetrical induction machines. We will not consider unbalanced operation; a detailed analysis of unbalanced operation is found in [1].

Voltage Equations and Winding Inductances

For the three-phase symmetrical induction machine,

$$v_{as} = r_s i_{as} + \frac{d\lambda_{as}}{dt} \tag{8.11-1}$$

$$v_{bs} = r_s i_{bs} + \frac{d\lambda_{bs}}{dt} \tag{8.11-2}$$

$$v_{cs} = r_s i_{cs} + \frac{d\lambda_{cs}}{dt} \tag{8.11-3}$$

$$v_{ar} = r_r i_{ar} + \frac{d\lambda_{ar}}{dt} \tag{8.11-4}$$

$$v_{br} = r_r i_{br} + \frac{d\lambda_{br}}{dt} \tag{8.11-5}$$

$$v_{cr} = r_r i_{cr} + \frac{d\lambda_{cr}}{dt} \tag{8.11-6}$$

where $r_s(r_r)$ is the resistance of each stator (rotor) winding. In matrix form,

$$\mathbf{v}_{abcs} = \mathbf{r}_s \mathbf{i}_{abcs} + p\boldsymbol{\lambda}_{abcs} \tag{8.11-7}$$

$$\mathbf{v}_{abcr} = \mathbf{r}_r \mathbf{i}_{abcr} + p\boldsymbol{\lambda}_{abcr} \tag{8.11-8}$$

where

$$(\mathbf{f}_{abcs})^T = [f_{as} \;\; f_{bs} \;\; f_{cs}] \tag{8.11-9}$$

$$(\mathbf{f}_{abcr})^T = [f_{ar} \;\; f_{br} \;\; f_{cr}] \tag{8.11-10}$$

and

$$\mathbf{r}_s = r_s \mathbf{I} \tag{8.11-11}$$

$$\mathbf{r}_r = r_r \mathbf{I} \tag{8.11-12}$$

As in the two-phase case, $\mathbf{f}$ can represent voltage, current, or flux linkage, the superscript T denotes the transpose, p is d/dt, and $\mathbf{I}$ a 3×3 identity matrix.

The flux-linkage equations may be written as

$$\begin{bmatrix} \boldsymbol{\lambda}_{abcs} \\ \boldsymbol{\lambda}_{abcr} \end{bmatrix} = \begin{bmatrix} \mathbf{L}_s & \mathbf{L}_{sr} \\ (\mathbf{L}_{sr})^T & \mathbf{L}_r \end{bmatrix} \begin{bmatrix} \mathbf{i}_{abcs} \\ \mathbf{i}_{abcr} \end{bmatrix} \tag{8.11-13}$$

Although the same notation is used for the inductance matrices as for the two-phase case, they are not the same. The self-inductances are all constant and can be expressed as in the case of the two-phase machine; however, mutual coupling exists between stator (rotor) phases. To determine this mutual inductance, let us consider the coupling between the *as* and *bs* windings, and let us imagine that we can rotate the *bs* winding 120° in the clockwise direction. If this were possible, then the *as* and *bs* windings would be on top of each other and, thus, tightly coupled with a mutual inductance of

$$L = L_{ms} \tag{8.11-14}$$

If now we rotate the *bs* winding back to its original position and if we assume that the mutual inductance varies in proportion to $\cos\phi_s$ as the *bs* winding is rotated, then

$$L_{asbs} = L_{ms}\cos\phi_s\big|_{\phi_s = 2\pi/3} = -\frac{1}{2}L_{ms} \tag{8.11-15}$$

It follows that

$$\mathbf{L}_s = \begin{bmatrix} L_{ss} & -\frac{1}{2}L_{ms} & -\frac{1}{2}L_{ms} \\ -\frac{1}{2}L_{ms} & L_{ss} & -\frac{1}{2}L_{ms} \\ -\frac{1}{2}L_{ms} & -\frac{1}{2}L_{ms} & L_{ss} \end{bmatrix} \tag{8.11-16}$$

$$\mathbf{L}_r = \begin{bmatrix} L_{rr} & -\frac{1}{2}L_{mr} & -\frac{1}{2}L_{mr} \\ -\frac{1}{2}L_{mr} & L_{rr} & -\frac{1}{2}L_{mr} \\ -\frac{1}{2}L_{mr} & -\frac{1}{2}L_{mr} & L_{rr} \end{bmatrix} \tag{8.11-17}$$

where $L_{ss} = L_{ls} + L_{ms}$ and $L_{rr} = L_{lr} + L_{mr}$. Also,

$$\mathbf{L}_{sr} = L_{sr}\begin{bmatrix} \cos\theta_r & \cos\left(\theta_r + \frac{2}{3}\pi\right) & \cos\left(\theta_r - \frac{2}{3}\pi\right) \\ \cos\left(\theta_r - \frac{2}{3}\pi\right) & \cos\theta_r & \cos\left(\theta_r + \frac{2}{3}\pi\right) \\ \cos\left(\theta_r + \frac{2}{3}\pi\right) & \cos\left(\theta_r - \frac{2}{3}\pi\right) & \cos\theta_r \end{bmatrix} \tag{8.11-18}$$

where

$$L_{sr} = \frac{N_s N_r}{\mathcal{R}_m} \tag{8.11-19}$$

All rotor variables may be referred to the stator windings by the following turns-ratios:

$$\mathbf{i}'_{abcr} = \frac{N_r}{N_s}\mathbf{i}_{abcr} \tag{8.11-20}$$

$$\mathbf{v}'_{abcr} = \frac{N_s}{N_r}\mathbf{v}_{abcr} \tag{8.11-21}$$

$$\boldsymbol{\lambda}'_{abcr} = \frac{N_s}{N_r}\boldsymbol{\lambda}_{abcr} \tag{8.11-22}$$

Thus, (8.11-7) and (8.11-8) become

$$\mathbf{v}_{abcs} = \mathbf{r}_s\mathbf{i}_{abcs} + p\boldsymbol{\lambda}_{abcs} \tag{8.11-23}$$

$$\mathbf{v}'_{abcr} = \mathbf{r}'_r\mathbf{i}'_{abcr} + p\boldsymbol{\lambda}'_{abcr} \tag{8.11-24}$$

where

$$\mathbf{r}'_r = \left(\frac{N_s}{N_r}\right)^2\mathbf{r}_r \tag{8.11-25}$$

The flux-linkage equations may now be written as

$$\begin{bmatrix}\boldsymbol{\lambda}_{abcs}\\ \boldsymbol{\lambda}'_{abcr}\end{bmatrix} = \begin{bmatrix}\mathbf{L}_s & \mathbf{L}'_{sr}\\ (\mathbf{L}'_{sr})^T & \mathbf{L}'_r\end{bmatrix}\begin{bmatrix}\mathbf{i}_{abcs}\\ \mathbf{i}'_{abcr}\end{bmatrix} \tag{8.11-26}$$

where, by definition,

$$\mathbf{L}'_{sr} = \frac{N_s}{N_r}\mathbf{L}_{sr} = \frac{L_{ms}}{L_{sr}}\mathbf{L}_{sr} \tag{8.11-27}$$

and

$$\mathbf{L}'_r = \begin{bmatrix} L'_{lr} + L_{ms} & -\frac{1}{2}L_{ms} & -\frac{1}{2}L_{ms}\\ -\frac{1}{2}L_{ms} & L'_{lr} + L_{ms} & -\frac{1}{2}L_{ms}\\ -\frac{1}{2}L_{ms} & -\frac{1}{2}L_{ms} & L'_{lr} + L_{ms}\end{bmatrix} \tag{8.11-28}$$

In (8.11-28),

$$L'_{lr} = \left(\frac{N_s}{N_r}\right)^2 L_{lr} \tag{8.11-29}$$

Torque

The electromagnetic torque, positive for motor action, may be expressed by using the second entry in Table 2.5-1. In particular,

$$\begin{aligned} T_e = -\frac{P}{2}L_{ms}\Bigg\{&\left[i_{as}\left(i'_{ar} - \frac{1}{2}i'_{br} - \frac{1}{2}i'_{cr}\right) + i_{bs}\left(i'_{br} - \frac{1}{2}i'_{ar} - \frac{1}{2}i'_{cr}\right) + i_{cs}\left(i'_{cr} - \frac{1}{2}i'_{br} - \frac{1}{2}i'_{ar}\right)\right]\sin\theta_r\\ &+ \frac{\sqrt{3}}{2}\left[i_{as}\left(i'_{br} - i'_{cr}\right) + i_{bs}\left(i'_{cr} - i'_{ar}\right) + i_{cs}\left(i'_{ar} - i'_{br}\right)\right]\cos\theta_r\Bigg\} \end{aligned} \tag{8.11-30}$$

The torque and rotor speeds are related by (8.4-5), which is repeated for convenience,

$$T_e = J\left(\frac{2}{P}\right)\frac{d\omega_r}{dt} + B_m\left(\frac{2}{P}\right)\omega_r + T_L \tag{8.11-31}$$

where J is the inertia of the rotor and, in some cases, the connected load. The units of J and the damping coefficient B_m are discussed following (8.4-5). The load torque T_L is positive for a torque load on the shaft of the induction machine, as indicated in Fig. 8.11-1.

Voltage Equations in the Arbitrary Reference Frame

It may be advantageous to review Sections 5.7 and 5.8 before starting this section. A change of variables that formulates a transformation of the three-phase variables of stationary (stator) circuits to the arbitrary reference frame may by expressed as

$$\mathbf{f}_{qd0s} = \mathbf{K}_s\mathbf{f}_{abcs} \tag{8.11-32}$$

where

$$(\mathbf{f}_{qd0s})^T = [f_{qs}\ f_{ds}\ f_{0s}] \tag{8.11-33}$$

$$(\mathbf{f}_{abcs})^T = [f_{as}\ f_{bs}\ f_{cs}] \tag{8.11-34}$$

$$\mathbf{K}_s = \frac{2}{3}\begin{bmatrix} \cos\theta & \cos\left(\theta - \frac{2}{3}\pi\right) & \cos\left(\theta + \frac{2}{3}\pi\right) \\ \sin\theta & \sin\left(\theta - \frac{2}{3}\pi\right) & \sin\left(\theta + \frac{2}{3}\pi\right) \\ \frac{1}{2} & \frac{1}{2} & \frac{1}{2} \end{bmatrix} \tag{8.11-35}$$

where the $\frac{2}{3}$ factor was introduced by Park [4].

$$\frac{d\theta}{dt} = \omega(t) \tag{8.11-36}$$

It can be shown that the inverse is

$$(\mathbf{K}_s)^{-1} = \begin{bmatrix} \cos\theta & \sin\theta & 1 \\ \cos\left(\theta - \frac{2}{3}\pi\right) & \sin\left(\theta - \frac{2}{3}\pi\right) & 1 \\ \cos\left(\theta + \frac{2}{3}\pi\right) & \sin\left(\theta + \frac{2}{3}\pi\right) & 1 \end{bmatrix} \tag{8.11-37}$$

In the above equations, f can represent either voltage, current, flux linkage, or electric charge. The superscript T denotes the transpose of a matrix. The s subscript indicates the variables, parameters, and transformation associated with stationary circuits. The angular position θ must be continuous; however, the angular velocity associated with the change of variables is unspecified. The frame of reference may rotate at any constant or varying angular velocity or it may remain stationary.

A change of variables which formulates a transformation of the three-phase variables of the rotor circuits to the arbitrary reference frame is

$$\mathbf{f}'_{qd0r} = \mathbf{K}_r \mathbf{f}'_{abcr} \tag{8.11-38}$$

where

$$\left(\mathbf{f}'_{qd0r}\right)^T = \left[f'_{qr} \;\; f'_{dr} \;\; f'_{0r}\right] \tag{8.11-39}$$

$$\left(\mathbf{f}'_{abcr}\right)^T = \left[f'_{ar} \;\; f'_{br} \;\; f'_{cr}\right] \tag{8.11-40}$$

$$\mathbf{K}_r = \frac{2}{3}\begin{bmatrix} \cos\beta & \cos\left(\beta - \frac{2}{3}\pi\right) & \cos\left(\beta + \frac{2}{3}\pi\right) \\ \sin\beta & \sin\left(\beta - \frac{2}{3}\pi\right) & \sin\left(\beta + \frac{2}{3}\pi\right) \\ \frac{1}{2} & \frac{1}{2} & \frac{1}{2} \end{bmatrix} \tag{8.11-41}$$

$$\beta = \theta - \theta_r \tag{8.11-42}$$

where the angular position is defined by (8.11-36) and

$$\frac{d\theta_r}{dt} = \omega_r(t) \tag{8.11-43}$$

The inverse is

$$(\mathbf{K}_r)^{-1} = \begin{bmatrix} \cos\beta & \sin\beta & 1 \\ \cos\left(\beta - \frac{2}{3}\pi\right) & \sin\left(\beta - \frac{2}{3}\pi\right) & 1 \\ \cos\left(\beta + \frac{2}{3}\pi\right) & \sin\left(\beta + \frac{2}{3}\pi\right) & 1 \end{bmatrix} \tag{8.11-44}$$

The r subscript indicates the variables, parameters, and transformation associated with rotating circuits. Transforming the voltage equations given by (8.11-7) and (8.11-8) yields

$$v_{qs} = r_s i_{qs} + \omega\lambda_{ds} + p\lambda_{qs} \tag{8.11-45}$$

$$v_{ds} = r_s i_{ds} - \omega\lambda_{qs} + p\lambda_{ds} \tag{8.11-46}$$

$$v_{0s} = r_s i_{0s} + p\lambda_{0s} \tag{8.11-47}$$

$$v'_{qr} = r'_r i'_{qr} + (\omega - \omega_r)\lambda'_{dr} + p\lambda'_{qr} \tag{8.11-48}$$

$$v'_{dr} = r'_r i'_{dr} - (\omega - \omega_r)\lambda'_{qr} + p\lambda'_{dr} \tag{8.11-49}$$

$$v'_{0r} = r'_r i'_{0r} + p\lambda'_{0r} \tag{8.11-50}$$

Transforming (8.11-26) yields

$$\lambda_{qs} = L_{ls} i_{qs} + \frac{3}{2} L_{ms}\left(i_{qs} + i'_{qr}\right) \tag{8.11-51}$$

$$\lambda_{ds} = L_{ls} i_{ds} + \frac{3}{2} L_{ms}\left(i_{ds} + i'_{dr}\right) \tag{8.11-52}$$

$$\lambda_{0s} = L_{ls} i_{0s} \tag{8.11-53}$$

$$\lambda'_{qr} = L'_{lr} i'_{qr} + \frac{3}{2} L_{ms}\left(i_{qs} + i'_{qr}\right) \tag{8.11-54}$$

$$\lambda'_{dr} = L'_{lr} i'_{dr} + \frac{3}{2} L_{ms}\left(i_{ds} + i'_{dr}\right) \tag{8.11-55}$$

$$\lambda'_{0r} = L'_{lr} i'_{0r} \tag{8.11-56}$$

Equations (8.11-45) through (8.11-56) suggest the equivalent circuits shown in Fig. 8.11-2.

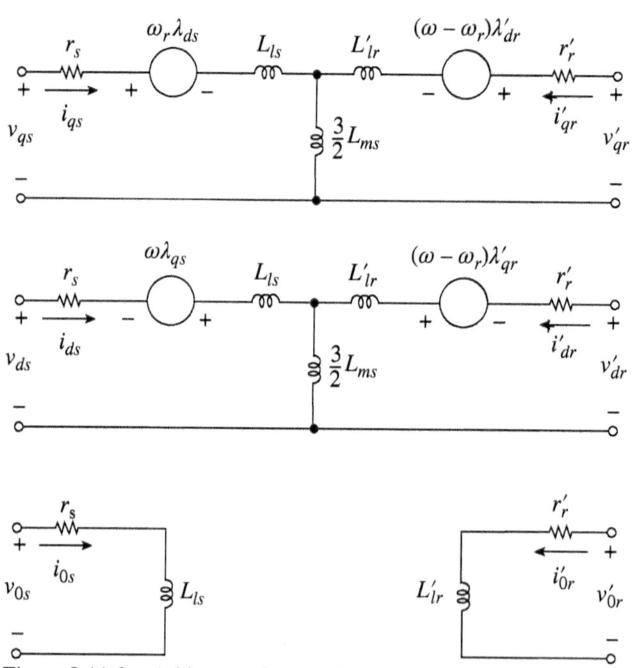

Figure 8.11-2 Arbitrary reference frame equivalent circuits for a three-phase symmetrical induction machine.

Other than the zero-voltage equations (8.11-47) and (8.11-50) and the $\frac{3}{2}$ factor, the voltage equations are the same as those for a two-phase induction machine. Moreover, if the three-phase induction machine is connected in wye without a neutral connection as shown in Fig. 8.11-1, the currents i_{0s} and i'_{0r} are zero for balanced or unbalanced operation since the sum of the three-phase currents is zero. Therefore, v_{0s} and v'_{0r} are zero since the sum of the three-phase stator and rotor flux linkages will be zero for symmetrical systems. We begin to see that the analysis and performance of the two-phase induction machine tells us a lot (in fact, nearly everything) about its bigger three-phase sister.

Torque Equation in Arbitrary Reference Frame Variables

Since the voltage equations for v_{0s} and v'_{0r} given by (8.11-47) and (8.11-50), respectively, do not contain a rotor speed term, the expression for torque can be determined from (8.7-1) if we realize that power input to the coupling field is $\frac{3}{2}$ of the sum of the instantaneous phase power. Hence, the analysis set forth in Section 8.7 can be applied to a three-phase induction machine with or without a neutral connection. We need only to multiply the torque expressions in Section 8.7 by $\frac{3}{2}$ and replace L_{ms} with $\frac{3}{2}L_{ms}$. Recall that $L_{ms}(X_{ms})$ is contained in $L_{ss}(X_{ss})$ and $L'_{rr}(X'_{rr})$.

SP8.11-1 Assume the stator and rotor windings of a symmetrical induction machine are wye connected. Express $\mathbf{L}_s$ (8.11-16) and $\mathbf{L}_r$ (8.11-17) as diagonal matrices. [$L_{ss}\mathbf{I}$, $L_{rr}\mathbf{I}$, where L_{ss} and L_{rr} are for a three-phase machine]

SP8.11-2 Express L_{ss} and L_{rr} in SP8.11-1. [$L_{ss} = L_{ls} + \frac{3}{2}L_{ms}$; $L_{rr} = L_{lr} + \frac{3}{2}L_{mr}$]

SP8.11-3 Express L'_{rr} from (8.11-55). $\left[L'_{rr} = L'_{lr} + \frac{3}{2}L_{ms}\right]$

8.12 PRINCIPLES OF FIELD ORIENTATION

In Chapter 6, the permanent-magnet ac machine was used in a brushless dc drive, where nearly instantaneous torque control was achieved by controlling the stator rotating magnetic field speed to be equal to the electrical angular velocity of the rotor and its magnitude and position relative to the permanent magnet of the rotor. This was accomplished by controlled switching of the drive inverter that supplied the stator applied voltages; however, we did not become involved with the design of the control. Instead, we analyzed the steady-state performance assuming that the sensing and control functions were performing perfectly, whereupon the control of torque was essentially instantaneous. The electric transients were thus negligible, that is, a change from one steady-state operating point to another was accomplished with a minimum of electric transients. Therefore, considering only steady-state performance was not as unrepresentative as one might have first thought.

The goal of field orientation of an induction machine is similar: fast torque response, which is accomplished with a controlled drive inverter [2,5,6]. Unfortunately, the necessary field orientation-based control is more involved in the case of the induction machine since, not only is a torque control involved but it is also necessary to control the stator variables to orient the maximum value of the rotor field, mmf_r^e, in the q axis, orthogonal to the d-axis component of the stator field, mmf_s^e. To do all this we have two things we can control, the stator applied voltage and the slip frequency. Although the principle of field orientation is quite straightforward, the implementation of the control is not. In particular, the voltage equations in the synchronously rotating reference frame are used to determine the necessary i_{qs}^e and i_{ds}^e to ensure field orientation, and these values are then transformed to the stationary reference frame to determine the stator currents needed for field orientation. This is achieved by controlling the output voltage of the drive inverter to shape the waveforms of the actual stator currents to those determined necessary to provide field orientation. In this section, we will not become involved with the details of implementing this control; instead, we will follow the same approach as in the case of the brushless dc drive, in that, we will assume that the control is functioning perfectly and focus on the steady-state performance of an induction machine with field orientation. In other words, in this section, our focus will be on what the control does, not how it does it.

The explanation of the control methodology is rather involved, requiring extensive use of reference frame theory and one can lose sight of the goal; however, the resulting phasor voltage equation is not complicated and similar in form to that for the brushless dc drive. If the reader wishes to bypass the reference frame derivation of the phasor voltage equation, the material up to (8.12-19) and (8.12-20) may be de-emphasized.

Reference Frame Derivation

The voltage and flux-linkage equations of a two-phase single-fed induction machine in the synchronously rotating reference frame are

$$v_{qs}^e = r_s i_{qs}^e + \omega_e \lambda_{ds}^e + p\lambda_{qs}^e \tag{8.12-1}$$

$$v_{ds}^e = r_s i_{ds}^e - \omega_e \lambda_{qs}^e + p\lambda_{ds}^e \tag{8.12-2}$$

$$0 = r_r' i_{qr}'^e + (\omega_e - \omega_r)\lambda_{dr}'^e + p\lambda_{qr}'^e \tag{8.12-3}$$

$$0 = r_r' i_{dr}'^e - (\omega_e - \omega_r)\lambda_{qr}'^e + p\lambda_{dr}'^e \tag{8.12-4}$$

where

$$\lambda_{qs}^e = L_{ss} i_{qs}^e + L_{ms} i_{qr}'^e \tag{8.12-5}$$

$$\lambda_{ds}^e = L_{ss} i_{ds}^e + L_{ms} i_{dr}'^e \tag{8.12-6}$$

$$\lambda_{qr}'^e = L_{rr}' i_{qr}'^e + L_{ms} i_{qs}^e \tag{8.12-7}$$

$$\lambda_{dr}'^e = L_{rr}' i_{dr}'^e + L_{ms} i_{ds}^e \tag{8.12-8}$$

The torque may be expressed as

$$T_e = \frac{P}{2}\left(\lambda'^e_{qr} i'^e_{dr} - \lambda'^e_{dr} i'^e_{qr}\right) \tag{8.12-9}$$

The aim is to select the applied stator voltages so that $\lambda'^e_{qr} = 0$; whereupon, (8.12-9) becomes

$$T_e = -\frac{P}{2}\lambda'^e_{dr} i'^e_{qr} \tag{8.12-10}$$

Please be careful here. Making λ'^e_{qr} zero does not mean that i'^e_{qr} will also be zero; instead, from (8.12-7) we see that with λ'^e_{qr} zero, then

$$i'^e_{qr} = -\frac{L_{ms}}{L'_{rr}} i^e_{qs} \tag{8.12-11}$$

Thus, if i^e_{qs} is (8.12-11), then $\lambda^e_{qr} = 0$ and T_e is (8.12-10).

From (8.12-10) we see that if λ'^e_{dr} is constant, then torque is proportional to i^e_{qs} from (8.12-11). We will work toward that goal but first let us control i'^e_{dr} to zero. Why? Well, if i'^e_{dr} is zero, the rotor poles will be positioned completely in the q axis. Now again, be careful; controlling i'^e_{dr} to zero does not mean that λ'^e_{dr} is zero; from (8.12-8) with $i'^e_{dr} = 0$ then

$$\lambda'^e_{dr} = L_{ms} i^e_{ds} \tag{8.12-12}$$

which will be constant if i^e_{ds} is held constant. If (8.12-11) and (8.12-12) are substituted into (8.12-10), the torque may be expressed as

$$T_e = \frac{P}{2}\frac{L^2_{ms}}{L'_{rr}} i^e_{qs} i^e_{ds} \tag{8.12-13}$$

Note that we have T_e in terms of stator-related currents and if i^e_{ds} is held constant, T_e is directly proportional to i^e_{qs}. This equation is used to control i^e_{qs} when T_e is commanded and with i^e_{ds} held constant, generally at its rated value.

We have eliminated (8.12-4) and since λ'^e_{qr} is zero, then so must be $p\lambda'^e_{qr}$, and if (8.12-11) is substituted for i'^e_{qr} and (8.12-12) for λ'^e_{dr} into (8.12-3), the angular velocity of the slip may be calculated as

$$(\omega_e - \omega_r)_{\text{calc}} = \frac{r'_r}{L'_{rr}}\frac{i^e_{qs}}{i^e_{ds}} \tag{8.12-14}$$

which we will also be denoted as $(\omega_s)_{\text{calc}}$ for compactness. This equation is used to control slip frequency. Note that (8.12-14), like (8.12-13), is in terms of stator-related currents. We have set forth the basic relationships for field orientation control, which in effect positions the rotor poles along the q axis in accordance with (8.12-11).

Steady-State Operation

Our purpose now is to consider the steady-state performance of an induction motor assuming that the field orientation is functioning properly. Although this idealized approach is an oversimplification of the control challenges involved, it helps to give insight to the basic features of field orientation; however, since the stator currents are commanded (controlled), the electric transients are minimized. Therefore, we will find that steady-state, "ideally" controlled operation and the actual field orientation drive with harmonics neglected are very similar.

Let us take a minute to express the steady-state torque assuming field orientation is functioning properly. From (5.5-9) and (5.5-10) we can express I_{qs}^e and I_{ds}^e, respectively, as

$$I_{qs}^e = \sqrt{2} I_s \cos \theta_{esi}(0) \tag{8.12-15}$$

$$I_{ds}^e = -\sqrt{2} I_s \sin \theta_{esi}(0) \tag{8.12-16}$$

where in (5.5-9) and (5.5-10), $\omega = \omega_e$ and $\theta(0) = 0$. Substituting (8.12-15) and (8.12-16) into (8.12-13) yields the expression for torque with field orientation as

$$\begin{aligned} T_e &= -\frac{P}{2}\frac{L_{ms}^2}{L'_{rr}}\left[\sqrt{2} I_s \cos \theta_{esi}(0)\right]\left[\sqrt{2} I_s \sin \theta_{esi}(0)\right] \\ &= -\frac{P}{2}\frac{L_{ms}^2}{L'_{rr}} I_s^2 \sin 2\theta_{esi}(0) \end{aligned} \tag{8.12-17}$$

The instantaneous phasors for the induction machine are given by (8.8-3) and (8.8-4). With field orientation and harmonics neglected, the flux linkages in the synchronous reference frame are zero or constants; whereupon, ideally the time rate of change is zero. Therefore, with ideal functioning field orientation the last two terms of (8.8-3) and (8.8-4) are zero and the instantaneous phasor equations become the steady-state phasor equations. The only dynamic feature is the relationship between rotor speed and torque given by (8.4-5).

We will work with the steady-state equations and modify V_{qs}^e and V_{ds}^e to account for the results of field orientation action on the $\overset{e}{qr}$ and $\overset{e}{dr}$ variables given earlier; thereby, making it necessary to work only with the stator phasor voltage equations. Substituting $I_{qr}'^e$, in terms of the I_{qs}^e from (8.12-11), into (8.12-5) yields

$$\lambda_{qs}^e = \left(\frac{L_{ss}}{L_{ms}} - \frac{L_{ms}}{L'_{rr}}\right) L_{ms} I_{qs}^e \tag{8.12-18}$$

We can now express steady-state V_{qs}^e and V_{ds}^e. From (8.12-1) and (8.12-2) with $p\lambda = 0$, λ_{ds}^e from (8.12-6), with $I_{dr}'^e$ zero, and (8.12-18) for λ_{qs}^e, we have

$$V_{qs}^e = r_s I_{qs}^e + \omega_e L_{ss} I_{ds}^e \tag{8.12-19}$$

$$V_{ds}^e = r_s I_{ds}^e - \omega_e \left(\frac{L_{ss}}{L_{ms}} - \frac{L_{ms}}{L'_{rr}}\right) L_{ms} I_{qs}^e \tag{8.12-20}$$

Let us take a minute to express $\widetilde{V}_{as}$ for steady-state operation. Substituting (8.12-19) and (8.12-20) into (8.8-1) yields $\sqrt{2}\,\widetilde{V}_{as}$ with field orientation, that is,

$$\begin{aligned}\sqrt{2}\,\widetilde{V}_{as} &= V_{qs}^{e} - jV_{ds}^{e} \\ &= r_s I_{qs}^{e} + \omega_e L_{ss} I_{ds}^{e} - j\left(r_s I_{ds}^{e} - \omega_e K_L L_{ms} I_{qs}^{e}\right) \\ &= r_s\left(I_{qs}^{e} - jI_{ds}^{e}\right) + \omega_e L_{ss} I_{ds}^{e} + j\omega_e K_L L_{ms} I_{qs}^{e}\end{aligned} \tag{8.12-21}$$

where

$$K_L = \frac{L_{ss}}{L_{ms}} - \frac{L_{ms}}{L'_{rr}} \tag{8.12-22}$$

If now we add and subtract $\omega_e K_L L_{ms} I_{ds}^{e}$ on the right-hand side of (8.12-21), we can express $\widetilde{V}_{as}$ as

$$\widetilde{V}_{as} = (r_s + j\omega_e K_L L_{ms})\widetilde{I}_{as} + \frac{1}{\sqrt{2}}\omega_e (L_{ss} - K_L L_{ms}) I_{ds}^{e} \tag{8.12-23}$$

Substituting (8.12-16) for I_{ds}^{e} into (8.12-23) yields

$$\widetilde{V}_{as} = (r_s + j\omega_e K_L L_{ms})\widetilde{I}_{as} - \omega_e (L_{ss} - K_L L_{ms}) I_s \sin\theta_{esi}(0) \tag{8.12-24}$$

Solving (8.12-13) for I_{qs}^{e} yields

$$I_{qs}^{e} = \frac{2}{P}\frac{L'_{rr}}{L_{ms}^{2}}\frac{T_e}{I_{ds}^{e}} \tag{8.12-25}$$

We are also going to use (8.12-14), but let us wait just a minute before we get to that.

Although we are not going to become involved in the details of the field orientation, the block diagram shown in Fig. 8.12-1 depicts the basic principles of the control. Therein, the input variables containing an asterisk are commanded variables. That is, if the machine is to operate with rated load torque and rated frequency, then T_e^* would be rated torque and I_{ds}^{e*} would be constant and generally selected to be the value of I_{ds}^{e} for rated conditions and maintained at that value regardless of T_e^*. Now, (8.12-14), (8.12-25), and (8.12-26), which we will get to in a moment, impose field orientation control on the currents, i_{as}^* and i_{bs}^*. The inverter forms the voltages v_{as} and v_{bs} to ensure i_{as} and i_{bs} are i_{as}^* and i_{bs}^*, respectively.

We will continue to use reference frame variable notation for compactness; hopefully, this will not be confusing for those who have not studied Chapter 5. Thus, we have I_{qs}^{e*} and I_{ds}^{e*} and, from (8.12-14), we can calculate $\omega_e - \omega_r$, which we are calling $(\omega_s)_{\text{calc}}$ in Fig. 8.12-1 (the slip angular velocity) and/or $(\omega_e - \omega_r)_{\text{calc}}$ since it is calculated from commanded inputs I_{qs}^{e*} and I_{ds}^{e*} and machine parameters. Now $(\omega_s)_{\text{calc}}$ is the slip necessary for field orientation with the

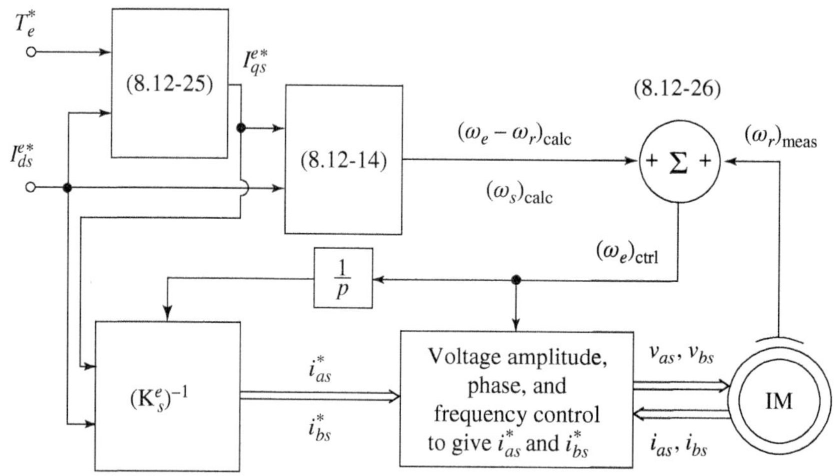

Figure 8.12-1 Block diagram depicting field-oriented control principles.
Note: $I_{qs}^e = \sqrt{2}I_s \cos\theta_{esi}(0)$ and $I_{ds}^e = -\sqrt{2}I_s \sin\theta_{esi}(0)$.

commanded stator-related currents. Now, $(\omega_r)_{meas}$ is the measured electrical angular velocity of the rotor and it is added to $(\omega_s)_{calc}$. Thus, from Fig. 8.12-1,

$$\begin{aligned}(\omega_e)_{ctrl} &= (\omega_s)_{calc} + (\omega_r)_{meas} \\ &= (\omega_e - \omega_r)_{calc} + (\omega_r)_{meas}\end{aligned} \tag{8.12-26}$$

where $(\omega_e)_{ctrl}$ is the electrical angular velocity of the controlled stator voltages applied to the machine (v_{as} and v_{bs} of Fig. 8.12-1) for the rotor magnetic field to be oriented in the q axis. There is no direct control of speed; however, if we are operating in the steady state and the torque load increases, $(\omega_r)_{meas}$ would decrease which would decrease ω_e until $(\omega_e - \omega_r)_{calc}$ was established. Again, $(\omega_e - \omega_r)_{calc}$ is the slip angular velocity, calculated from the commanded T_e^* and I_{ds}^{e*} by (8.12-14). Assuming the parameters remain constant with temperature change, then $(\omega_e - \omega_r)_{calc}$ will change only if T_e^* and/or I_{ds}^{e*} are changed.

It is important to note that the commanded variables are in the synchronously rotating reference frame $\left(I_{qs}^e \text{ and } I_{ds}^e\right)$. These commanded variables are transformed to the stationary reference frame by $\left(\mathbf{K}_s^e\right)^{-1}$ to make up the waveforms of i_{as} and i_{bs} or i_{as}, i_{bs}, and i_{cs} for a three-phase system. The phase and amplitude of the stator voltages are then controlled by the drive inverter in order to produce the desired stator currents.

TABLE 8.12-1 Steady-State Voltage and Torque Equations for Field Orientation of Induction Machines

For Two Phase

$$V_{qs}^e = r_s I_{qs}^e + \omega_e L_{ss} I_{ds}^e \qquad \tilde{V}_{as} = (r_s + j\omega_e K_L L_{ms})\tilde{I}_{as} - \omega_e (L_{ss} - K_L L_{ms}) I_s \sin\theta_{esi}(0)$$

$$V_{ds}^e = r_s I_{ds}^e - \omega_e K_L L_{ms} I_{qs}^e$$

$$T_e = \frac{P}{2}\frac{L_{ms}^2}{L'_{rr}} I_{qs}^e I_{ds}^e$$

$$T_e = -\frac{P}{2}\frac{L_{ms}^2}{L'_{rr}}\left[\sqrt{2} I_s \cos\theta_{esi}(0)\right]\left[\sqrt{2} I_s \sin\theta_{esi}(0)\right]$$

$$= -\frac{P}{2}\frac{L_{ms}^2}{L'_{rr}} I_s^2 \sin 2\theta_{esi}(0)$$

$$K_L = \frac{L_{ss}}{L_{ms}} - \frac{L_{ms}}{L'_{rr}}$$

For Three Phase

Replace L_{ms} with $\frac{3}{2}L_{ms}$ and multiply T_e by $\frac{3}{2}$.

Example 8F Operation with Field Orientation

The parameters of the two-phase two-pole 5-hp 110-V 60-Hz induction machine are

$$r_s = 0.295\ \Omega \qquad L_{ms} = 35.15\ \text{mH} \qquad r'_r = 0.201\ \Omega$$
$$L_{ls} = 0.944\ \text{mH} \qquad L'_{lr} = 0.944\ \text{mH}$$

The machine is equipped with a field orientation-based controller. Calculate operating conditions for rated and $\frac{1}{2}$ rated torque from stall to rated frequency. The rated torque is 10 N · m and the rated power factor angle is 30°. From (8.12-17), in the steady state,

$$T_e = -\frac{P}{2}\frac{L_{ms}^2}{L'_{rr}} I_s^2 \sin 2\theta_{esi}(0)$$

$$10 = -\left(\frac{2}{2}\right)\frac{(35.15)^2 \times 10^{-3}}{(35.15 + 0.944)} I_s^2 \sin 2(-30°) \tag{8F-1}$$

Solving for I_s yields

$$I_s = \left[\frac{(10)(35.15 + 0.944)}{(35.15)^2 \times 10^{-3} \sin 60°}\right]^{\frac{1}{2}} = 18.37\ \text{A} \tag{8F-2}$$

Therefore,

$$\tilde{I}_{as} = 18.37\underline{/-30°}\ \text{A} \tag{8F-3}$$

Now,

$$\begin{aligned} L_{ss} &= L_{ls} + L_{ms} \\ &= (0.944 + 35.15) \times 10^{-3} \\ &= 36.094 \text{ mH} \end{aligned} \tag{8F-4}$$

From (8.12-22),

$$K_L = \frac{L_{ss}}{L_{ms}} - \frac{L_{ms}}{L'_{rr}} = \frac{36.094}{35.15} - \frac{35.15}{36.094} = 0.053 \tag{8F-5}$$

From (8.12-24), and calculating at rated frequency,

$$\begin{aligned} \tilde{V}_{as} &= (r_s + j\omega_e K_L L_{ms})\tilde{I}_{as} - \omega_e (L_{ss} - K_L L_{ms}) I_s \sin\theta_{esi}(0) \\ &= \left[0.295 + j377(0.053)(35.15 \times 10^{-3})\right] 18.37\underline{/-30^\circ} \\ &\quad - 377(36.094 - 0.053 \times 35.15) \times 10^{-3}(18.37)(-0.5) \\ &= 129.7 + j8.5 = 130\underline{/3.74^\circ}\text{V} \end{aligned} \tag{8F-6}$$

This voltage exceeds the machine's rated value (110-V); however, the current is essentially rated. We will continue; the rotor speed is calculated from (8.12-14),

$$\omega_e - \omega_r = \frac{r'_r}{L'_{rr}} \frac{I_{qs}^{e*}}{I_{ds}^{e*}} \tag{8F-7}$$

where

$$\begin{aligned} I_{qs}^e &= \sqrt{2} I_s \cos\theta_{esi}(0) \\ &= \sqrt{2} 18.37 \cos 30^\circ = 22.5 \text{ A} \end{aligned} \tag{8F-8}$$

$$\begin{aligned} I_{ds}^e &= -\sqrt{2} 18.37 \sin\theta_{esi}(0) \\ &\quad -\sqrt{2} 18.37 \sin(-30^\circ) = 12.99 \text{ A} \end{aligned} \tag{8F-9}$$

From (8F-7),

$$\omega_e - \omega_r = \frac{0.201}{36.09 \times 10^{-3}} \frac{22.5}{12.99} = 9.6 \text{ rad/s} \tag{8F-10}$$

$$-\omega_r = -\omega_e + 9.6$$

$$\omega_r = 377 - 9.6 = 367.4 \text{ rad/s} \tag{8F-11}$$

Let us continue with $\overset{e}{qs}$ and $\overset{e}{ds}$ variables to calculate $\tilde{V}_{as}$ again. From (8.12-19) and (8.12-20) or Table 8.12-1,

$$\begin{aligned} V_{qs}^e &= r_s I_{qs}^e + \omega_e L_{ss} I_{ds}^e \\ &= (0.295)(22.5) + 377(36.09 \times 10^{-3})12.99 \\ &= 6.638 + 176.74 = 183.38 \text{ V} \end{aligned} \tag{8F-12}$$

$$
\begin{aligned}
V^e_{ds} &= r_s I^e_{ds} - \omega_e K_L L_{ms} I^e_{qs} \\
&= (0.295)(12.99) - (377)(0.053)(35.15 \times 10^{-3})(22.5) \\
&= 3.8 - 15.8 = -12.0 \text{ V}
\end{aligned} \tag{8F-13}
$$

Then,

$$
\begin{aligned}
\widetilde{V}_{as} &= \frac{1}{\sqrt{2}}\left(V^e_{qs} - jV^e_{ds}\right) \\
&= \frac{1}{\sqrt{2}}(183.38 + j12) = 130\angle 3.74°\text{V}
\end{aligned} \tag{8F-14}
$$

Let us continue the calculation of $\widetilde{V}_{as}$ for T_e^* equal to its rated value (10 N · m) and one-half rated (5 N · m) over the speed range from $\omega_e = 377$ to zero (stall). To expedite the calculations, we will assume ω_e of the stator applied voltages is rated (377 rad/s), $\frac{1}{2}\omega_e(185.5\text{ rad/s})$, $\frac{1}{4}\omega_e(94.3\text{ rad/s})$, and $\omega_e = (\omega_s)_{\text{calc}}$, with $I^e_{ds} = -\sqrt{2}I_s \sin\theta_{esi}(0)$ maintained constant at rated (12.99 A). The results of these calculations are given in Table 8F-1 and the Operating Points are shown on a T_e versus ω_r plot in Fig. 8F-1.

TABLE 8F-1 Tabulation of $\widetilde{V}_{as}$ and $\widetilde{I}_{as}$ for Different T_e^* and Stator Frequency

Operating Point	T_e^* (N·m)	$\omega_e - \omega_r$ (rad/s)	ω_e (rad/s)	ω_r (rad/s)	$\widetilde{V}_{as}$ (V)	$\widetilde{I}_{as}$ (A)
1	10	9.6	377	367.4	$130\angle 3.74°$	$18.37\angle{-30°}$
2	$\frac{10}{2}$	4.8	377	372.2	$125\angle 1.3°$	$12.2\angle{-49.1°}$
3	10	9.6	$\frac{377}{2}$	178.9	$67.2\angle 2.46°$	$18.37\angle{-30°}$
4	$\frac{10}{2}$	4.8	$\frac{377}{2}$	183.7	$62.55\angle 0.42°$	$12.2\angle{-49.1°}$
5	10	9.6	$\frac{377}{4}$	84.7	$35.92\angle 0°$	$18.37\angle{-30°}$
6	$\frac{10}{2}$	4.8	$\frac{377}{4}$	89.4	$33.59\angle 0°$	$12.2\angle{-49.1°}$
7	10	9.6	9.6	0	$8.24\angle{-17.1°}$	$18.37\angle{-30°}$
8	$\frac{10}{2}$	4.8	4.8	0	$6.1\angle{-24.9°}$	$12.2\angle{-49.1°}$

The change in $|\widetilde{V}_{as}|$ and $\theta_{esv}(0)$, which is the shift in the applied voltage relative to the q axis, is shown in Fig. 8F-2 for $T_e^* =$ 10 N · m and in Fig. 8F-3 for $T_e^* =$ 5 N · m.

In each case we see that $\widetilde{I}'_{ar}$ is oriented orthogonal to the d axis and the torque is 10 N · m and this has been done by controlling $\widetilde{V}_{as}$ and $\omega_e - \omega_r$. The phasor diagram shown in Fig. 8F-4 should be compared to Fig. 8C-2 which is for uncontrolled induction machine operation.

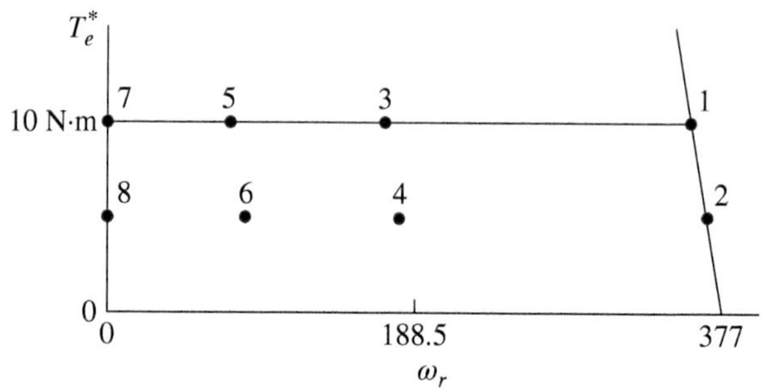

Figure 8F-1 Plot of Operating Points given in Table 8F-1 on T_e versus ω_r characteristics.

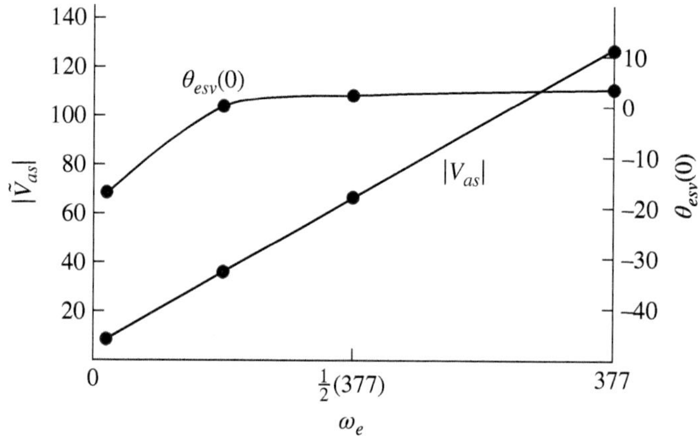

Figure 8F-2 Amplitude of $\tilde{V}_{as}$ and phase angle $\theta_{esv}(0)$ relative to the q axis for T_e^* of 10 N · m.

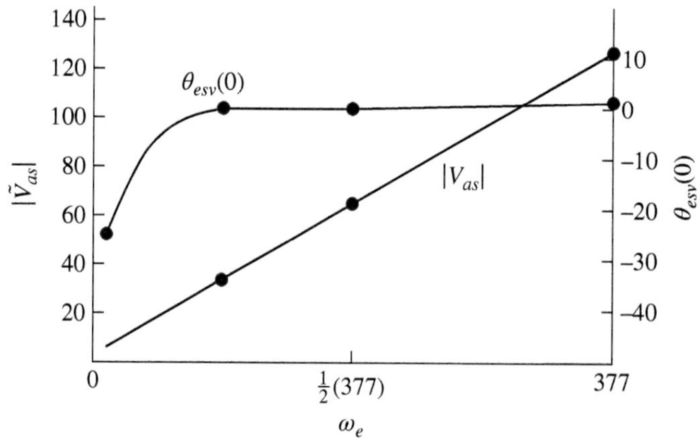

Figure 8F-3 Same as Fig. 8F-2 for T_e^* of 5 N · m.

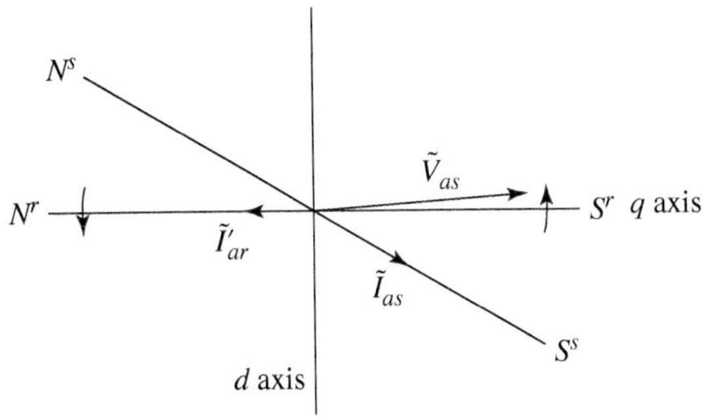

Figure 8F-4 Phasor diagram for Operating Point 1.

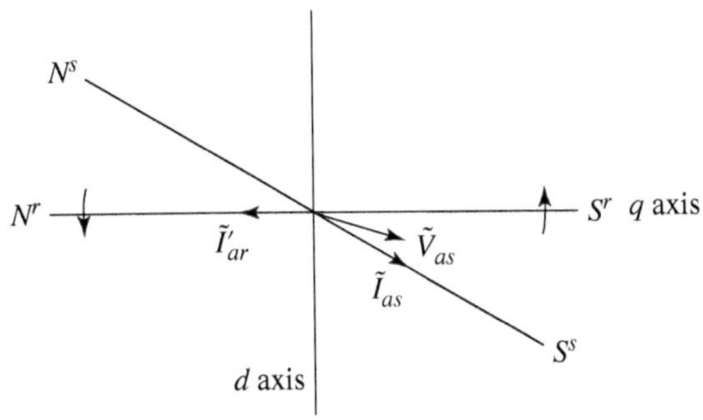

Figure 8F-5 Phasor diagram for Operating Point 7.

The phasor diagram for Operating Point 1 is shown in Fig. 8F-4 and for Operating Point 7 in Fig. 8F-5.

Example 8G Torque Control

Let us consider the dynamics of an induction motor drive with field orientation control given in Example 8F. The drive is originally operating at Operating Point 1 of Fig. 8F-1. The load line is

$$T_L = K\omega_r^2 \tag{8G-1}$$

where for $T_e = 10\,\mathrm{N \cdot m}$ at Operating Point 1 with $\omega_r = 367.4\,\mathrm{rad/s}$ from Table 8F-1.

$$K = \frac{10}{(367.4)^2} = 7.41 \times 10^{-5}\,\mathrm{N \cdot m \cdot s^2} \tag{8G-2}$$

This mode of operation is shown in Fig. 8G-1.

The commanded torque is changed from 10 N · m to 5 N · m. The motor slows down and after the mechanical dynamics subside, the new Operating Point is at 2 on Fig. 8G-1. At this point the new speed from (8G-1) is

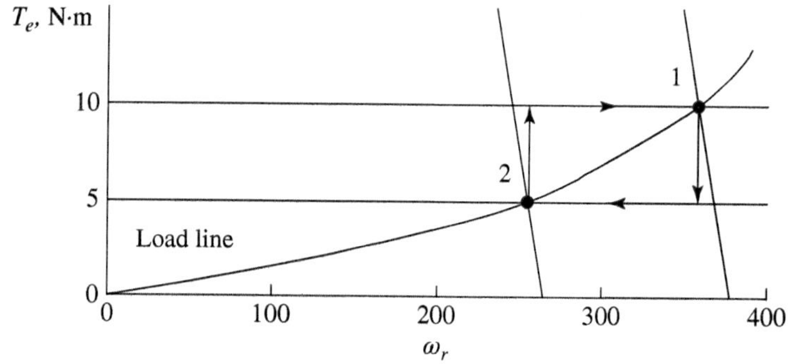

Figure 8G-1 Operation of induction motor drive with field orientation for step change in T_e^*.

$$T_L = K\omega_r^2$$

$$5 = 7.41 \times 10^{-5}\omega_r^2$$

$$\omega_r = \left(\frac{5}{7.41 \text{ x } 10^{-5}}\right)^{\frac{1}{2}} = 259.8 \text{ rad/s} \tag{8G-3}$$

We could liken this scenario to reducing the accelerator of an electric car by one half. Now, from Operating Point 2, the accelerator is depressed and we return to Operating Point 1 as shown in Fig. 8G-1. For this purpose let us calculate the applied voltage of Operating Point 2.

$$\tilde{V}_{as} = (r_s + j\omega_e K_L L_{ms})\tilde{I}_{as} - \omega_e(L_{ss} - K_L L_{ms})I_s \sin\theta_{esi}(0) \tag{8G-4}$$

Now, for $T_e^* = 5\text{ N}\cdot\text{m}$ $\omega_e - \omega_r = 4.8$ rad/s, therefore,

$$\omega_e = 259.8 + 4.8 = 264.6 \text{ rad/s} \tag{8G-5}$$

Also from Example 8F $\tilde{I}_{as} = 12.2\underline{/-49.1^\circ}$ and for a two-phase motor $r_s = 0.295\,\Omega$, $L_{ss} = 36.094$ mH, $L_{ms} = 35.15$ mH, and $K_L = 0.053$.

Substituting into (8G-4),

$$\begin{aligned}\tilde{V}_{as} = &\left[0.295 + j264.6(0.053)(35.151 \times 10^{-3})\right] 12.2\,\underline{/-49.1^\circ} \\ &- 264.6(36.094 - 0.053 \times 35.15) \times 10^{-3}(12.2)(\sin -49.1^\circ) \\ &= (0.5744\,\underline{/59.1^\circ})(12.2\,\underline{/-49.1^\circ}) + 83.51\underline{/0^\circ} \\ &= 90.4 + j1.22 = 90.4\,\underline{/0.77^\circ}\end{aligned} \tag{8G-6}$$

SP8.12-1 Verify ω_r for Operating Point 8 given in Table 8F-1.

8.13 RECAPPING

The section showing the free acceleration of an induction motor in the stationary, rotor, and synchronously rotating reference frames illustrated the fact that reference frame theory comes from Tesla's rotating magnetic field. Clearly, the change of variables assures that the substitute variables portray Tesla's rotating magnetic fields as viewed from the selected reference frame. This is a feature of the transformation that we have been talking about since Chapter 5.

With the assumption that the stator and rotor of an induction machine are symmetrical magnetically, rotor position-dependent magnetic coupling can be eliminated by transforming the stator and rotor variables to substitute variables in any frame of reference. This feature of the induction machine enables steady-state analysis to be carried out in any asynchronous frame of reference using the standard equivalent circuit and dc circuit analysis in the synchronously rotating reference frame. This is unlike the synchronous-type machines, where the rotor-dependent magnetic coupling is eliminated only in the rotor reference frame.

The dynamic and steady-state positioning of the stator and rotor poles developed from Tesla's rotating magnetic fields provide an excellent instructional tool. In fact, the instantaneous phasor becomes a means of visualizing dynamic and steady-state operation during motor and generator action and provides a clear and direct means of portraying induction motor operation with field orientation.

8.14 PROBLEMS

1. Consider the two-pole, two-phase induction machine shown in Fig. 8.2-1. The device is operating as a motor at $\omega_r = 95\pi$ rad/s with $I'_{ar} = \cos 5\pi t$ and $I'_{br} = -\sin 5\pi t$. Determine the angular velocity and direction of mmf$_r$ relative to (a) an observer on the rotor and (b) an observer on the stator. Also determine (c) the angular velocity of the stator currents and (d) the direction of rotation of the rotor.

2. The windings shown in Fig. 8.15-1 are sinusoidally distributed and the device is symmetrical. The amplitude of the stator-to-rotor mutual inductance is L_{sr}. Express all mutual inductances as functions of L_{sr} and θ_r.

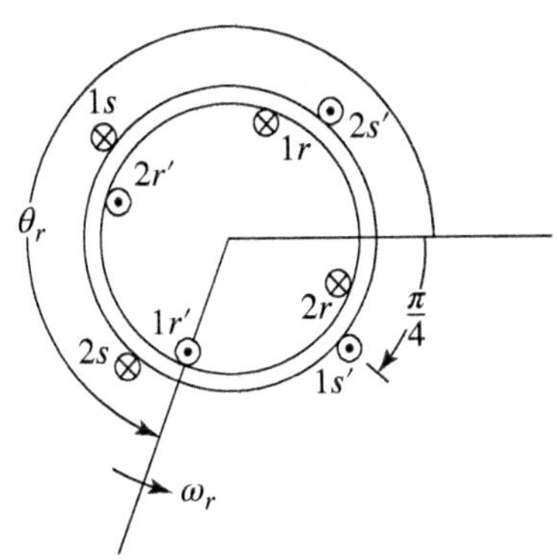

Figure 8.15-1 Coupled windings.

3. The rotor windings of the two-pole, two-phase symmetrical induction machine shown in Fig. 8.2-1 are open-circuited. $I_{as} = \sqrt{2}I_s \cos \omega_e t$ and $I_{bs} = -\sqrt{2}I_s \sin \omega_e t$. The rotor is driven at $\omega_r = \omega_e$ in the counterclockwise direction. Express V_{ar}.

4. Derive the expression for the electromagnetic torque T_e given by (8.4-3), starting with (8.4-1) and (8.4-2).

5. Synchronously rotating variables may be related to rotor reference frame variables by $\mathbf{f}^e_{qds} = {}^r\mathbf{K}^e\mathbf{f}^r_{qds}$. Determine ${}^r\mathbf{K}^e$ in terms of $\mathbf{K}^e_s$ and $\mathbf{K}^r_s$.

6. Repeat SP8.8-4b without approximation.

7. A four-pole two-phase induction machine has the following parameters: $r_s = 0.3\ \Omega$, $L_{ls} = 1$ mH, $L_{ms} = 20$ mH, $r'_r = 0.2\ \Omega$, and $L'_{lr} = 1$ mH. The device is supplied from a 60-Hz source; the rotor speed is $\omega_r = 360$ rad/s. In this mode of operation $\tilde{I}_{as} = 28.8\angle -36.1°$ and $\tilde{I}'_{ar} = 23.9\angle -173.2°$. Calculate (a) T_e, (b) the total ohmic loss in the rotor windings, (c) the mechanical power delivered to the load, and (d) express I_{as}, I'_{ar}, I^s_{qs}, and I'^s_{qr}.

8. Consider Problem 7. Calculate $\tilde{V}_{as}$. Draw the phasor diagram and locate the poles. Show all voltage drops associated with the equivalent circuit shown in Fig. 8C-1.

9. Select two identical capacitors so that when they are connected in parallel with each phase (one capacitor per phase) of the induction machine described in Example 8E, the no-load capacitor–induction machine combination operates at unity power factor. Assume the capacitors are ideal (zero resistance).

10. Determine the phasor diagram and locate the poles for the two-phase 5-hp induction machine given in Example 8F for (a) stall and (b) with a slip of 0.001.

11. Consider Table 8F-1. Verify the calculations for Operating Point 5 and draw the phasor diagram.

12. Show that

$$\tilde{I}'_{ar} = -\frac{jX_{ms}}{\frac{r'_r}{s} + jX'_{rr}}\tilde{I}'_{as}$$

is equal to (8.12-11). ■

8.15 REFERENCES

1. P. C. Krause, *Analysis of Electric Machinery*, McGraw-Hill Book Company, New York, 1986.
2. P. C. Krause, O. Wasynczuk, S. D. Sudhoff, and S. D. Pekarek, *Analysis of Electric Machinery and Drive Systems*, 3rd Edition, John Wiley & Sons, IEEE Press, New York, 2013.
3. P. C. Krause, O. Wasynczuk, and S. D. Sudhoff, *Analysis of Electric Machinery*, 1st Edition, IEEE Press, 1994.

4. R. H. Park, "Two-Reaction Theory of Synchronous Machines – Generalized Method of Analysis – Part I," *AIEE Trans.*, Vol. 84, July 1929, pp. 716–727.
5. A. M. Trzynadlowski, *The Field Orientation Principle in Control of Induction Motors*, Kluwer Academic Publishers, 1994.
6. F. Blaschke, *Das Verfahren der Feldorientierung zur Regelung der Drehfeld-maschine*, PhD thesis, TU Braunschweig, 1974.

CHAPTER 9

STEPPER MOTORS

9.1 INTRODUCTION

Stepper motors are electromechanical motion devices that are used primarily to convert information in digital form to mechanical motion. Although stepper motors were used as early as the 1920s, their use has skyrocketed with the advent of the digital computer. Whenever stepping from one position to another is required, whether the application is industrial, military, or medical, the stepper motor is often the motor of choice [1]. Stepper motors come in various sizes and shapes but most fall into two types – the variable-reluctance stepper motor and the permanent-magnet stepper motor. Both types are considered in this chapter. We shall find that the operating principle of the variable-reluctance stepper motor is much the same as that of the reluctance machine that we have already discussed in earlier chapters, and the permanent-magnet stepper motor is similar in principle to the permanent-magnet synchronous machine.

9.2 BASIC CONFIGURATIONS OF MULTISTACK VARIABLE-RELUCTANCE STEPPER MOTORS

There are two general types of variable-reluctance stepper motors: single-stack and multistack. As a first approximation, the behavior of both types may be described from similar equations. Actually, the principle of operation of variable-reluctance stepper motors is the same as the reluctance machine that we considered in earlier chapters; only the mode of operation differs. There are, however, some new terms to define, and it is necessary for us to extend some of our previous definitions to fit the stepper motor. First, we will look at the multistack device in some detail, followed by a brief discussion of the single-stack, variable-reluctance stepper motor.

In its most basic form, the multistack variable-reluctance stepper motor consists of three or more single-phase reluctance motors on a common shaft with their

Electromechanical Motion Devices: Rotating Magnetic Field-Based Analysis and Online Animations, Third Edition. Paul Krause, Oleg Wasynczuk, Steven D. Pekarek, and Timothy O'Connell.

Published 2020 by John Wiley & Sons, Inc.

stator magnetic axes displaced from each other. The rotor of an elementary three-stack device is shown in Fig. 9.2-1. It has three cascaded two-pole rotors with a minimum-reluctance path of each aligned at the angular displacement θ_{rm}. In stepper motor language, each of the two-pole rotors is said to have two teeth. Now, visualize that each of these rotors has its own, separate, single-phase stator with the magnetic axes of the stators displaced from each other. In Fig. 9.2-1 we have labeled the individual rotors *a, b,* and *c*. The corresponding stators are shown in Fig. 9.2-2; the stator with the *as* winding is associated with the *a* rotor, the *bs* winding with the *b* rotor, and so on. There are several things to note. First, we see that each of the single-phase stators has two poles, much the same in configuration as the dc machine, with the stator winding wound around both poles. In particular, positive current flows into as_1 and out as_1'; as_1' is connected to as_2 so that positive current flows into as_2 and out as_2'. Although we have shown only one circle for as_1, …, as_2', we realize that each would represent several turns, and that the number of turns from as_1 to as_1' (indicated by $N_s/2$ in Fig. 9.2-2) is the same as from as_2 to as_2'. Let us note one more thing. Heretofore, we have referenced θ_{rm} (or θ_r) from the *as* axis to the maximum-reluctance path of a salient-pole rotor. In Fig. 9.2-2, θ_{rm} is referenced to the minimum-reluctance path of the rotor. Since this is more or less standard in stepper motor analysis, we will deviate from the convention we have established for reluctance and synchronous machines.

Each stack is often called a phase. In other words, a three-stack machine is a three-phase machine. This nomenclature can be misleading since we generally

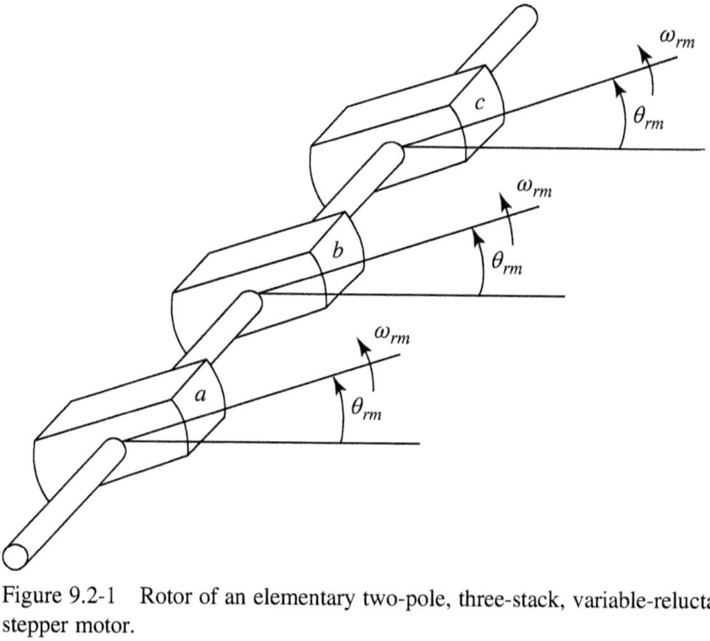

Figure 9.2-1 Rotor of an elementary two-pole, three-stack, variable-reluctance stepper motor.

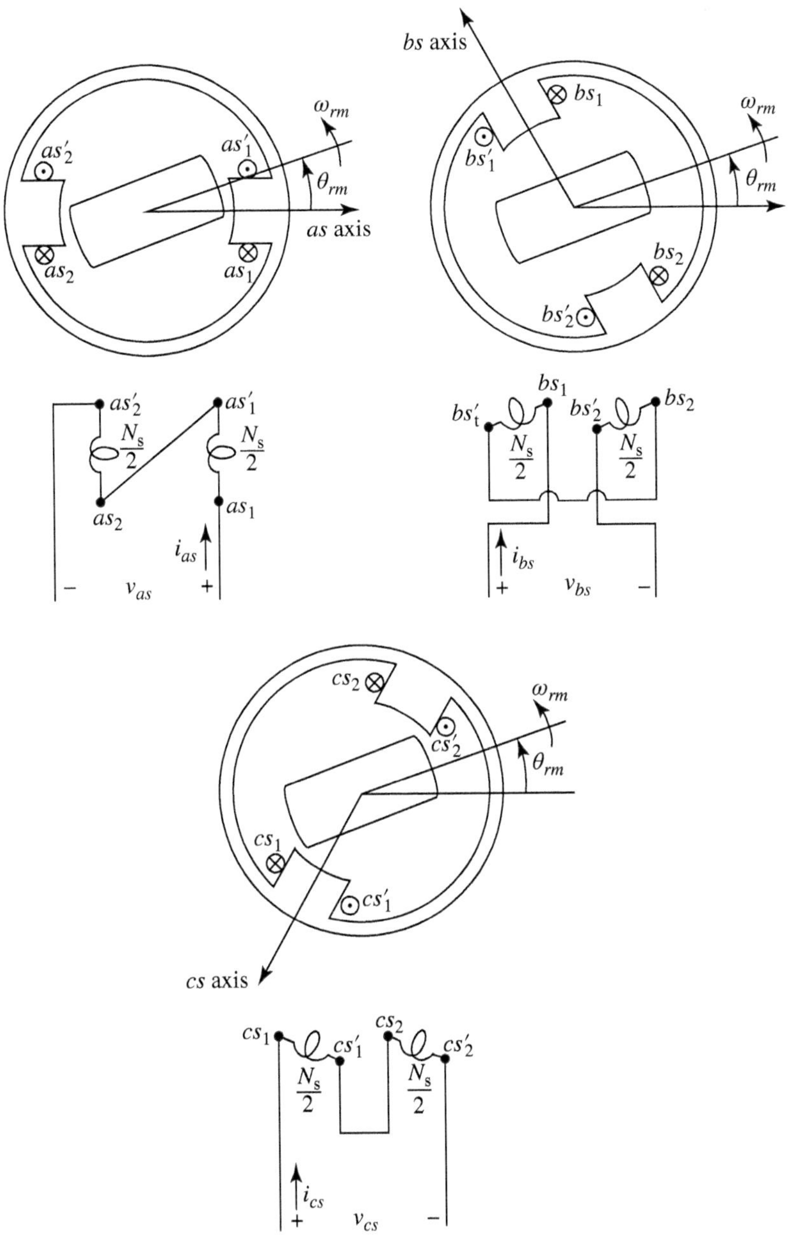

Figure 9.2-2 Stator configuration for an elementary two-pole, three-stack, variable-reluctance stepper motor.

think of a three-phase ac device when we hear the words three-phase machine. We will find that a stepper motor is a discrete device, operated by switching a dc voltage from one stator winding to the other. Although more than three stacks (phases) may be used, perhaps as many as seven, three-stack variable-reluctance stepper motors are quite common. Our previous meaning of phase must be modified to accommodate the stepper motor.

Before writing any equations, let us see if we can gain some insight in regard to the operation of this device. To start, let the *bs* and *cs* windings in Fig. 9.2-2 be open-circuited, and let us apply a dc voltage to the *as* winding, whereupon we will assume that a constant i_{as} is immediately established. Now, since the magnetic systems of the three single-phase stators are separate, flux set up by one winding does not link the other windings. Hence, with only the *as* winding energized, flux exists only in the *as* axis. We know from our work back in Chapter 2 that the minimum-reluctance path of the *a* part of the rotor (see Fig. 9.2-2) will align with the *as* axis. That is, at equilibrium with zero load torque, θ_{rm}, in all parts of Fig. 9.2-2 is the same: either zero or 180°. Let us say it is zero to make our discussion easier. (What would the rotor do if we could instantaneously reverse the direction of i_{as}?)

Stepper motors are used to convert digital or discrete information into a change in angular position. Let us see how positioning (stepping) is achieved. For this, let us instantaneously de-energize the *as* winding and immediately establish a direct current in the *bs* winding. The minimum-reluctance path of the rotor will align itself with the *bs* axis. To do this, the rotor would rotate clockwise from $\theta_{rm} = 0$ to $\theta_{rm} = -60°$. Note that by advancing the mmf from the positive *as* axis to the positive *bs* axis, 120° counterclockwise, we have caused a 60° clockwise rotation of the rotor. There must be something wrong here. We recall from our work with rotating magnetic fields in Chapter 4 that, with the magnetic axes as shown in Fig. 9.2-2, an *abc* sequence of balanced sinusoidal currents will yield operation at synchronous speed with the rotor rotating counterclockwise. Therefore, it would seem that rotating the air-gap mmf from the positive *as* axis to the positive *bs* axis would cause rotation in the counterclockwise direction; however, two of the three phases are not involved. In the case of variable-reluctance stepper motors, we will find that the direction of stepping can be either in the same or opposite direction of the rotation of the air-gap mmf depending upon the number of phases (stacks), the number of poles created by the stator windings, and the number of rotor teeth.

If, instead of energizing the *bs* winding, we energize the *cs* winding in Fig. 9.2-2, the rotor would have stepped counterclockwise from $\theta_{rm} = 0$ to $\theta_{rm} = 60°$. Thus, applying a dc voltage separately in the sequence *as, bs, cs, as,* ... produces 60° steps in the clockwise direction, whereas the sequence *as, cs, bs, as,* ... produces 60° steps in the counterclockwise direction. We need at least three stacks to achieve controlled stepping in both directions.

Before defining some stepper motor terms, let us think of one more thing. What if we energized the *as* and *bs* windings with the same current? That is, assume that initially the *as* winding is energized with $\theta_{rm} = 0$ and the *bs* winding is energized without de-energizing the *as* winding. What happens? Well, the rotor rotates clockwise from $\theta_{rm} = 0$ to $\theta_{rm} = -30°$. We have reduced our step length by one half. This is referred to as *half-step operation*.

It is time to define terms. Let RT denote the number of rotor teeth per stack and ST the number of stator teeth per stack. The elementary device shown in Figs. 9.2-1 and 9.2-2 has two poles, two rotor teeth, and two stator teeth per stack; thus, $RT = ST = 2$. In fact, RT always equals ST in a multistack variable-reluctance stepper motor. The number of stacks (phases) is denoted as N; here $N = 3$. Now, the tooth pitch, which we will denote as TP, is the angular displacement between rotor teeth. In this case, $TP = 180°$. We can write

$$TP = \frac{2\pi}{RT} \tag{9.2-1}$$

We have one more term to define: the step length, denoted as SL. It is the angular rotation of the rotor as we change the excitation (dc voltage) from one phase to the other. In this case, the step length is 60°, $SL = 60°$. If we energize each stack separately, then going from *as* to *bs* to *cs* back to *as* causes the rotor to rotate one tooth pitch. In other words, the number of stacks (phases) times the step length is a tooth pitch. That is,

$$TP = N \cdot SL \tag{9.2-2}$$

We can substitute (9.2-1) into (9.2-2) and obtain

$$SL = \frac{TP}{N} = \frac{2\pi}{RT \cdot N} \tag{9.2-3}$$

We shall find use for all of these new terms as we go along.

Although the elementary device shown in Figs. 9.2-1 and 9.2-2 offers a good starting point in our analysis of stepper motors, it has limited application owing to its large step length. Let us consider the four-pole, three-stack, variable-reluctance stepper motor with four rotor teeth, as illustrated in Fig. 9.2-3. Here, $RT = 4$ and $N = 3$; therefore, from (9.2-1) the tooth pitch is $TP = 2\pi/RT = 90°$. From (9.2-2), the step length is $SL = TP/N = 30°$ and an *as, bs, cs, as,* … sequence produces 30° steps in the clockwise direction.

The device shown in Fig. 9.2-4 is a four-pole, three-stack variable-reluctance stepper motor with eight rotor teeth. In this case, $RT = 8$ and $N = 3$, thus, $TP = 45°$ and $SL = 15°$; however, in this device an *as, bs, cs, as,* … sequence produces 15° steps in the counterclockwise direction. The pattern is clear; by increasing the number of rotor teeth, we reduce the step length. The step lengths of multistack variable-reluctance stepping motors typically range from 2° to 15°.

There appears to be an inconsistency in Fig. 9.2-4. In particular, θ_{rm} is referenced from the *as* axis to a position between rotor teeth. Earlier in this section, we established that, in the case of stepper motors, we would reference θ_{rm} from the *as* axis to the minimum-reluctance path of the rotor, whereupon the reluctance of the

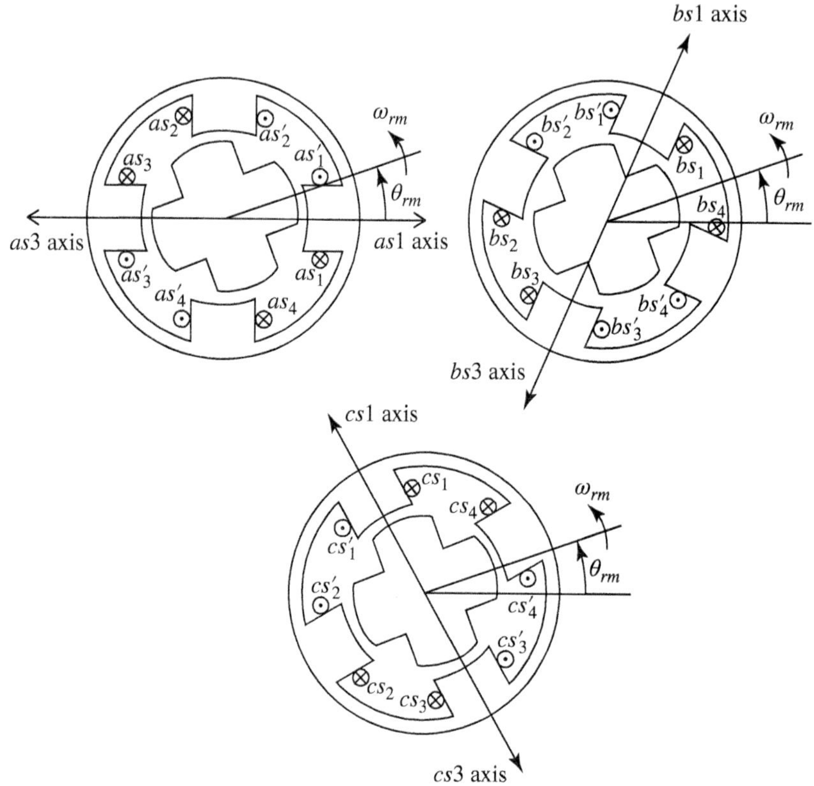

Figure 9.2-3 Four-pole, three-stack, variable-reluctance stepper motor with four rotor teeth.

magnetic system associated with the *as* winding would be minimum when $\theta_{rm} = 0$. At first glance it appears that we have violated this stepper motor convention; however, when θ_{rm} is zero in Fig. 9.2-4, the reluctance of the magnetic system associated with the *as* winding is minimum. Hence, we must reference θ_{rm} from a position between rotor teeth to maintain the convention that we established earlier in this section. A cutaway view of a four-pole, three-stack variable-reluctance stepper motor with 16 rotor teeth is shown in Fig. 9.2-5.

SP9.2-1 Calculate the step length for an eight-pole, three-stack, variable-reluctance stepper motor with 16 rotor teeth. [$SL = 7.5°$]

SP9.2-2 Consider the two-pole, two-phase reluctance motor shown in Fig. 7.6-1. Calculate (a) *TP*, (b) *SL*, and (c) determine the direction of rotation when a dc voltage is switched from the *as* winding to the *bs* winding. [(a) $TP = 180°$; (b) $SL = 90°$; (c) either ccw or cw]

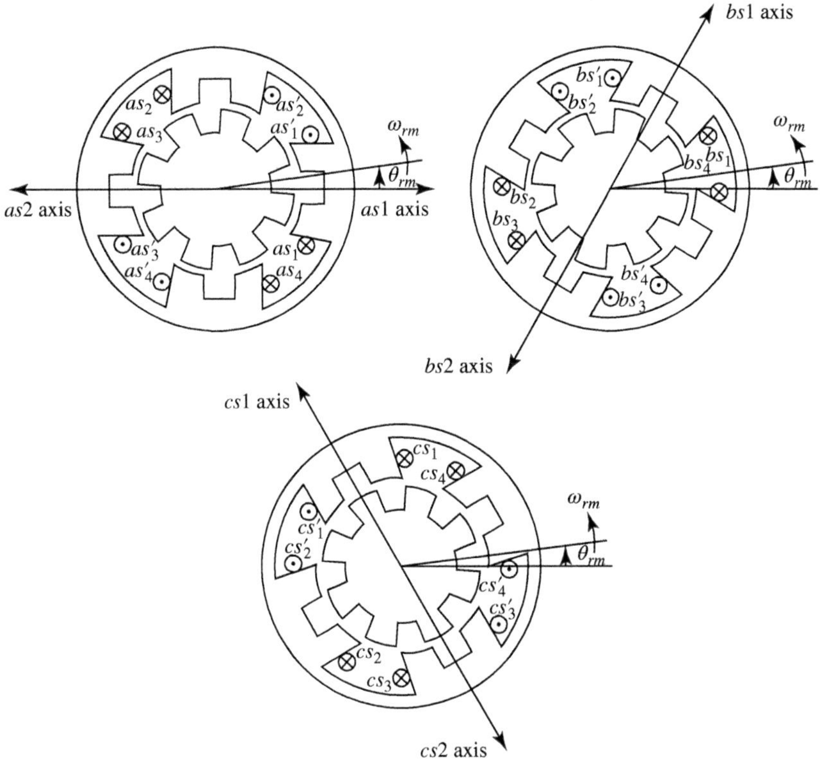

Figure 9.2-4 Four-pole, three-stack, variable-reluctance stepper motor with eight rotor teeth.

Figure 9.2-5 Cutaway view of four-pole, three-stack, variable-reluctance stepper motor with 16 rotor teeth. (Courtesy of Warner Electric.)

9.3 EQUATIONS FOR MULTISTACK VARIABLE-RELUCTANCE STEPPER MOTORS

The voltage equations for a three-stack, variable-reluctance stepper motor may be written as

$$v_{as} = r_s i_{as} + \frac{d\lambda_{as}}{dt} \tag{9.3-1}$$

$$v_{bs} = r_s i_{bs} + \frac{d\lambda_{bs}}{dt} \tag{9.3-2}$$

$$v_{cs} = r_s i_{cs} + \frac{d\lambda_{cs}}{dt} \tag{9.3-3}$$

In matrix form,

$$\mathbf{v}_{abcs} = \mathbf{r}_s \mathbf{i}_{abcs} + p\boldsymbol{\lambda}_{abcs} \tag{9.3-4}$$

where p is the operator d/dt and, for voltages, currents, and flux linkages

$$(\mathbf{f}_{abcs})^T = [f_{as} \quad f_{bs} \quad f_{cs}] \tag{9.3-5}$$

with

$$\mathbf{r}_s = \begin{bmatrix} r_s & 0 & 0 \\ 0 & r_s & 0 \\ 0 & 0 & r_s \end{bmatrix} \tag{9.3-6}$$

Since magnetic coupling does not exist between phases, we can write the flux linkages as

$$\begin{bmatrix} \lambda_{as} \\ \lambda_{bs} \\ \lambda_{cs} \end{bmatrix} = \begin{bmatrix} L_{asas} & 0 & 0 \\ 0 & L_{bsbs} & 0 \\ 0 & 0 & L_{cscs} \end{bmatrix} \begin{bmatrix} i_{as} \\ i_{bs} \\ i_{cs} \end{bmatrix} \tag{9.3-7}$$

For the purpose of expressing the self-inductances L_{asas}, L_{bsbs}, and L_{cscs}, let us first consider the elementary two-pole device illustrated in Fig. 9.2-2. With our work from Chapter 1, we can write as a first approximation,

$$L_{asas} = L_{ls} + L_A + L_B \cos 2\theta_{rm} \tag{9.3-8}$$

$$L_{bsbs} = L_{ls} + L_A + L_B \cos 2\left(\theta_{rm} - \frac{2}{3}\pi\right) \tag{9.3-9}$$

$$L_{cscs} = L_{ls} + L_A + L_B \cos 2\left(\theta_{rm} - \frac{4}{3}\pi\right) \tag{9.3-10}$$

From our previous work, we are aware that L_{ls} is the leakage inductance, whereas L_A and L_B are constants with $L_A > L_B$. For constant rotor speed, the rotor displacement can be expressed as

$$\theta_{rm} = \omega_{rm}t + \theta_{rm}(0) \tag{9.3-11}$$

We will use θ_{rm}, the actual rotor displacement, rather than θ_r, the electrical angular displacement. Although θ_{rm} and θ_r are related, $\theta_r = (P/2)\theta_{rm}$, where P is the number of poles, we will find it more convenient to use θ_{rm} in the analysis of stepper motors. We see that (9.3-8) is similar to (1.7-29) or (2.7-3), with θ_{rm} referenced to the minimum-reluctance path of the rotor. Equation (9.3-9) is easily developed once we realize that the self-inductance of the *bs* winding is the same as that of the *as* winding. However, since θ_{rm} is referenced from the *as* axis, the angular displacement to the *bs* axis from the *as* axis must be subtracted from θ_{rm} so that, when $\theta_{rm} = \frac{2}{3}\pi$, the argument of (9.3-9) is zero and (9.3-9) with $\theta_{rm} = \frac{2}{3}\pi$ becomes the same as (9.3-8) with $\theta_{rm} = 0$. Following this same line of reasoning, we would determine that the angular displacement of (9.3-10) is $-\frac{4}{3}\pi$. However, since $\cos 2\left(\theta_{rm} - \frac{4}{3}\pi\right) = \cos 2\left(\theta_{rm} + \frac{2}{3}\pi\right)$, we can use $\frac{2}{3}\pi$ as the angular displacement for L_{cscs}. It is obvious that we can express (9.3-8) through (9.3-10) in various forms. By adding or subtracting the appropriate multiple of 2π to or from the arguments in (9.3-8) through (9.3-10), we can also express the inductances as

$$L_{asas} = L_{ls} + L_A + L_B \cos 2\theta_{rm} \tag{9.3-12}$$

$$L_{bsbs} = L_{ls} + L_A + L_B \cos 2\left(\theta_{rm} + \frac{1}{3}\pi\right) \tag{9.3-13}$$

$$L_{cscs} = L_{ls} + L_A + L_B \cos 2\left(\theta_{rm} + \frac{2}{3}\pi\right) \tag{9.3-14}$$

The self-inductances of the four-pole, three-stack, variable-reluctance device with four rotor teeth shown in Fig. 9.2-3 can be approximated as

$$L_{asas} = L_{ls} + L_A + L_B \cos 4\theta_{rm} \tag{9.3-15}$$

$$L_{bsbs} = L_{ls} + L_A + L_B \cos 4\left(\theta_{rm} + \frac{1}{6}\pi\right) \tag{9.3-16}$$

$$L_{cscs} = L_{ls} + L_A + L_B \cos 4\left(\theta_{rm} + \frac{2}{6}\pi\right) \tag{9.3-17}$$

Although we are using the same L_{ls}, L_A, and L_B to denote constants, we realize that these are not equal for the various machines.

For the four-pole, three-stack, variable-reluctance stepper motor with eight rotor teeth shown in Fig. 9.2-4, we can approximate the self-inductances as

$$L_{asas} = L_{ls} + L_A + L_B \cos 8\theta_{rm} \tag{9.3-18}$$

$$L_{bsbs} = L_{ls} + L_A + L_B \cos 8\left(\theta_{rm} - \frac{1}{12}\pi\right) \tag{9.3-19}$$

$$L_{cscs} = L_{ls} + L_A + L_B \cos 8\left(\theta_{rm} - \frac{2}{12}\pi\right) \tag{9.3-20}$$

It may be observed that for the devices shown in Figs. 9.2-2 and 9.2-3, where we have previously noted that a counterclockwise rotation of the stator mmf and stepping are in opposite directions, the inductances may be expressed as

$$L_{asas} = L_{ls} + L_A + L_B \cos\left(RT\theta_{rm}\right) \tag{9.3-21}$$

$$L_{bsbs} = L_{ls} + L_A + L_B \cos\left[RT(\theta_{rm} + SL)\right] \tag{9.3-22}$$

$$L_{cscs} = L_{ls} + L_A + L_B \cos\left[RT(\theta_{rm} + 2SL)\right] \tag{9.3-23}$$

For the device shown in Fig. 9.2-4, where rotation of the stator mmf and stepping are in the same direction, the self-inductances may be expressed as

$$L_{asas} = L_{ls} + L_A + L_B \cos\left(RT\theta_{rm}\right) \tag{9.3-24}$$

$$L_{bsbs} = L_{ls} + L_A + L_B \cos\left[RT(\theta_{rm} - SL)\right] \tag{9.3-25}$$

$$L_{cscs} = L_{ls} + L_A + L_B \cos\left[RT(\theta_{rm} - 2SL)\right] \tag{9.3-26}$$

For each additional phase (stack), the self-inductance is offset by the appropriate multiple of SL. An expression for the electromagnetic torque may be obtained from Table 2.5-1:

$$T_e = \frac{\partial W_c(\mathbf{i}, \theta_{rm})}{\partial \theta_{rm}} \tag{9.3-27}$$

Since we are assuming a linear magnetic system, the field energy and coenergy are equal. Thus, since the mutual inductances are zero,

$$W_c = \frac{1}{2}L_{asas}i_{as}^2 + \frac{1}{2}L_{bsbs}i_{bs}^2 + \frac{1}{2}L_{cscs}i_{cs}^2 \tag{9.3-28}$$

Substituting the self-inductances given by (9.3-21) through (9.3-23) into (9.3-28) and taking the partial derivative with respect to θ_{rm} yields

$$T_e = -\frac{RT}{2}L_B\left\{i_{as}^2 \sin\left(RT\theta_{rm}\right) + i_{bs}^2 \sin\left[RT(\theta_{rm} + SL)\right] + i_{cs}^2 \sin\left[RT\left(\theta_{rm} + 2SL\right)\right]\right\} \tag{9.3-29}$$

An alternate form of (9.3-29) using the tooth pitch TP is

$$T_e = -\frac{RT}{2}L_B\left\{i_{as}^2 \sin\left(\frac{2\pi}{TP}\theta_{rm}\right) + i_{bs}^2 \sin\left[\frac{2\pi}{TP}\left(\theta_{rm} + \frac{TP}{N}\right)\right] + i_{cs}^2 \sin\left[\frac{2\pi}{TP}\left(\theta_{rm} + \frac{2TP}{N}\right)\right]\right\} \tag{9.3-30}$$

It is important to note that (9.3-29) and (9.3-30) are written for rotation of the stator mmf and stepping of the rotor in opposite directions. For stepping in the same direction, the sign preceding both *SL*s in (9.3-29) and both (*TP/N*)s in (9.3-30) must be changed. Note also that the magnitude of the torque is proportional to the number of rotor teeth *RT*.

The torque and rotor angular position are related as

$$T_e = J\frac{d^2\theta_{rm}}{dt^2} + B_m\frac{d\theta_{rm}}{dt} + T_L \tag{9.3-31}$$

where J is the total inertia in kg·m^2 and B_m is a damping coefficient associated with the mechanical rotational system in N·m·s/rad. The electromagnetic torque T_e is positive in the counterclockwise direction (positive direction of θ_{rm}), whereas the load torque T_L is positive in the clockwise direction.

SP9.3-1 The stator currents of a three-stack, variable-reluctance machine are $i_{as} = I$, $i_{bs} = -I$, and $i_{cs} = 0$. Determine the no-load rotor position. [$\theta_{rm} = \pm TP/6$]

SP9.3-2 Repeat SP9.3-1 with $i_{as} = i_{bs} = i_{cs}$. [T_e is zero for all values of θ_{rm}.]

9.4 OPERATING CHARACTERISTICS OF MULTISTACK VARIABLE-RELUCTANCE STEPPER MOTORS

It is instructive to take a little closer look at the operating characteristics of a multistack variable-reluctance stepper motor from the standpoint of idealized, pseudo-steady-state conditions. For this purpose, let us consider the expression for torque given by (9.3-30) for a three-stack motor with opposite directions of rotation of the stator mmf and stepping. In particular,

$$T_e = -\frac{RT}{2}L_B\left\{i_{as}^2\sin\left(\frac{2\pi}{TP}\theta_{rm}\right) + i_{bs}^2\sin\left[\frac{2\pi}{TP}\left(\theta_{rm} + \frac{TP}{3}\right)\right] + i_{cs}^2\sin\left[\frac{2\pi}{TP}\left(\theta_{rm} + \frac{2TP}{3}\right)\right]\right\} \tag{9.4-1}$$

In Fig. 9.4-1, the three terms of (9.4-1) are plotted separately for equal, constant (steady-state) currents. Let us assume that there is no load torque, $T_L = 0$, and $i_{as} = I$, whereas i_{bs} and i_{cs} are zero. Only the first term of (9.4-1) is present; that is, only the steady-state torque due to i_{as} exists. The stable steady-state rotor position would be at $\theta_{rm} = 0$ denoted as point 1 in Fig. 9.4-1. Now, let us assume that i_{as} is instantaneously decreased from I to zero while i_{bs} is increased from zero to I. Hence, the steady-state torque plot due to i_{as} would instantaneously disappear from Fig. 9.4-1 and the torque due to i_{bs} would immediately appear. Now, we know that this cannot happen in practice since there would be electrical transients involved, but we are neglecting all transients in this discussion. Since the torque at point 2 is negative, the rotor will rotate in the clockwise direction. We will then proceed along the i_{bs} torque plot until we have reached point 3. Note that we have moved one step length in the clockwise direction. If, instead of energizing the *bs* winding after de-energizing the *as* winding, we energized the *cs* winding, then the torque at

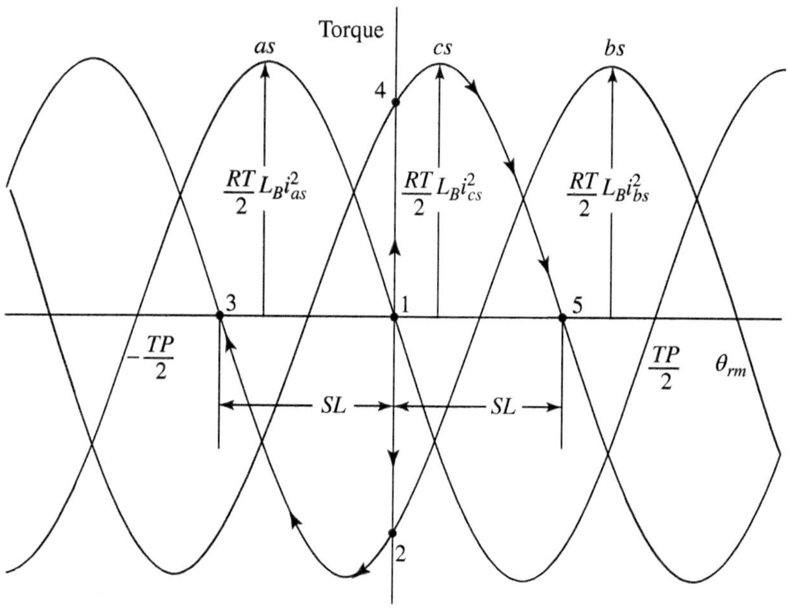

Figure 9.4-1 Stepping operation of a three-stack, variable-reluctance stepper motor without load torque – steady-state torque-angle plots.

point 4 would appear. This is a positive T_e, so the rotor will rotate in the counterclockwise direction and we will ride along the torque angle plot to point 5. Please realize that not only are we neglecting the electrical transients in this discussion but we are also neglecting the mechanical transients. Normally, there would be a damped oscillation about the new operating point, in this example, either point 3 or 5.

Half-step operation is depicted in Fig. 9.4-2. To explain this, let us again start at point 1, where $T_L = 0$ and only the *as* winding is energized $(i_{as} = I)$. Instantaneously, the *bs* winding is energized and $i_{bs} = I$. Now, both i_{as} and i_{bs} are I and only the *as* + *bs* torque plot, shown in Fig. 9.4-2, exists. Immediately, the torque at point 2 appears and the rotor starts to rotate in the clockwise direction, coming to rest at point 3. The rotor has moved $SL/2$ clockwise.

Stepping action with a load torque is shown in Fig. 9.4-3. Assume that initial operation is at point 1 with $i_{as} = I$ and $i_{bs} = i_{cs} = 0$. Recall that T_e is positive in the counterclockwise direction, whereas T_L is positive in the clockwise direction, and stable operation occurs when $T_e = T_L$. Thus, at point 1, $T_e = T_L$. The *as* winding is de-energized while the *bs* winding is energized. Immediately, the negative T_e at point 2 appears and the rotor will start to move clockwise to point 3. If the *cs* winding is energized rather than the *bs* winding, the torque at point 4 would appear and the rotor would move to point 5. Note that the step length is still the same in both directions. However, the rotor will move more rapidly in the clockwise direction than in the counterclockwise direction since the load torque is in the clockwise

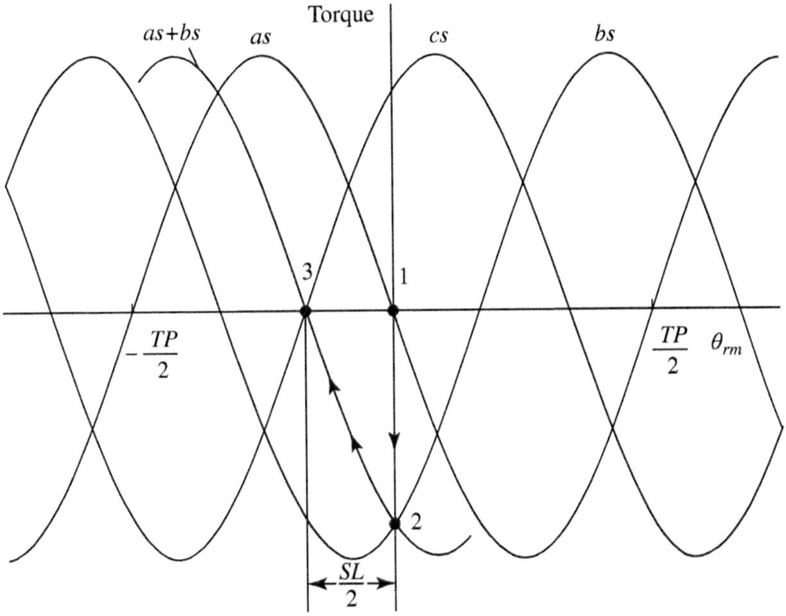

Figure 9.4-2 Half-step operation of a three-stack, variable-reluctance stepper motor – steady-state torque-angle plots.

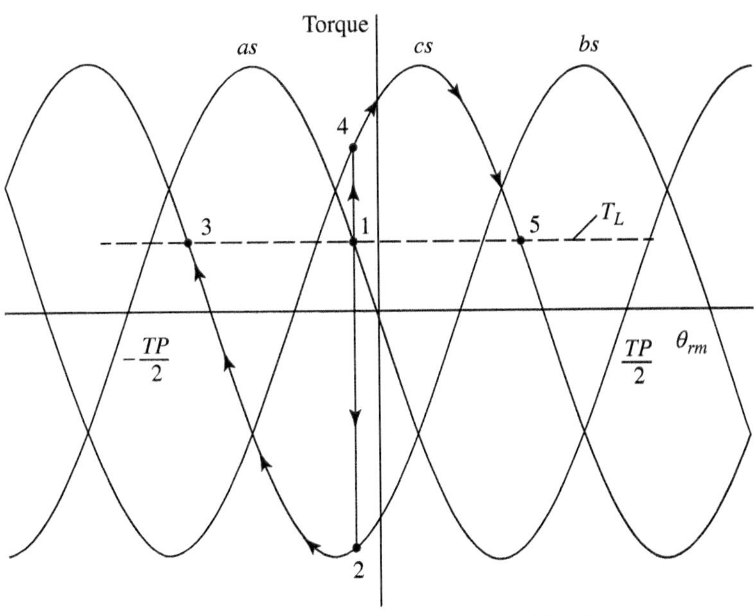

Figure 9.4-3 Stepping operation of a three-stack, variable-reluctance stepper motor with load torque – steady-state torque-angle plots.

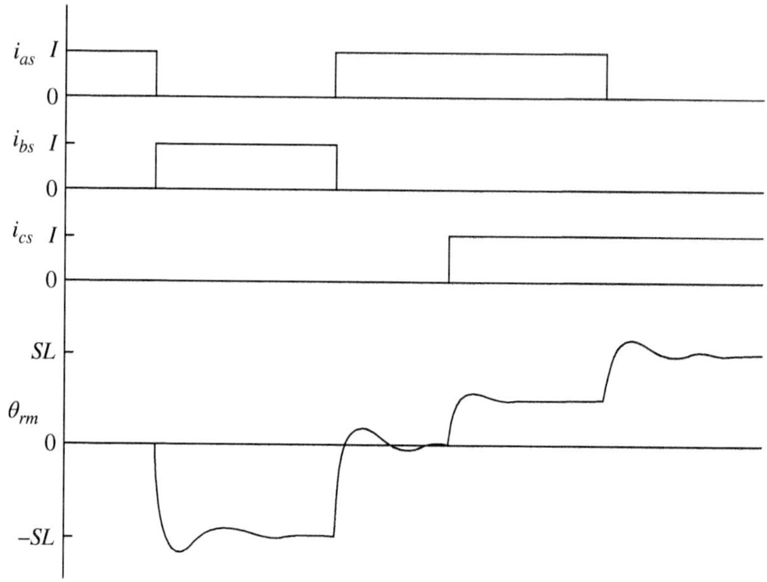

Figure 9.4-4 Stepping operation depicting θ_{rm} versus time – no load torque.

direction. In other words, there is a larger torque to accelerate the rotor in the clockwise direction than in the counterclockwise direction.

The plots of i_{as}, i_{bs}, i_{cs}, and θ_{rm} shown in Fig. 9.4-4 allow us to view stepping operation from another standpoint. Initially, there is no load torque and $i_{as} = I$. The current i_{as} is stepped off and i_{bs} is stepped on. The rotor rotates clockwise to $\theta_{rm} = -SL$. Here, we have indicated the presence of a damped mechanical oscillation, which was not shown in the steady-state torque angle plots. Next, i_{bs} is switched off and i_{as} is switched back on. The rotor ends up back at $\theta_{rm} = 0$. Next, we see half-step operation; i_{as} remains at I while i_{cs} is switched to I. The rotor advances to $\frac{1}{2}SL$. When i_{as} is switched to zero, the rotor again advances by $\frac{1}{2}SL$ to $\theta_{rm} = SL$.

SP9.4-1 In Fig. 9.4-3, the load torque is such that the initial operating point with $i_{as} = I$ and $i_{bs} = i_{cs} = 0$ is at $\theta_{rm} = -TP/8$. The current in the *as* winding is switched to zero and the current in the *cs* winding is switched to I. Determine the final value of θ_{rm}. In which direction will the rotor rotate? [$\theta_{rm} = -TP/8 - 2SL$; cw]

SP9.4-2 It is desirable to step from $\theta_{rm} = 0$ to $\theta_{rm} = -SL/3$ for the device shown in Fig. 9.2-3. Assume that we have the facility to control the winding currents. Let $i_{as} = I$; determine i_{bs}. [$i_{bs} = 0.81I$]

9.5 SINGLE-STACK VARIABLE-RELUCTANCE STEPPER MOTORS

As its name suggests, the single-stack, variable-reluctance stepper motor has only one stack and all stator phases are arranged on this single stack. A three-phase, single-stack, variable-reluctance stepper motor is shown in Fig. 9.5-1. Here, it appears that we have taken the three two-pole, single-phase stators shown in Fig. 9.2-2 and squeezed them into one stack. The magnetic axes of the stator windings are displaced 120° as in the case of the three-phase machines considered in earlier chapters; however, the stepper motor generally has stator teeth or poles that protrude rather than a circular inner stator surface.

Recall that in the case of the multistack variable-reluctance motor, the number of rotor and stator teeth per stack is the same. In the case of the single-stack stepper motor, RT is never equal to ST. If, for example, the rotor shown in Fig. 9.5-1 had the same number of teeth as the stator, then, when two diagonally opposite rotor teeth are aligned with two diagonally opposite stator teeth, all diagonally opposite rotor teeth would be aligned with diagonally opposite stator teeth and stepping action could not occur. The equations that we derived for the tooth pitch TP and step length SL for the multistack variable-reluctance stepper motor also apply for the single-stack stepper motor. For the two-pole, three-phase stepper motor shown in Fig. 9.5-1, $RT = 4$ and, thus, $TP = 2\pi/RT = 90°$ and $SL = TP/N = 30°$. Note that the sequence *as, bs, cs, as, …* produces a counterclockwise stepping of the rotor.

Two other types of three-phase, single-stack, variable-reluctance stepper motors are shown in Figs. 9.5-2 and 9.5-3. The two-pole device shown in Fig. 9.5-2 has six stator teeth and eight rotor teeth. $TP = 45°$ and $SL = 15°$, and an *as, bs, cs, as, …* sequence produces a clockwise stepping of the rotor. For the four-pole, three-phase device shown in Fig. 9.5-3, $ST = 12$ and $RT = 8$. Thus, $TP = 45°$ and $SL = 15°$. The step length is the same as for the stator with six teeth (Fig. 9.5-2); however, counterclockwise stepping of the rotor occurs with the

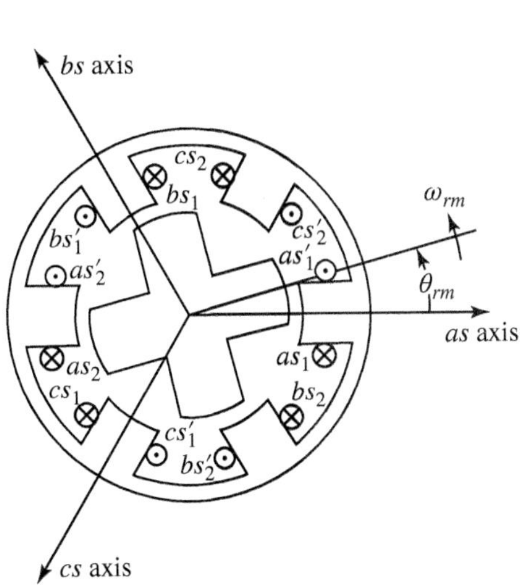

Figure 9.5-1 Two-pole, three-phase, single-stack, variable-reluctance stepper motor with six stator teeth and four rotor teeth.

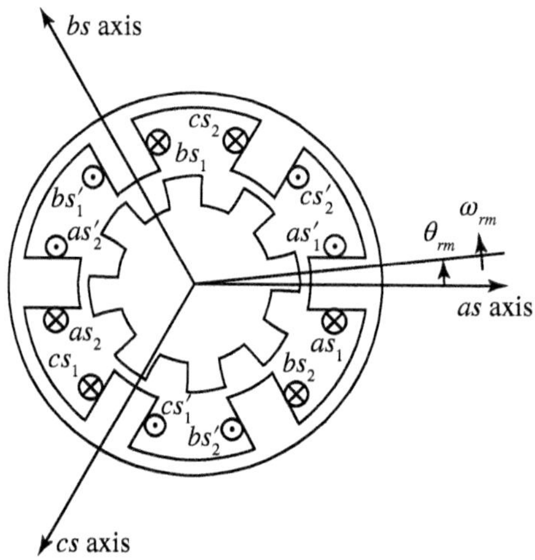

Figure 9.5-2 Two-pole, three-phase, single-stack, variable-reluctance stepper motor with six stator teeth and eight rotor teeth.

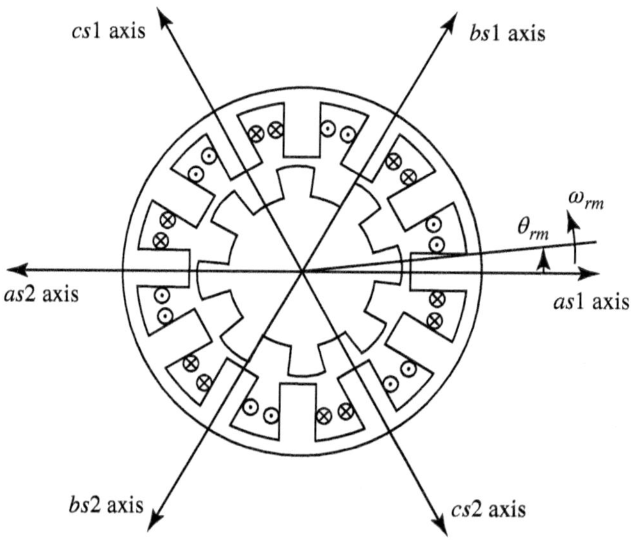

Figure 9.5-3 Four-pole, three-phase, single-stack, variable-reluctance stepper motor with 12 stator teeth and eight rotor teeth.

sequence *as, bs, cs, as,* .... In Fig. 9.5-3, the labeling of the coil sides of the windings is omitted because of lack of space.

The expressions given for the self-inductances of the three-stack (phase) variable-reluctance stepper motor, (9.3-21) through (9.3-26), also apply to the three-phase, single-stack, variable-reluctance stepper motor. Therefore, it would appear

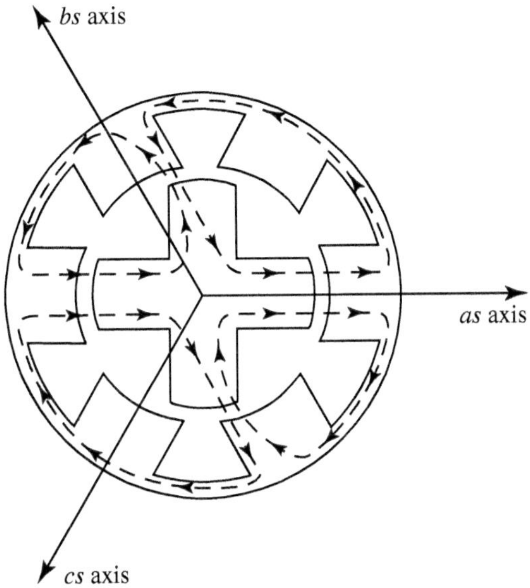

Figure 9.5-4 Two-pole, three-phase, single-stack, variable-reluctance stepper motor given in Fig. 9.5-1 with $\theta_{rm} = 0$.

that the operation of the single-stack and multistack variable-reluctance stepper motors may be described by the same set of equations. Although this perception is essentially valid from an idealized point of view, it is not valid in the practical world. We see from Figs. 9.5-1 through 9.5-3 that the stator windings share the same magnetic system. Hence, there is a possibility of mutual coupling between stator phases. For the purposes of discussion, let us consider Fig. 9.5-4, which is Fig. 9.5-1 with $\theta_{rm} = 0$. The dashed lines shown therein depict the flux linking the *bs* winding due to positive current flowing in the *as* winding. If we assume that the reluctance of the iron is small so that it can be neglected, the flux linkages cancel, whereupon mutual coupling would not exist between stator phases. From an idealized standpoint, this is a valid line of reasoning; from a practical standpoint, it is not.

Stepper motors are generally designed to operate at current levels that saturate the iron of the machine. Hence, owing to the increased reluctance of the saturated iron, less flux will be circulating around the longer paths through iron than through the short paths. Hence, a net mutual flux would exist between stator phases. For the case depicted in Fig. 9.5-4, there would be a net flux in the direction of the positive *bs* axis as a result of the saturation of the iron. Albeit relatively small in amplitude, a mutual inductance does exist in the practical application of single-stack, variable-reluctance stepper motors. This complicates the analysis of these devices far beyond that which we care to deal with in this text. Instead, for our first look at stepper motors, we will consider it sufficient to neglect saturation and the mutual coupling it causes in single-stack, variable-reluctance stepper motors.

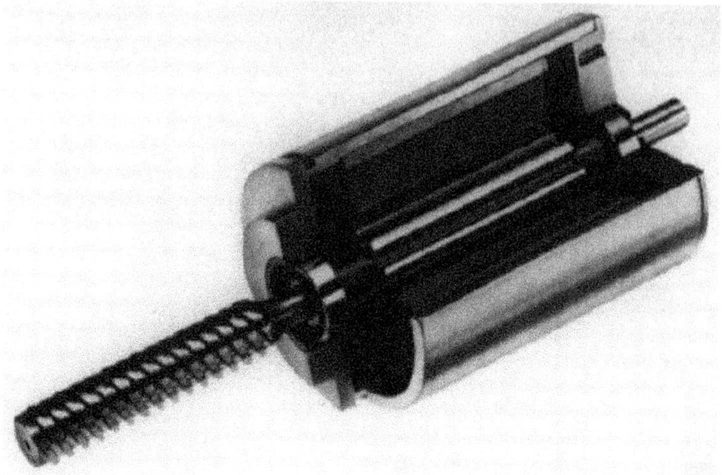

Figure 9.5-5 Single-stack, 15°/step, variable-reluctance stepper motor. (Courtesy of Warner Electric.)

A single-stack, variable-reluctance stepper motor is shown in Fig. 9.5-5. This device has a 15° step length and is equipped with an integral lead screw for translational motion.

SP9.5-1 Express the number of stator teeth possible for an N-phase, single-stack, variable-reluctance stepper motor. [$ST = n(2N)$, where $n = 1, 2, 3, \ldots$]

SP9.5-2 The rotor in Fig. 9.5-1 is replaced by the rotor from Fig. 9.5-3. Determine (a) TP, (b) SL, and (c) the direction of rotation with an as, bs, cs, as, ... sequence. [(a) $TP = 45°$; (b) $SL = 15°$; (c) cw]

9.6 BASIC CONFIGURATION OF PERMANENT-MAGNET STEPPER MOTORS

The permanent-magnet stepper motor is quite common. Actually, it is a permanent-magnet ac machine and it may be operated either as a stepping motor or as a continuous-speed device. Here, we will concern ourselves only with its application as a stepping motor since continuous-speed operation is similar to the operation of a permanent-magnet ac machine covered in Chapter 6.

A two-pole, two-phase, permanent-magnet stepper motor with five rotor teeth is shown in Fig. 9.6-1. Most permanent-magnet stepper motors have more than two poles and more than five rotor teeth; some may have as many as eight poles and as many as 50 rotor teeth. Nevertheless, the elementary device shown in Fig. 9.6-1 is sufficient to illustrate the principle of operation of the permanent-magnet stepper motor. The axial cross-sectional view shown in Fig. 9.6-1b illustrates the permanent magnet that is mounted on the rotor. The permanent magnet magnetizes the iron end caps that are also mounted on the rotor and are slotted to

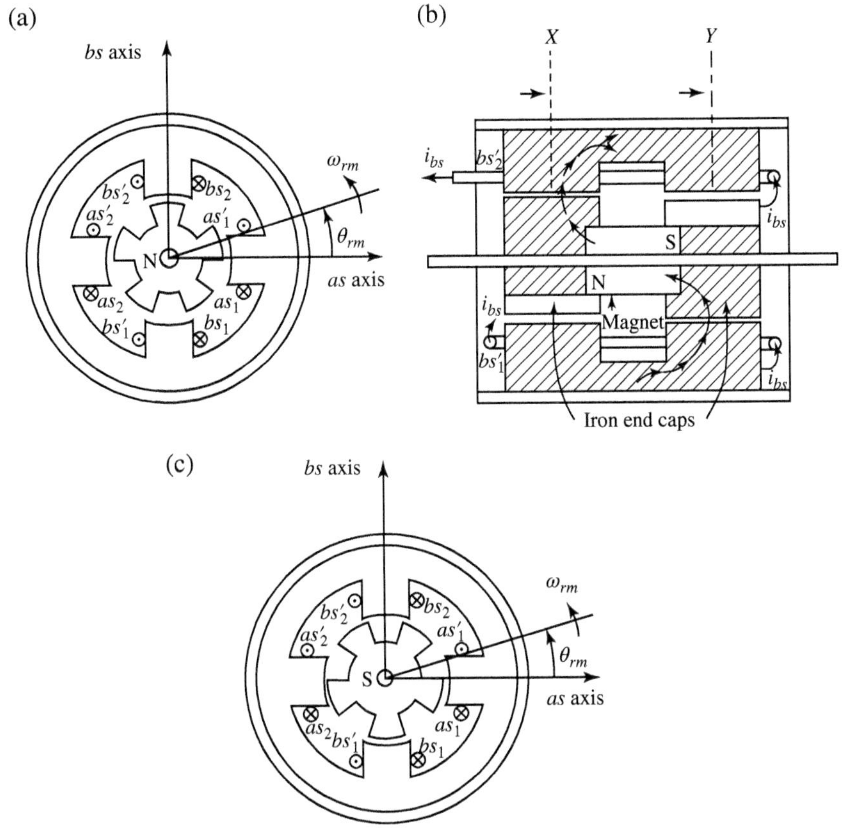

Figure 9.6-1 Two-pole, two-phase, permanent-magnet stepper motor: (a) axial view at X; (b) side cross-sectional view; (c) axial view at Y.

form the rotor teeth. The view looking from left to right at X is shown in Fig. 9.6-1a. Figure 9.6-1c is the view from left to right at Y. The left end cap shown in Fig. 9.6-1a is magnetized as a north pole; the right end cap shown in Fig. 9.6-1c is magnetized as a south pole. Note that the rotor teeth of the left end cap are displaced one-half a tooth pitch from the teeth on the right end cap. Also, note that the stator windings are wound over the full axial length of the device; a part of the bs winding is shown in Fig. 9.6-1b.

Let us trace the main path of flux linking the bs winding for the rotor position shown in Fig. 9.6-1. This path is depicted by dashed lines in Fig. 9.6-1b. However, it is necessary to visualize the drawing in three dimensions. Flux leaves the left end cap through the rotor tooth at the top of the rotor that is aligned with the stator tooth that has the bs_2 part of the bs winding. The flux travels up through the stator tooth in the stator iron. The flux then splits and travels around the circumference of the stator and returns to the south pole of the rotor through the stator tooth positioned at the bottom in Fig. 9.6-1c, on which the bs_1 winding is wound. The main flux linking the *as* winding for the rotor position shown in Fig. 9.6-1 would enter the stator tooth on which the as_1 winding is wound from the rotor tooth on the right of Fig. 9.6-1a. The flux would travel around the circumference of the stator and return to the rotor through the stator pole upon which the as_2 winding is wound, Fig. 9.6-1c.

Stepping action can be explained by first assuming that the *bs* winding is open-circuited and a constant positive current is flowing in the *as* winding. As a result of this current, a south pole is established at the stator tooth on which the as_1 winding is wound, and a stator north pole is established at the stator tooth on which the as_2 winding is wound. The rotor would be positioned at $\theta_{rm} = 0$. Now, let us simultaneously de-energize the *as* winding while energizing the *bs* winding with a positive current. The rotor will move one step length in the counterclockwise direction. To continue stepping in the counterclockwise direction, the *bs* winding is de-energized and the *as* winding is energized with a negative current. That is, counterclockwise stepping occurs with a current sequence of i_{as}, i_{bs}, $-i_{as}$, $-i_{bs}$, i_{as}, ... Clockwise rotation is achieved by i_{as}, $-i_{bs}$, $-i_{as}$, i_{bs}, ...

The tooth pitch can be calculated from (9.2-1); however, the step length for a permanent-magnet stepper motor cannot be calculated from (9.2-3). As we have mentioned, counterclockwise rotation of the device shown in Fig. 9.6-1 is achieved by a sequence of i_{as}, i_{bs}, $-i_{as}$, $-i_{bs}$, i_{as}, ... We see that it takes four switchings (steps) to advance the rotor one tooth pitch. Thus,

$$TP = 2N \cdot SL \tag{9.6-1}$$

where N is the number of phases. Substituting (9.2-1) into (9.6-1) and solving for SL yields

$$SL = \frac{\pi}{RT \cdot N} \tag{9.6-2}$$

For the device shown in Fig. 9.6-1, $RT = 5$ and $N = 2$. From (9.6-2), $SL = 18°$.

Recall that in the case of variable-reluctance stepper motors, it is unnecessary to reverse the direction of the current in the stator windings to achieve rotation; therefore, the stator voltage source need only be unidirectional. However, in the case of a permanent-magnet stepper motor, it is necessary for the phase currents to flow in both directions to achieve rotation. Generally, stepper motors are supplied from a dc voltage source; hence, the electronic interface between the phase windings and the dc source must be bidirectional; that is, it must have the capability of applying a positive and negative voltage to each phase winding. This requirement markedly increases the cost of the electronic interface and its associated controls relative to a unidirectional source. As an alternative, permanent-magnet stepper motors are often equipped with what is referred to as *bifilar windings*. Rather than only one winding on each stator tooth, there are two identical windings with one wound opposite to the other, each having separate independent external terminals. With this type of winding configuration, the direction of the magnetic field established by the stator windings is reversed, not by changing the direction of the current but by reversing the sense of the winding through which current is flowing. If, for example, the device shown in Fig. 9.6-1 is equipped with bifilar windings, there would be another *as* winding and another *bs* winding with separate, independent, external terminals wound opposite on the stator teeth to the windings shown. Although this increases the size and weight of the stepper motor, it eliminates the need for a bidirectional electronic interface. When this permanent-magnet stepper motor is equipped with bifilar windings as just described, it is (perhaps, inappropriately) called a four-phase device. Actually it has four windings, but it is still a two-phase device magnetically. Although we are not going to consider the bifilar-wound machine in detail, one should be aware of this somewhat ambiguous nomenclature. More specifically, care should be taken when using (9.6-2) to calculate the step length. The number of phases N in (9.6-2) is the number of phases magnetically rather than the number of windings. A cutaway view of a permanent-magnet stepper motor is shown in Fig. 9.6-2.

Figure 9.6-2 Cutaway view of a permanent-magnet stepper motor. (Courtesy of Sanyo Denki.)

SP9.6-1 Consider the device shown in Fig. 9.6-1. The load torque is zero. Initially $i_{as} = I$ and $i_{bs} = 0$. From this condition, the following sequence occurs: $i_{as} = 0$ and $i_{bs} = I$, then $i_{as} = -I$ and $i_{bs} = I$. Determine the initial, intermediate, and final positions. [$\theta_{rm} = 0$, 18°, 27°]

SP9.6-2 A four-pole, two-phase, permanent-magnet stepper motor has 18 rotor teeth. Calculate *TP* and *SL*. [*TP* = 20°; *SL* = 5°]

9.7 EQUATIONS FOR PERMANENT-MAGNET STEPPER MOTORS

The voltage equations for a two-phase, permanent-magnet stepper motor may be written as

$$v_{as} = r_s i_{as} + \frac{d\lambda_{as}}{dt} \tag{9.7-1}$$

$$v_{bs} = r_s i_{bs} + \frac{d\lambda_{bs}}{dt} \tag{9.7-2}$$

In matrix form,

$$\mathbf{v}_{abs} = \mathbf{r}_s \mathbf{i}_{abs} + p\boldsymbol{\lambda}_{abs} \tag{9.7-3}$$

where p is the operator d/dt, and for voltages, currents, and flux linkages

$$(\mathbf{f}_{abs})^T = [f_{as} \quad f_{bs}] \tag{9.7-4}$$

with

$$\mathbf{r}_s = \begin{bmatrix} r_s & 0 \\ 0 & r_s \end{bmatrix} \tag{9.7-5}$$

The flux linkages may be expressed as

$$\lambda_{as} = L_{asas} i_{as} + L_{asbs} i_{bs} + \lambda_{asm} \tag{9.7-6}$$

$$\lambda_{bs} = L_{bsas} i_{as} + L_{bsbs} i_{bs} + \lambda_{bsm} \tag{9.7-7}$$

In matrix form,

$$\boldsymbol{\lambda}_{abs} = \mathbf{L}_s \mathbf{i}_{abs} + \boldsymbol{\lambda}'_m \tag{9.7-8}$$

where

$$\mathbf{L}_s = \begin{bmatrix} L_{asas} & L_{asbs} \\ L_{bsas} & L_{bsbs} \end{bmatrix} \tag{9.7-9}$$

$$\boldsymbol{\lambda}'_m = \begin{bmatrix} \lambda_{asm} \\ \lambda_{bsm} \end{bmatrix} \tag{9.7-10}$$

From Fig. 9.6-1, we can write, as a first approximation,

$$\boldsymbol{\lambda}'_m = \lambda'_m \begin{bmatrix} \cos(RT\,\theta_{rm}) \\ \sin(RT\,\theta_{rm}) \end{bmatrix} \tag{9.7-11}$$

where λ_m' is the amplitude of the flux linkages established by the permanent magnet as viewed from the stator phase windings. In other words, the magnitude of λ_m' is proportional to the magnitude of the open-circuit sinusoidal voltage induced in each stator phase winding. In (9.7-11),

$$\frac{d\theta_{rm}}{dt} = \omega_{rm} \tag{9.7-12}$$

Those who have read Chapter 6 on the permanent-magnet ac machine will recognize the similarity between (9.7-11) and (6.3-9). Also, the procedure for calculating λ_m' in the case of the stepper motor is identical to that illustrated in Example 6A.

From the idealized standpoint, the self-inductance of the stator phases of the device shown in Fig. 9.6-1 is essentially constant, and the reluctance seen by the permanent magnet is also constant, independent of rotor position. However, in practice, both the self-inductances and the reluctance vary with rotor position due to saturation of the stator iron and the differences from the idealized configuration that occur when shaping the poles. We shall disregard these departures from the idealized case and assume constant self-inductances and a constant reluctance seen by the permanent magnet independent of rotor position. When doing so, we are neglecting the reluctance torques caused by variation in self-inductances and the permanent magnet, both of which attempt to place the rotor in its minimum-reluctance position. The latter torque is often referred to as the *detent* or *retention torque,* since it exists whether or not the stator windings are excited, and, if the load torque is not too large, this detent torque will preserve the rotor position during a power failure. Nevertheless, the reluctance torques are small relative to the torque produced by the interaction of the permanent magnet and the stator currents and, although we are not looking at the complete picture when we neglect the reluctance torques, this approximation is certainly adequate for our first look at the permanent-magnet stepper motor.

With the assumption of constant self-inductances, we can write

$$L_{asas} = L_{ls} + L_{ms} = L_{ss} \tag{9.7-13}$$

$$L_{bsbs} = L_{ls} + L_{ms} = L_{ss} \tag{9.7-14}$$

Following a line of reasoning similar to that used in the case of the single-stack variable-reluctance stepper motor, it can be shown that stator mutual inductances do not exist if saturation is neglected. Thus,

$$\mathbf{L}_s = \begin{bmatrix} L_{ss} & 0 \\ 0 & L_{ss} \end{bmatrix} \tag{9.7-15}$$

An expression for the electromagnetic torque may be obtained by taking the partial derivative of the coenergy with respect to θ_{rm}, as given in Table 2.5-1 and (9.3-27). Since the stator mutual inductances are zero, the coenergy may be expressed as

$$W_c = \frac{1}{2}L_{asas}i_{as}^2 + \frac{1}{2}L_{bsbs}i_{bs}^2 + \lambda_{asm}i_{as} + \lambda_{bsm}i_{bs} + W_{pm} \tag{9.7-16}$$

where L_{asas} and L_{bsbs} are given by (9.7-13) and (9.7-14), respectively, and λ_{asm} and λ_{bsm} are given by (9.7-10) and (9.7-11). The term W_{pm} is related to the energy associated with the permanent magnet. Since we are neglecting variations in the self-inductances and in W_{pm}, taking the partial derivative of W_c with respect to θ_{rm} yields

$$T_e = -RT\,\lambda_m'[i_{as}\sin(RT\,\theta_{rm}) - i_{bs}\cos(RT\,\theta_{rm})] \tag{9.7-17}$$

The terms of (9.7-17) are plotted in Fig. 9.7-1, wherein it is assumed that constant currents are present in both phase windings. Each term of (9.7-17) is identified in Fig. 9.7-1. In particular, $\pm T_{eam}$ is the torque due to the interaction of the permanent magnet and $\pm i_{as}$, and $\pm T_{ebm}$ is due to the interaction of the permanent magnet and $\pm i_{bs}$.

The reluctance of the permanent magnet is large, approaching that of air. Since the flux established by the phase currents flows through the magnet, the reluctance of the flux path is relatively large. Hence, the variation in the reluctance due to rotation of the rotor is small and, consequently, the amplitudes of the reluctance torques are small relative to the torque produced by the interaction between the permanent magnet and the phase currents. For this reason, the reluctance torques are generally neglected, as we have done here, and the self-inductances are assumed to be constant. Therefore, the voltage equations for the permanent-magnet stepper motor become those of the permanent-magnet ac machine that we considered in Chapter 6, except for the use of θ_{rm} instead of θ_r and the difference in the referencing of θ_{rm} (to the minimum reluctance rather than the maximum reluctance). These equations are set forth in Section 9.8.

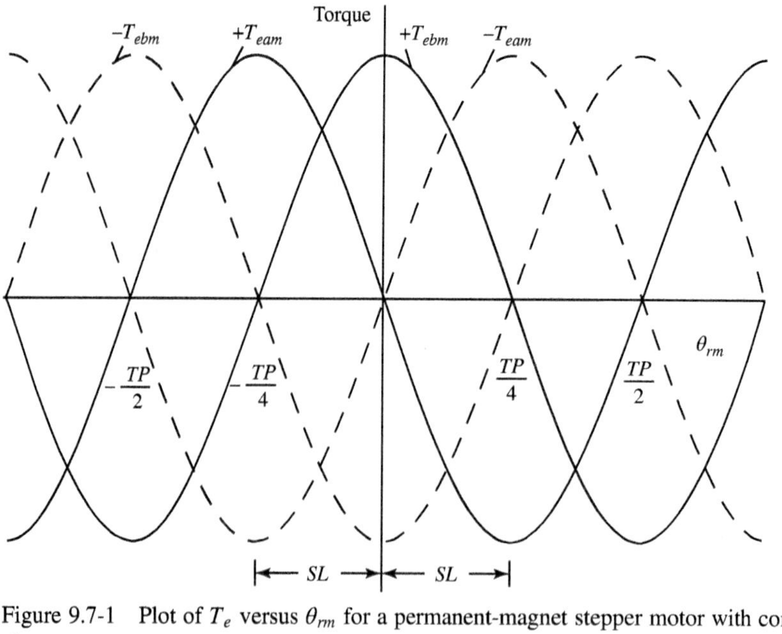

Figure 9.7-1 Plot of T_e versus θ_{rm} for a permanent-magnet stepper motor with constant phase currents.

Although a discussion of the stepping action of a permanent-magnet stepper motor using the steady-state torque-angle characteristics is appropriate, this explanation would be essentially a repeat of that given in Section 9.4 for the variable-reluctance devices. We will not do this. Instead, we will ask a few questions to help emphasize this similarity.

SP9.7-1 Express λ_{asm} in terms of SL rather than RT. Determine the number of step lengths in a period for the device shown in Fig. 9.6-1. $\{\lambda_{asm} = \lambda'_m \cos[\pi/(SL \cdot N)\theta_{rm}];\ 4\}$

SP9.7-2 Consider Fig. 9.7-1. The load torque is zero. Initially, $i_{as} = I$ and $i_{bs} = 0$. Then, $i_{as} = 0$ and $i_{bs} = -3I$. Finally, $i_{as} = -\frac{1}{2}I$ and $i_{bs} = 0$. Determine the three positions. $[\theta_{rm} = 0,\ -TP/4,\ -TP/2]$

9.8 EQUATIONS OF PERMANENT-MAGNET STEPPER MOTORS IN ROTOR REFERENCE FRAME – RELUCTANCE TORQUES NEGLECTED

Stepper motors often operate for a time in a continuous rotational mode, especially when connected through a speed-reduction gear to the member being positioned. In some cases, the phase voltages may be stepped as fast as 500–1500 steps per second. When the permanent-magnet stepper motor is operated in this mode, it behaves very similar to either a permanent-magnet ac machine or a brushless dc machine, depending upon the type of control employed. It is, therefore, advantageous to set forth the voltage equations of the permanent-magnet stepper motor in the rotor reference frame. This transformation is of advantage only if the self-inductances are assumed constant and the detent torque is neglected.

A change of variables that, in effect, transforms the stator variables to the rotor reference frame is

$$\begin{bmatrix} f^r_{qs} \\ f^r_{ds} \end{bmatrix} = \begin{bmatrix} -\sin(RT\theta_{rm}) & \cos(RT\theta_{rm}) \\ \cos(RT\theta_{rm}) & \sin(RT\theta_{rm}) \end{bmatrix} \begin{bmatrix} f_{as} \\ f_{bs} \end{bmatrix} \tag{9.8-1}$$

or

$$\mathbf{f}^r_{qds} = \mathbf{K}^r_s \mathbf{f}_{abs} \tag{9.8-2}$$

where f can represent either voltage, current, or flux linkage and θ_{rm} is defined by (9.7-12). It follows that

$$\mathbf{f}_{abs} = \left(\mathbf{K}^r_s\right)^{-1} \mathbf{f}^r_{qds} \tag{9.8-3}$$

where $\left(\mathbf{K}^r_s\right)^{-1} = \mathbf{K}^r_s$. The s subscript denotes stator variables and the r superscript indicates that the transformation is to a reference frame rotating with the rotor.

If you have studied Chapters 6 and 7, you will recall that therein the stator variables of a permanent-magnet synchronous machine (Chapter 6) and the stator variables of a synchronous machine (Chapter 7) were transformed to the rotor reference frame by $\mathbf{K}^r_s$. However, the $\mathbf{K}^r_s$ used in Chapters 6 and 7, (6.5-4) or (7.3-29), is different from that given by (9.8-1). In fact, the $\mathbf{K}^r_s$ given by (9.8-1) is the $\mathbf{K}^r_s$

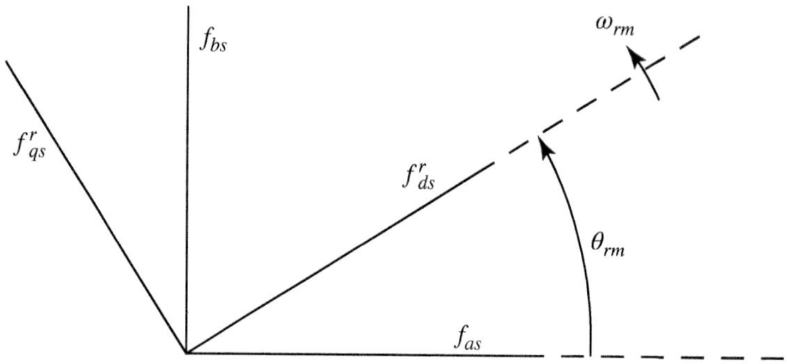

Figure 9.8-1 Trigonometric interpretation of change of stator variables.

given in Chapters 6 and 7 rotated by $\frac{\pi}{2}$ (see Fig. 9.8-1). We expect that the $\overset{r}{q}s$- and $\overset{r}{d}s$-equations will be similar to those in Chapters 6 and 7; however, we will go through the transformation.

Substituting (9.8-3) into (9.7-3) yields

$$(\mathbf{K}_s^r)^{-1}\mathbf{v}_{qds}^r = \mathbf{r}_s(\mathbf{K}_s^r)^{-1}\mathbf{i}_{qds}^r + p\left[(\mathbf{K}_s^r)^{-1}\boldsymbol{\lambda}_{qds}^r\right] \tag{9.8-4}$$

Multiplying (9.8-4) by $\mathbf{K}_s^r$ and simplifying yields

$$\mathbf{v}_{qds}^r = \mathbf{r}_s\mathbf{i}_{qds}^r + RT\ \omega_{rm}\boldsymbol{\lambda}_{dqs}^r + p\boldsymbol{\lambda}_{qds}^r \tag{9.8-5}$$

where

$$\left(\boldsymbol{\lambda}_{dqs}^r\right)^T = \begin{bmatrix}\lambda_{ds}^r & -\lambda_{qs}^r\end{bmatrix} \tag{9.8-6}$$

The last two terms of (9.8-5) are obtained from the last term of (9.8-4); in particular,

$$\mathbf{K}_s^r p\left[(\mathbf{K}_s^r)^{-1}\boldsymbol{\lambda}_{qds}^r\right] = \mathbf{K}_s^r\left[p(\mathbf{K}_s^r)^{-1}\right]\boldsymbol{\lambda}_{qds}^r + \mathbf{K}_s^r(\mathbf{K}_s^r)^{-1}p\boldsymbol{\lambda}_{qds}^r \tag{9.8-7}$$

It is left to the reader to show that the right-hand side of (9.8-7) reduces to the last two terms of (9.8-5).

The stator flux linkages are expressed as

$$\boldsymbol{\lambda}_{abs} = \mathbf{L}_s\mathbf{i}_{abs} + \boldsymbol{\lambda}_m' \tag{9.8-8}$$

where $\mathbf{L}_s$ is given by (9.7-15) and $\boldsymbol{\lambda}'_m$ is defined by (9.7-11). Substituting the change of variables into (9.8-8) yields

$$(\mathbf{K}_s^r)^{-1}\boldsymbol{\lambda}_{qds}^r = \mathbf{L}_s(\mathbf{K}_s^r)^{-1}\mathbf{i}_{qds}^r + \boldsymbol{\lambda}_m' \tag{9.8-9}$$

Premultiplying by $\mathbf{K}_s^r$ and substituting (9.7-15) for $\mathbf{L}_s$ and (9.7-11) for $\boldsymbol{\lambda}_m'$ yield

$$\boldsymbol{\lambda}_{qds}^r = \begin{bmatrix} L_{ss} & 0 \\ 0 & L_{ss}\end{bmatrix}\begin{bmatrix} i_{qs}^r \\ i_{ds}^r\end{bmatrix} + \lambda_m'^r\begin{bmatrix}0\\1\end{bmatrix} \tag{9.8-10}$$

which is identical in form to (6.5-5). To be consistent with our previous notation, we have added the superscript r to λ'_m. We see from (9.8-10) that in our new system of variables, the flux linkage created by the permanent magnet appears constant. Hence, our fictitious circuits are fixed relative to the permanent magnet and, therefore, fixed in the rotor. We have accomplished the goal of eliminating flux linkages that vary with θ_{rm}.

In expanded form, the voltage equations are

$$v_{qs}^r = r_s i_{qs}^r + RT\,\omega_{rm}\lambda_{ds}^r + p\lambda_{qs}^r \tag{9.8-11}$$

$$v_{ds}^r = r_s i_{ds}^r - RT\,\omega_{rm}\lambda_{qs}^r + p\lambda_{ds}^r \tag{9.8-12}$$

where

$$\lambda_{qs}^r = L_{ss} i_{qs}^r \tag{9.8-13}$$

$$\lambda_{ds}^r = L_{ss} i_{ds}^r + \lambda'^r_m \tag{9.8-14}$$

Substituting (9.8-13) and (9.8-14) into (9.8-11) and (9.8-12), and since λ'^r_m is constant, $p\lambda'^r_m = 0$.

$$v_{qs}^r = (r_s + pL_{ss})i_{qs}^r + RT\,\omega_{rm}L_{ss}i_{ds}^r + RT\,\omega_{rm}\lambda'^r_m \tag{9.8-15}$$

$$v_{ds}^r = (r_s + pL_{ss})i_{ds}^r - RT\,\omega_{rm}L_{ss}i_{qs}^r \tag{9.8-16}$$

The above equations may be written in matrix form as

$$\begin{bmatrix} v_{qs}^r \\ v_{ds}^r \end{bmatrix} = \begin{bmatrix} r_s + pL_{ss} & RT\,\omega_{rm}L_{ss} \\ -RT\,\omega_{rm}L_{ss} & r_s + pL_{ss} \end{bmatrix} \begin{bmatrix} i_{qs}^r \\ i_{ds}^r \end{bmatrix} + \begin{bmatrix} RT\,\omega_{rm}\lambda'^r_m \\ 0 \end{bmatrix} \tag{9.8-17}$$

The expression for the electromagnetic torque, with the reluctance torques neglected, is obtained by expressing i_{as} and i_{bs} in (9.7-17) in terms of i_{qs}^r. In particular,

$$T_e = RT\,\lambda'^r_m i_{qs}^r \tag{9.8-18}$$

In Section 6.5, the qs^r and ds^r voltage equations are derived for a permanent-magnet ac machine. These equations are similar to those which we have derived in this section with the exception that ω_r is replaced by $RT\omega_{rm}$. Also, the expression for the electromagnetic torque given by (9.8-18) is identical in form to (6.5-18) if $P/2$ in (6.5-18) is replaced by the rotor teeth RT.

When operated in the continuous rotational mode, the fundamental component of the voltages applied to the stator windings of the permanent-magnet stepper motor to provide ccw rotation of mmf$_s$ can be expressed as

$$v_{as} = -\sqrt{2}v_s \sin\theta_{esv} \tag{9.8-19}$$

$$v_{bs} = \sqrt{2}v_s \cos\theta_{esv} \tag{9.8-20}$$

These equations form a balanced two-phase set, where the amplitude v_s may be a function of time. One may question the form of these voltages since in the past we have selected v_{as} as the $\cos\theta_{esv}$. Recall that, in the case of the stepper motor, we referenced θ_{rm} as being different from that of the synchronous machine. Actually, $\theta_{rm} = \theta_r + \frac{\pi}{2}$. We have just pointed out that the equations given in Chapter 6 are identical to those for the permanent-magnet stepper motor with ω_r replaced with $RT\omega_{rm}$. To apply the material given in Chapter 6 to the permanent-magnet stepper motor, it is necessary to select the instantaneous values of v_{as} and v_{bs} so that v_{qs}^r and v_{ds}^r are the same as in Chapter 6. This is the case if v_{as} and v_{bs} are defined by (9.8-19) and (9.8-20) with

$$\frac{d\theta_{rm}}{dt} = \omega_{rm} \tag{9.8-21}$$

and $\theta_{esv}(0)$ is the time-zero position of the applied voltages.

Now then, when the permanent-magnet stepper motor is operated in the continuous rotational mode between target positions, it is operated either in the open-loop control mode or the closed-loop control mode. In the so-called open-loop control mode, stator voltages of a constant stepping rate are applied until the desired position is reached. In the case of the closed-loop control scheme, the rotor position of the permanent-magnet stepper motor is detected, and this information is supplied to the driving inverter that generates the voltages applied to the stator phases. In this case, the frequency of the stator voltages is made to correspond to the rotor speed and the stepper motor is operated as a permanent-magnet ac machine between target positions. Therefore, when the permanent-magnet stepper motor is operated in the closed-loop control mode, it is a permanent-magnet ac machine and ω_e is made equal to $RT\ \omega_{rm}$. It would appear that we need only to replace ω_r with $RT\ \omega_{rm}$ in all voltage equations given in Chapter 6 and they would apply to the permanent-magnet stepper motor with a closed-loop control. There is an exception; however, when converting between phasors and $\overset{r}{qs}$ and $\overset{r}{ds}$ as quantities, it is necessary to replace $\tilde{F}_{as}$ with $\tilde{F}_{bs}$ since $\theta_{rm} = \theta_r + \frac{\pi}{2}$. In particular, we see from (9.8-19) and (9.8-20) that, during steady-state closed-loop control operation, F_{as} and F_{bs} are expressed as

$$F_{as} = -\sqrt{2}F_s \sin\left[RT\ \omega_{rm}t + \theta_{esf}(0)\right] \tag{9.8-22}$$

$$F_{bs} = \sqrt{2}F_s \cos\left[RT\ \omega_{rm}t + \theta_{esf}(0)\right] \tag{9.8-23}$$

wherein ω_e has been set equal to $RT\ \omega_{rm}$. Substituting F_{as} and F_{bs} into the transformation equation, (9.8-1), yields

$$F_{qs}^r = \sqrt{2}F_s \cos\left[\theta_{esf}(0) - RT\ \theta_{rm}(0)\right] \tag{9.8-24}$$

$$F_{ds}^r = -\sqrt{2}F_s \sin\left[\theta_{esf}(0) - RT\ \theta_{rm}(0)\right] \tag{9.8-25}$$

Now, from (9.8-23),

$$\begin{aligned}\widetilde{F}_{bs} &= F_s e^{j\theta_{esf}(0)} \\ &= F_s \cos\theta_{esf}(0) + jF_s \sin\theta_{esf}(0)\end{aligned} \tag{9.8-26}$$

If we let $\theta_{rm}(0) = 0$, then (9.8-26) may be expressed in terms of (9.8-24) and (9.8-25) as

$$\sqrt{2}\widetilde{F}_{bs} = F_{qs}^r - jF_{ds}^r \tag{9.8-27}$$

This is (5.5-16) with $\widetilde{F}_{as}$ replaced by $\widetilde{F}_{bs}$. Thus, for the permanent-magnet stepper motor during closed loop operation, (6.6-5) becomes

$$\widetilde{V}_{bs} = (r_s + jRT\,\omega_{rm}L_{ss})\widetilde{I}_{bs} + \widetilde{E}_b \tag{9.8-28}$$

where $\widetilde{E}_b$ is $\widetilde{E}_a$ given by (6.6-6) with ω_r replaced by $RT\,\omega_{rm}$; in particular,

$$\widetilde{E}_b = \frac{1}{\sqrt{2}}RT\,\omega_{rm}\lambda_m'^r e^{j0} \tag{9.8-29}$$

The stepper motor can be operated with or without phase shift, just as the permanent-magnet ac motor, and with $\theta_{rm}(0) = 0$ the phase shift for maximum torque per volt is (6.9-18) and maximum torque per Amp (6.9-19) with ω_r replaced with $RT\ \omega_{rm}$.

SP9.8-1 A permanent-magnet stepper motor with $RT = 7$ is operated at $\omega_{rm} = 100$ rad/s. $\widetilde{V}_{bs} = (10/\sqrt{2})\underline{/0^\circ}$, $\widetilde{I}_{bs} = (1/\sqrt{2})\underline{/-60.3^\circ}$, $\theta_{rm}(0) = 0$, $r_s = 4\ \Omega$, and $L_{ss} = 0.01$ H. Calculate $\lambda_m'^r$. [$\lambda_m'^r = 0.00277$ V·s/rad]

SP9.8-2 The stepper motor given in SP9.8-1 is operated at $\omega_{rm} = 100$ rad/s with $\widetilde{I}_{bs} = (1/\sqrt{2})\underline{/-20^\circ}$ and $\theta_{rm}(0) = 0$. Calculate ϕ_v. [$\phi_v = 32.8°$]

9.9 RECAPPING

We have considered the two types of electromechanical stepping devices used most often: the variable-reluctance and the permanent-magnet stepper motors. Although we have considered an important operating mode of the stepper motor, there are numerous aspects of stepper motor operation that we are unable to cover in an introductory text. For example, the types and control of the voltage source and stability problems that can occur during continuous rotation of the stepper motor are beyond the scope of this textbook. Nevertheless, the information presented in this chapter sets the stage for study of these more advanced topics.

9.10 PROBLEMS

1. Sketch the configuration of a two-pole, four-stack, variable-reluctance stepper motor with two rotor teeth. Use *as, bs, cs,* and *ds* to denote the phase windings. Calculate *TP, SL,* and give the excitation sequence for ccw rotation.

2. For Problem 1, express the self-inductances and the torque using SL in the arguments.

3. Those who have read Chapter 7 on synchronous and reluctance machines may wonder if the transformation can be used in the case of the variable-reluctance stepper motor to transform the stator circuits to fictitious circuits with constant inductances fixed in the rotor. The answer is no, and the reason is that mutual inductances must exist between stator phases. To show this, take a two-pole, two-phase reluctance machine, assume $L_{asbs} = L_{bsas} = 0$, and evaluate $\mathbf{K}_s^r \mathbf{L}_s (\mathbf{K}_s^r)^{-1}$.

4. The four-pole, three-stack, variable-reluctance stepper motor shown in Fig. 9.2-3 is to be operated at a continuous speed of 30 rad/s. Neglect electrical transients and sketch the current i_{as}, indicating the times at which it is zero and nonzero.

5. A four-pole, five-stack, variable-reluctance stepper motor has eight rotor teeth, as shown in Fig. 9.2-4. Its magnetic axes are arranged *as, bs, cs, ds,* and *es,* in the counterclockwise direction. Express the self-inductances with the constant angular displacement in terms of step length.

6. Express the self-inductances for the single-stack, variable-reluctance stepper motor shown in Fig. 9.5-1 with the constant angular displacement in terms of step length.

7. A two-phase, permanent-magnet stepper motor has 50 rotor teeth. When the rotor is driven by an external mechanical source at $\omega_{rm} = 100$ rad/s, the measured open-circuit phase voltage is 25 V, peak to peak. Calculate λ_m and SL. If $i_{as} = 1$ A, $i_{bs} = 0$, express T_e.

8. Consider the two-phase, permanent-magnet stepper motor of Fig. 9.6-1. Sketch i_{as} and i_{bs} versus time for the excitation sequence $i_{as}, i_{bs}, -i_{as}, -i_{bs}, i_{as}, \ldots$. Denote the time between steps as T_s and the stepping rate as $f_s = 1/T_s$. Establish a relationship between the fundamental frequency (ω_e) of i_{as} and i_{bs}, and the stepping rate f_s. Relate ω_{rm} to ω_e and to f_s.

9. A two-phase, permanent-magnet stepper motor has 50 rotor teeth. The parameters are $\lambda_m' = 0.00226$ V·s/rad, $r_s = 10\ \Omega$, and $L_{ss} = 1.1$ mH. The applied stator voltages form a balanced two-phase set with $V_s = 10$ V, $\omega_e = 314$ rad/s. Establish the steady-state rotor speed ω_{rm} and the maximum electromagnetic torque $T_{eM/V}$ that can be developed at this speed.

10. The permanent-magnet stepper motor shown in Fig. 9.6-1 is equipped with bifilar windings. Assume that the *cs* winding (*ds* winding) is wound on the same stator pole as the *as* winding (*bs* winding) but opposite to it, and also assume that the windings are tightly coupled (no leakage inductances). Express $v_{as} - v_{cs}$ and $v_{bs} - v_{ds}$. ■

9.11 REFERENCE

1. P. P. Acarnley, *Stepping Motors: A Guide to Modern Theory and Practice*, Peter Peregrinus Ltd. for the Institution of Electrical Engineers, Stevenage, England, 1984.

CHAPTER 10

POWER ELECTRONICS

10.1 INTRODUCTION

Power electronic circuits, which have enabled the ever-expanding electrification of, among others, consumer electronics, the transportation sector, and wireless technologies, form the backbone of electric drives. Power-electronic-drive-controlled machines are impacting industrial systems; propulsion systems in air-, marine-, and ground-based vehicles; aircraft actuation systems; and many others. This chapter serves to open the door to advanced study in this area by establishing the fundamental principles of switching circuits and then analyzing a few of the most common power electronic circuits, in particular those used in electric drives.

In Section 10.2, we introduce several fundamental concepts relevant to power conversion that can be applied across a broad range of basic and advanced circuits. In Sections 10.3 through 10.5, these concepts are used to introduce dc–dc converters, ac–dc rectifiers, and dc–ac inverters, respectively. This introduction includes the common buck and boost dc–dc converters, half- and full-wave rectifiers, and the single-phase inverter. The three-phase inverter is covered in Chapter 6 as part of a brushless dc drive.

10.2 SWITCHING-CIRCUIT FUNDAMENTALS

Power electronics embodies electronic circuits and devices that control the flow of energy (i.e., power) between sources and loads. Ideally, power electronic circuits accomplish their goal as efficiently as possible so that energy is transferred without loss.

Power electronic devices are everywhere, often doing their jobs silently and diligently behind the scenes while other, more "glamorous" electrical devices

Electromechanical Motion Devices: Rotating Magnetic Field-Based Analysis and Online Animations, Third Edition. Paul Krause, Oleg Wasynczuk, Steven D. Pekarek, and Timothy O'Connell.

Published 2020 by John Wiley & Sons, Inc.

get all the attention. In fact, most power converters are designed to be hidden or inconspicuous; those black "bricks" that come with computers are not exactly eye-catching. However, their role in today's world is irreplaceable and cannot be overstated. Because electrical energy is produced and consumed over a wide range of voltages and frequencies, the applications for power electronic devices are countless.

As you read this chapter, there is a good chance you are surrounded by the impacts of power electronic devices. If you use a smart phone, a power electronic converter charged it by transporting power from the ac grid to its dc battery. The next time you ride in an automobile, you will be surrounded by power electronic devices moving energy from the engine or battery to the car's numerous electrical accessories. These are just a few of the ways you interact with power electronics daily.

Power Conversion Principles

It is useful to consider the generic power electronic circuit shown in Fig. 10.2-1. This device creates an interface between two circuits, facilitating the flow of power between them when a direct connection is inefficient or impossible. For example, a dc–ac inverter can act as an interface between a dc battery and an ac electric motor. However, most rotating electric machinery requires ac currents and voltages to operate; thus, unless it is a dc machine, a motor cannot operate directly from a battery. An inverter is required, converting the battery's stored energy at dc to ac that can be used by the motor.

Historically, the connected circuits denoted "Circuit 1" and "Circuit 2" in Fig. 10.2-1 have been characterized as *source* and *load*, respectively, where power-producing circuits are sources and power-consuming circuits are loads. However, modern trends in the power field have blurred the line between sources and loads, as many devices are both producers and consumers of power. For example, the battery in a cellular telephone is a source when it is powering the phone and a load when it is being charged. Likewise, the electric machine in a hybrid electric vehicle can act as both a load (motor mode) and source (generator mode) depending on the state of operation of the vehicle. In general, power electronic circuits can, and often do, facilitate bidirectional power flow between circuits, as indicated by the double arrows in Fig. 10.2-1.

There are two key principles of power conversion, as related to Fig. 10.2-1 [1]. First, power conversion circuits should consume as little power as possible. An ideal power device is one that is 100% efficient and consumes none of the

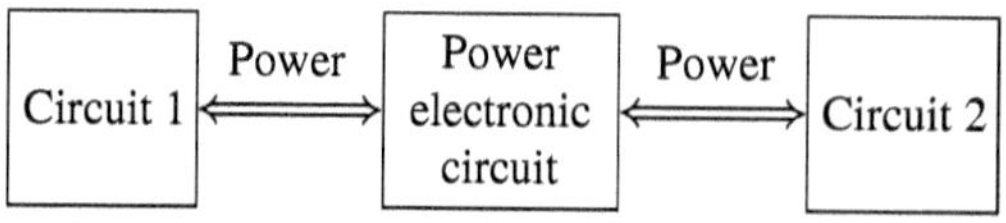

Figure10.2-1 A generic power electronic circuit.

power flowing through it. Second, power conversion circuits should *transfer* much more power than they consume. This allows a relatively small power device to interface between two large devices that process a lot of power. If the battery and motor in the above example are each rated for 1 kW of electric power during normal operation, a good 1 kW inverter interfacing between them might be 98% efficient. Thus, although the inverter can transfer 1 kW from battery to motor, it does this so efficiently that it only consumes 20 W in the process.

Power conversion circuits often serve as an interface between dc sources/loads at different voltages. A naïve (and poor) choice of circuit for this application is the resistive voltage divider. Although the voltage divider is simple and a great option for logic-level signals, it can be shown with some quick calculations that it violates both of the power conversion principles put forth in the previous paragraph (it consumes a lot of power and is inefficient). Thus, a voltage divider is not a power electronic circuit. A better choice is a dc–dc (pronounced "dc to dc") converter, which will be covered in more detail in Section 10.3. Before we get there, however, recognize that this application – interfacing between two circuits at different voltages – seems well-suited to the function provided by a transformer. Unfortunately, as shown in Chapter 1, a transformer requires ac voltage to function; dc voltage into the transformer's primary winding will neither induce a changing magnetic field in its core nor a voltage in its secondary winding. Dc voltage only leads to saturation of the transformer's magnetic core, rendering it useless for power transfer. In fact, achieving transformer action with dc signals is one of the core applications of power electronics, as will be made evident in the next several sections.

Fourier Analysis

To proceed, it is useful to expand the common definition of the term "dc" – i.e., a signal with constant value – to something broader. Power conversion circuits often switch between two or more configurations in a periodic manner at a known frequency, and Fourier series are well-suited for analyzing the periodic signals that thus arise. A review of basic Fourier analysis is in order.

A periodic function with period T is one for which $f(t) = f(t + T)$. Whereas in other chapters f is a general variable that represents transient voltage, current, or flux linkage, here f represents any time-domain periodic function. Although there are some esoteric periodic signals that cannot be represented by a Fourier series, any periodic signal found in a real circuit, including those with step discontinuities, will meet the mathematical conditions necessary. Thus, any periodic signal encountered in a power conversion circuit can be expressed as a sum of sinusoids, also known as the *Fourier series*:

$$f(t) = a_0 + \sum_{n=1}^{\infty}[a_n \cos(n\omega t) + b_n \sin(n\omega t)] \tag{10.2-1}$$

where $\omega = 2\pi/T$ is the radian frequency, and a_n and b_n are the series coefficients found by computing the following integrals over an interval $[t_0, t_0 + T]$:

$$a_0 = \frac{1}{T}\int_{t_0}^{t_0 + T} f(t)dt \tag{10.2-2}$$

$$a_n = \frac{2}{T}\int_{t_0}^{t_0 + T} f(t)\cos(n\omega t)dt \tag{10.2-3}$$

$$b_n = \frac{2}{T}\int_{t_0}^{t_0 + T} f(t)\sin(n\omega t)dt \tag{10.2-4}$$

Through algebraic manipulation, the Fourier series can be written in an alternate compact form that is sometimes easier to work with. The sines and cosines in (10.2-1) are combined into phase-shifted cosine terms to give the *compact Fourier series*:

$$f(t) = a_0 + \sum_{n=1}^{\infty} c_n \cos(n\omega t + \theta_n) \tag{10.2-5}$$

The coefficients of the compact series are related to those of the standard series by

$$c_n = \sqrt{a_n^2 + b_n^2} \tag{10.2-6}$$

and the phase angles are

$$\theta_n = -\tan^{-1}\left(\frac{b_n}{a_n}\right) \tag{10.2-7}$$

The time average of a transient signal $f(t)$ over an interval $[t_0, t_0 + T]$, denoted $\widehat{f}$, is given by

$$\widehat{f} = \frac{1}{T}\int_{t_0}^{t_0 + T} f(t)dt \tag{10.2-8}$$

Comparing this to (10.2-2) we see that $a_0 = \widehat{f}$: the Fourier coefficient a_0 is the average value (i.e. the dc component or dc offset) of the periodic signal. To develop a method of dc–dc conversion, we consider any periodic signal with a nonzero dc component to be a candidate dc signal. For these, the dc component of the Fourier series can be "pulled out" by filtering out the ac components, which are the Fourier series terms with coefficients a_n and b_n (or c_n in the compact series). This broadens the family of dc signals we can consider from "signals that are constant" to "signals with a dc component." This broader classification gives us considerable flexibility as we look for a method of dc–dc conversion.

Although the Fourier series equations may appear daunting, under certain circumstances often arising in power conversion circuits, they can be simplified considerably [2]. For example, if $f(t)$ has no dc component, then $a_0 = 0$:

$$a_0 = 0: \quad \text{when } \widehat{f} = 0 \tag{10.2-9}$$

If, after removing the dc component, the remaining function is odd, then the cosine terms will drop out[1]:

$$\begin{aligned}&\text{Odd function} \Rightarrow \text{sines only}\\ &\qquad \Rightarrow a_n = 0, \quad \forall n \geq 1: \quad \text{when } f(t) - a_0 = -[f(-t) - a_0]\end{aligned} \tag{10.2-10}$$

Likewise, if the function is even (i.e., it has y-axis symmetry), then the sine terms will drop out:

$$\begin{aligned}&\text{Even function} \Rightarrow \text{cosines only}\\ &\qquad \Rightarrow b_n = 0, \quad \forall n \geq 1: \quad \text{when } f(t) = f(-t)\end{aligned} \tag{10.2-11}$$

Finally, if after removing the dc component, the remaining function has half-wave symmetry, it will contain no even harmonics:

$$\begin{aligned}&\text{Half-wave symmetry} \Rightarrow \text{no even harmonics}\\ &\qquad \Rightarrow a_n, \ b_n = 0 \ \forall n \text{ even}:\\ &\qquad \text{when } f\left(t + \frac{T}{2}\right) - a_0 = -[f(t) - a_0]\end{aligned} \tag{10.2-12}$$

Examples of periodic functions which exhibit some of these properties are shown in Fig. 10.2-2 and described in Example 10A.

Example 10A Fourier Series Properties of Periodic Signals
In this example, we describe the general features of the each of the periodic functions shown in Fig. 10.2-2 using (10.2-9) through (10.2-12).

1. This signal has no average value; thus, from (10.2-9), $a_0 = 0$. It is an odd function; thus, from (10.2-10), $a_n = 0, \ \forall n \geq 1$ and the series is made up of sines only. It is half-wave symmetric; thus, from (10.2-12), its Fourier series contains no even harmonics.

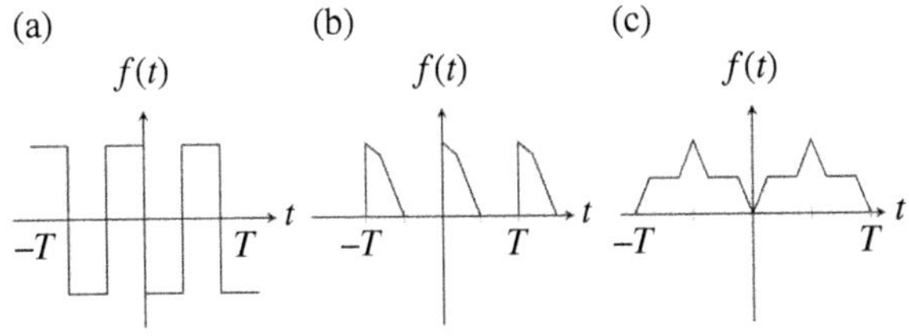

Figure 10.2-2 Periodic functions with various Fourier series properties. (a)–(c) are as described in parts 1–3 of Example 10A, respectively.

[1] In this chapter, the ∀ symbol stands for "for all."

2. The signal has an average value; thus, from (10.2-9), $a_0 \neq 0$. In fact, from (10.2-2), $a_0 = \widehat{f}$. The signal is neither even nor odd; thus, from (10.2-10) and (10.2-11), the series contain sines and cosines. The signal has no half-wave symmetry; thus, from (10.2-12), its Fourier series contains even and odd harmonics.
3. The signal has an average value; thus, from (10.2-9), $a_0 \neq 0$. It is an even function; thus, from (10.2-11), $b_n = 0, \quad \forall n \geq 1$ and the series is made up of cosines only. After removing the dc component, the signal is half-wave symmetric; thus, from (10.2-12), its Fourier series contains no even harmonics.

Switches and Switching Functions

In power electronics, a *switch* is a device used to regulate the flow of energy between sources and loads. Switches can either be controlled or uncontrolled. Common examples of controlled and uncontrolled switches, respectively, are the transistor and diode.

A transistor found in power circuits is typically a bipolar junction transistor (BJT), metal-oxide-semiconductor field-effect transistor (MOSFET), or insulated-gate bipolar transistor (IGBT). Of these, the IGBT has become popular for electric drives because of its simple voltage-controlled switching, high power and current capabilities, and fast switching speeds. It is the industrial workhorse device. However, new wide bandgap silicon-carbide (SiC) and gallium-nitride (GaN) MOSFETS, which can switch must faster than traditional silicon-based devices, are beginning to compete with IGBTs as their voltage and power ratings and reliability increase. Diodes can be simple p–n junction or MOS devices. Although conceptually simpler than transistors, diodes vary widely in their current-carrying and voltage-blocking capabilities depending on their construction. Every power electronic circuit has at least one switch, and it is the operation and the location of its switches that determine the circuit's behavior.

A real switch can be approximated as an *ideal switch*, a concept whose circuit representation is shown in Fig. 10.2-3. An ideal switch is lossless, carries no current and maintains an open circuit between its terminals when it is open/off, and has no voltage drop and maintains a short circuit between its terminals when it is closed/on. The real devices described in the previous paragraph are, of course, not ideal. However, they can be approximated well to first-order with the ideal

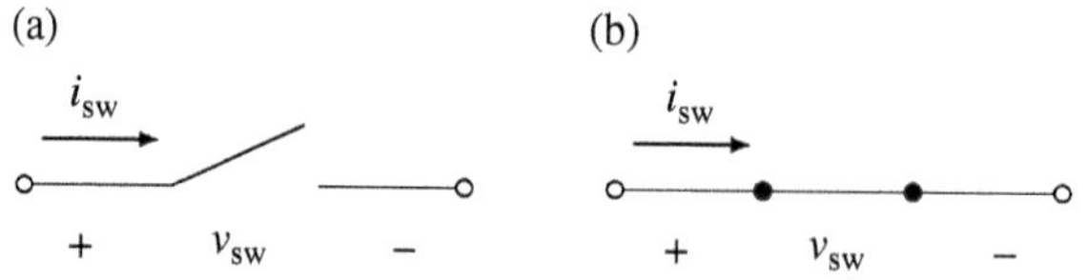

Figure 10.2-3 Ideal switch in (a) open/off and (b) closed/on configurations.

assumptions. Thus, to avoid unnecessary complication, we limit our discussion and subsequent analyses to ideal switches.

The ideal switch voltage, v_{sw}, and current, i_{sw}, behaviors are summarized as follows:

$$\begin{aligned} \text{Switch open} &\Rightarrow i_{sw} = 0 \\ \text{Switch closed} &\Rightarrow v_{sw} = 0 \end{aligned} \tag{10.2-13}$$

This behavior is described mathematically using a binary switch variable, q, as

$$q = \begin{cases} 0, & \text{Switch open} \\ 1, & \text{Switch closed} \end{cases} \tag{10.2-14}$$

Note that this q is a variable and is different from the label of the q-axis introduced in Chapters 4 and 5. To avoid confusion, we could use a different variable name here; however, in an attempt at consistency with traditional power electronic circuit analysis notation, we will not. Through the remainder of the text it should be clear from context which q is being used.

A power electronic switch does not remain in one state for long; rather, it is opened and closed rapidly (kHz to MHz frequencies) in order to transfer power in a desired fashion. Thus, the switch variable q is most often a transient *switching function, $q(t)$*, which opens and closes the switch at a fixed rate known as the *switching frequency, f_{sw}*. The inverse of the switching frequency is the *switching period, T_{sw}*. The fraction of the switching period that the switch is on is the *duty ratio, k*. The switching frequency is chosen based on many factors, among these are efficiency, circuit ripple and/or distortion requirements, acoustic noise, and device ratings.

A common switching function encountered in the design of power electronic converters is shown in Fig. 10.2-4. By utilizing the even symmetry of the waveform (defined by (10.2-11)) and solving for the Fourier series coefficients using (10.2-2) through (10.2-4), we can write the compact Fourier series of the switching function as

$$q(t) = k + \frac{2}{\pi}\sum_{n=1}^{\infty}\frac{\sin(n\pi k)}{n}\cos(n\omega_{sw}t - n\theta(0)) \tag{10.2-15}$$

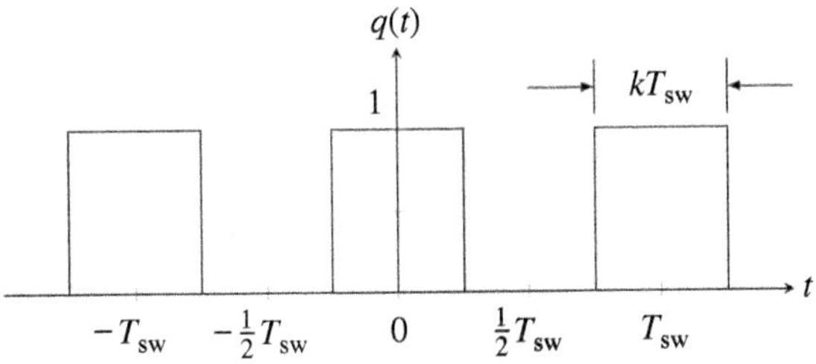

Figure 10.2-4 General periodic switching function $q(t)$.

where $\omega_{sw} = 2\pi f_{sw}$ and $\theta(0)$ is the phase shift angle of the fundamental component at time zero. Notice that the switching function's average value is its duty ratio, k. Also, note that the square wave in Fig. 10.2-4 is an even function centered at $t = 0$, which means that its fundamental has $\theta(0) = 0$. However, the Fourier series in (10.2-15) is general and can easily be adjusted for shifted square waves by setting $\theta(0)$ appropriately.

While switching functions are used to control switches, uncontrolled devices, such as diodes, are not actively controlled. Rather, ideal diodes obey the following:

$$\begin{aligned} &\text{Diode off}: \quad i_D = 0 \Rightarrow v_D < 0 \\ &\text{Diode on}: \quad v_D = 0 \Rightarrow i_D > 0 \end{aligned} \tag{10.2-16}$$

where v_D and i_D are the diode voltage and current, respectively, as shown in Fig. 10.2-5. In power electronic circuits, controlled and uncontrolled switches are often used in tandem, so that if the controlled switch is opened and closed according to $q(t)$, the uncontrolled switch will obey $1-q(t)$. When one switch is open, the other is closed, and vice versa. An uncontrolled switch is simpler to use and often cheaper than a similarly rated controlled device; however, a controlled device fed with a $1-q(t)$ switching function can be used to perform the same function. This is often done in high-performance circuits, where precise control of all switches is important.

A final word on switches is in order. Although throughout this chapter we assume that all switches are ideal, this is only an approximation. Although sufficient for our introductory treatment, the switching rules in (10.2-13) do not hold for real devices. In fact, IGBTs, MOSFETs, diodes, and other real switches have finite turn-on and turn-off transition times. During these transitions (typically less than 1 μs), both v_{sw} and i_{sw} are nonzero concurrently as they transition to their final state. Because $v_{sw} \times i_{sw} \neq 0$, power is lost in the switch. This *switching loss* is proportional to f_{sw} and negatively impacts a power converter's efficiency. In addition, due to both solid-state physical effects and ohmic losses, the voltage across a real switch is nonzero when the switch is fully on and conducting current. This *conduction loss*, which also negatively impacts a converter's efficiency, is proportional to duty cycle – the longer a switch is on, the higher its conduction loss. Because they affect efficiency and generate heat, switching and conduction losses are important considerations in the design of high-performance power converters. For a thorough treatment of this topic, interested students should consult a dedicated, comprehensive power electronics text, such as [1, 3, 4].

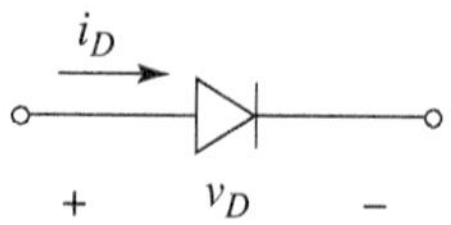

Figure 10.2-5 Ideal diode circuit representation.

Example 10B Fourier Series of a Switching Function

Let us now calculate the Fourier series of the switching function given in Fig. 10B-1. Note that the frequency of the function in this illustrative example is much smaller than would be found in a typical power converter.

Regardless of frequency, the steps to find the Fourier series are the same. The series we seek is given by (10.2-15). We must find the series parameters by comparing Fig. 10B-1 to Fig. 10.2-4. By inspection, we see that

$$k = \frac{\pi/2}{2\pi} = 1/4 \tag{10B-1}$$

and

$$T_{\text{sw}} = 2\pi \tag{10B-2}$$

Thus,

$$\omega_{\text{sw}} = \frac{2\pi}{T_{\text{sw}}} = 1 \text{ rad/s} \tag{10B-3}$$

The center of the second pulse in the example, indicated with the dotted vertical line, is at $t = \pi/4$, which is 1/8 of the switching period. Thus,

$$\begin{aligned}\theta(0) &= \omega_{\text{sw}} \times \frac{T_{\text{sw}}}{8} \\ &= (1 \text{ rad/s}) \times \left(\frac{2\pi}{8}\text{s}\right) = \frac{\pi}{4} \text{ rad}\end{aligned} \tag{10B-4}$$

We can now write the Fourier series using (10B-1) through (10B-4) and (10.2-15):

$$q(t) = \frac{1}{4} + \frac{2}{\pi}\sum_{n=1}^{\infty} \frac{\sin(n\pi/4)}{n} \cos(nt - n\pi/4). \tag{10B-5}$$

Before concluding this example, note that harmonics that are multiples of four (which is the inverse of k) are zero because of the sine coefficient term. This relationship between duty ratio and harmonics is a general one, and it is important in advanced power electronic circuit analysis and design.

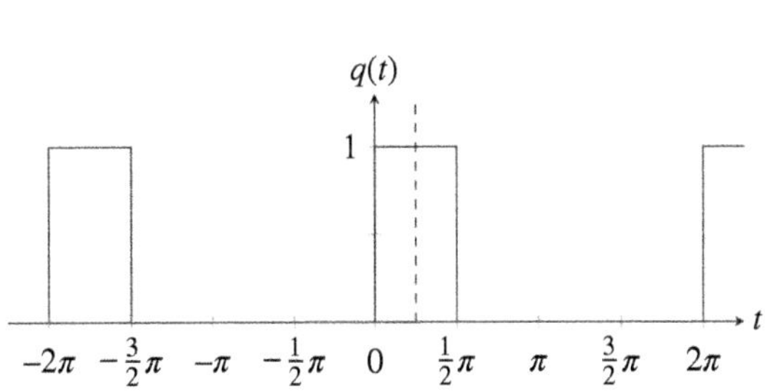

Figure 10B-1 Switching function.

Energy Storage Elements

Power circuits also rely on *energy storage elements*, namely, inductors and capacitors. In fact, with components from just these three fundamental categories – switches, inductors (possibly coupled), and capacitors – one can construct the majority of the most commonly used power electronic circuit topologies. Two important concepts, based on the principle of conservation of energy, govern the use of energy storage devices in power circuits.

It is instructive to consider the switched inductor circuit shown in Fig. 10.2-6a. When the switch is closed, the inductor current will ramp up linearly according to $di/dt = V_{dc}/L$. Because the energy stored in the inductor's magnetic field is $\frac{1}{2}Li^2$, it will increase quadratically. Theoretically, if the switch remains closed, the current will ramp up indefinitely and the inductor will continue to store energy at an ever-increasing rate. Of course, in a real inductor the conductors will overheat and melt before the current gets too high, but the takeaway is the same: it is a bad idea to close a dc source into an inductor indefinitely. Rather, it is better to keep the average current constant and transfer energy into and out of the inductor cyclically. In the circuit shown, this would require additional circuitry, but we will leave that detail out for now. If the energy put into the inductor during charging equals the energy taken out while discharging, the inductor will not overheat. Simple analysis (see Example 10C) shows that this steady-state operation is achieved if and only if the average inductor voltage, $\widehat{v}_L$, is zero. To summarize, the first key principle of energy storage devices is

$$\widehat{v}_L = 0 \tag{10.2-17}$$

Analogous arguments can be applied to the switched capacitor circuit shown in Fig. 10.2-6b. When the switch is opened, the capacitor voltage will increase linearly according to $dv/dt = I_{dc}/C$. The energy stored in the capacitor's electric field is $\frac{1}{2}Cv^2$, and it will ramp up quadratically with the voltage. Theoretically, the voltage will ramp up indefinitely and the capacitor will continue to store energy at an ever-increasing rate. Of course, as with a real inductor, in a real capacitor physical reality will kick in; the voltage will eventually reach the breakdown voltage of the capacitor separator and arc across the capacitor terminals, creating a short circuit. Thus, it is a bad idea to close a dc current source into a capacitor indefinitely. Instead, energy should be transferred into and out of the capacitor periodically, so that the average voltage remains constant. If the energy put into the

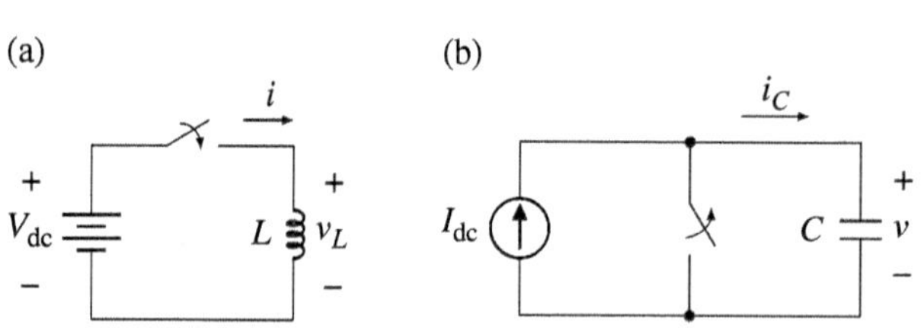

Figure 10.2-6 Switched (a) inductor and (b) capacitor circuits.

capacitor during charging is the same as the energy extracted during discharge, the capacitor will not overcharge and short-circuit. One can show that this steady-state operation can be achieved if and only if the average capacitor current, $\widehat{i_C}$, is zero. To summarize, the second key principle of energy storage devices in power electronic circuits is

$$\widehat{i_C} = 0 \tag{10.2-18}$$

We now have all the basic tools to begin an introductory analysis of several power electronic switching circuits. Together with basic circuit theory, these tools enable us to analyze dc–dc, ac–dc, and dc–ac circuits in the following sections.

Example 10C Average Inductor Voltage for Energy Balance
In this example, we show that the average inductor voltage must be zero for energy balance in the switched inductor circuit in Fig. 10.2-6a.

The inductor voltage, v_L, is expressed as

$$v_L = L\frac{di}{dt} \tag{10C-1}$$

With constant inductance, the average inductor voltage is

$$\widehat{v_L} = L\widehat{\frac{di}{dt}} \tag{10C-2}$$

If $\widehat{\frac{di}{dt}} \neq 0$, then the inductor current magnitude will continually increase. Since the stored inductor energy is $\frac{1}{2}Li^2$, this represents an unbalanced energy condition. Therefore, for energy balance

$$\widehat{\frac{di}{dt}} = 0 \tag{10C-3}$$

Combining (10C-2) and (10C-3) for energy balance reveals that

$$\widehat{v_L} = 0 \tag{10C-4}$$

SP10.2-1 (a) Find the voltage conversion efficiency $\eta = P_{\text{out}}/P_{\text{in}}$ for a voltage divider circuit with $r_1 = 1\,\Omega$, $r_2 = 10\,\Omega$, and $r_{\text{out}} = 100\,\Omega$ (r_1 is in series with the parallel combination of r_2 and r_{out}). (b) Is this a good power conversion circuit? [(a) $\eta = 0.0817 = 8.17\%$; (b) No, approximately 92% of the input power is consumed in the circuit]

SP10.2-2 Using (10.2-9) through (10.2-12) describe the Fourier series of the square wave switching function defined by $q(t) = \frac{1}{2}[1 + sign(\sin(t))]$. [dc component with $a_0 = 1/2$; odd function, so no cosines; half-wave symmetry, so no even harmonics]

SP10.2-3 Show that $\widehat{i_C} = 0$ is necessary for energy balance in a capacitor. [See Example 10C and use analogous arguments]

10.3 DC–DC CONVERSION

Let us consider the ideal dc–dc conversion circuit shown conceptually in Fig. 10.3-1. Power is to be transferred from the dc source, V_{hi}, to the lower voltage dc load, V_{lo}. Although this motivating example transfers power from a higher voltage source to a lower voltage load, the analysis concepts that follow are equally applicable for low-to-high-voltage conversion circuits. Through examining the general Fourier series equations, we have seen that a periodic signal with a dc component can be filtered to give a constant value dc signal. Thus, if we can create a periodic signal from V_{hi} that has an average value of V_{lo}, we can filter out the harmonic content and be left with the desired output. A simple periodic signal that meets the requirements is a square wave with magnitude V_{hi} and duty ratio $k = V_{lo}/V_{hi}$. k has a value between 0 and 1, and the length of time, per period, that the square wave is high is kT_{sw}, as is indicated in Fig. 10.3-2.

This leads us to a dc–dc conversion strategy: transfer power from V_{hi} in square pulses with $k = V_{lo}/V_{hi}$ and filter the result to leave only the dc component, V_{lo}. Because a low-pass filter is needed to eliminate all frequencies above dc, the higher the frequency of the pulses, the simpler and more effective this filter can be.

Buck Converter

A circuit that accomplishes the dc–dc conversion strategy laid out above is the *buck converter*, shown in Fig. 10.3-3. This circuit has one controlled switch – a transistor – and one uncontrolled switch – a diode. Assuming L and C are "large" (we will see what this means a little later) and that the circuit is in steady state, we can approximate as constants the output voltage, v_{out}, and the inductor current, i_L. When the transistor is closed, the diode is reverse-biased and does not conduct current; the circuit is in Configuration A, as shown in Fig. 10.3-4, and current flows

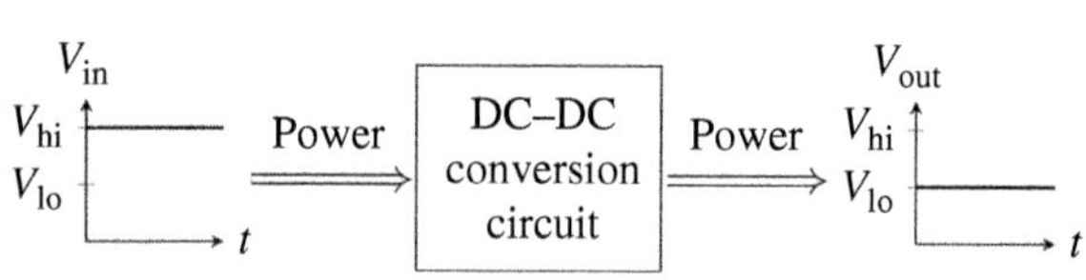

Figure 10.3-1 Ideal dc–dc conversion circuit.

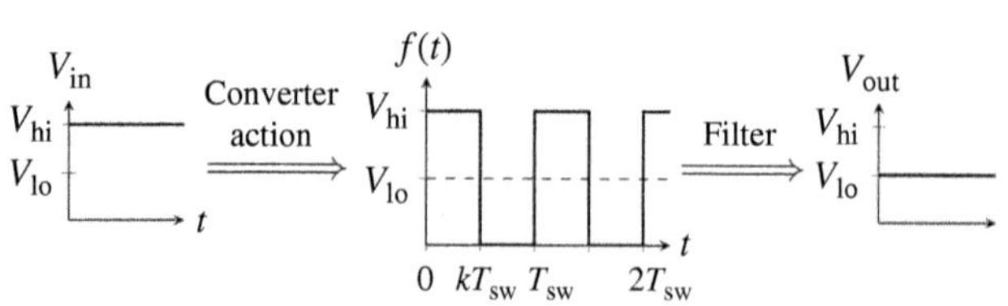

Figure 10.3-2 Filtered square wave voltage dc–dc voltage conversion strategy.

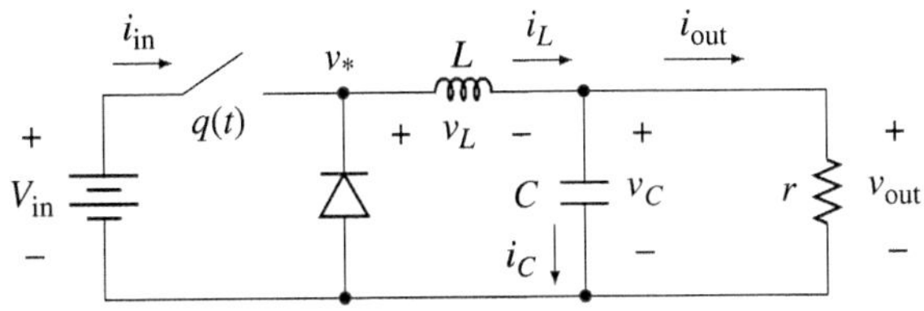

Figure 10.3-3 Buck converter circuit.

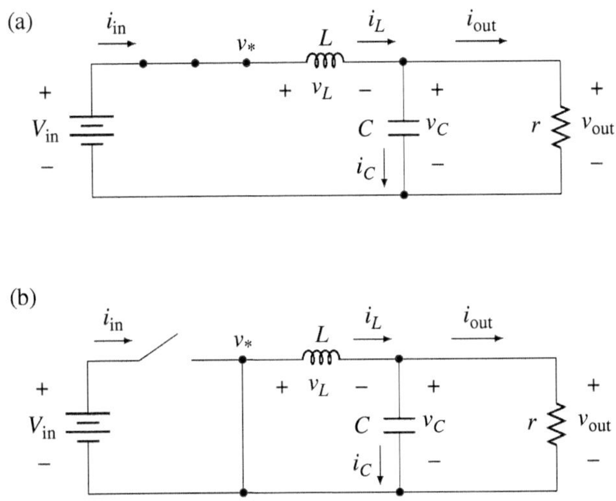

Figure 10.3-4 Buck converter in (a) Configuration A, with $q(t) = 1$, and (b) Configuration B, with $q(t) = 0$.

through the inductor from the input to the output. When the transistor opens, the inductor current cannot change instantaneously and Kirchoff's Current Law (KCL) requires that this current have a closed path in which to flow. The inductor's current decreases, inducing a negative inductor voltage. This forward-biases the diode, turning it on; the circuit switches to Configuration B. An appropriate switching function controls the transistor, switching the circuit between these two configurations. Let us now analyze Configurations A and B in more detail.

In Configuration A, $i_{in} = i_L$, $v_* = V_{in}$, and $v_L = V_{in} - v_{out}$. In Configuration B, the input voltage is detached from the circuit, $v_* = 0$, $i_{in} = 0$, and $v_L = -v_{out}$. If we assume that the transistor is switched at a high rate so that the circuit rapidly oscillates between Configurations A and B, an average behavior will emerge. We will first analyze this, and then add the high-frequency details afterward. In terms of the switching function,

$$\begin{aligned} v_*(t) &= q(t)V_{in} + (1 - q(t)) \times 0 \\ &= q(t)V_{in} \end{aligned} \tag{10.3-1}$$

The square wave voltage, v_*, can be described using (10.2-15) with $\theta(0) = \pi/2$. The time average of v^* over one switching period is

$$\begin{aligned}\widehat{v}_* &= \frac{1}{T_{sw}}\int_0^{T_{sw}} v_* dt \\ &= \frac{1}{T_{sw}}\left(\int_0^{kT_{sw}} V_{in} dt + \int_{kT_{sw}}^{T_{sw}} 0 \times dt\right) \\ &= \frac{1}{T_{sw}}\int_0^{kT_{sw}} V_{in} dt = kV_{in}\end{aligned} \tag{10.3-2}$$

Now, because the average voltage across the inductor is zero (10.2-17), we have found the average output voltage:

$$\widehat{v}_{out} = \widehat{v}_* = kV_{in} \tag{10.3-3}$$

The buck converter's voltage *conversion ratio* is then

$$\frac{\widehat{v}_{out}}{V_{in}} = k. \tag{10.3-4}$$

Equations (10.3-3) and (10.3-4) equivalently describe the average behavior of the buck converter, showing that the average output voltage is a fraction of the input voltage and is determined by the duty ratio k. Look again at Fig. 10.3-2. The equation we have derived encompasses all the features of that figure. The input voltage is first "chopped" by the controlled switch into the square wave v^*. A buck converter is sometimes called a *chopper* because of this. With rapid enough switching and "large" L and C, v^* is filtered to leave its average value as kV_{in}. Thus, $\widehat{v}_{out}/V_{in} = k$ and dc voltage transformation has been accomplished!

Let us now consider the current and power flows in this circuit. Because the average current in the output capacitor is zero (10.2-18),

$$\widehat{i}_{out} = \widehat{i}_L \tag{10.3-5}$$

Further, the input current can be expressed in terms of the switching function as

$$\begin{aligned}i_{in}(t) &= q(t)i_L + (1 - q(t)) \times 0 \\ &= q(t)i_L\end{aligned} \tag{10.3-6}$$

The time-average input current over one switching period is

$$\begin{aligned}\widehat{i}_{in} &= \frac{1}{T_{sw}}\int_0^{T_{sw}} i_{in}(t) dt \\ &= \frac{1}{T_{sw}}\left(\int_0^{kT_{sw}} i_L dt + \int_{kT_{sw}}^{T_{sw}} 0 \times dt\right) \\ &= \frac{1}{T_{sw}}\int_0^{kT_{sw}} i_L dt = k\widehat{i}_L\end{aligned} \tag{10.3-7}$$

In the last step, we have used our assumption that i_L is approximately constant, and we have set $i_L = \widehat{i}_L$. The average power into the converter, $\widehat{P}_{\text{in}}$, is

$$\widehat{P}_{\text{in}} = V_{\text{in}}\widehat{i}_{\text{in}} = V_{\text{in}}k\widehat{i}_L \tag{10.3-8}$$

and the average power out of the converter, $\widehat{P}_{\text{out}}$, is approximated as

$$\widehat{P}_{\text{out}} = \widehat{(v_{\text{out}}i_{\text{out}})} = \widehat{v}_{\text{out}}\widehat{i}_L \tag{10.3-9}$$

To set the average of a product equal to the product of two averages, which is in general not a valid mathematical step, we have used our assumption that v_{out} and i_L are constant and equal to $\widehat{v}_{\text{out}}$ and $\widehat{i}_L$, respectively. Under this assumption (10.3-9) is valid. Substituting the buck converter voltage relationship in (10.3-3) into (10.3-8) and comparing to (10.3-9) yields

$$\widehat{P}_{\text{in}} = \widehat{v}_{\text{out}}\widehat{i}_L = \widehat{P}_{\text{out}} \tag{10.3-10}$$

showing that the average power is conserved. The ideal buck converter transfers all of the average input power to the output and consumes none, making it an exemplary power conversion device.

Thus far, we have left out a few important details. Namely, what constitutes "large" L and C, and what is "fast" switching? We will look at these details now; however, before we do it is worth reviewing the power conversion strategy embodied in Fig. 10.3-2 and the discussion and equations leading to (10.3-3). Similar strategies and concepts are used over and over in power electronic converters, and a firm understanding now will go a long way in your understanding of more advanced devices.

Let us now look at the steady-state operation of the buck converter circuit. We will examine the inductor current and capacitor voltage more closely to gain insight into the relationship between L, C, and f_{sw}. Thus far, we have assumed constant v_{out} and i_L. To analyze the inductor current in more detail, we will continue to assume that the output capacitance is large so that the capacitor can be approximated as a dc voltage source with voltage V_{out}. Fig. 10.3-5 shows the approximate circuit.

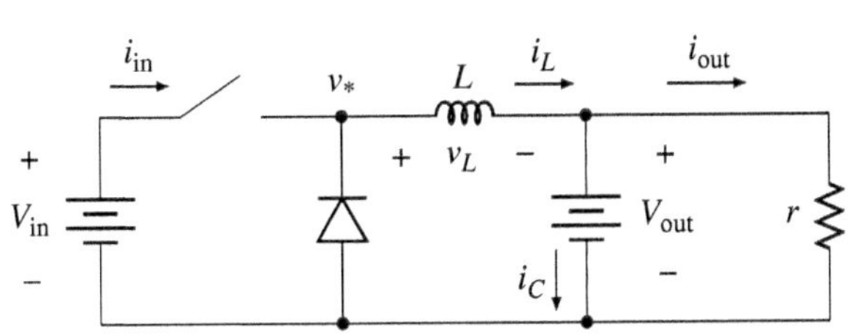

Figure 10.3-5 Buck converter with "large" output capacitance approximated as a voltage source.

The inductor voltage can be expressed using (10.3-1) as

$$\begin{aligned} v_L &= v_* - V_{out} \\ &= q(t)V_{in} - V_{out} \end{aligned} \tag{10.3-11}$$

Because the inductor voltage is also Ldi_L/dt, we have

$$\frac{di_L}{dt} = \frac{q(t)V_{in} - V_{out}}{L} \tag{10.3-12}$$

When the transistor is on and the circuit is in Configuration A, the inductor current ramps up linearly with slope $(V_{in} - V_{out})/L$; when the transistor is off and the circuit is in Configuration B, the current ramps down linearly with slope $-V_{out}/L$. The triangular-shaped periodic waveform with average $\widehat{i}_L = I_L$ shown in Fig. 10.3-6 is the result. The peak-to-peak ripple magnitude of this waveform is ΔI, and its period is T_{sw}.

We can calculate ΔI by taking the magnitude of the result of integrating (10.3-12) with respect to t either from 0 to kT_{sw} or from kT_{sw} to T_{sw}. In either case the result is

$$\begin{aligned} \Delta I &= \frac{V_{out}}{L}(T_{sw} - kT_{sw}) \\ &= \frac{V_{out}}{L \times f_{sw}}(1-k) \end{aligned} \tag{10.3-13}$$

Notice that the ripple magnitude is a function of L and f_{sw}. For a given switching frequency, if a maximum ripple ΔI_{max} is specified, the minimum inductance necessary is

$$L_{min} = \frac{V_{out}(1-k)}{f_{sw} \times \Delta I_{max}}. \tag{10.3-14}$$

According to (10.3-14), if $L \geq L_{min}$, then $\Delta I \leq \Delta I_{max}$.

We have thus derived a minimum threshold for determining a "large" L, based on the switching frequency and a specified maximum allowable ripple

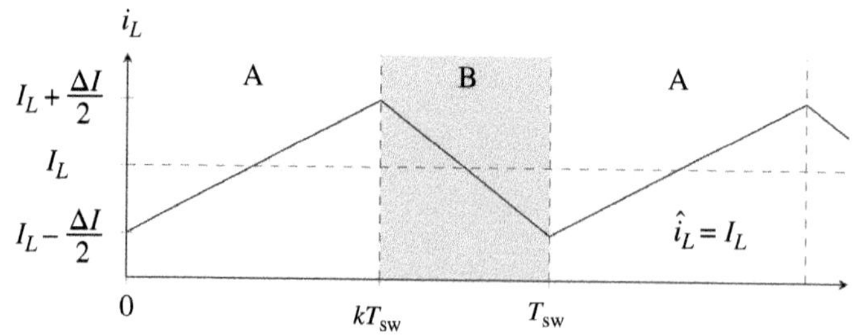

Figure 10.3-6 Buck converter inductor current waveform over one and a half switching periods.

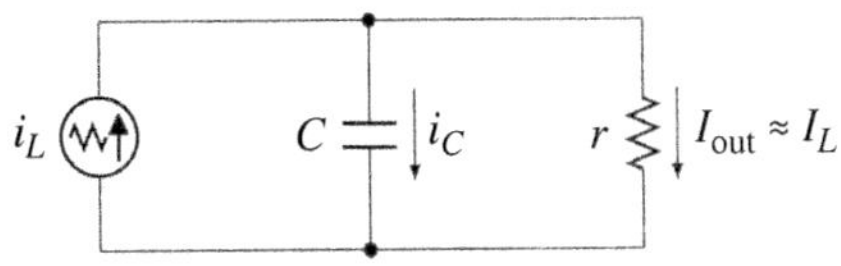

Figure 10.3-7 Buck converter output stage used to calculate C_{min}.

current. But, what about C? Thus far, C has been assumed large so that the capacitor could be considered a dc voltage source. In reality, the capacitor voltage will have an average value with some triangular ripple, similar to the inductor current waveform. To analyze the capacitor, we will use the circuit shown in Fig. 10.3-7. The inductor current is represented as a current source with average $\widehat{i_L} = I_L$ and triangular ripple of magnitude $\pm\ \Delta I/2$. The capacitor is in parallel with the resistive load. Because, according to (10.2-18) the capacitor has no average current in the steady state, I_L flows through the resistor and the ripple current flows through the capacitor. In reality, a small fraction of the ripple current will also flow through the resistive load; however, if the capacitor impedance at the ripple frequency is much smaller than the load impedance, which it will be in normal circumstances, then the majority of the ripple current will flow through the capacitor and our assumption is valid. In fact, in many instances, with a reasonable amount of output capacitance the inductance in the conductors from the converter to the load is large enough to validate our assumption, regardless of the load impedance.

The capacitor voltage, v_C, will rise and fall according to

$$\frac{dv_C}{dt} = \frac{i_C(t)}{C} \tag{10.3-15}$$

where i_C is the ripple current from the inductor. The capacitor current is shown in Fig. 10.3-8, which is similar to Fig. 10.3-6 except that the dc offset has been removed.

The variable δ_{mid} is introduced to mark the midway point between kT_{sw} and T_{sw}:

$$\delta_{mid} \triangleq \frac{T_{sw} - kT_{sw}}{2} \tag{10.3-16}$$

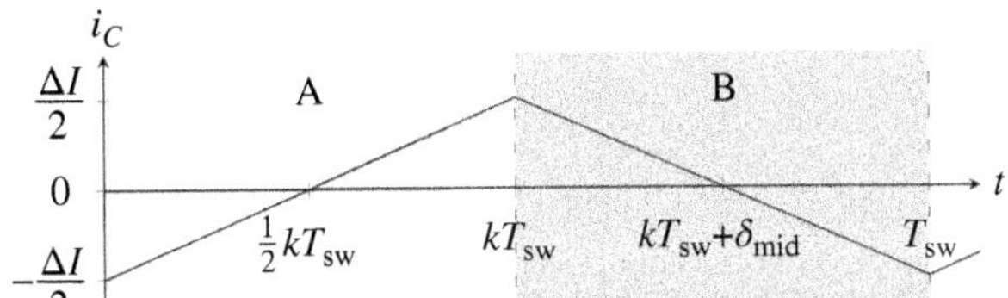

Figure 10.3-8 Buck converter approximate capacitor current.

The capacitor charges when $i_C > 0$ and discharges when $i_C < 0$. Thus, the capacitor peak-to-peak voltage ripple ΔV is found by calculating the voltage rise that occurs between $kT_{sw}/2$ and $kT_{sw} + \delta_{mid}$. Because the ripple is symmetrical about the t-axis, its magnitude can also be found by calculating the voltage drop that occurs when i_C is negative; however, because this requires integrating over a split interval, we choose the positive region for simplicity. Specifically,

$$\begin{aligned} \Delta V &= \int_{kT_{sw}/2}^{kT_{sw}+\delta_{mid}} \frac{dv_C}{dt} dt \\ &= \frac{1}{C} \frac{\left[\left(\frac{kT_{sw}}{2} \times \frac{\Delta I}{2}\right) + \left(\delta_{mid} \times \frac{\Delta I}{2}\right)\right]}{2} \\ &= \frac{\Delta I}{8C}(kT_{sw} + 2\delta_{mid}) \end{aligned} \tag{10.3-17}$$

Using (10.3-16) to simplify, this becomes

$$\begin{aligned} \Delta V &= \frac{\Delta I \times T_{sw}}{8C} \\ &= \frac{\Delta I}{8C \times f_{sw}} \end{aligned} \tag{10.3-18}$$

The voltage ripple magnitude is a function of C, f_{sw}, and the current ripple magnitude. For a given switching frequency and current ripple, then, the minimum capacitance C_{min} necessary to keep the output voltage ripple below a desired limit ΔV_{max} is

$$C_{min} = \frac{\Delta I}{8 f_{sw} \Delta V_{max}} \tag{10.3-19}$$

According to (10.3-19), if $C \geq C_{min}$, then $\Delta V \leq \Delta V_{max}$. Thus, we have derived a minimum threshold for determining a "large" C, based on the switching frequency, the allowable voltage ripple, and the inductor current ripple.

The last remaining detail is to discuss the switching frequency. Specifically, what is meant by "fast" switching? There are several ways to answer this question; we will only discuss a few of the most important ones here. First, notice from (10.3-14) and (10.3-19) that both L_{min} and C_{min} are inversely proportional to f_{sw}. Thus, all else being equal, a higher switching frequency reduces the size of the converter's energy storage components. In a general sense, then, "fast" switching is that which allows the circuit to have reasonably sized components. Next, notice from (10.3-13) and (10.3-18) that f_{sw} is also inversely proportional to both ΔI and ΔV. All else being equal, then, faster switching reduces ripple magnitude. Thus, "fast" switching keeps the circuit ripple below a specified threshold. A third way to consider the question is by considering the Fourier series of v_*, which, according to (10.3-1) is simply (10.2-15) multiplied by V_{in}. According to our dc–dc conversion process shown in Fig. 10.3-2, we would like to low-pass filter the square wave v_* to leave only its dc component as V_{out}. Any ripple on the output is unwanted. Together, the inductor and capacitor serve as a low-pass filter to remove it. The higher the switching frequency, the farther it will be in the frequency domain from the dc component and the easier it will be for the filter components to remove it and higher harmonics. Thus, "fast" switching is that which allows the low-pass *LC* output filter to work effectively.

Of course, all of these qualitative fast switching descriptions are related and essentially describe the same physical circuit behavior; however, we provide three alternative ways of thinking about the concept in order to facilitate a deeper understanding. Other considerations that influence the switching frequency include audible noise – humans hear frequencies up to around 20 kHz, so switching frequencies are often chosen higher than this to avoid converter "buzzing," – switch hardware limitations, losses – real switches have small losses each time they switch; the higher the frequency, the higher these losses – and switch gate control hardware limits. These advanced topics are considered in later courses.

Example 10D Buck Converter Design

Here, we design a buck converter to power a 5 V, 25 W resistive load from a 12 V battery. First, we choose appropriate values for f_{sw}, k, L, and C so that the inductor's peak-to-peak current ripple should be no more than 40 percent of its average current and the output peak-to-peak voltage ripple should be no more than 4 percent of the nominal value. Then, for the chosen circuit components, we calculate the expected ΔI and ΔV and compare these to the specifications.

The duty ratio is given by (10.3-4) and is

$$k = \frac{\widehat{v}_{out}}{V_{in}} = \frac{5}{12} \tag{10D-1}$$

To avoid the audible range but remain low enough for average switch hardware to accommodate, we will choose $f_{sw} = 30$ kHz. Once the design is complete, we can verify that our components are reasonably sized. If they are not, we can adjust the switching frequency and component design accordingly. The output power is 25 W; thus

$$\widehat{i}_{out} = \frac{25\text{ W}}{5\text{ V}} = 5\text{ A} \tag{10D-2}$$

and from (10.3-5), $\widehat{i}_L = 5$ A. From the specification,

$$\Delta I_{max} = 0.4 \times \widehat{i}_L = 2\text{ A} \tag{10D-3}$$

thus, from (10.3-14),

$$L_{min} = \frac{5 \times (1 - 5/12)}{(30 \times 10^3) \times 2} = 48.61\ \mu\text{H} \tag{10D-4}$$

Let us choose $L = 50\ \mu$H. Equation (10.3-13) now gives the expected current ripple as

$$\Delta I = \frac{5 \times (1 - 5/12)}{(30 \times 10^3) \times (50 \times 10^{-6})} = 1.94\text{ A} \tag{10D-5}$$

which is 39 percent of the nominal current, within the specification. From the specification,

$$\Delta V_{max} = 0.04 \times 5\ \text{V} = 0.2\ \text{V} \tag{10D-6}$$

and from (10.3-19),

$$C_{min} = \frac{1.94}{8 \times (30 \times 10^3) \times 0.2} = 40.4\ \mu\text{F}. \tag{10D-7}$$

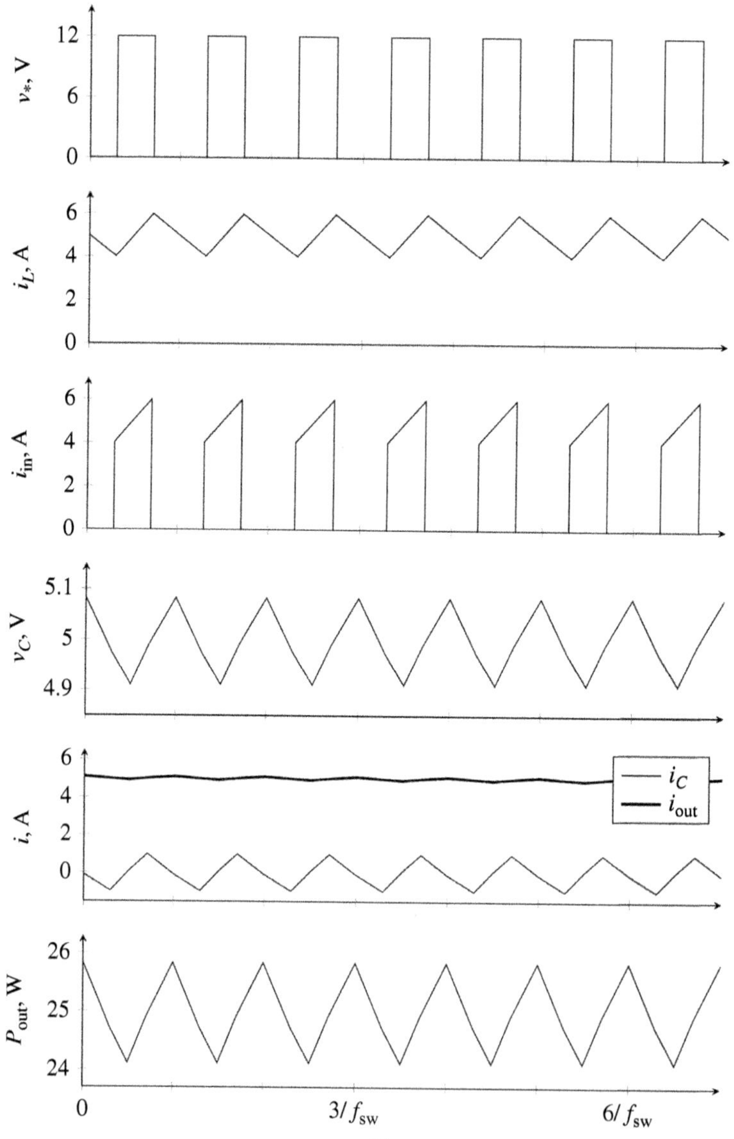

Figure 10.3-9 Circuit waveforms from the 25 W buck converter designed in Example 10D.

Because it is a standard capacitor size, we will choose $C = 47$ μF. Based on this choice, (10.3-18) gives the expected voltage ripple as

$$\Delta V = \frac{1.94}{8 \times \left(47 \times 10^{-6}\right) \times \left(30 \times 10^{3}\right)} = 0.17 \text{ V} \tag{10D-8}$$

which is 3.4 percent of the nominal output voltage, within the specification. The chosen components are reasonably sized for this type of circuit and the circuit achieves its design goals. Thus, the switching frequency does not need to be adjusted.

The circuit designed in Example 10D has been simulated and several waveforms are shown in Fig. 10.3-9 to illustrate the behavior. Notice that $v_* = q(t)V_{\text{in}}$ is as designed, with $k = 5/12$. The period is 33.33 μs, which corresponds to the 30 kHz switching frequency. The inductor current has an average of 5 A and peak-to-peak ripple of approximately 2 A. The input current is equal to the inductor current when the transistor is on and zero otherwise. The output capacitor voltage has an average of 5 V and peak-to-peak ripple of approximately 0.18 V. Notice that the output voltage ripple is out of phase with the inductor current ripple; this is because the capacitor voltage rises when the inductor current is greater than its average, 5 A, and discharges when it is less than its average. The capacitor and output (resistor) current are plotted together to show that the majority of the ripple in i_L goes through the capacitor while the dc current goes exclusively through the resistor. Finally, the output resistor power is plotted to show the average power is 25 W, as designed, with small peak-to-peak ripple of roughly 2 W, or 8 percent.

Boost Converter

To highlight the power electronic conversion concepts we have covered thus far, we will briefly introduce a companion circuit to the buck converter: the dc–dc *boost converter*. The basic boost converter circuit is shown in Fig. 10.3-10. Its purpose is the inverse of the buck converter: it transfers power from a lower voltage dc source to a higher voltage dc load. In contrast to the buck converter, in the boost converter the inductor is always connected in series with the input voltage source. By controlling the transistor with an appropriate switching function, the inductor can be charged and discharged while maintaining an average current. This current is alternately disconnected from and connected to the load through the uncontrolled

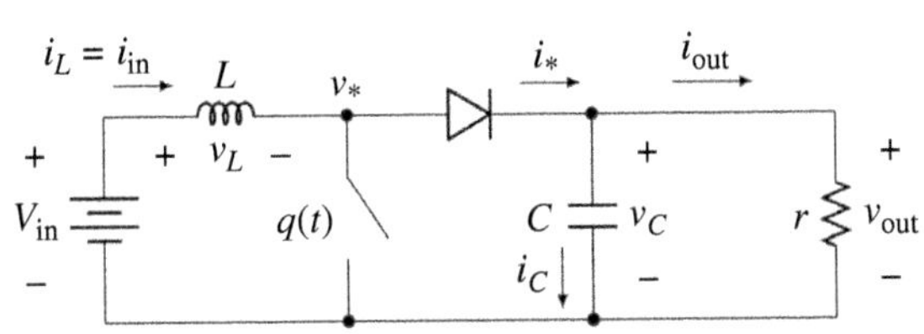

Figure 10.3-10 Boost converter circuit.

(diode) switch. The load current is filtered with the output capacitor. With sufficiently "large" L and C, the inductor and capacitor can be approximated as dc current and voltage sources, respectively, to facilitate design and analysis.

Like the buck converter, the boost converter is periodically switched between two circuit configurations. When the transistor is closed, the diode is reverse-biased and turns off. In this circuit configuration, the inductor's voltage is V_{in} and its current is ramped up as energy is transferred from the source to the inductor's magnetic field. The transistor is then opened, which causes the diode to turn on in order to provide a path for the inductor current to the load. Since $\widehat{v}_{out} > V_{in}$, the inductor voltage is now negative and its current ramps down as the energy transferred to it from the source is now transferred to the load. The inductor voltage is thus

$$v_L(t) = q(t)V_{in} + (1 - q(t)) \times (V_{in} - \widehat{v}_{out}) \tag{10.3-20}$$

Because the average inductor voltage is zero in the steady state (10.2-17), we have

$$\begin{aligned}
\widehat{v}_L &= \frac{1}{T_{sw}} \int_0^{T_{sw}} [q(t)V_{in} + (1 - q(t)) \times (V_{in} - \widehat{v}_{out})]dt \\
&= \frac{1}{T_{sw}} \left[\int_0^{kT_{sw}} V_{in} dt + \int_{kT_{sw}}^{T_{sw}} (V_{in} - \widehat{v}_{out}) dt \right] \\
&= \frac{1}{T_{sw}} [V_{in} \times kT_{sw} + (V_{in} - \widehat{v}_{out}) \times (T_{sw} - kT_{sw})] = 0
\end{aligned} \tag{10.3-21}$$

which is simplified to

$$\frac{\widehat{v}_{out}}{V_{in}} = \frac{1}{1-k} \tag{10.3-22}$$

Equation (10.3-22) describes the steady-state voltage conversion ratio of the boost converter. It is analogous to (10.3-4) for the buck converter. In the boost converter, the average output voltage is greater than or equal to the input voltage, and the conversion ratio is controlled by k. When $k = 0$, the transistor is never on and the input and average output voltages are equal. At any other value of k, the average output voltage is higher than the input. For example, for $k = 0.5$, $\widehat{v}_{out} = 2V_{in}$ and for $k = 0.75$, $\widehat{v}_{out} = 4V_{in}$. Note that according to (10.3-22) $\widehat{v}_{out}$ tends toward infinity as k approaches unity. This is because as k increases, more energy is transferred to the inductor while the transistor is in the on-state, and the fraction of the switching period for this energy to be transferred to the load when the transistor is off, $1 - k$, decreases. In order for this increasing energy to be transferred to the load in a decreasing time, it must be done in short pulses of current at high voltage. In the extreme as k approaches unity, these pulses become infinitesimally small while the voltage becomes infinitely large. Although this description is true in the ideal circuit we have analyzed, in any real converter this is not the case. Series resistance in the inductor will limit the amount of energy that can be transferred to it, bounding the output voltage. This is addressed in Problem 6. More generally, the converter's

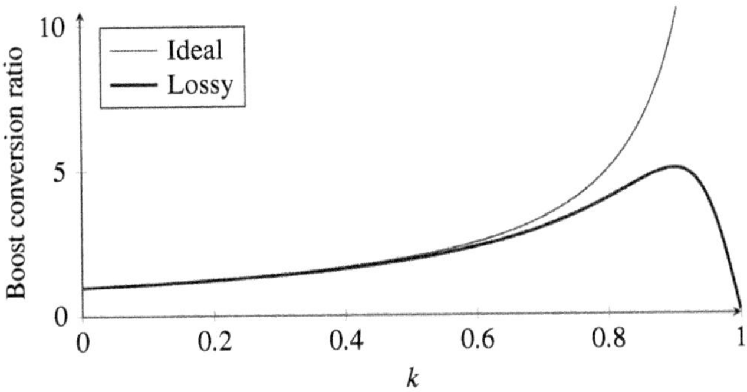

Figure 10.3-11 Boost converter conversion ratio ($\widehat{v}_{out}/V_{in}$) for ideal (lossless) and lossy circuits.

losses will dominate the behavior at higher duty ratios, bounding the output voltage to a finite value. For example, in Fig. 10.3-11 the conversion ratio of an ideal boost converter (10.3-22) is compared to that of a boost converter with resistive losses in its inductor equal to 1 percent of the load resistance. As can be seen, in the lossy circuit the boost ratio peaks at 5 with $k = 0.9$, then falls off to zero at $k = 1$.

Analysis of the circuit currents and powers in the ideal boost converter reveals that

$$i_{in} = i_L \tag{10.3-23}$$

$$\begin{aligned} \widehat{i}_{out} &= (1-k)\widehat{i}_L \\ &= (1-k)\widehat{i}_{in} \end{aligned} \tag{10.3-24}$$

$$\begin{aligned} \widehat{P}_{in} &= V_{in}\widehat{i}_{in} \\ &= V_{in}\frac{\widehat{i}_{out}}{1-k} = \frac{V_{in}}{1-k}\widehat{i}_{out} \\ &= \widehat{v}_{out}\widehat{i}_{out} \end{aligned} \tag{10.3-25}$$

Assuming v_{out} and i_{out} are approximately constant and equal to $\widehat{v}_{out}$ and $\widehat{i}_{out}$, respectively, we have

$$\widehat{P}_{out} = \left(\widehat{v_{out}i_{out}}\right) = \widehat{v}_{out}\widehat{i}_{out} \tag{10.3-26}$$

Thus, $\widehat{P}_{in} = \widehat{P}_{out}$ and power is conserved.

The boost converter power conversion strategy is hinted at by (10.3-24), and is summarized as follows. In the steady state, the inductor can be approximated as a dc current source; we ignore small ripple current in this discussion. Because $\widehat{P}_{in} = \widehat{P}_{out}$ and $\widehat{v}_{out} > V_{in}$, this requires that $\widehat{i}_{out} < \widehat{i}_{in}$. This dc current reduction is accomplished by first "chopping" the input current to create the square wave diode current i_*. The dc value of i_* is then obtained by filtering; the whole process being analogous to the voltage reduction strategy used in the buck converter and shown in Fig. 10.3-2. In the boost converter, the circuit reduces dc current the same way the buck converter reduces dc voltage. In both circuits, because power is conserved, when output voltage or current decreases, its corresponding current or voltage must increase.

Circuit waveforms from a 25-W, 5-V to 12-V boost converter with the same inductor and capacitor values as the buck converter in Example 10D are shown in Fig. 10.3-12. The waveforms are arranged so that when compared side-by-side

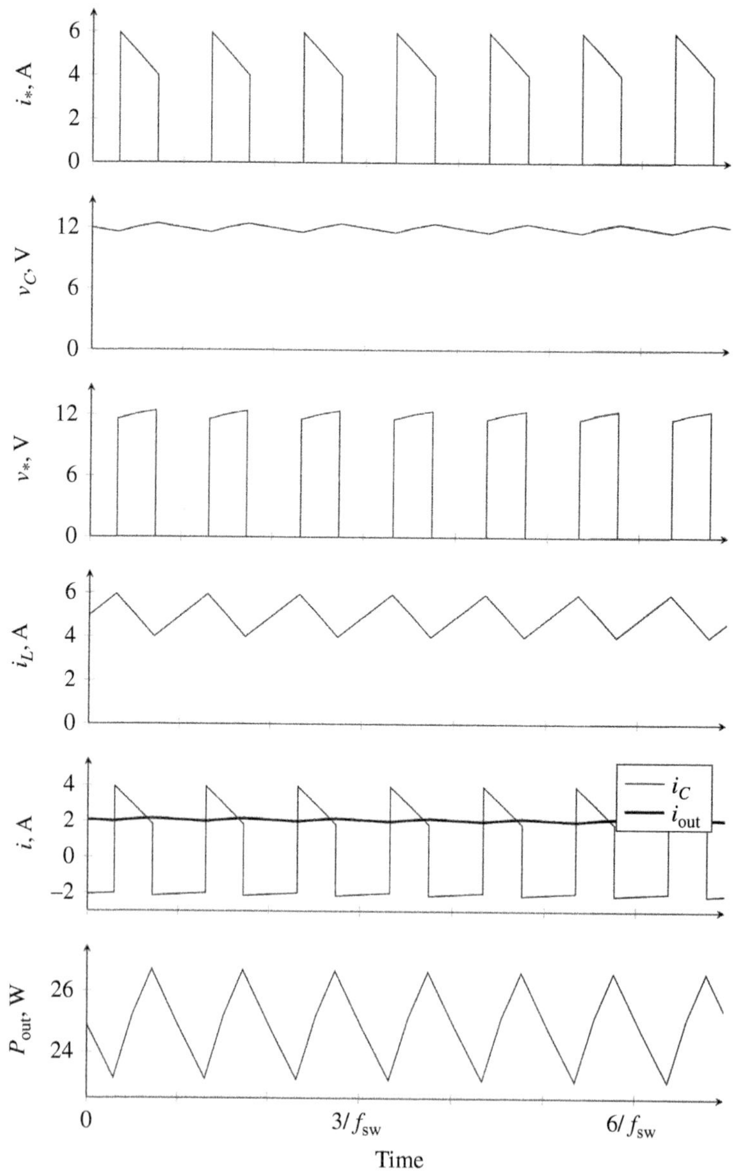

Figure 10.3-12 Circuit waveforms from ideal 25 W boost converter with $L = 50\ \mu H$, $C = 47\ \mu F$, $f_{sw} = 30\ kHz$, $V_{in} = 5\ V$, $\widehat{v}_{out} = 12\ V$, and $k = 7/12$.

with Fig. 10.3-9, the analogous conversion strategies are evident. For example, the top boost waveform, i_*, is analogous to the top buck waveform, v_*. Each of these is the "chopped" version of the related dc input value.

Choosing appropriate L, C, and f_{sw} in boost converter design requires similar analysis as was done for the buck converter. Inductor current ripple, capacitor voltage ripple, and switch frequency considerations are all used to choose appropriate values. Although we will not explicitly cover this here, you should now have the necessary knowledge and tools to accomplish the task.

Advanced Circuit Topologies

Numerous other dc–dc converter topologies exist. Buck–boost and boost–buck converters combine elements of the buck and boost converters to allow both voltage increase and decrease in a single converter. A flyback converter also allows voltage increase and decrease, and uses a coupled inductor to achieve isolation and to add flexibility in choosing the output voltage polarity. Multistage series and parallel configurations are also common when efficiency, ripple, or other design specifications are restrictive. Although the topologies of these circuits are more elaborate than the simple buck and boost circuits introduced here, the basic concepts we have introduced can be applied to their analysis.

SP10.3-1 For a buck converter with $f_{sw} = 25$ kHz, what is T_{sw}? [$T_{sw} = 40$ µs]

SP10.3-2 For the buck converter in SP10.3-1, if the transistor is on for 10 µs of each switch period, what is the duty ratio k? [$k = 0.25$]

SP10.3-3 For the buck converter in SP10.3-1 and SP10.3-2, (a) what is the average output voltage, $\hat{v}_{out}$, if the input voltage, $V_{in} = 28$ V? (b) What is the buck voltage conversion ratio? [(a) $\hat{v}_{out} = 7$ V; (b) $\dfrac{\hat{v}_{out}}{V_{in}} = k = 0.25$]

SP10.3-4 (a) Assume the boost conversion ratio for a lossy boost converter is $\dfrac{\hat{v}_{out}}{V_{in}} = \dfrac{(1-k)r}{(1-k)^2 r + r_L}$ (you are asked to derive this in Problem 6). (a) Find the maximum achievable boost ratio k_{max}. Assume $r = 1\ \Omega$ and $r_L = 0.01\ \Omega$. (b) What is k at this point? (c) Compare your results to the plot in Fig. 10.3-11. [(a) $\hat{v}_{out}/V_{in} = 5$; (b) $k_{max} = 0.9$; (c) results should align with the plot]

10.4 AC–DC CONVERSION

We now shift gears to ac–dc conversion circuits, commonly called *rectifiers*. We analyze two common rectifiers – the half- and full-wave, respectively – to introduce the key concepts. The analysis tools introduced in this section are broadly applicable to advanced rectifier circuits.

Half-Wave Rectifier

The single-phase, *half-wave rectifier* is shown in Fig. 10.4-1. An ac voltage source $V_{in}(t) = V_p \sin(\omega_{in} t)$, where $\omega_{in} = 2\pi f_{in}$ is the radian frequency, is connected in series with a diode and a load. In the steady state, the current is i_{load} and the diode

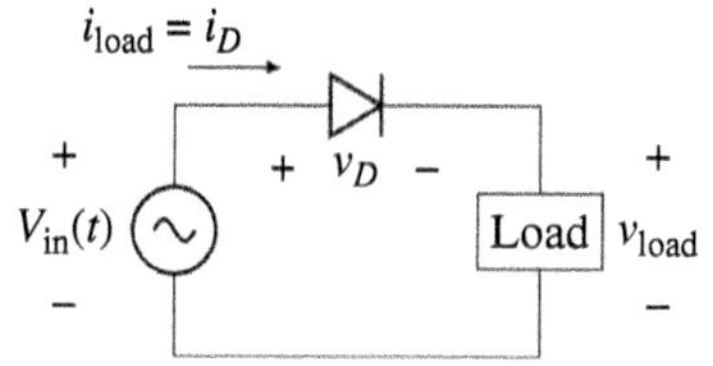

Figure 10.4-1 Half-wave rectifier circuit.

and load voltages are v_D and v_{load}, respectively, as shown. In our analysis, the load will consist of various arrangements of passive elements. The diode obeys the rules given in (10.2-16).

The *trial method* [1] is an effective strategy to analyze this diode circuit over one excitation period. Because the input is periodic, the circuit behavior will be as well. At each instant of time, assume a diode state (on or off); if the diode is on, set $v_D = 0$ and analyze the circuit to determine the direction of the diode current i_D; if the diode is off, set $i_D = 0$ and analyze the circuit to determine the polarity of v_D. Except at the instants when the diode is changing state, one and only one diode state will be self-consistent, as described by (10.2-16); the other state will be impossible. Once the diode state is known, the circuit can be solved. Formally, this process must be applied at each instant of time to calculate the circuit's full transient behavior. However, in practice, the behavior can be bracketed into a few discrete time regions. For example, with sinusoidal input the time domain can be divided into three regions: positive, negative, and zero input, respectively. Within each of these regions the diode state is constant.

Assuming a resistive load r, the diode trial method can be used to solve the circuit in Fig. 10.4-1. Within the interval $t \in (0, T/2)$, $V_{in} > 0$. First, assume the diode is off and set $i_D = i_{load} = 0$. With $i_{load} = 0$, $v_{load} = i_{load} \times r = 0$ and $v_D = V_{in} - v_{load} = V_{in} > 0$. But, if the diode is off, v_D must be negative; thus, this state is inconsistent and cannot be true. Thus, the diode is on. To see this, set $v_D = 0$ and solve the circuit. In this case, $v_{load} = V_{in} > 0$, and $i_{load} = v_{load}/r > 0$. $i_{load} > 0$ *is* consistent with the diode being on, so this is the diode's correct state. Following the same procedure for the interval $t \in (T/2, T)$ when $V_{in} < 0$, we find that the diode must be off. Finally, the instants when $V_{in} = 0$ must be examined. Unfortunately, the trial method tells us nothing about the diode state in these instants, because both the on and off states lead to inconsistent results. This simply means that the diode is changing state at these instants, transitioning from on to off at $t = T/2$ and from off to on at $t = T$. Because the diode is transitioning at these points, we cannot assign it a definitive state. The full diode trial method analysis procedure is summarized in Table 10.4-1.

From Table 10.4-1, the operation of the circuit is clear. When the input voltage is positive the diode is on, the load voltage equals the input voltage, and positive current flows through the load. As the input voltage crosses zero, the diode turns off. When the input voltage is negative, the diode is off, no current flows, and the load voltage is zero. The diode switches back on as the voltage crosses zero again one half period later, and the process continues periodically. Half-wave

TABLE 10.4-1 Summary of Diode Trial Method for Half-Wave Rectifier with Resistive Load

t	V_{in}	Diode State Assumed	Test	Result	OK?
$(0, T/2)$	>0	On	$v_D = 0$	$i_D > 0$	Yes
		Off	$i_D = 0$	$v_D > 0$	No
$(T/2, T)$	<0	On	$v_D = 0$	$i_D < 0$	No
		Off	$i_D = 0$	$v_D < 0$	Yes
$T/2$ or T	=0	On	$v_D = 0$	$i_D = 0$	No
		Off	$i_D = 0$	$v_D = 0$	No

rectifier input and output voltage waveforms for $f_{in} = 60$ Hz are shown in Fig. 10.4-2. This plot shows why the circuit is named the way it is: half of the input sine wave is allowed to pass through to the output; the negative half of the wave is blocked.

The average value of this rectified signal is easily calculated as

$$\begin{aligned}\widehat{v}_{load} &= \frac{1}{T}\int_0^{T/2} V_p \sin(\omega_{in}t)dt \\ &= \frac{-V_p}{\omega_{in}T}\left[\cos\left(\frac{\omega_{in}T}{2}\right) - \cos(0)\right] \\ &= \frac{V_p}{\pi}\end{aligned} \quad (10.4\text{-}1)$$

For the example in Fig. 10.4-2, $\widehat{v}_{load} = 10/\pi \approx 3.18$ V, as plotted. Because it has a nonzero average value, the half-wave rectifier's load voltage is, in fact, a dc signal because it fits our expanded definition of dc from Section 10.2. Accordingly, a Fourier series can be constructed for this waveform, and its dc component is V_p/π. The half-wave rectifier accomplishes the goal of converting the ac $V_{in}(t)$ to the dc v_{load}, with average value $\widehat{v}_{load} = V_p/\pi$. Additional waveforms for circuit current and load power are plotted in Fig. 10.4-3. Notice that current and power only flow to the load when the input voltage is positive and the diode is on.

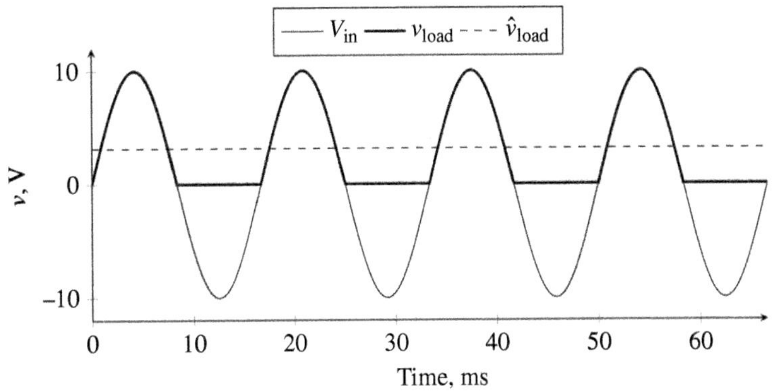

Figure 10.4-2 Voltage waveforms for a half-wave rectifier with resistive load; $V_p = 10$ V, $r = 1\ \Omega$ and $f_{in} = 60$ Hz.

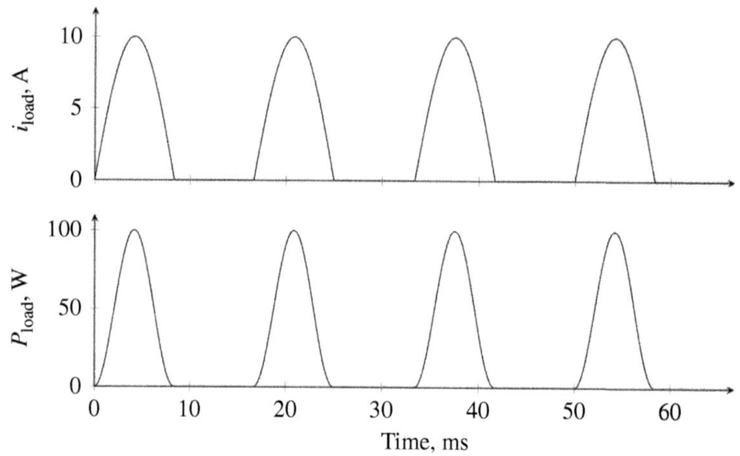

Figure 10.4-3 Load current and power waveforms for the half-wave rectifier with resistive load.

The analysis is complicated somewhat when the load contains inductors and capacitors, which allow the load to maintain its current and/or voltage independent of the source. This changes the diode switching behavior compared to the resistive load case. However, by modifying the diode trial method's resistive load results and using some basic circuit analysis, we can meet the challenge.

First, we will examine a parallel rC load. This type of load might represent, for example, a bus for dc loads (e.g., a battery charging circuit) or a dc link feeding additional power electronic circuitry. If the capacitance is "large" so that the load time constant is much greater than the input voltage period (i.e., $\tau = rC \gg T$), the capacitor will charge up when the diode is on, but will maintain its charge and slowly discharge approximately linearly through the resistor when the diode is off. Here, we only examine the large capacitance case, because if C is small, the capacitor will discharge quickly when the diode is off and the load behavior will not differ significantly from the resistive case. With large output capacitance the diode will be reverse-biased and remain off until the input voltage is greater than the decaying capacitor voltage, at which point the diode will turn on again. In essence, with respect to the resistive load circuit, the introduction of the capacitor delays the turn-on time of the diode until the input voltage is greater than the capacitor voltage and accelerates the turn-off time to when the diode current crosses zero (when the capacitor and resistor currents balance exactly), which occurs while the input voltage is still positive. Because the source current leads the source voltage in this capacitive circuit, the diode current always crosses zero before the source voltage.

Calculating the diode turn-on and turn-off times t_{on} and t_{off}, respectively, requires some calculus [1], which we will omit. However, it suffices to understand conceptually the differences between the resistive and rC load circuits. Waveforms for a half-wave rectifier with a parallel rC load with $r = 1\ \Omega$ and $C = 47$ mF ($\tau = 47$ ms) and source parameters $V_p = 10$ V and $f_{in} = 60$ Hz ($T \approx 16.67$ ms) are given in Figs. 10.4-4 and 10.4-5. For this circuit, $\tau/T \approx 2.82$. The capacitor's effect is clear: compared to a resistive load, $\widehat{v}_{load}$ is much higher and the ripple about

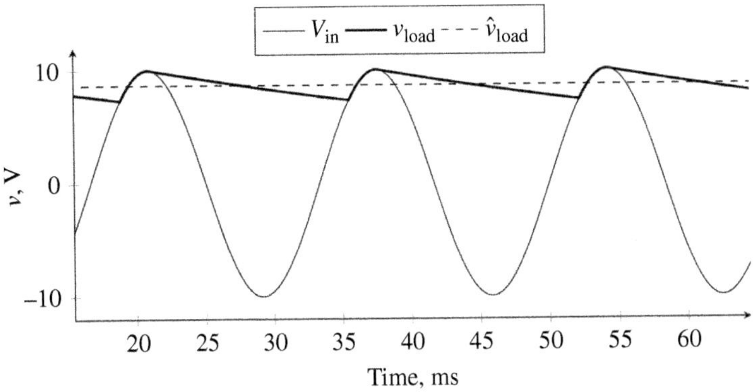

Figure 10.4-4 Steady-state voltage waveforms from a half-wave rectifier with parallel rC load; $V_p = 10$ V, $r = 1\ \Omega$, $C = 47$ mF, and $f_{in} = 60$ Hz.

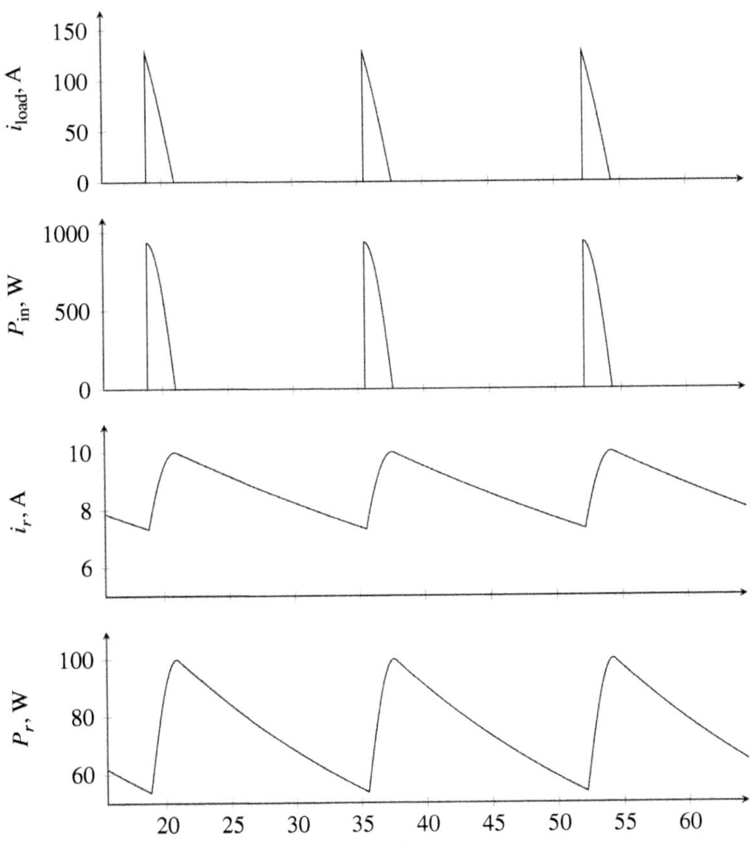

Figure 10.4-5 Source current and power, and load (resistor) current and power waveforms for the half-wave rectifier with parallel rC load.

the average is much lower. The capacitor acts as both a low-pass filter and a charge storage device increasing and smoothing the resistor voltage.

In Fig. 10.4-5, it is shown that the parallel rC load draws current in short, large bursts when the diode is on. Compared to Fig. 10.4-3, here both the current and power pulses from the source are over ten times larger than in the resistive load circuit. This is common in diode capacitor circuits. Filters are often used to reduce these spikes in order to improve power quality, as is described later in this section.

Next, let us consider a series rL load. This might represent, for example, the field winding of either a synchronous or dc machine. If the inductance is "small"[2] so that the load time constant is much smaller than the input voltage period (i.e., $\tau = L/r \ll T$), the source current will lag the voltage by a small angle much less than $\pi/2$. The current zero-crossing, and the corresponding diode turn-off, will lag the voltage zero-crossing. For a short time during each cycle the source voltage will be negative while the current is positive: energy will flow from the load to the source. Once the diode turns off, the load conditions are identical to the off-state conditions of the resistive load circuit; thus, the diode will turn on at the same times. However, the turn-off will be delayed, which will affect the average rectified load voltage.

As with the rC load, explicitly calculating the diode turn-off times requires nontrivial calculations that will not be shown here. However, a firm conceptual understanding of the impact of the inductor's addition is our goal. This is achieved through an understanding of the previous discussion and by studying Figs. 10.4-6 and 10.4-7, which plot waveforms from a half-wave rectifier with series rL load, the same parameters as the resistive circuit and $L = 3$ mH ($\tau = 3$ ms). For this circuit, $\tau/T \approx 0.18$. Comparing Fig. 10.4-6 to its resistive (Fig. 10.4-2) and parallel rC (Fig. 10.4-4) counterparts, and comparing Fig. 10.4-7 to Figs. 10.4-3 and 10.4-5 shows how the addition of energy storage elements impacts the performance of the half-wave rectifier.

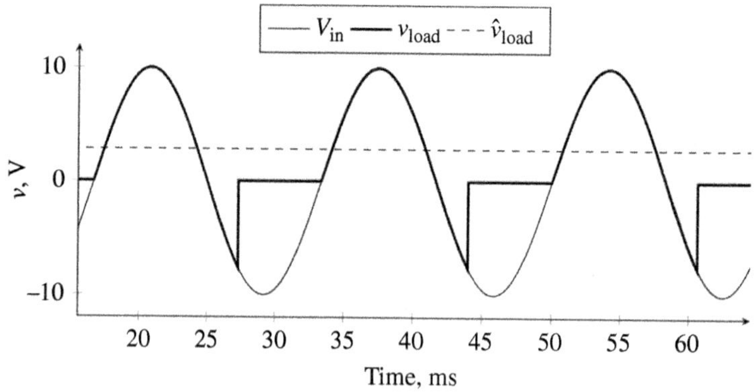

Figure 10.4-6 Steady-state voltage waveforms for a half-wave rectifier with series rL load; $V_p = 10$ V, $r = 1\ \Omega$, $L = 3$ mH, and $f_{in} = 60$ Hz.

[2] We only consider the small inductance case, because with large L the inductor current will not go to zero during the half cycle when the input voltage is negative; thus, the diode will never turn off, and rectification will not occur.

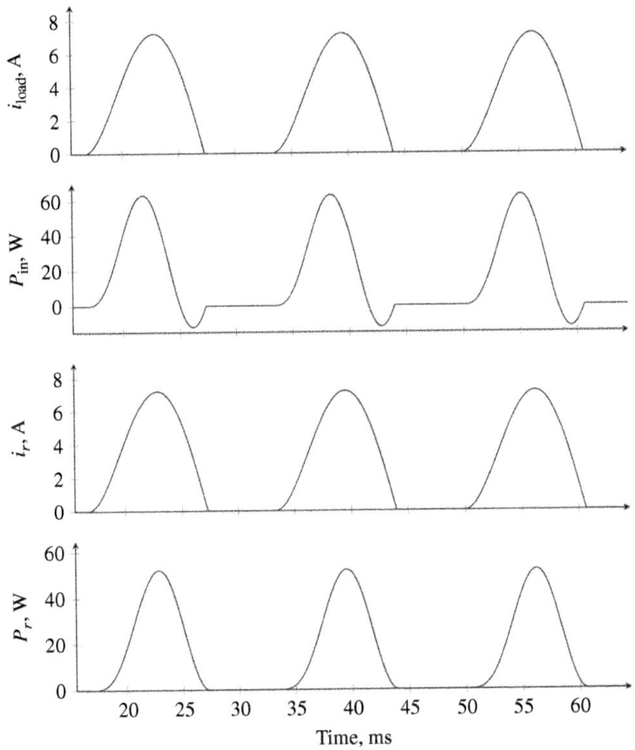

Figure 10.4-7 Source current and power, and load (resistor) current and power waveforms for the half-wave rectifier with series rL load.

Example 10E Average Power in a Half-Wave Rectifier

We now calculate the average power delivered by the half-wave rectifier to a resistive load, r.

The relevant circuit waveforms are shown in Figs. 10.4-2 and 10.4-3. The load power, P_{load}, is expressed as

$$P_{load} = \frac{v_{load}^2}{r} \tag{10E-1}$$

The average power delivered to the load is

$$\widehat{P}_{load} = \frac{\widehat{\left(v_{load}^2\right)}}{r} \tag{10E-2}$$

which is calculated as

$$\begin{aligned} \widehat{P}_{load} &= \frac{1}{r}\frac{1}{T_{in}}\int_0^{T_{in}/2} V_p^2 \sin^2(\omega_{in} t)dt \\ &= \frac{1}{r}\frac{V_p^2}{T_{in}}\int_0^{T_{in}/2} \sin^2(\omega_{in} t)dt \end{aligned} \tag{10E-3}$$

We use the identity

$$\sin^2(\omega_{\text{in}}t) = \frac{1}{2}[1 - \cos(2\omega_{\text{in}}t)] \tag{10E-4}$$

to integrate (10E-3), giving

$$\widehat{P}_{\text{load}} = \frac{V_p^2}{4r} \tag{10E-5}$$

For comparison, note that from (10.4-1),

$$\begin{aligned} \frac{(\widehat{v}_{\text{load}})^2}{r} &= \frac{(V_p/\pi)^2}{r} \\ &= \frac{V_p^2}{\pi^2 r} \neq \widehat{P}_{\text{load}} \end{aligned} \tag{10E-6}$$

This illustrates a key fact to remember when averaging transient signals. The average of a product of two signals (i.e., $(\widehat{v_{\text{load}}^2})$) is not necessarily equal to the product of the averages of the two signals (i.e., $(\widehat{v}_{\text{load}})^2$). This is important to remember when calculating average power.

Full-Wave Rectifier

From Fig. 10.4-2 it is apparent that the average load voltage obtained from the half-wave rectifier can be doubled by flipping the clipped negative humps, making them positive, and applying them to the load as well. In other words, we would like to create $v_{\text{load}} = |V_{\text{in}}|$. This is accomplished by adding three diodes and forming a *full-wave rectifier* (also called a *full-bridge rectifier*). The circuit is shown in Fig. 10.4-8.

The full-wave rectifier can be analyzed using the trial method: we first assume a resistive load and analyze all the diode state combinations, both for positive, negative, and zero input voltage. Because there are now four diodes, there are $2^4 = 16$ switch combinations for each of the three input voltage regions, leaving 48 combinations to analyze. Although this seems daunting at first, many of the combinations can be eliminated quickly.

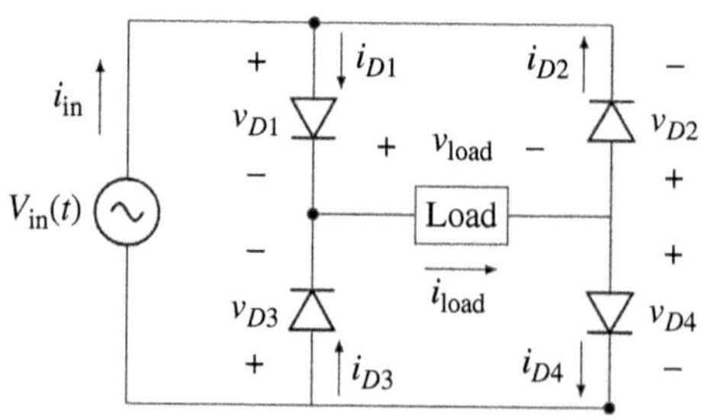

Figure 10.4-8 Full-wave rectifier circuit.

Similarly to the half-wave rectifier, when $V_{\text{in}} = 0$, none of the 16 diode states are self-consistent. These are the instants when the diodes change state; each diode will switch from its state before the zero-crossing to its new state after the zero-crossing. Of the remaining 32 configurations, several are quickly eliminated due to Kirchoff's Voltage Law (KVL) and KCL violations. Any configuration with only one diode on violates KCL, because there is not a closed loop for current to flow. This fact eliminates eight configurations (four with $V_{\text{in}} > 0$ and four with $V_{\text{in}} < 0$). Likewise, any configuration with diodes 1 and 3 on simultaneously and/or diodes 2 and 4 on simultaneously violates KVL, because these configurations short the input source and also require negative current to flow through one of the diodes. This eliminates 14 more configurations. In the two cases with all four diodes off, no load current can flow, which means that $v_{D1} = -v_{D2}$ and $v_{D3} = -v_{D4}$. This is inconsistent with all off diode voltages being negative, eliminating these two cases. There remain only eight configurations to examine.

For simplicity, the collective diode state assumed is represented by a four-digit binary number, where 1 = on and 0 = off. For example, the diode state with diodes 2 and 3 on and diodes 1 and 4 off is represented as 0110. After applying the trial method to the remaining eight cases, only two cases – when $V_{\text{in}} > 0$ and diodes 1 and 4 are on, and when $V_{\text{in}} < 0$ and diodes 2 and 3 are on – are consistent. The two key circuit configurations resulting from the trial method for the full-wave rectifier are given in Table 10.4-2. The full analysis is left to the reader as an end-of-chapter problem.

The full-wave rectifier operation is now evident. As time increases from 0 to $T/2$, the circuit is in diode state 1001, the positive half cycle of the sinusoid is applied across the load, and current flows through diodes 1 and 3 and the load. The diodes switch state as the voltage crosses zero at $t = T/2$. The input voltage sinusoid is negative from $T/2$ to T, and $-V_{\text{in}}$ (which is positive) is applied across the load. Current flows through diodes 3 and 4 and the load. As the input voltage crosses zero again, the diodes switch back to their original state, and the cycle continues.

The diode trial method thus far presented may seem tedious; however, once you work through it for a few circuits, you will develop an intuition for how diode circuits work. The tedium of checking each and every switch state can then be bypassed as you quickly identify the acceptable configurations. Many introductory

TABLE 10.4-2 Summary of Defining States in the Diode Trial Method for a Full-Wave Rectifier with Resistive Load

			Test Result				
t	V_{in}	Diode State	D1	D2	D3	D4	OK?
$(0, \frac{T}{2})$	>0	1001	$i_{D1} > 0$	$v_{D2} < 0$	$v_{D3} < 0$	$i_{D4} > 0$	Yes
$(\frac{T}{2}, T)$	<0	0110	$v_{D1} < 0$	$i_{D2} > 0$	$i_{D3} > 0$	$v_{D4} < 0$	Yes
$\frac{T}{2}$ or T	=0	Indeterminate diode transition state					

texts bypass any discussion of this rigorous method in favor of simply stating how the full-wave rectifier operates; however, a thorough understanding of the trial method is very useful when you encounter a complicated circuit containing diodes. Thus, time invested now will pay dividends down the road.

Input, load, and average voltage waveforms for the full-wave rectifier with resistive load are shown in Fig. 10.4-9. Performing the averaging calculation in (10.4-1) on v_{load} gives

$$\hat{v}_{\text{load}} = \frac{2V_p}{\pi} \tag{10.4-2}$$

which is twice the average value of the half-wave rectifier output. Figure 10.4-10 contains load current and power waveforms. Comparing Figs. 10.4-9 and 10.4-10 to Figs. 10.4-2 and 10.4-3, respectively, reveals that the full-wave rectifier output

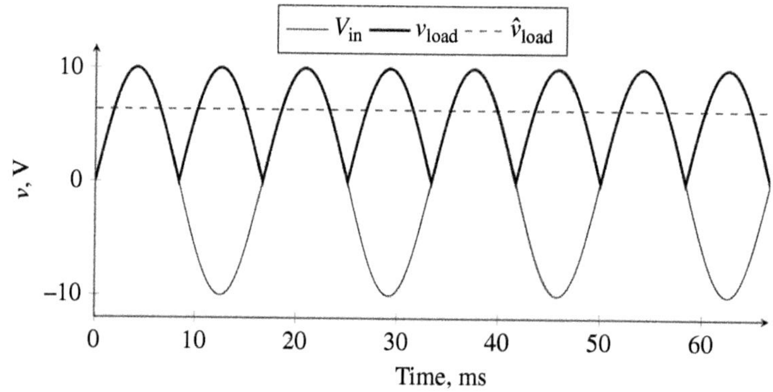

Figure 10.4-9 Voltage waveforms from a full-wave rectifier with resistive load; $V_p = 10$ V, $r = 1\ \Omega$, and $f_{\text{in}} = 60$ Hz.

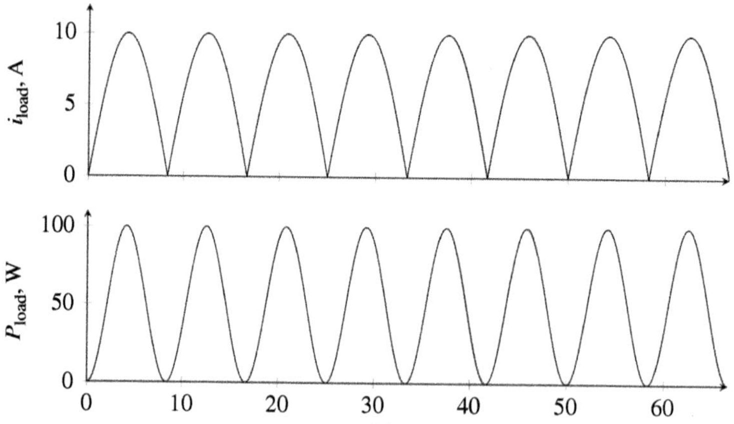

Figure 10.4-10 Load current and power waveforms for the full-wave rectifier with resistive load.

waveforms have double the frequency of the half-wave rectifier. The compact Fourier series representation of the full-wave rectified waveform shown in Fig. 10.4-9, labeled $v_{\text{rec}}(t)$, can be calculated using (10.2-1) through (10.2-7) and the fact that the waveform is even to give

$$v_{\text{rec}}(t) = \frac{2V_p}{\pi} + \frac{4V_p}{\pi}\sum_{n=1}^{\infty}\frac{\cos(n\pi)}{1-4n^2}\cos\left[n(2\omega_{\text{in}}t - 2\theta(0))\right] \tag{10.4-3}$$

where $V_p = 10$ V, $\omega_{\text{in}} = 2\pi \times 60$ rad/s, and $\theta(0) = \pi/2$. Notice that the dc component of the Fourier series matches the average value in (10.4-2), and the series' fundamental frequency is twice the frequency of V_{in}. When compared to the half-wave rectifier output, the double frequency harmonics of the full-wave rectifier are relatively easier to filter.

The effect of adding parallel capacitance to the full-wave rectifier output is similar to the half-wave rectifier. However, since the full-wave rectified voltage has twice the frequency of the half-wave, the filtered voltage has a higher average and reduced ripple magnitude. With a capacitor in parallel with the load, the circuit is often called a *classical rectifier*, because this is a classical (i.e., common) charging circuit for batteries. Waveforms for the full-wave rectifier shown in Fig. 10.4-8 with 47 mF parallel load capacitance are shown in Figs. 10.4-11 and 10.4-12. Comparing these figures to the half-wave rectifier waveforms in Figs. 10.4-4 and 10.4-5, respectively, clearly shows the double frequency and reduced load ripple.

Adding inductance in series with the full-wave rectifier output has effects similar to the half-wave rectifier, with one key difference. Because there are four switches instead of one, with an inductive load the full-wave rectifier always has an available current path, so there are no diode switching delays and the current need not decay to zero for the diodes to switch states. This is illustrated in Figs. 10.4-13 and 10.4-14. Comparing these to Figs. 10.4-6 and 10.4-7, respectively, shows that

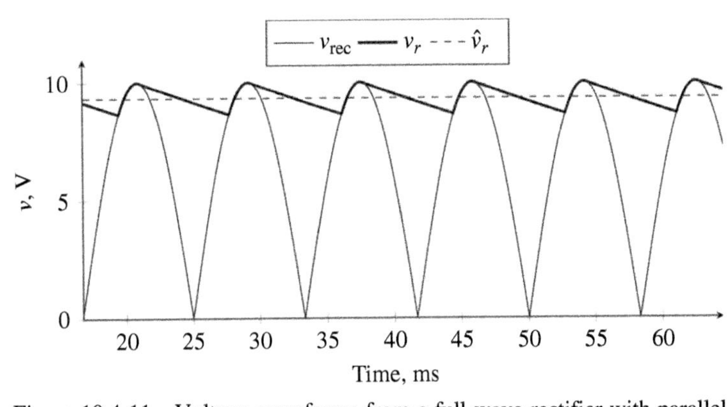

Figure 10.4-11 Voltage waveforms from a full-wave rectifier with parallel rC load; $V_p = 10$ V, $r = 1\ \Omega$, $C = 47$ mF, and $f_{\text{in}} = 60$ Hz.

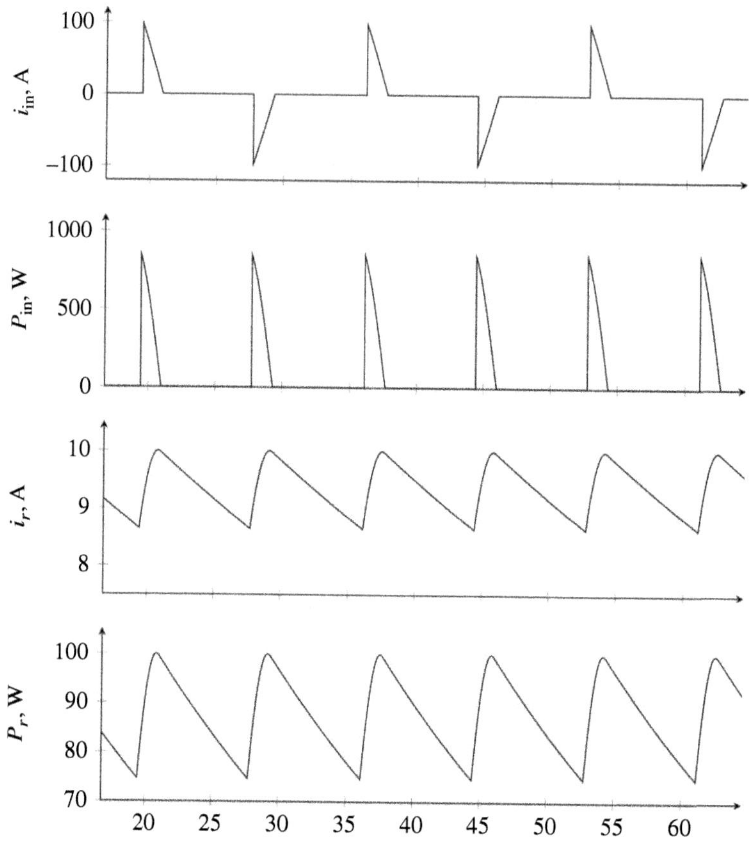

Figure 10.4-12 Source current and power, and load (resistor) current and power waveforms for the full-wave rectifier with parallel rC load.

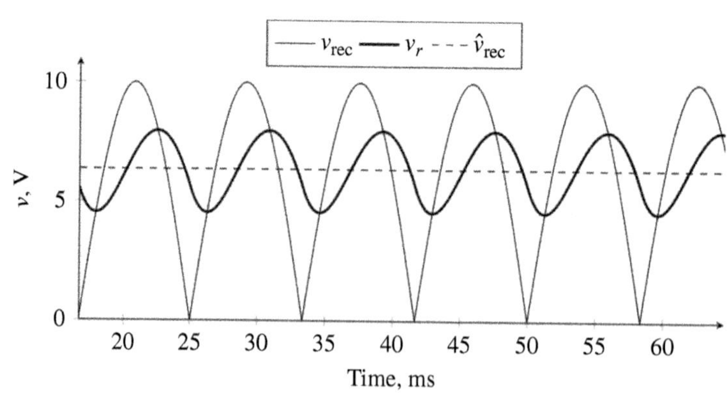

Figure 10.4-13 Steady-state voltage waveforms from a full-wave rectifier with series rL load; $V_p = 10$ V, $r = 1\ \Omega$, $L = 3$ mH, and $f_{in} = 60$ Hz.

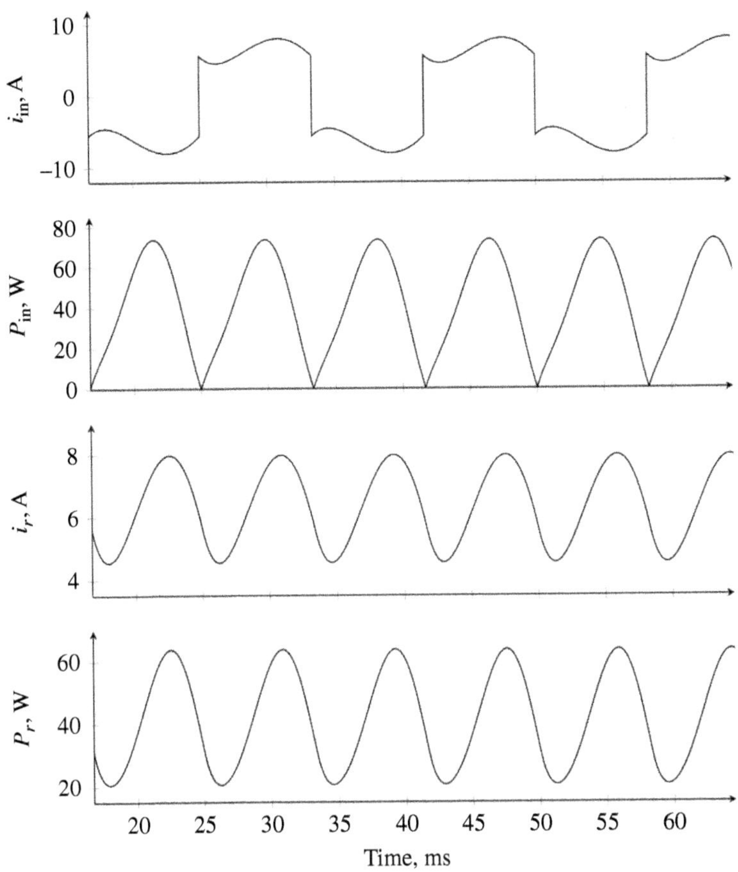

Figure 10.4-14 Source current and power, and load (resistor) current and power waveforms for the full-wave rectifier with series rL load.

the rectified voltage is not altered by diode switching delays, the input power is always non-negative, and the load current remains positive. The inductor acts as a low-pass filter that smooths the resistor voltage.

The full-wave rectifier contains four diodes and provides a path for the load current in each of its allowed diode switch states. This facilitates the implementation of more sophisticated load filtering to achieve the desired load power quality. To see this, we consider the LC low-pass filter shown in Fig. 10.4-15. The input to the filter is $v_{\text{rec}}(t) = |V_{\text{in}}(t)|$, which consists of a dc component equal to $2V_p/\pi$ and harmonics at multiples of $2\omega_{\text{in}}$. To achieve true ac-to-dc conversion, all the harmonics above dc should be filtered out. A straightforward steady-state analysis of the circuit (this is left to the reader in SP10.4-2) reveals that the filter transfer function $G_{\text{filt}}(\omega)$ is

$$G_{\text{filt}}(\omega) = \frac{v_r(\omega)}{v_{\text{rec}}(\omega)} = \frac{r}{j\omega L + r(1 - \omega^2 LC)} \tag{10.4-4}$$

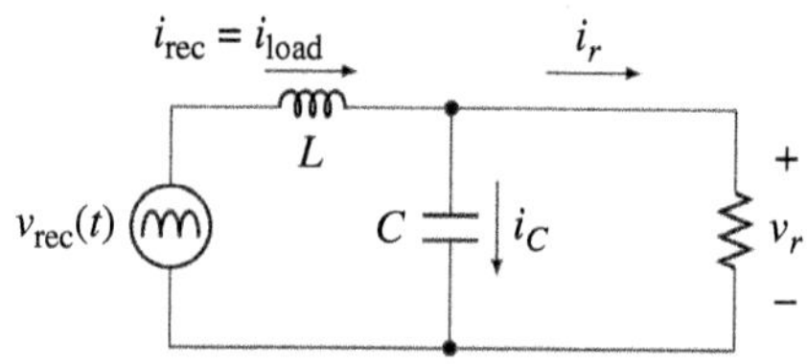

Figure 10.4-15 Full-wave rectifier with LC output filter and resistive load.

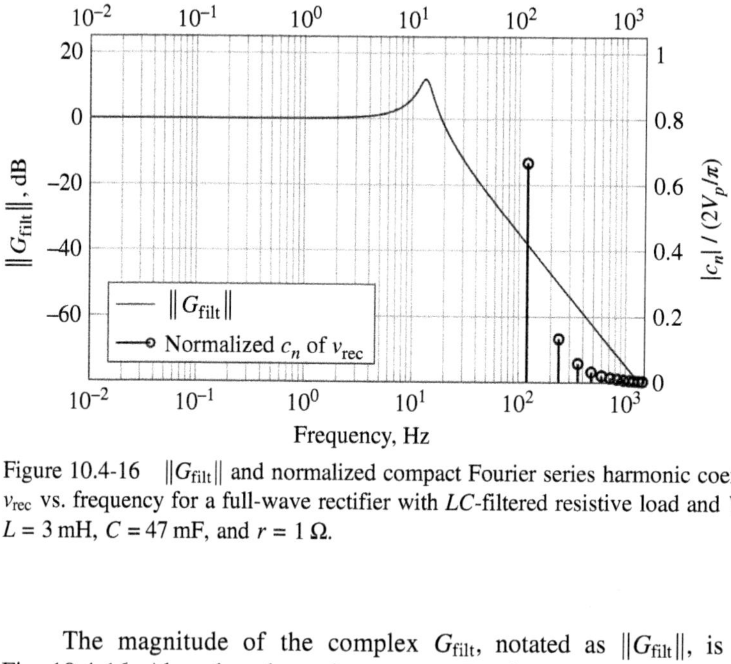

Figure 10.4-16 $\|G_{filt}\|$ and normalized compact Fourier series harmonic coefficients of v_{rec} vs. frequency for a full-wave rectifier with LC-filtered resistive load and $V_p = 10$ V, $L = 3$ mH, $C = 47$ mF, and $r = 1\ \Omega$.

The magnitude of the complex G_{filt}, notated as $\|G_{filt}\|$, is plotted in Fig. 10.4-16. Also plotted are the compact Fourier series harmonic coefficients (i.e., the c_n values of $v_{rec}(t)$ in (10.4-3)), normalized to the dc component $\frac{2V_p}{\pi}$. With $L = 3$ mH and $C = 47$ mF, the harmonics of $v_{rec}(t)$ are well to the right of the resonance frequency, $f_{res} = \frac{1}{2\pi\sqrt{LC}} \approx 13.4$ Hz, in the filter's capacitive region. The harmonics are thus filtered out effectively, leaving the dc component to dominate the filter output.

Voltage waveforms for the filtered full-wave rectifier are shown in Fig. 10.4-17. The load voltage v_r closely matches $\widehat{v}_{rec}$, highlighting the LC low-pass filter effects and clearly showing the ac–dc converter action of this circuit. In general, L and C should be chosen so that the harmonics of the rectified waveform are located well into the capacitive region of the filter to the right of the resonance frequency. If the harmonics fall to the left of the resonance frequency, the filter's attenuation will be minimal. This is the essence of power electronics, as we have also seen with the dc–dc converters earlier in this chapter. Power conversion

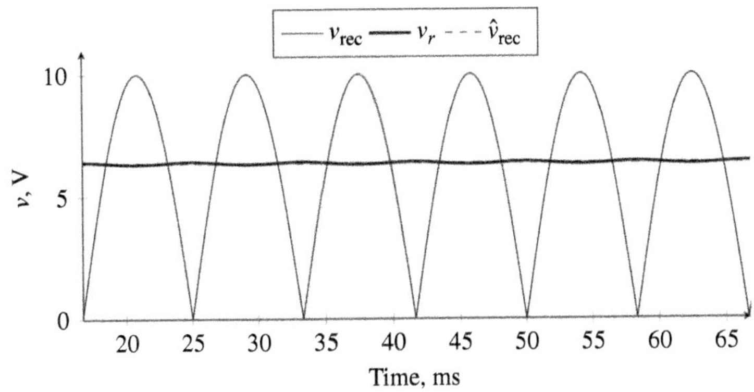

Figure 10.4-17 Steady-state voltage waveforms from a full-wave rectifier with *LC*-filtered resistive load; $V_p = 10\,\text{V}$, $r = 1\,\Omega$, $L = 3\,\text{mH}$, $C = 47\,\text{mF}$, and $f_{in} = 60\,\text{Hz}$.

is achieved by chopping up an input signal with switches and filtering the chopped up signal before applying it to a load. In the buck converter, the input voltage was chopped into a square wave and filtered; in the boost converter, the input current was chopped into a square wave and filtered; in the full- and half-wave rectifiers, the input voltage sinusoid is chopped into half-waves with diodes, and the waves are filtered and applied to a dc load.

SP10.4-1 (a) Calculate the average power delivered by the full-wave rectifier to a resistive load. The input voltage is $V_{in}(t) = V_p \sin(\omega_{in} t)$, where $T_{in} = 2\pi/\omega_{in}$ is the input period. Relevant circuit waveforms are shown in Figs. 10.4-9 and 10.4-10. (b) Compare your answer in (a) to the result in Example 10E. [(a)$\widehat{P}_{load} = V_p^2/2r$; (b) The average power in the full-wave rectifier is twice that in the half-wave rectifier]

SP10.4-2 Derive (10.4-4) using Fig. 10.4-15.

SP10.4-3 Why does the Fourier series of v_{rec} in (10.4-3) have a fundamental frequency of $2\omega_{in}$? [$|V_{in}|$ has twice the frequency of V_{in} (the absolute value of a sinusoid has twice the frequency of the sinusoid itself)]

10.5 DC–AC CONVERSION

We conclude our introduction to power electronic circuits with dc–ac conversion circuits, called *inverters*. Here, we analyze the single-phase dc–ac inverter; the basic three-phase inverter is covered in Chapter 6 as part of a brushless dc drive.

Single-Phase Inverter

A single-phase dc–ac inverter is shown in Fig. 10.5-1. This circuit converts energy from the dc source, V_{in}, to the ac load using a bridge of four controlled transistor switches, notated with the capital letter S. Although there are numerous switch control methods that accomplish the task, we will cover one here and leave the others to an advanced course.

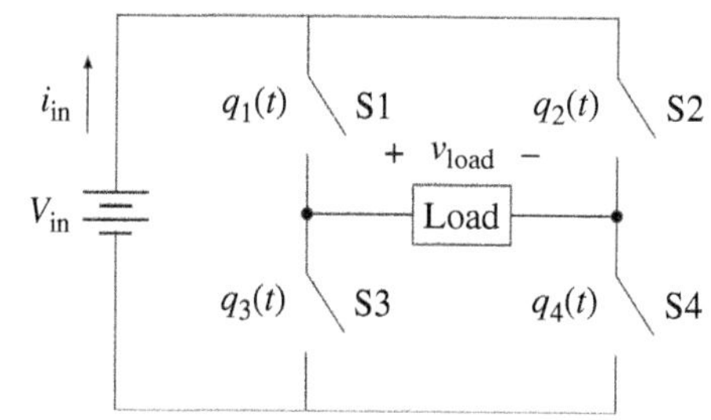

Figure 10.5-1 Single-phase dc–ac inverter circuit.

It is fairly straightforward to identify the allowed and useful switch states. First, switches 1 and 3 and 2 and 4, respectively, can never be on simultaneously, lest they short the source. Thus, their respective switching functions obey $q_1(t) = 1 - q_3(t)$ and $q_2(t) = 1 - q_4(t)$. Second, if switches 1 and 2 or 3 and 4, respectively, are on simultaneously, the load is essentially short-circuited and no power will be transferred. Further, if only one (or no) switches are on, there will not be a continuous path for current to flow through the load. Thus, the only allowable switch states that transfer power from the source to the load are $(q_1, q_2, q_3, q_4) = (1, 0, 0, 1)$ and $(0, 1, 1, 0)$. If the circuit is switched between these two states alternately with equal time in each state, the square wave voltage shown in Fig. 10.5-2 will be applied across the load.

Although typically when we think of an ac waveform we think of a sinusoid, the inverter voltage square wave shown in Fig. 10.5-2 also qualifies. In fact, our earlier Fourier analysis showed that this periodic waveform can be expressed as a sum of sinusoids: the fundamental and its harmonics. The series in (10.2-15) can be easily modified to represent the inverter square wave (see SP10.5-1). If we apply appropriate filtering, we can remove the harmonics and "pull out" the fundamental frequency of the square wave, leaving a pure sinusoid. In practice, such filtering is difficult to achieve because the harmonics are relatively close to the fundamental in the frequency domain. In later courses, more advanced switching methods based on *pulse width modulation* (PWM) are used to separate the unwanted harmonics from the fundamental and make it easier to filter them. For the purposes of our

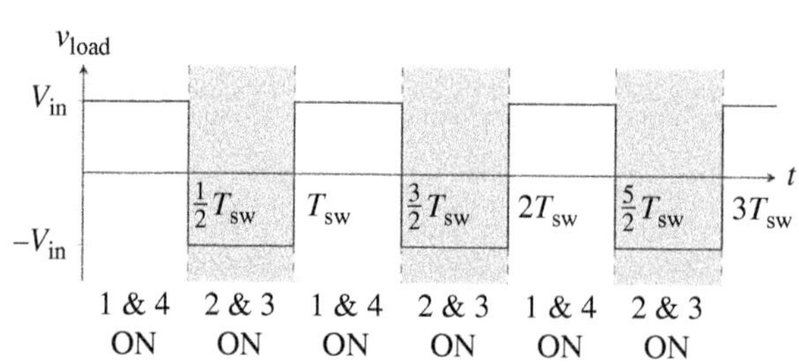

Figure 10.5-2 Single-phase inverter voltage.

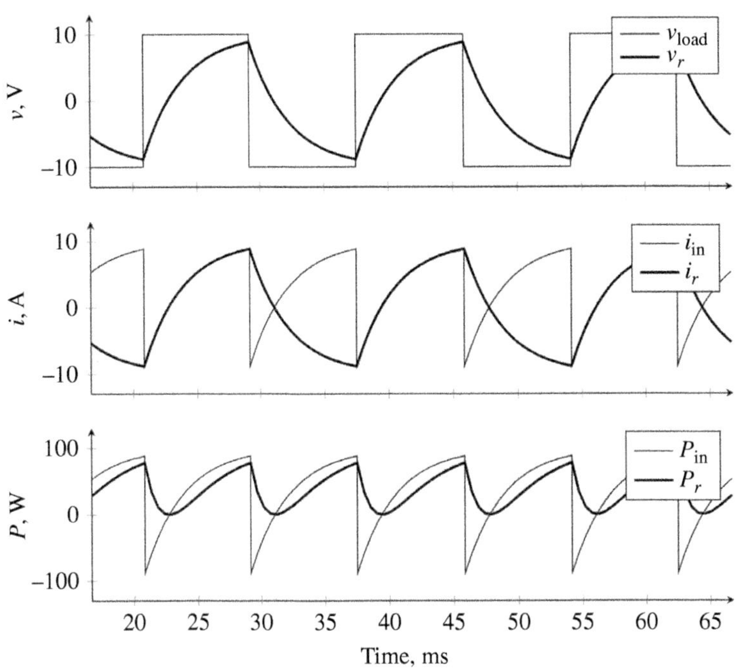

Figure 10.5-3 Steady-state voltage, current, and power waveforms from a single-phase inverter with series *rL* load; V_{in} = 10 V, r = 1 Ω, L = 3 mH, and f_{in} = 60 Hz.

introductory analysis, we will limit ourselves to the square wave switching scheme and analyze the circuit with both inductive and resonant *LC* load filters applied.

Voltage, current, and power waveforms are shown in Fig. 10.5-3 for the single-phase inverter with a series *rL* load and V_{in} = 10 V, r = 1 Ω, L = 3 mH, and f_{in} = 60 Hz. The load *rL* time constant is τ = 3 ms, which is about one-sixth of the switching period T_{sw} (16.7 ms), and the response is as expected: a sequence of *rL* charges and decays. The load current is slightly more sinusoidal than the square wave current that would result without the inductive filtering, but we are still far from a sinusoid. Luckily, we can do better.

A load waveform closer to a sinusoid can be formed by adding a capacitor in series with the series *rL* load and choosing *L* and *C* to resonate at the inverter switch frequency. This creates a series *resonant filter*. For example, by adding series capacitance *C* to the circuit characterized by the waveforms in Fig. 10.5-3 such that $\dfrac{1}{2\pi\sqrt{LC}}$ = 60 Hz, the waveforms in Fig. 10.5-4 result. Although the shapes of these load voltage and current waveforms are more sinusoidal than their counterparts in Fig. 10.5-3, they are not strictly sinusoidal. Luckily, though, in many applications these are acceptable ac waveforms. For example, numerous low-cost consumer single-phase dc–ac inverters create similar waveforms.

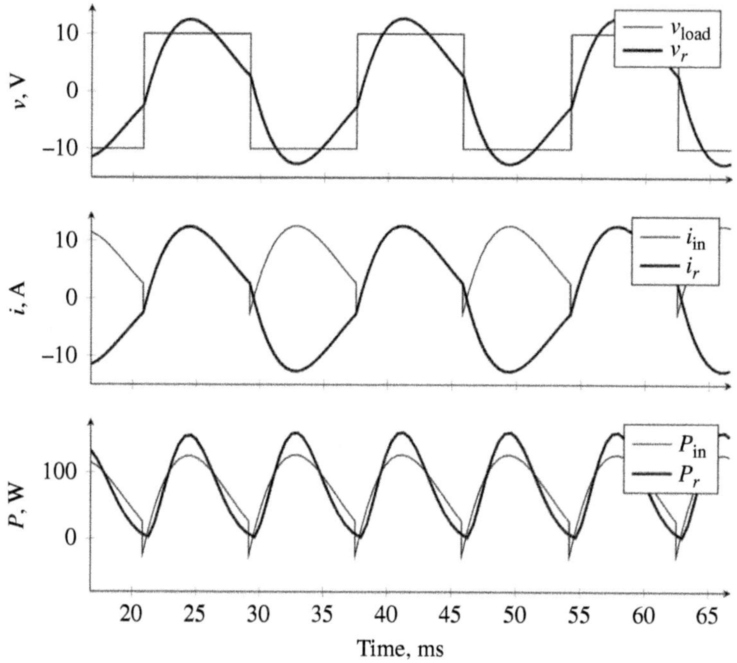

Figure 10.5-4 Steady-state voltage, current, and power waveforms for the single-phase inverter with series resonant rLC load; $V_{in} = 10\,\text{V}$, $r = 1\,\Omega$, $L = 3\,\text{mH}$, $C = 2.345\,\text{mF}$, and $f_{in} = 60\,\text{Hz}$.

Example 10F Power Calculation for Single-Phase Inverter

It is instructive to calculate the average power delivered to a resistive load by the single-phase inverter in Fig. 10.5-1. Because the load is resistive, the load current is simply the square wave in Fig. 10.5-2 scaled by the load resistance. The power delivered to the load is calculated as

$$P_{\text{load}} = i_{\text{load}}^2 \times r \tag{10F-1}$$

The average power is

$$\widehat{P}_{\text{load}} = \left(\widehat{i_{\text{load}}^2 \times r}\right) = \left(\widehat{i_{\text{load}}^2}\right) \times r \tag{10F-2}$$

Notice that this is *not* equivalent to $\left(\widehat{i}_{\text{load}}\right)_r^2$, which is zero. We can calculate the load current's mean square, denoted $(\cdot)_{\text{ms}}$, as

$$\begin{aligned}\widehat{P}_{\text{load}} &= r \times \left(i_{\text{load}}\right)_{\text{ms}} = \frac{r}{T_{\text{sw}}}\int_0^{T_{\text{sw}}} i_{\text{load}}^2 dt \\ &= \frac{r}{T_{\text{sw}}}\int_0^{T_{\text{sw}}} \left(\frac{V_{\text{in}}}{r}\right)^2 dt \\ &= \frac{V_{\text{in}}^2}{r}\end{aligned} \tag{10F-3}$$

We conclude this section by noting that the single-phase inverter can be expanded in a straightforward manner to a three-phase inverter, a device which is common in electric drives. This device, often referred to as a "six-step" inverter, is covered in Chapter 6, where it is used to drive the permanent-magnet ac machine as a brushless dc machine. The concepts covered here with respect to the single-phase inverter apply to that device.

SP10.5-1 Modify (10.2-15) to find the Fourier series of the single-phase inverter waveform shown in Fig. 10.5-2. [$v_{\text{load}}(t) = \dfrac{4V_{\text{in}}}{\pi}\sum_{n=1}^{\infty}\dfrac{\sin(n\pi/2)}{n}\cos(n\omega_{\text{sw}}t - n\pi/2)$]

SP10.5-2 Qualitatively describe any changes to the waveforms in Fig. 10.5-3 if L is reduced to 0.3 mH (90 percent reduction) [v_r and i_r would look much more like square waves; i_{in} and P_{in} would have much faster rise times; P_r would have less ripple (would look closer to a constant value)]

10.6 RECAPPING

This chapter has presented power electronic circuit analysis at an introductory level. In Section 10.2, we established the fundamentals of switching circuits, including power conversion principles, useful Fourier analysis techniques, ideal switches and switching functions, and energy storage elements. These are important concepts that can be applied across a broad range of basic and advanced circuits. In Sections 10.3 through 10.5, respectively, we applied these concepts to introduce and analyze buck and boost dc–dc converters, half- and full-wave rectifiers, and the single-phase inverter. You might be asking, "What about ac–ac conversion?" As we have seen in Chapter 1, ac transformers work great for this, provided the transformation is at a fixed frequency. Since this is the case more often than not, power electronic converters are not as prevalent in this application. The exception is when the frequency must be changed. In that case, a transformer will not suffice and an ac–ac power electronic circuit – the cycloconverter is one example – is required. We will leave the cycloconverter and other ac–ac conversion methods as topics for advanced study.

With full mastery of the techniques introduced in this chapter, the student should be well-prepared to analyze the machine drive circuits in this text, and to undertake advanced study beyond this course.

10.7 PROBLEMS

1. Show that the compact Fourier series of the periodic triangle wave, $tri(t)$, with period T defined over the interval $[0, T]$ as

$$tri(t) = \begin{cases} 1 - \dfrac{4t}{T}, & \forall t \in [0, T/2) \\ -1 + \dfrac{4(t - T/2)}{T}, & \forall t \in [T/2, T) \end{cases}$$

is the following:

$$tri(t) = \frac{8}{\pi^2} \sum_{n=1,3,5,\ldots}^{\infty} \frac{1}{n^2} \cos\left(\frac{2\pi n}{T} t\right)$$

It may be helpful to first calculate the Fourier series in the standard form, simplifying as appropriate using the rules in (10.2-9) through (10.2-12), and then to convert to the compact form. In addition, it may be useful to consider the relationship between the triangle wave and a square wave and to use the Fourier series of the square wave to obtain that of the triangle wave.

2. Show that the compact Fourier series of the symmetric square wave switching function defined by $q(t) = \frac{1}{2}[1 + sign(\sin(t))]$ is the following:

$$q(t) = \frac{1}{2} + \frac{2}{\pi} \sum_{n=1,3,5,\ldots}^{\infty} \frac{(-1)^{\frac{n-1}{2}}}{n} \sin(nt)$$

It may help to sketch the function, compare it to Fig. 10.2-4, and use (10.2-15). Some simplification of the series may be necessary to match the given solution.

3. Repeat the buck converter design in Example 10D if instead the battery voltage is 24 V and the inductor peak-to-peak current ripple should be no more than 20 percent of the average current. All other specifications and assumptions are the same. Note that your results can vary from the solution, depending on your design choices.

4. For the buck converter you designed in Problem 3, sketch the inductor current waveform over three switching periods. Be sure to label key points on the x- and y-axes to explain your sketch.

5. You wish to interface a photovoltaic array, which produces dc current at 35 V, to a dc 140 V, 2.8 kW load. (a) Suggest a dc–dc converter for this application. (b) Based on your choice of converter, find k, assuming no losses. (c) Calculate the load current. (d) Assuming no losses in the converter, calculate the photovoltaic array output current and power.

6. Show that the boost conversion ratio for the nonideal boost converter shown in Fig. 10.7-1 is given by $\frac{\hat{v}_{out}}{V_{in}} = \frac{(1-k)r}{(1-k)^2 r + r_L}$. This result produces the lossy curve in Fig. 10.3-11. [*Hint*: First, write the power balance equation relating average input, output, and loss power. Then, express each power in terms of the average inductor current, $\hat{i}_L$, and other circuit values.]

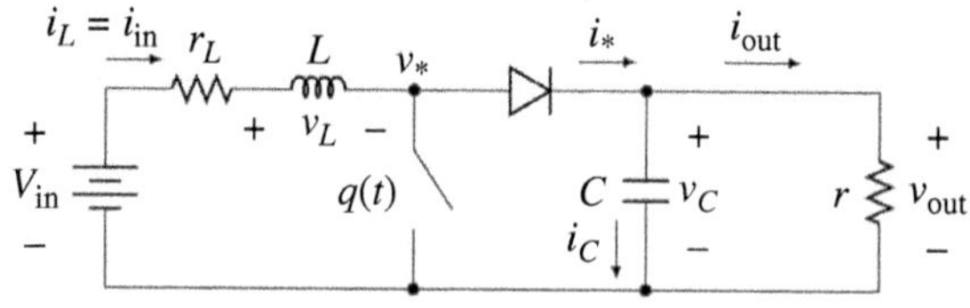

Figure 10.7-1 Boost converter circuit with resistive losses in the inductor; for Problem 6.

7. For the diode circuit in Fig. 10.7-2 with input voltage $v_{in}(t)$ shown in Fig. 10.7-3, perform the diode trial method by completing the following steps: (a) Create and fill in a trial method table similar to Table 10.4-1. (b) Based on your results in (a), sketch $v_{load}(t)$ over a few periods. (c) Calculate $\widehat{P}_{load}$.

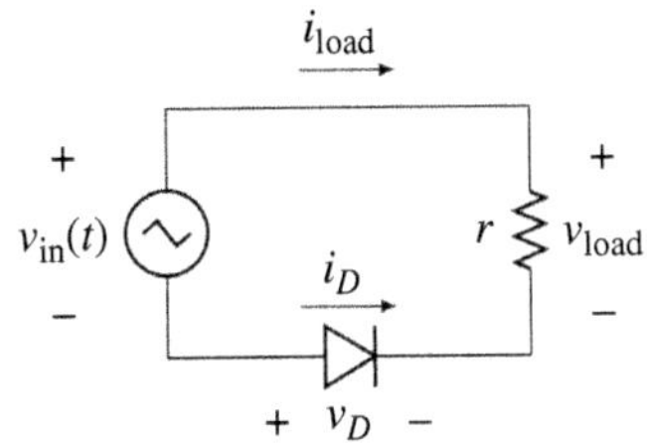

Figure 10.7-2 Diode circuit for Problem 7.

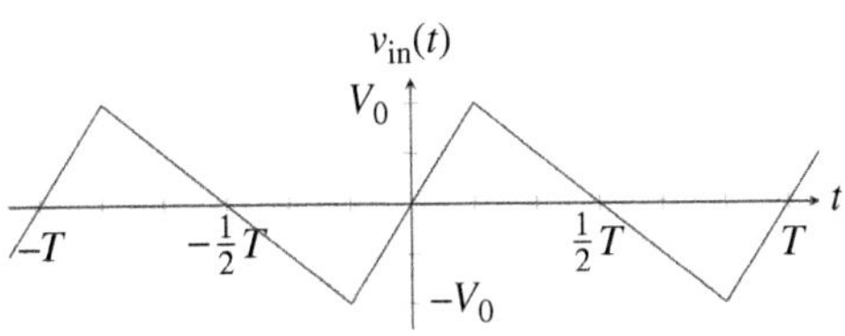

Figure 10.7-3 Input voltage for diode circuit; for Problem 7.

8. Conduct the diode trial method and complete a table similar to Table 10.4-1 for the full-wave rectifier. [*Hint*: Only the states shown in Table 10.4-2 are viable. Thus, your table will fill in all of the unviable states.]

9. A single-phase inverter with V_{in} = 10 V is supplying a series rL load with r = 1 Ω and L = 3 mH. The switching frequency, f_{sw}, is 60 Hz and relevant circuit waveforms are shown in Fig. 10.5-3. Find the load current phasor and average power for (a) the fundamental and (b) the third harmonic. The series you found in SP10.5-1 will be useful. ■

10.8 REFERENCES

1. P. T. Krein, *Elements of Power Electronics*, Oxford University Press, 1998.
2. G. M. Masters, *Renewable and Efficient Electric Power Systems*, 2nd Edition, Wiley-IEEE Press, 2013.
3. R. W. Erickson and D. Maksimovic, *Fundamentals of Power Electronics*, 2nd Edition, Springer, 2001.
4. N. Mohan, T. M. Undeland, and W. P. Robbins, *Power Electronics: Converters, Applications, and Design*, John Wiley & Sons, 1989.

APPENDIX A

ABBREVIATIONS, CONSTANTS, CONVERSIONS, AND IDENTITIES

Term	Abbreviation
alternating current	ac
ampere	A
ampere-turn	At
coulomb	C
direct current	dc
electromotive force	emf
foot	ft
gauss	G
gram	g
henry	H
hertz	Hz
horsepower	hp
inch	in
joule	J
kilogram	kg
kilovar	kvar
kilovolt	kV
kilovoltampere	kVA
kilowatt	kW
magnetomotive force	mmf
maxwell	Mx
megawatt	MW
meter	m
microfarad	μF
millihenry	mH

(*continued*)

Electromechanical Motion Devices: Rotating Magnetic Field-Based Analysis and Online Animations, Third Edition. Paul Krause, Oleg Wasynczuk, Steven D. Pekarek, and Timothy O'Connell.

Published 2020 by John Wiley & Sons, Inc.

(*continued*)

Term	Abbreviation
newton	N
newton meter	N·m
oersted	Oe
pound	lb
poundal	pdl
power factor	pf
pulse-width modulation	PWM
radian	rad
revolution per minute	r/min (rpm)
root mean square	rms
second	s
voltampere reactive	var
volt	V
voltampere	VA
watt	W
weber	Wb

Constants and Conversion Factors

permeability of free space	$\mu_0 = 4\pi \times 10^{-7}$ Wb/A·m
permittivity of free space	$\epsilon_0 = 8.854 \times 10^{-12}$ $C^2/N{\cdot}m^2$
acceleration of gravity	$g = 9.807$ m/s^2
length	1 m = 3.218 ft = 39.37 in.
mass	1 kg = 0.0685 slug = 2.205 lb (mass)
force	1 N = 0.225 lb = 3.6 oz
torque	1 N·m = 0.738 lb·ft
energy	1 J (W·s) = 0.738 lb·ft
power	1 W = 1.341×10^{-3} hp
moment of inertia	1 kg·m^2 = 0.738 slug·ft^2 = 23.7 lb·ft^2
magnetic flux	1 Wb = 10^8 Mx (lines)
magnetic flux density	1 Wb/m^2 = 10,000 G = 64.5 klines/in^2
magnetizing force	1 At/m = 0.0254 At/in = 0.0126 Oe

Trigonometric Identities

(I-1)	$e^{j\alpha} = \cos\alpha + j\sin\alpha$
(I-2)	$a\cos x + b\sin x = \sqrt{a^2+b^2}\cos(x+\varphi) \qquad \varphi = \tan^{-1}(-b/a)$
(I-3)	$\cos^2 x + \sin^2 x = 1$
(I-4)	$\sin 2x = 2\sin x\cos x$
(I-5)	$\cos 2x = \cos^2 x - \sin^2 x = 2\cos^2 x - 1 = 1 - 2\sin^2 x$
(I-6)	$\cos x\cos y = \frac{1}{2}\cos(x+y) + \frac{1}{2}\cos(x-y)$
(I-7)	$\sin x\sin y = \frac{1}{2}\cos(x-y) - \frac{1}{2}\cos(x+y)$
(I-8)	$\sin x\cos y = \frac{1}{2}\sin(x+y) + \frac{1}{2}\sin(x-y)$
(I-9)	$\cos(x \pm y) = \cos x \cos y \mp \sin x \sin y$
(I-10)	$\sin(x \pm y) = \sin x \cos y \pm \cos x \sin y$
(I-11)	$\cos^2 x + \cos^2\left(x-\frac{2}{3}\pi\right) + \cos^2\left(x+\frac{2}{3}\pi\right) = \frac{3}{2}$
(I-12)	$\sin^2 x + \sin^2\left(x-\frac{2}{3}\pi\right) + \sin^2\left(x+\frac{2}{3}\pi\right) = \frac{3}{2}$
(I-13)	$\sin x\cos x + \sin\left(x-\frac{2}{3}\pi\right)\cos\left(x-\frac{2}{3}\pi\right) + \sin\left(x+\frac{2}{3}\pi\right)\cos\left(x+\frac{2}{3}\right) = 0$
(I-14)	$\cos x + \cos\left(x-\frac{2}{3}\pi\right) + \cos\left(x+\frac{2}{3}\pi\right) = 0$
(I-15)	$\sin x + \sin\left(x-\frac{2}{3}\pi\right) + \sin\left(x+\frac{2}{3}\pi\right) = 0$
(I-16)	$\sin x\cos y + \sin\left(x-\frac{2}{3}\pi\right)\cos\left(y-\frac{2}{3}\pi\right) + \sin\left(x+\frac{2}{3}\pi\right)\cos\left(y+\frac{2}{3}\pi\right) = \frac{3}{2}\sin(x-y)$
(I-17)	$\sin x\sin y + \sin\left(x-\frac{2}{3}\pi\right)\sin\left(y-\frac{2}{3}\pi\right) + \sin\left(x+\frac{2}{3}\pi\right)\sin\left(y+\frac{2}{3}\pi\right) = \frac{3}{2}\cos(x-y)$
(I-18)	$\cos x\sin y + \cos\left(x-\frac{2}{3}\pi\right)\sin\left(y-\frac{2}{3}\pi\right) + \cos\left(x+\frac{2}{3}\pi\right)\sin\left(y+\frac{2}{3}\pi\right) = -\frac{3}{2}\sin(x-y)$
(I-19)	$\cos x\cos y + \cos\left(x-\frac{2}{3}\pi\right)\cos\left(y-\frac{2}{3}\pi\right) + \cos\left(x+\frac{2}{3}\pi\right)\cos\left(y+\frac{2}{3}\pi\right) = \frac{3}{2}\cos(x-y)$
(I-20)	$\sin x\cos y + \sin\left(x+\frac{2}{3}\pi\right)\cos\left(y-\frac{2}{3}\pi\right) + \sin\left(x-\frac{2}{3}\pi\right)\cos\left(y+\frac{2}{3}\pi\right) = \frac{3}{2}\sin(x+y)$
(I-21)	$\sin x\sin y + \sin\left(x+\frac{2}{3}\pi\right)\sin\left(y-\frac{2}{3}\pi\right) + \sin\left(x-\frac{2}{3}\pi\right)\sin\left(y+\frac{2}{3}\pi\right) = -\frac{3}{2}\cos(x+y)$
(I-22)	$\cos x\sin y + \cos\left(x+\frac{2}{3}\pi\right)\sin\left(y-\frac{2}{3}\pi\right) + \cos\left(x-\frac{2}{3}\pi\right)\sin\left(y+\frac{2}{3}\pi\right) = \frac{3}{2}\sin(x+y)$
(I-23)	$\cos x\cos y + \cos\left(x+\frac{2}{3}\pi\right)\cos\left(y-\frac{2}{3}\pi\right) + \cos\left(x-\frac{2}{3}\pi\right)\cos\left(y+\frac{2}{3}\pi\right) = \frac{3}{2}\cos(x+y)$

APPENDIX **B**

ANIMATIONS

This book is accompanied by a companion website, https://engineering.purdue.edu/KWPO/, that includes links to animations supporting the material in the text. These animations include: DC Machines (Chapter 3), Tesla's Rotating Magnetic Field (Chapter 4), Introduction to Reference Frame Theory (Chapter 5), Permanent-Magnet AC and Brushless DC Machines (Chapter 6), Synchronous Machines (Chapter 7), and Symmetrical Induction Machines (Chapter 8). The reader is encouraged to visit the web site, which provides additional details regarding the available animations.

Electromechanical Motion Devices: Rotating Magnetic Field-Based Analysis and Online Animations, Third Edition. Paul Krause, Oleg Wasynczuk, Steven D. Pekarek, and Timothy O'Connell.

Published 2020 by John Wiley & Sons, Inc.

INDEX

Electromechanical Motion Devices: Rotating Magnetic Field-Based Analysis and Online Animations, Third Edition. Paul Krause, Oleg Wasynczuk, Steven D. Pekarek, and Timothy O'Connell.

Published 2020 by John Wiley & Sons, Inc.

IEEE Press Series on Power Engineering

Series Editor: **M. E. El-Hawary,** Dalhousie University, Halifax, Nova Scotia, Canada

The mission of IEEE Press Series on Power Engineering is to publish leading- edge books that cover the broad spectrum of current and forward-looking technologies in this fast-moving area. The series attracts highly acclaimed authors from industry/ academia to provide accessible coverage of current and emerging topics in power engineering and allied fields. Our target audience includes the power engineering professional who is interested in enhancing their knowledge and perspective in their areas of interest.

1. *Electric Power Systems: Design and Analysis, Revised Printing*
Mohamed E. El-Hawary

2. *Power System Stability*
Edward W. Kimbark

3. *Analysis of Faulted Power Systems*
Paul M. Anderson

4. *Inspection of Large Synchronous Machines: Checklists, Failure Identification, and Troubleshooting*
Isidor Kerszenbaum

5. *Electric Power Applications of Fuzzy Systems*
Mohamed E. El-Hawary

6. *Power System Protection*
Paul M. Anderson

7. *Subsynchronous Resonance in Power Systems*
Paul M. Anderson, B.L. Agrawal, and J.E. Van Ness

8. *Understanding Power Quality Problems: Voltage Sags and Interruptions*
Math H. Bollen

9. *Analysis of Electric Machinery*
Paul C. Krause, Oleg Wasynczuk, and S.D. Sudhoff

10. *Power System Control and Stability, Revised Printing*
Paul M. Anderson and A.A. Fouad

Electromechanical Motion Devices: Rotating Magnetic Field-Based Analysis and Online Animations, Third Edition. Paul Krause, Oleg Wasynczuk, Steven D. Pekarek, and Timothy O'Connell.

Published 2020 by John Wiley & Sons, Inc.

11. *Principles of Electric Machines with Power Electronic Applications, Second Edition*
Mohamed E. El-Hawary

12. *Pulse Width Modulation for Power Converters: Principles and Practice*
D. Grahame Holmes and Thomas A. Lipo

13. *Analysis of Electric Machinery and Drive Systems, Second Edition*
Paul C. Krause, Oleg Wasynczuk, and Scott D. Sudhoff

14. *Risk Assessment for Power Systems: Models, Methods, and Applications*
Wenyuan Li

15. *Optimization Principles: Practical Applications to the Operation of Markets of the Electric Power Industry*
Narayan S. Rau

16. *Electric Economics: Regulation and Deregulation*
Geoffrey Rothwell and Tomas Gomez

17. *Electric Power Systems: Analysis and Control*
Fabio Saccomanno

18. *Electrical Insulation for Rotating Machines: Design, Evaluation, Aging, Testing, and Repair, Second Edition*
Greg C. Stone, Ian Culbert, Edward A. Boulter, and Hussein Dhirani

19. *Signal Processing of Power Quality Disturbances*
Math H. J. Bollen and Irene Y. H. Gu

20. *Instantaneous Power Theory and Applications to Power Conditioning*
Hirofumi Akagi, Edson H. Watanabe, and Mauricio Aredes

21. *Maintaining Mission Critical Systems in a 24/7 Environment*
Peter M. Curtis

22. *Elements of Tidal-Electric Engineering*
Robert H. Clark

23. *Handbook of Large Turbo-Generator Operation and Maintenance, Second Edition*
Geoff Klempner and Isidor Kerszenbaum

24. *Introduction to Electrical Power Systems*
Mohamed E. El-Hawary

25. *Modeling and Control of Fuel Cells: Distributed Generation Applications*
M. Hashem Nehrir and Caisheng Wang

26. *Power Distribution System Reliability: Practical Methods and Applications*
Ali A. Chowdhury and Don O. Koval

27. *Introduction to FACTS Controllers: Theory, Modeling, and Applications*
Kalyan K. Sen and Mey Ling Sen

28. *Economic Market Design and Planning for Electric Power Systems*
James Momoh and Lamine Mili

29. *Operation and Control of Electric Energy Processing Systems*
James Momoh and Lamine Mili

30. *Restructured Electric Power Systems: Analysis of Electricity Markets with Equilibrium Models*
Xiao-Ping Zhang

31. *An Introduction to Wavelet Modulated Inverters*
S. A. Saleh and M. Azizur Rahman

32. *Control of Electric Machine Drive Systems*
Seung-Ki Sul

33. *Probabilistic Transmission System Planning*
Wenyuan Li

34. *Electricity Power Generation: The Changing Dimensions*
Digambar M. Tagare

35. *Electric Distribution Systems*
Abdelhay A. Sallam and Om P. Malik

36. *Practical Lighting Design with LEDs*
Ron Lenk and Carol Lenk

37. *High Voltage and Electrical Insulation Engineering*
Ravindra Arora and Wolfgang Mosch

38. *Maintaining Mission Critical Systems in a 24/7 Environment, Second Edition*
Peter M. Curtis

39. *Power Conversion and Control of Wind Energy Systems*
Bin Wu, Yongqiang Lang, Navid Zargari, and Samir Kouro

40. *Integration of Distributed Generation in the Power System*
Math H. J. Bollen and Fainan Hassan

41. *Doubly Fed Induction Machine: Modeling and Control for Wind Energy Generation*
Gonzalo Abad, Jesús López, Miguel Rodrigues, Luis Marroyo, and Grzegorz Iwanski

42. *High Voltage Protection for Telecommunications*
Steven W. Blume

43. *Smart Grid: Fundamentals of Design and Analysis*
James Momoh

44. *Electromechanical Motion Devices, Second Edition*
Paul Krause, Oleg Wasynczuk, and Steven Pekarek

45. *Electrical Energy Conversion and Transport: An Interactive Computer-Based Approach, Second Edition*
George G. Karady and Keith E. Holbert

46. *ARC Flash Hazard and Analysis and Mitigation*
J. C. Das

47. *Handbook of Electrical Power System Dynamics: Modeling, Stability, and Control*
Mircea Eremia and Mohammad Shahidehpour

48. *Analysis of Electric Machinery and Drive Systems, Third Edition*
Paul Krause, Oleg Wasynczuk, Scott Sudhoff, and Steven Pekarek

49. *Extruded Cables for High-Voltage Direct-Current Transmission: Advances in Research and Development*
Giovanni Mazzanti and Massimo Marzinotto

50. *Power Magnetic Devices: A Multi-Objective Design Approach*
S. D. Sudhoff

51. *Risk Assessment of Power Systems: Models, Methods, and Applications, Second Edition*
Wenyuan Li

52. *Practical Power System Operation*
Ebrahim Vaahedi

53. *The Selection Process of Biomass Materials for the Production of Bio-Fuels and Co-Firing*
Najib Altawell

54. *Electrical Insulation for Rotating Machines: Design, Evaluation, Aging, Testing, and Repair, Second Edition*
Greg C. Stone, Ian Culbert, Edward A. Boulter, and Hussein Dhirani

55. *Principles of Electrical Safety*
Peter E. Sutherland

56. *Advanced Power Electronics Converters: PWM Converters Processing AC Voltages*
Euzeli Cipriano dos Santos Jr. and Edison Roberto Cabral da Silva

57. *Optimization of Power System Operation, Second Edition*
Jizhong Zhu

58. *Power System Harmonics and Passive Filter Designs*
J. C. Das

59. *Digital Control of High-Frequency Switched-Mode Power Converters*
Luca Corradini, Dragan Maksimovic, Paolo Mattavelli, and Regan Zane

60. *Industrial Power Distribution, Second Edition*
Ralph E. Fehr, III

61. *HVDC Grids: For Offshore and Supergrid of the Future*
Dirk Van Hertem, Oriol Gomis-Bellmunt, and Jun Liang

62. *Advanced Solutions in Power Systems: HVDC, FACTS, and Artificial Intelligence*
Mircea Eremia, Chen-Ching Liu, and Abdel-Aty Edris

63. *Operation and Maintenance of Large Turbo-Generators*
Geoff Klempner and Isidor Kerszenbaum

64. *Electrical Energy Conversion and Transport: An Interactive Computer-Based Approach*
George G. Karady and Keith E. Holbert

65. *Modeling and High-Performance Control of Electric Machines*
John Chiasson

66. *Rating of Electric Power Cables in Unfavorable Thermal Environment*
George J. Anders

67. *Electric Power System Basics for the Nonelectrical Professional*
Steven W. Blume

68. *Modern Heuristic Optimization Techniques: Theory and Applications to Power Systems*
Kwang Y. Lee and Mohamed A. El-Sharkawi

69. *Real-Time Stability Assessment in Modern Power System Control Centers*
Savu C. Savulescu

70. *Optimization of Power System Operation*
Jizhong Zhu

71. *Insulators for Icing and Polluted Environments*
Masoud Farzaneh and William A. Chisholm

72. *PID and Predictive Control of Electric Devices and Power Converters Using MATLAB®/Simulink®*
Liuping Wang, Shan Chai, Dae Yoo, Lu Gan, and Ki Ng

73. *Power Grid Operation in a Market Environment: Economic Efficiency and Risk Mitigation*
Hong Chen

74. *Electric Power System Basics for the Nonelectrical Professional, Second Edition*
Steven W. Blume

75. *Energy Production Systems Engineering*
Thomas Howard Blair

76. *Model Predictive Control of Wind Energy Conversion Systems*
Venkata Yaramasu and Bin Wu

77. *Understanding Symmetrical Components for Power System Modeling*
J. C. Das

78. *High-Power Converters and AC Drives, Second Edition*
Bin Wu and Mehdi Narimani

79. *Current Signature Analysis for Condition Monitoring of Cage Induction Motors: Industrial Application and Case Histories*
William T. Thomson and Ian Culburt

80. *Introduction to Electric Power and Drive Systems*
Paul Krause, Oleg Wasynczuk, Timothy O'Connell, and Maher Hasan

81. *Instantaneous Power Theory and Applications to Power Conditioning, Second Edition*
Hirofumi Akagi, Edson Hirokazu Watanabe, and Mauricio Aredes

82. *Practical Lighting Design with LEDs, Second Edition*
Ron Lenk and Carol Lenk

83. *Introduction to AC Machine Design*
Thomas A. Lipo

84. *Advances in Electric Power and Energy Systems: Load and Price Forecasting*
Mohamed E. El-Hawary

85. *Electricity Markets: Theories and Applications*
Jeremy Lin and Fernando H. Magnago

86. *Multiphysics Simulation by Design for Electrical Machines, Power Electronics, and Drives*
Marius Rosu, Ping Zhou, Dingsheng Lin, Dan Ionel, Mircea Popescu, Frede Blaabjerg, Vandana Rallabandi, and David Staton

87. *Modular Multilevel Converters: Analysis, Control, and Applications*
Sixing Du, Apparao Dekka, Bin Wu, and Navid Zargari

88. *Electrical Railway Transportation Systems*
Morris Brenna, Federica Foiadelli, and Dario Zaninelli

89. *Energy Processing and Smart Grid*
James A. Momoh

90. *Handbook of Large Turbo-Generator Operation and Maintenance, Third Edition*
Geoff Klempner and Isidor Kerszenbaum

91. *Advanced Control of Doubly Fed Induction Generator for Wind Power Systems*
Dehong Xu, Frede Blaabjerg, Wenjie Chen, and Nan Zhu

92. *Electric Distribution Systems, Second Edition*
Abdelhay A. Sallam and Om P. Malik

93. *Power Electronics in Renewable Energy Systems and Smart Grid: Technology and Applications*
Bimal K. Bose

94. *Power System and Control and Stability, Third Edition*
Vijay Vittal, James D. McCalley, Paul M. Anderson, and A. A. Fouad

95. *Electromechanical Motion Devices: Rotating Magnetic Field-Based Analysis and Online Animations, Third Edition*
Paul Krause, Oleg Wasynczuk, Steven D. Pekarek, and Timothy O'Connell